T0313175

Modeling, Control, Estimation, and Optimization for Microgrids

Modeling, Control, Estimation, and Optimization for Microgrids

A Fuzzy-Model-Based Method

Zhixiong Zhong

CRC Press is an imprint of the
Taylor & Francis Group, an informa business

MATLAB® is a trademark of The MathWorks, Inc. and is used with permission. The MathWorks does not warrant the accuracy of the text or exercises in this book. This book's use or discussion of MATLAB® software or related products does not constitute endorsement or sponsorship by The MathWorks of a particular pedagogical approach or particular use of the MATLAB® software.

CRC Press
Taylor & Francis Group
6000 Broken Sound Parkway NW, Suite 300
Boca Raton, FL 33487-2742

Printed on acid-free paper

International Standard Book Number-13: 978-1-138-49165-6 (Hardback)

Visit the Taylor & Francis Web site at
http://www.taylorandfrancis.com

and the CRC Press Web site at
http://www.crcpress.com

Dedication

To my parent Yongxiang Zhong and Fengmei Hong, my wife Yanyu Hong, and my sons Hongli Zhong and Ceyi Zhong.

Contents

PART IV Cyber-Physical Control Framework for Microgrids

List of Figures

List of Tables

Preface

Microgrids provide appealing solutions for integrating renewable energy sources into power grids. They have attracted increasing interest in recent years because of environmental concerns and shortages of traditional energy sources (natural gas, oil, and coal). Microgrids are small-scale electrical distribution systems consisting of power converters that link generation, storage, and distribution facilities. Microgrids utilizing renewable energy sources present little or no inertia and thus are more instable than conventional power grids with synchronous generators. They represent the main building blocks of future "smart" grids.

Renewable energy generators such as solar photovoltaic (PV) systems and wind turbines have been widely studied. However, their high installation costs make maximum power point tracking (MPPT) control a major factor. MPPT is impacted by changes in external conditions and intrinsic characteristics of systems. It is therefore difficult to implement traditional MPPT controls such as Newton-like extremum seeking, adaptive perturbation, and close observation. Current control methods lack strict convergence analysis capability and guarantee only near-maximum power.

Power converters based on pulsewidth modulation (PWM) techniques are commonly used to control microgrids. PWM converters exhibit several desirable characteristics such as high efficiency, constant frequency operation, few components, and high conversion ratios. However, they also exhibit nonlinear dynamic behaviors and are thus difficult to control.

Traditional power generators are large-scale nonlinear systems. Simplified linear models in synchronous generators have been investigated for a long time. Their inherent simplicity of design means that they provide asymptotic stability on a small scale and attenuate the impacts of small-scale disturbances. Conversely, power networks are large-scale spatially distributed systems with multiple locations and require complex monitoring and control systems to guarantee safety and stability while providing needed power. This is typically done by a centralized control framework utilizing a single controller that has powerful processing capabilities for handling complex data measurements. Centralized control systems with higher sampling rates reduced overall system reliability and increased system sensitivity to a single point of failure. Distributed control systems are becoming more commonly used in interconnected power systems.

The coexistence of multiple energy resources with varying dynamic properties such as inertial levels and dispatch characteristics has raised concerns about the stability, control, and efficiency of microgrids. Their flexibility can be ensured by operating various types of generators as power demands change. Maintaining the power balance in the face of changing demands represent a challenging problem in all areas of the power industry. Another vital issue is the need to maintain power generation security in the present climate of increasing numbers of cyber attacks that can produce catastrophic results.

This book consists of eleven chapters divided into four parts. Part I (Chapters 1 through 3) explains fuzzy modelling and local controls of microgrid components. Part II (Chapters 4 through 6) discusses centralized, decentralized, and distributed fuzzy control schemes. Part III (Chapters 7 and 8) describes operational and optimization aspects of microgrid energy management. Part IV (Chapters 9 through 11) details various aspects of cyber-physical system (CPS) controls for microgrids and devotes an entire chapter to time delay switch (TDS) attacks.

Chapter 1 addresses the stability analysis and control synthesis for stand-alone nonlinear solar power systems via the T-S fuzzy-model-based approach, nonlinear photovoltaic (PV) systems with DC/AC loads, and the reformulation of the MPPT problems of PV systems in the framework of descriptor systems. Other topics presented are a robust fuzzy observer for state feedback control and finite-time stabilization via sliding mode control of descriptor systems. Two numerical examples demonstrate the effectiveness of the proposed method.

Chapter 2 develops a novel MTTP method for stand-alone wind power generators via the T-S fuzzy-model-based approach. It covers nonlinear wind power systems, the MPPT problem in relation to wind power systems, and a fuzzy observer for state feedback control under partial state measurement. A fuzzy sliding mode controller for descriptor systems is proposed. The chapter concludes with two numerical examples that demonstrate effectiveness.

Chapter 3 focuses on the development of a model framework for lead-acid batteries, lithium (Li)-ion batteries, and supercapacitors; it also covers the reformulation of the original partial differential equation (PDE)-based battery model to a fuzzy-based version to precisely characterize its charge and discharge operations. The chapter also explains a fuzzy state of charge (SOC) estimation approach for various types of batteries and supercapacitors and includes a numerical example demonstrating the effectiveness of the proposed model.

Centralized fuzzy control is the topic of Chapter 4. It examines methods for tracking voltage synchronization of PV installations and wind systems and the concept that all generator subsystems act as a single entity to achieve synchronization through a communication network. It details network-based controls with sampled data measurement and time-triggered zero order hold (ZOH), and a numerical example concludes the chapter.

Chapter 5 investigates the problems of tracking voltage synchronization of multiple PV and wind turbine systems. Again, the subsystems must act as a single group to achieve synchronization through local information exchange. Use of sampled data measurement and time-triggered ZOH in decentralized fuzzy control systems is examined. A numerical example is provided.

Chapter 6 focuses on distributed fuzzy controls and tracking voltage synchronization of microgrids. Sampled data measurement and time-triggered ZOH and a numerical example are also discussed.

Chapter 7 studies detailed models and fuzzy logical formulations of microgrid systems consisting of PV, wind turbines, and energy storage facilities. Three switching models are proposed to maintain power balance.

Optimization is the subject of Chapter 8. It details power management strategies. Renewable energy sources are utilized as distributed connections to a common bus in a microgrid. To maintain the power balance, the power management system determines the operating modes of the energy systems based on measured currents. The chapter also covers maintenance of system stability and transient and steady performances of microgrids through optimization using reachable set estimation and finite time control.

Chapter 9 continues the discussion of using renewable energy sources routed through a common bus in a microgrid. The sources communicate with each other via networks. Network-induced delays are introduced along with proposals for solving stability and control problems. The chapter includes a numerical example.

Chapter 10 covers some of the same topics explained in earlier chapters in Part IV in relation to event-triggered fuzzy control. New concepts for solving stability analysis and control synthesis problems are proposed and a numerical example is provided.

Chapter 11 is a timely discussion of the vulnerability of power generation systems to time delay switch (TDS) attacks with a focus on nonlinear power networks. Its intent is to establish an effective method to deal with such attacks, particularly through better monitoring of system states by an augmented observer that ensures finite-time boundedness (FTB) and compensation control. A numerical example that demonstrates the effectiveness of the method is included.

The research described in this book represents a fresh approach to explaining fuzzy-model-based control approaches to microgrid applications. This book covers the author's long-term research results, teaching, and practical experience focusing on microgrid control and operation. It is intended for use by engineers and operators in all areas of power grid and microgrid planning, control, and operations, and also by students, and academic researchers. It describes microgrid dynamics, modelling, and control issues from introductory through advanced levels. It can serve as a text for both undergraduate and post-graduate electrical engineering students in courses on microgrids, smart grids, and modern power system controls.

The author has been inspired over the years by many scientists who undoubtedly give their help on this book; in particular by Hao Ying, Chih-Min Lin, Lixian Zhang, Michael V. Basin, and Hak-Keung Lam. This work would not have been possible without my colleagues and the excellent professional environments at Fujian Provincial Key Laboratory of Information Processing and Intelligent Control, and Digital Fujian IoT Laboratory of Intelligent Production. To all these people I extend our sincere thanks. I also sincerely thank Marc Gutierrez, Nick Mould, and Arun for giving me the opportunity to publish my book with Taylor & Francis CRC Press, and the editorial and production team at Taylor & Francis for their valuable help. We gratefully acknowledge the financial support of the the Central Government Guides Local Science and Technology Development Projects (grant no. 2019L3009), the Advanced Research Program of Minjiang University (grant no. K-30404307; MJY18003), the Science and Technology Planning Project of Fuzhou City (grant no. 2017-G-106; 2019-G-49; 2018-G-98; 2018-G-96), the Fujian Industrial Technology Development and Application Plan Project (grant no. 2019h0025), the Industrial Robot Application Fujian University Engineering Research Center (grant no. MJUKF-IRA1802), and the scientific research project of Xiamen City (grant no. 3502Z20189033). I also thank my families for their continual understanding, patience, and support. Many researchers have made significant contribution to microgrid applications. Owing to the structural arrangement and length limitation of the book, many of their published results are not included or even not cited. I would extend my apologies to these researchers.

Zhixiong Zhong
Minjiang University
Fuzhou, China

Part I

Fuzzy Modeling and Local Control for Microgrid Components

Preview

Due to increasing demands for electricity, the costs of conventional power sources (coal, petroleum, and other fossil fuels) have increased and their stocks are finite. The accelerating demands threaten the security of energy supplies worldwide and will worsen environmental pollution. Renewal energy sources such as wind power and photovoltaic (PV) systems are promising alternatives because they are freely available, environmentally friendly, and accrue fewer operational and maintenance costs.

Microgrids are key components used by modern power systems to integrate renewable energy sources. They control and coordinate renewable energy sources, balance loads and also monitor protective devices and communication networks. They utilize maximum power point tracking (MPPT) to achieve optimum energy generation. This is an important factor because outputs of wind turbines and PV arrays must be regulated based on load demands. Furthermore, weather conditions cause fluctuations of wind and solar power that create variations of bus voltages and impact power flows through transmission systems. Conventional linear controllers based on small-signal analysis decrease system performance and transient events such as islanding, maintenance, and load variations may result in instability.

Modern electrical utilities utilize model-based techniques to analyze operations but these models often fail to generate sufficient details that would allow them to effectively coordinate microgrid management of renewable energy sources. Fuzzy logic control (FLC) has been applied successfully to control complex nonlinear and even nonanalytic systems. The Takagi-Sugeno (T-S) model is regarded as a likely candidate for controlling complex nonlinear systems. The T-S fuzzy models can approximate nonlinear system operations with good precision and aid utilities to develop linear methods to resolve control problems.

Many theoretical findings focused on stability analysis and control synthesis by T-S fuzzy systems have been published over the years. More recently, reports of uses of T-S models to achieve MPPT for wind turbines and PV arrays have appeared in the literature.

Chapter 1 investigates the modelling procedure and local control of AC-DC PV systems and presents a fuzzy-based approach for controlling the dynamics of PV power generation. The chapter also discusses a singular system approach to achieving MPPT. Since PV panel output is very sensitive to solar activity and temperature and these factors impacted MPPT, a robust fuzzy observer-based MPPT and a finite-time MPPT via sliding mode strategy are proposed. Chapter 2 extends some of these results to wind power generators. Chapter 3 covers fuzzy modelling and state of charge (SOC) estimation design on energy storage sytems (lead-acid and lithium ion batteries and supercapacitors).

1 Fuzzy Modeling and Control of Photovoltaic (PV) Power

Solar energy has experienced dramatic growth in the past few decades. It has been predicted that, the global capacity of solar power will reach 980 GW by 2020 [1]. A solar PV system directly converts solar irradiation into electricity. The main drawbacks of PV systems include high device cost and low energy conversion efficiency. In order to reduce the cost of energy, it is crucial to maintain the PV operation at its maximum efficiency at all times. However, the maximum power point of PV power depends on the changes in its intrinsic characteristics and external disturbances, such as aging of the device, irradiance intensity, and temperature conditions. Therefore, it is difficult to ensure the achievement of MPPT control for solar PV systems.

Nowadays, several MPPT techniques and their implementations are reported in the open literature [2]. Traditional MPPT control is based on perturb and observe, incremental conductance, fuzzy logic, and maximum power voltage-based methods. Unfortunately, the maximum power produced by the PV array changes with solar radiation and cell temperature so that most of MPPT methods lack strict convergence analysis and only provide near-maximum power. Although the works propose nonlinear MPPT control with guaranteed stability, these approaches are realized with difficulty due to the use of either the discontinuous control law or the time derivative of the PV voltage and current. In addition, no result reported in current literature has dealt with the robust MPPT control problem for PV power systems with partial state measurement, parametric uncertainty, and disturbance. It is always difficult for users to select an MPPT technique implementing a particular application. Until 2007, only a few papers discussed MPPT techniques. But many new MPPT techniques such as the Newton-like extremum seeking technique [4], the distributed MPPT [3], and the adaptive perturbation and observation [5], have been reported since then.

In the last three decades, the DC-DC converters have been widely used in the PV systems. The buck, boost and buck-boost circuits are three basic configurations for the DC-DC converters [6]. The duty ratio determines the switching action via pulsewidth modulation, which implements the control of the DC-DC converters, and exhibits a nonlinear dynamic behavior. Moreover, in most cases the approximated linear models based on a single operating point are not limited to minimum phase types for the DC-DC converters. For a specific operating condition, there is usually a unique maximum power point on the P-V characteristics. Maximum power point tracking (MPPT) control of the PV system aims to locate the MPP for online operation regardless of the change of the PV intrinsic and environment uncertainties. Although the linear controller is easier to design and implement, it is difficult to ensure MPPT performance in all the operating conditions [6]. Recently, it has been shown that nonlinear systems can be described by several local linear systems blending IF-THEN fuzzy rules [7, 8]. More recently, a T-S fuzzy-model-based approach has been developed for the MPPT control of PV systems with the DC-DC converters [9].

In this chapter, the stability analysis and control synthesis are developed for standalone solar power nonlinear systems via the T-S fuzzy-model-based approach. First, the nonlinear PV powers with DC-AC load are represented in the T-S fuzzy model. Then, the MPPT problem of the considered PV system is reformulated into the framework of descriptor systems. A robust fuzzy observer for state feedback control is proposed under partial state measurement. We further consider the finite-time stabilization via the sliding mode control in the framework of descriptor systems. Finally, two numerical examples are provided to show the effectiveness of the proposed method.

1.1 MODELING OF PV POWER

1.1.1 MODELING OF PV POWER WITH DC LOAD

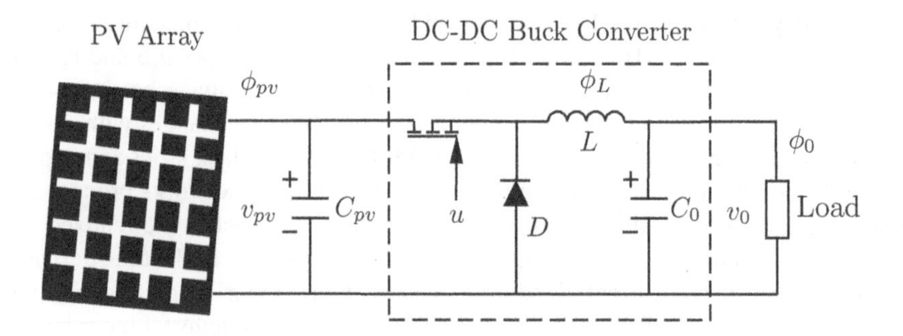

Figure 1.1 Solar PV power with DC-DC buck converter.

Consider a solar PV power system using the DC-DC buck converter as shown in Figure 1.1. Its dynamic model can be represented by the following differential equations [9],

$$\begin{cases} \dot{v}_{pv} = \frac{1}{C_{pv}}\left(\phi_{pv} - \phi_L u\right), \\ \dot{\phi}_L = \frac{1}{L}R_0\left((\phi_0 - \phi_L) - R_L\phi_L - v_0\right) + \frac{1}{L}\left(V_D + v_{pv} - R_M\phi_L\right)u - \frac{V_D}{L}, \\ \dot{v}_0 = \frac{1}{C_0}\left(\phi_L - \phi_0\right), \end{cases} \tag{1.1}$$

where v_{pv}, ϕ_L, and v_0 denote the PV array voltage, the current of the inductance L, and the voltage of the capacitance C_0, respectively; R_0, R_L, and R_M are the resistances on the capacitance C_0, on the inductance L, and on the power MOSFET, respectively; V_D is the forward voltage of power diode; ϕ_0 is the measurable load current; u is the duty ratio using the pulsewidth-modulated signal to control the switching MOSFET.

Note that a nonlinear system can be described by several local linear systems blending IF-THEN fuzzy rules at any given accuracy [7]. Here, define $x(t) =$

$\begin{bmatrix} v_{pv} & \phi_L & v_O \end{bmatrix}^T$, and choose $z_1 = \frac{\phi_{pv}}{v_{pv}}, z_2 = \phi_L, z_3 = \frac{\phi_O}{\phi_L}, z_4 = \frac{V_D}{v_{pv}}$, and $z_5 = v_{pv}$ as fuzzy premise variables. Thus, it follows from (1.1) that the PV power nonlinear system is represented by the following T-S model,

Plant Rule $\mathscr{R}^l$: **IF** z_1 is $\mathscr{F}_1^l$ and z_2 is $\mathscr{F}_2^l$ and $\cdots$ and z_5 is $\mathscr{F}_5^l$, **THEN**

$$\dot{x}(t) = A_l x(t) + B_l u(t), l \in \mathscr{L} := \{1, 2, \ldots, r\} \tag{1.2}$$

where $\mathscr{R}^l$ denotes the l-th fuzzy inference rule; r is the number of inference rules; $\mathscr{F}_\theta^l(\theta = 1, 2, \ldots, 5)$ is the fuzzy set; $x(t) \in \Re^{n_x}$ and $u(t) \in \Re^{n_u}$ denote the system state and control input, respectively; n_x, and n_u can be determined from the context; $z(t) \triangleq [z_1, z_2, z_3, z_4, z_5]$ are the measurable variables; $\{A_l, B_l\}$ is the l-th local model as below:

$$A_l = \begin{bmatrix} \frac{1}{C_{pv}} \mathscr{F}_1^l & 0 & 0 \\ -\frac{\mathscr{F}_4^l}{L} & \frac{R_O}{L} \mathscr{F}_3^l - \frac{R_O}{L} - \frac{R_O R_L}{L} & -\frac{R_O}{L} \\ 0 & \frac{1 - \mathscr{F}_3^l}{C_O} & 0 \end{bmatrix}, B_l = \begin{bmatrix} -\frac{\mathscr{F}_2^l}{C_{pv}} \\ \frac{1}{L}\left(V_D + \mathscr{F}_5^l - R_M \mathscr{F}_2^l\right) \\ 0 \end{bmatrix}. \tag{1.3}$$

By denoting $\mu_l[z(t)]$ as the normalized membership function, one gets

$$\mu_l[z(t)] := \frac{\prod_{\phi=1}^5 \mu_{\phi l}[z_\phi(t)]}{\sum_{\varsigma=1}^r \prod_{\phi=1}^5 \mu_\phi^\varsigma [z_\phi(t)]} \geq 0, \sum_{l=1}^r \mu_l[z(t)] = 1. \tag{1.4}$$

Here, we denote $\mu_l \triangleq \mu_l[z(t)]$ for brevity.

By fuzzy blending, the global T-S fuzzy dynamic model is given by

$$\dot{x}(t) = A(\mu)x(t) + B(\mu)u(t), \tag{1.5}$$

where $A(\mu) := \sum_{l=1}^r \mu_l A_l, B(\mu) := \sum_{l=1}^r \mu_l B_l$.

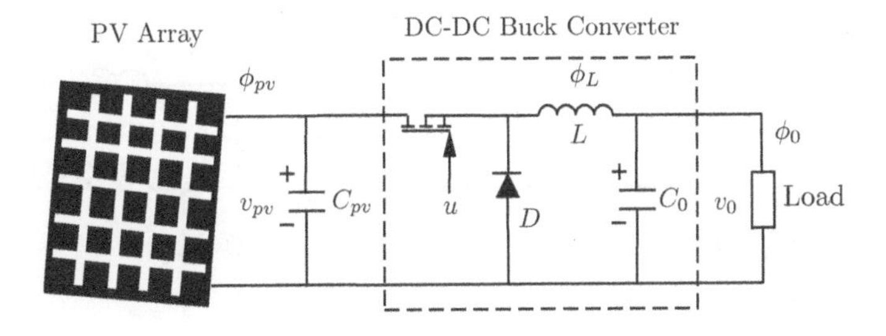

Figure 1.2 Solar PV power with DC-DC boost converter.

Now, consider a solar PV power system using the DC-DC boost converter as shown in Figure 1.2. Its dynamic model can be represented by the following differential equations [14]

$$\begin{cases} \dot{\phi}_{pv} = -\frac{1}{L}(1-u)v_{dc} + \frac{1}{L}v_{pv}, \\ \dot{v}_{dc} = \frac{1}{C_0}(1-u)\phi_{pv} - \frac{1}{C_0}\phi_0, \end{cases} \tag{1.6}$$

where $u \in [0,1]$ denotes the duty ratio, and ϕ_{pv} and v_{dc} stand for the inductor current and the output voltage, respectively. It should be noted that the duty ratio u determines the switching action via the pulsewidth modulation.

Define $x(t) = \begin{bmatrix} \phi_{pv} & v_{dc} \end{bmatrix}^T$, and choose $z_1 = \frac{v_{pv}}{\phi_{pv}}$, $z_2 = \frac{\phi_0}{v_{dc}}$, $z_3 = v_{dc}$, $z_4 = \phi_{pv}$, as fuzzy premise variables. Similar to the procedure in (1.2) and (1.3), and it follows from (1.6) that the PV power nonlinear system is represented by the following T-S model,

Plant Rule $\mathscr{R}^l$: **IF** z_1 is $\mathscr{F}_1^l$ and z_2 is $\mathscr{F}_2^l$ and z_3 is $\mathscr{F}_3^l$ and z_4 is $\mathscr{F}_4^l$, **THEN**

$$\dot{x}(t) = A_l x(t) + B_l u(t), l \in \mathscr{L} := \{1,2,\ldots,r\} \tag{1.7}$$

where

$$A_l = \begin{bmatrix} \frac{1}{L}\mathscr{F}_1^l & -\frac{1}{L} \\ \frac{1}{C_0} & -\frac{1}{C_0}\mathscr{F}_2^l \end{bmatrix}, B_l = \begin{bmatrix} \frac{1}{L}\mathscr{F}_3^l \\ -\frac{1}{C_0}\mathscr{F}_4^l \end{bmatrix}. \tag{1.8}$$

By fuzzy blending, the global T-S fuzzy dynamic model is obtained by

$$\dot{x}(t) = A(\mu)x(t) + B(\mu)u(t), \tag{1.9}$$

where $A(\mu) := \sum_{l=1}^{r} \mu_l A_l, B(\mu) := \sum_{l=1}^{r} \mu_l B_l$.

1.1.2 MODELING OF PV POWER WITH AC LOAD

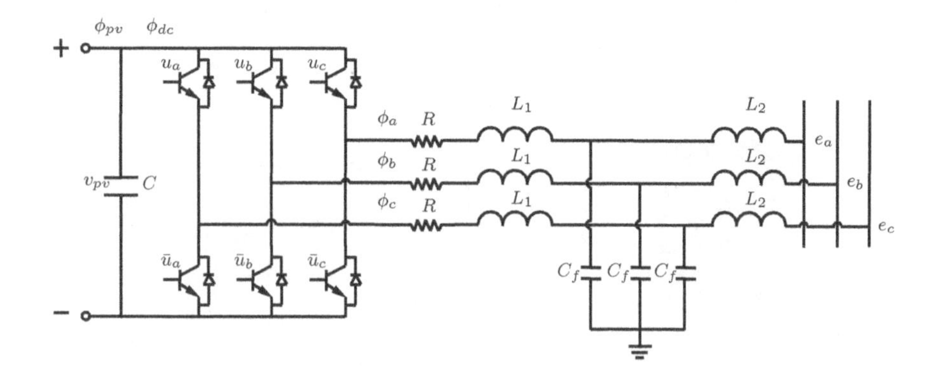

Figure 1.3 Solar PV power using DC-AC converter.

Consider a solar PV power system using the DC-AC converter as shown in Figure 1.3. Its dynamic model can be represented by the following differential equations [15],

$$C_{pv}\dot{v}_{pv} = \phi_{pv} - \phi_{dc}, \tag{1.10}$$

where C_{pv} denotes the value of the DC-bus capacitor; v_{pv}, ϕ_{pv}, and ϕ_{dc} denote the PV output voltage, the output DC currents, and the input DC current of the interlinking DC-AC converter, respectively.

The DC-AC converter model in the synchronous dq frame of the grid voltage can be depicted as

$$\begin{cases} L_1\dot{\phi}_{invd} = e_d - \phi_{invd}R_1 - [v_{cd} + (\phi_{invd} - i_d)R_d] - \omega L_1\phi_{invq}, \\ L_1\dot{\phi}_{invq} = e_q - \phi_{invq}R_1 - [v_{cq} + (\phi_{invq} - i_q)R_d] + \omega L_1\phi_{invd}, \\ C_f\dot{v}_{cd} = \phi_{invd} - i_d - \omega C_f v_{cq}, \\ C_f\dot{v}_{cq} = \phi_{invq} - i_q + \omega C_f v_{cd}, \\ L_2\dot{\phi}_d = v_{cd} + (\phi_{invd} - \phi_d)R_d - \phi_d R_2 - u_d - \omega L_2\phi_q, \\ L_2\dot{\phi}_q = v_{cq} + (\phi_{invq} - \phi_q)R_d - \phi_q R_2 + \omega L_2\phi_d, \end{cases} \tag{1.11}$$

where e_d and e_q are the d- and q-axis components of inverter output voltage; u_d is the d-axis component of grid voltage; ω is the fundamental angular frequency; L_1 and L_2 are the filter inductances (see Figure 1.3); R_1 and R_2 denote the equivalent resistances describing the system loss and the parasitic resistances of the filter inductances, respectively; C_f and R_d are the filter capacitance and passive damping resistance, respectively; ϕ_{invd} and ϕ_{invq} denote the active and reactive components of the converter side current, respectively; ϕ_d and ϕ_q are the active and reactive components of the grid side current, respectively; v_{cd} and v_{cq} are the d- and q- axis components of filter capacitance voltage, respectively.

Neglecting the conversion loss of the converters, the active power pg transferred between the DC subgrid and the AC grid can be expressed by [16]

$$\begin{aligned} P &= v_{PV}\phi_{dc} \\ &= 1.5u_d\phi_d. \end{aligned} \tag{1.12}$$

Note that the LCL filter is used to primarily reduce the high frequency current ripple, and the bandwidth of the closed-loop control system is always designed to be less than the resonant frequency of the LCL filter. Therefore, in the latter the influence of the filter capacitors can be neglected for the DC-bus control system design. Based on the above discussion and (1.10)-(1.12), and considering only the active current dynamic equation, the DC-bus voltage system can be described by [17]

$$\begin{cases} \dot{v}_{pv} = \frac{1}{C_{pv}}(\phi_{pv} - \frac{1.5u_d}{v_{pv}}\phi_d), \\ \dot{\phi}_d = -\frac{R_1}{L_1}\phi_d - \omega\phi_q + \frac{1}{L_1}e_d, \\ \dot{\phi}_q = \omega\phi_d - \frac{R_1}{L_1}\phi_q + \frac{1}{L_1}e_q. \end{cases} \tag{1.13}$$

In that case, choose $z_1 = \frac{\phi_{pv}}{v_{pv}}, z_2 = \frac{u_d}{v_{pv}}$, and $z_3 = \omega$ as the fuzzy premise variables. Similar to the fuzzy processes in (1.9), the PV power nonlinear system using the DC-AC converter is represented by

$$\dot{x}(t) = A(\mu)x(t) + B(\mu)u(t), \tag{1.14}$$

where $x(t) = \begin{bmatrix} v_{pv} & \phi_d & \phi_q \end{bmatrix}^T, A(\mu) := \sum_{l=1}^{r} \mu_l A_l, B(\mu) := \sum_{l=1}^{r} \mu_l B_l$, and

$$A_l = \begin{bmatrix} \frac{1}{C_{pv}}\mathscr{F}_1^l & \frac{-1.5}{C_{pv}}\mathscr{F}_2^l & 0 \\ 0 & -\frac{R_1}{L_1} & -\mathscr{F}_3^l \\ 0 & \mathscr{F}_3^l & -\frac{R_1}{L_1} \end{bmatrix}, B = \begin{bmatrix} 0 & 0 \\ \frac{1}{L_1} & 0 \\ 0 & \frac{1}{L_1} \end{bmatrix}, u(t) = \begin{bmatrix} e_d \\ e_q \end{bmatrix}. \tag{1.15}$$

1.2 CONTROL OF PV POWER

In this section, based on the fuzzy model (1.5) and (1.14), the stability analysis and controller design will be given, respectively.

1.2.1 STABILITY ANALYSIS OF PV POWER

Consider a T-S fuzzy controller, which shares the same premise variables in (1.9), is proposed as below:

Controller Rule $\mathscr{R}^l$: **IF** z_1 is $\mathscr{F}_1^l$ and z_2 is $\mathscr{F}_2^l$ and $\cdots$ and z_4 is $\mathscr{F}_4^l$, **THEN**

$$u(t) = K_l x(t), l \in \mathscr{L} \tag{1.16}$$

where $K_l \in \Re^{n_u \times n_x}$ are controller gains to be designed.

Likewise, the total T-S fuzzy controller is given by

$$u(t) = K(\mu)x(t), \tag{1.17}$$

where $K(\mu) := \sum_{l=1}^{r} \mu_l K_l$.

Note: The fuzzy controller in (1.17) is proposed for the PV power with the DC-DC buck converter in (1.5). However, the corresponding results can be easily extended to the PV power with the DC-DC boost converter in (1.9) and (1.14).

It follows from (1.5) and (1.17) that the closed-loop PV power fuzzy system is given by

$$\dot{x}(t) = (A(\mu) + B(\mu)K(\mu))x(t). \tag{1.18}$$

Based on the closed-loop fuzzy control system in (1.18), the result on stability analysis is proposed as below:

Theorem 1.1: Stability Analysis of PV Power Fuzzy System

Consider the PV power fuzzy system with the DC-DC buck converter in (1.5), and a T-S fuzzy controller in the form of (1.17). Then, the asymptotic stability of the closed-loop fuzzy control system is achieved, if the following condition is satisfied:

$$\mathrm{Sym}(P(A(\mu) + B(\mu)K(\mu))) < 0, \tag{1.19}$$

where $\mathrm{Sym}(\star) = (\star) + (\star)^T$, $0 < P = P^T \in \Re^{n_x \times n_x}$, $K(\mu) \in \Re^{n_u \times n_x}$. ■

Proof. Consider the following Lyapunov functional:

$$V(t) = x^T(t)Px(t), \tag{1.20}$$

where $0 < P = P^T \in \Re^{n_x \times n_x}$.

By differentiating $V(t)$ shown in (1.20) with respect to time and using the closed-loop control system in (1.18), it yields

$$\dot{V}(t) = \mathrm{Sym}\left(P\left(A(\mu) + B(\mu)K(\mu)\right)\right). \tag{1.21}$$

The inequality in (1.19) is obtained directly.

Note: A quadratic Lyapunov function $V(t) = x^T(t)Px(t)$ is considered in (1.20). It is clear that if $P \equiv \sum_{l=1}^{r} \mu_l P_l$, the function in (1.20) turns to the fuzzy-basis-dependent Lyapunov function $V(t) = x^T(t)P(\mu)x(t)$. However, it requires that the time-derivative of μ_l is known a priori, which may be unpractical for the considered system.

1.2.2 CONTROL SYNTHESIS OF PV POWER

Based on the result on Theorem 1.1, the fuzzy controller gains can be calculated as below:

Theorem 1.2: Controller Design of PV Power Fuzzy System

For the PV power fuzzy system with DC-DC buck converter in (1.5), a T-S fuzzy controller in the form of (1.17) can be used to stabilize its closed-loop control system, if the following LMIs (MATLAB®) are satisfied:

$$\Sigma_{ll} < 0, l \in \mathscr{L} \tag{1.22}$$

$$\Sigma_{ls} + \Sigma_{sl} < 0, 1 \leq l < s \leq r \tag{1.23}$$

where $\Sigma_{ls} = \mathrm{Sym}(A_l X + B_l \bar{K}_s)$, $0 < X = X^T \in \Re^{n_x \times n_x}$, $\bar{K}_s \in \Re^{n_u \times n_x}$.

In that case, the controller gains can be calculated by

$$K_l = \bar{K}_l X^{-1}. \tag{1.24}$$

■

Proof. By performing the congruence transformation to (1.19) by $X = P^{-1}$, and define $\bar{K}_l = K_l X$. Then, by extracting fuzzy premise variable, the controller design result can be directly obtained. Thus, the proof is completed.

1.3 MPPT FUZZY CONTROL OF PV POWER

In order to maximize the efficiency of the PV power-generation systems, the subsection will propose an MPP tracking (MPPT) technique based on the descriptor system approach. First, the electric characteristic of the PV arrays is considered as below [13]:

$$\begin{cases} \phi_{pv} = n_p I_{ph} - n_p I_{rs} \left(e^{\gamma v_{pv}} - 1 \right), \\ P_{pv} = \phi_{pv} v_{pv}, \end{cases} \tag{1.25}$$

where n_p and n_s are the number of the parallel and series cells, respectively; $\gamma = q/(n_s \phi KT)$ is of the electronic charge $q = 1.6 \times 10^{-19}$ C, the Boltzmann's constant $K = 1.3805 \times 10^{-23}$J/°K, the cell temperature T, and the ideal $p-n$ junction characteristic factor $\phi = 1-5$; I_{ph} and I_{rs} are the light-generated current and the reverse saturation current, respectively. Here, the series resistances and their intrinsic shunt are neglected.

According to the array power (1.25) and by taking the partial derivative of P_{pv} with respect to the PV voltage v_{pv}, one gets [13]

$$\frac{dP_{pv}}{dv_{pv}} = \phi_{pv} - n_p \gamma I_{rs} v_{pv} e^{\gamma v_{pv}}. \tag{1.26}$$

1.3.1 MODELING OF MPPT OF PV POWER WITH DC LOAD

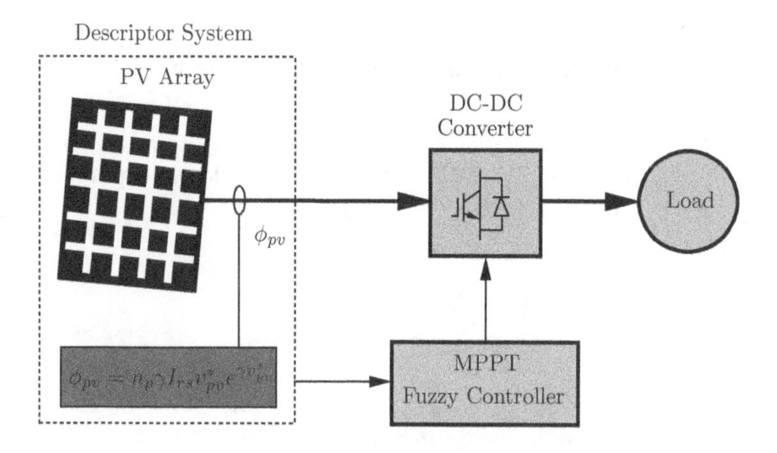

Figure 1.4 MPPT fuzzy control for PV power system.

To achieve the MPPT performance, it needs $\frac{dP_{pv}}{dv_{pv}} = 0$. Here, the proposed descriptor system approach is shown in Figure 1.4. First we measure the PV array current ϕ_{pv}, and we solve the equation $\phi_{pv} - n_p \gamma I_{rs} v^*_{pv} e^{\gamma v^*_{pv}} = 0$ to obtain the reference PV array voltage v^*_{pv}. When the condition $v_{pv} \to v^*_{pv}$ holds, the closed-loop PV power control system achieves the maximum power tracking. Thus we further define the output $e_{pv} = v_{pv} - v_{pv}^*$ and introduce the virtual state variable v_{pv}^*, we get

$$\left\{\begin{array}{l} \dot{\phi}_L = \frac{1}{L}R_0\left((\phi_0 - \phi_L) - R_L\phi_L - v_0\right) + \frac{1}{L}\left(V_D + v_{PV} - R_M\phi_L\right)u - \frac{V_D}{L}, \\ \dot{v}_0 = \frac{1}{C_0}\left(\phi_L - \phi_0\right), \\ \dot{e}_{pv} = \frac{1}{C_{PV}}\left(\phi_{PV} - \phi_L u\right) - \dot{v}_{pv}^*, \\ 0\dot{v}_{pv}^* = \phi_{PV} - n_p\gamma I_{rs}v_{pv}^* e^{\gamma v_{pv}^*}. \end{array}\right. \tag{1.27}$$

Here, choose $z_1 = \frac{\phi_0}{\phi_L}, z_2 = \phi_L, z_3 = v_{pv}, z_4 = \frac{V_D}{v_{pv}^*}, z_5 = \frac{\phi_{pv}}{e_{pv}}, z_6 = \frac{\dot{v}_{pv}^*}{v_{pv}^*}$ and $z_7 = e^{\gamma v_{pv}^*}$ as the fuzzy premise variables. The PV power fuzzy system is given by the following T-S model,

$$E\dot{x}(t) = A(\mu)x(t) + B(\mu)u(t), \tag{1.28}$$

where $A(\mu) := \sum_{l=1}^{r} \mu_l A_l, B(\mu) := \sum_{l=1}^{r} \mu_l B_l$, and

$$A_l = \begin{bmatrix} \frac{R_0}{L}\mathscr{F}_1^l - \frac{R_0}{L} - \frac{R_0R_L}{L} & -\frac{R_0}{L} & 0 & -\frac{\mathscr{F}_4^l}{L} \\ \frac{1-\mathscr{F}_1^l}{C_0} & 0 & 0 & 0 \\ 0 & 0 & \frac{\mathscr{F}_5^l}{C_{PV}} & -\mathscr{F}_6^l \\ 0 & 0 & \mathscr{F}_5^l & -n_p\gamma I_{rs}\mathscr{F}_7^l \end{bmatrix},$$

$$E = \begin{bmatrix} 1 & 0 & 0 & 0 \\ 0 & 1 & 0 & 0 \\ 0 & 0 & 1 & 0 \\ 0 & 0 & 0 & 0 \end{bmatrix}, B_l = \begin{bmatrix} \frac{1}{L}\left(V_D + \mathscr{F}_3^l - R_M\mathscr{F}_2^l\right) \\ 0 \\ -\frac{\mathscr{F}_2^l}{C_{PV}} \\ 0 \end{bmatrix}. \tag{1.29}$$

Recall the solar PV power system using the DC-DC boost converter as shown in (1.9). Now, by introducing the virtual state variable $\varepsilon_{pv} = v_{pv} - v_{pv}^*$, it follows from (1.25) and (1.26) that the considered PV system with MPPT control problem is reformulated into the following descriptor system:

$$\left\{\begin{array}{l} \dot{\phi}_{pv} = -\frac{1}{L}\left(1-u\right)v_{dc} + \frac{1}{L}v_{pv}, \\ \dot{v}_{dc} = \frac{1}{C_0}\left(1-u\right)\phi_{pv} - \frac{1}{C_0}\phi_0, \\ 0\cdot\dot{\varepsilon}_{pv} = \phi_{pv} - n_p\gamma I_{rs}e^{\gamma v_{pv}^*}\varepsilon_{pv} - n_p\gamma I_{rs}v_{pv}e^{\gamma v_{pv}^*}. \end{array}\right. \tag{1.30}$$

Define $x(t) = \begin{bmatrix} \phi_{pv} & v_{dc} & \varepsilon_{pv} \end{bmatrix}^T$, and choose $z_1 = \frac{v_{pv}}{\phi_{pv}}, z_2 = \frac{\phi_0}{v_{dc}}, z_3 = e^{\gamma v_{pv}^*}, z_4 = \frac{v_{pv}}{\phi_{pv}}e^{\gamma v_{pv}^*}, z_5 = v_{dc}$, and $z_6 = \phi_{pv}$ as fuzzy premise variables. The PV nonlinear system in (4) is represented by the following descriptor T-S model:

Plant Rule $\mathscr{R}^l$: **IF** z_1 is $\mathscr{F}_1^l$, and z_2 is $\mathscr{F}_2^l$, and, $\cdots$, and z_6 is $\mathscr{F}_6^l$, **THEN**

$$E\dot{x}(t) = A_l x(t) + B_l u(t), l \in \mathscr{L} := \{1, 2, \ldots, r\} \tag{1.31}$$

where

$$A_l = \begin{bmatrix} \frac{\mathscr{F}_1^l}{L} & -\frac{1}{L} & 0 \\ \frac{1}{C_0} & -\frac{\mathscr{F}_2^l}{C_0} & 0 \\ 1 - n_p \gamma I_{rs} \mathscr{F}_4^l & 0 & -n_p \gamma I_{rs} \mathscr{F}_3^l \end{bmatrix},$$

$$E = \begin{bmatrix} 1 & 0 & 0 \\ 0 & 1 & 0 \\ 0 & 0 & 0 \end{bmatrix}, B_l = \begin{bmatrix} \frac{\mathscr{F}_5^l}{L} \\ -\frac{\mathscr{F}_6^l}{C_0} \\ 0 \end{bmatrix}. \tag{1.32}$$

Note: It is worth noting that analyzing the MPPT properties of the PV power falls within the framework of descriptor fuzzy systems, as shown in (1.28) and (1.31).

1.3.2 MODELING OF MPPT OF PV POWER WITH AC LOAD

Referring back to Figure 1.4, we consider the MPPT of the PV power with the AC load. Define the output $e_{pv} = v_{pv}^* - v_{pv}$ and introduce the virtual state variable v_{pv}^*,

$$\begin{cases} \dot{v}_{pv} = \frac{1}{C_{pv}}(\phi_{pv} - \frac{1.5u_d}{v_{pv}}\phi_d), \\ \dot{\phi}_d = -\frac{R_1}{L_1}\phi_d - \omega\phi_q + \frac{1}{L_1}e_d, \\ \dot{\phi}_q = \omega\phi_d - \frac{R_1}{L_1}\phi_q + \frac{1}{L_1}e_q, \\ 0\dot{e}_{pv} = \phi_{pv} - n_p\gamma I_{rs}e^{\gamma(e_{pv}+v_{pv})}e_{pv} - n_p\gamma I_{rs}e^{\gamma(e_{pv}+v_{pv})}v_{pv}. \end{cases} \tag{1.33}$$

Define $x(t) = \begin{bmatrix} v_{pv} & \phi_d & \phi_q & e_p & e_u \end{bmatrix}^T$, and choose $z_1 = \frac{\phi_{pv}}{v_{pv}}, z_2 = \frac{u_d}{v_{pv}}, z_3 = \omega$, and $z_4 = e^{\gamma(e_{pv}+v_{pv})}$ as the fuzzy premise variables. Thus, it follows from (1.33) that the PV power nonlinear system is represented by

Plant Rule $\mathscr{R}^l$: **IF** z_1 is $\mathscr{F}_1^l$ and z_2 is $\mathscr{F}_2^l$ and z_3 is $\mathscr{F}_3^l$ and z_4 is $\mathscr{F}_4^l$, **THEN**

$$\dot{x}(t) = A_l x(t) + B_l u(t), l \in \mathscr{L} := \{1, 2, \ldots, r\} \tag{1.34}$$

where $\mathscr{R}^l$ denotes the l-th fuzzy inference rule; r is the number of inference rules; $\mathscr{F}_\theta^l(\theta = 1, 2, \ldots, 4)$ is the fuzzy set; $x(t) \in \Re^{n_x}$ and $u(t) \in \Re^{n_u}$ denote the system state and control input, respectively; $z(t) \triangleq [z_1, z_2, z_3, z_4]$ are the measurable variables; $\{A_l, B_l\}$ is the l-th local model as below:

$$A_l = \begin{bmatrix} \frac{1}{C_{pv}}z_1^l & -\frac{1.5}{C_{pv}}z_2^l & 0 & 0 \\ 0 & -\frac{R_1}{L_1} & -\omega & 0 \\ 0 & z_3^l & -\frac{R_1}{L_1} & 0 \\ z_1^l - n_p\gamma I_{rs}z_4^l & 0 & 0 & -n_p\gamma I_{rs}z_4^l \end{bmatrix},$$

$$E = \begin{bmatrix} 1 & 0 & 0 & 0 \\ 0 & 1 & 0 & 0 \\ 0 & 0 & 1 & 0 \\ 0 & 0 & 0 & 0 \end{bmatrix}, B_l = \begin{bmatrix} 0 & 0 \\ \frac{1}{L_1} & 0 \\ 0 & \frac{1}{L_1} \\ 0 & 0 \end{bmatrix}. \tag{1.35}$$

Denoting as $\mathscr{F}^l := \prod_{\phi=1}^{4} \mathscr{F}_\phi^l$ the inferred fuzzy set, and $\mu_l[z(t)]$ as the normalized membership function, yields

$$\mu_l[z(t)] := \frac{\prod_{\phi=1}^{4} \mu_{\phi l}[z_\phi(t)]}{\sum_{\varsigma=1}^{r} \prod_{\phi=1}^{4} \mu_\phi^\varsigma [z_\phi(t)]} \geq 0, \sum_{l=1}^{r} \mu_l[z(t)] = 1. \tag{1.36}$$

Now, $\mu_l \triangleq \mu_l[z(t)]$ is defined for brevity.

By fuzzy blending, the T-S fuzzy dynamic model is obtained by

$$E\dot{x}(t) = A(\mu)x(t) + B(\mu)u(t), \tag{1.37}$$

where $A(\mu) := \sum_{l=1}^{r} \mu_l A_l, B(\mu) := \sum_{l=1}^{r} \mu_l B_l$.

1.3.3 MPPT CONTROLLER DESIGN

Consider a T-S fuzzy controller, which shares the same premise variables in (1.28), as follows:

Controller Rule $\mathscr{R}^l$: **IF** z_1 is $\mathscr{F}_1^l$ and z_2 is $\mathscr{F}_2^l$ and $\cdots$ and z_7 is $\mathscr{F}_7^l$, **THEN**

$$u(t) = K_l x(t), l \in \mathscr{L} \tag{1.38}$$

where $K_l \in \Re^{n_u \times n_x}$ are the controller gains to be designed.

Likewise, the global T-S fuzzy controller is given by

$$u(t) = K(\mu)x(t), \tag{1.39}$$

where $K(\mu) := \sum_{l=1}^{r} \mu_l K_l$.

Note: In this section, the fuzzy controller is proposed for the MPPT of PV power with DC load in (1.28). However, the corresponding results can be easily extended to the MPPT of PV power with AC load in (1.37).

Based on the descriptor fuzzy system in (1.28), the MPPT fuzzy control problem can be solved as below:

Theorem 1.3: Stability Analysis for MPPT of PV System

Consider the PV power fuzzy system with the DC-DC buck converter in (1.5) using the T-S fuzzy controller laws in the form of (1.34). The MPPT control system of the PV generation is asymptotically achieved if the following inequalities hold:

$$E^T P = P^T E \geq 0, \tag{1.40}$$

$$P^T \bar{A}(\mu) + \bar{A}^T(\mu) P < 0, \tag{1.41}$$

where $\bar{A}(\mu) = A(\mu) + B(\mu)K(\mu)$, $P \in \Re^{n_x \times n_x}$, $K(\mu) \in \Re^{n_u \times n_x}$. ■

Proof. Consider the following Lyapunov function,

$$V(t) = x^T(t) E^T P x(t), \tag{1.42}$$

with $E^T P = P^T E \geq 0$.

Then, define $\bar{A}(\mu) = A(\mu) + B(\mu)K(\mu)$ and take the time derivative of $V(t)$ along the trajectory of the descriptor system in (1.28) with the fuzzy controller in (1.39),

$$\begin{aligned}\dot{V}(t) &= 2\dot{x}^T(t) E^T P x(t) \\ &= 2\left[\bar{A}(\mu)x(t)\right]^T P x(t) \\ &= x^T(t)\left(P^T \bar{A}(\mu) + \bar{A}^T(\mu)P\right)x(t).\end{aligned} \tag{1.43}$$

It directly obtains the inequalities in (1.40) and (1.41), and thus the proof is completed.

Note: It is well-known that the strict LMI conditions are nice results, which can be easily checked with MATLAB's LMI toolbox. However, the conditions in Theorem 1.3 are not all of strict LMI form due to matrix equality constraint $E^T P = P^T E \geq 0$ as shown in (1.40). This may cause the difficulty in solving the MPPT problem.

Now, define

$$P = \begin{bmatrix} P_{(1)} & 0 \\ P_{(2)} & P_{(3)} \end{bmatrix}, \tag{1.44}$$

where $0 < P_{(1)} = P_{(1)}^T \in \Re^{(n_x-1)\times(n_x-1)}, P_{(2)} \in \Re^{1\times(n_x-1)}$, $P_{(3)}$ is a scalar. It is easy to see that the inequality $E^T P = P^T E \geq 0$ holds.

In order to derive an LMI-based result, we define $X = P^{-1}$, that is

$$X = \begin{bmatrix} X_{(1)} & 0 \\ X_{(2)} & X_{(3)} \end{bmatrix}, \tag{1.45}$$

where $0 < X_{(1)} = X_{(1)}^T \in \Re^{(n_x-1)\times(n_x-1)}, X_{(2)} \in \Re^{1\times(n_x-1)}$, $X_{(3)}$ is a scalar.

Now, the strict LMI result on the MPPT controller design can be proposed as below:

Theorem 1.4: Controller Design for MPPT of PV with AC Load

Consider the PV power fuzzy system with DC-DC buck converter in (1.5) using the T-S fuzzy controller laws in the form of (1.34). The MPPT control system of PV generation is asymptotically achieved if the following LMIs hold:

$$\Sigma_{ll} < 0, l \in \mathscr{L} \tag{1.46}$$

$$\Sigma_{ls} + \Sigma_{sl} < 0, 1 \leq l < s \leq r \tag{1.47}$$

where $\Sigma_{ls} = \text{Sym}(A_l X + B_l \bar{K}_s), X = \begin{bmatrix} X_{(1)} & 0 \\ X_{(2)} & X_{(3)} \end{bmatrix}, 0 < X_{(1)} = X_{(1)}^T \in \Re^{(n_x-1)\times(n_x-1)}$, $X_{(2)} \in \Re^{1\times(n_x-1)}$, $X_{(3)}$ is a scalar.

In that case, the controller gains can be calculated by

$$K_l = \bar{K}_l X^{-1}, l \in \mathscr{L} \tag{1.48}$$

■

Proof. By performing a congruence transformation to (1.41) by $X = P^{-1}$, one gets

$$\text{Sym}\left(A(\mu)X + B(\mu)K(\mu)X\right) < 0. \tag{1.49}$$

By defining $\bar{K}_l = K_l X$, and extracting the fuzzy premise variables, the strict LMI result on controller design can be obtained in (1.46) and (1.47), and thus the proof is completed.

1.4 ROBUST MPPT FUZZY OBSERVER-BASED CONTROL

The operation at the maximum power point (MPP) of PV power panels is very sensitive to solar irradiance and cell temperature as shown in Figure 1.5. The cell temperature is easy to obtained by sensors, but this is not the case for solar irradiance because solar irradiance sensors are expensive and difficult to calibrate [18]. Nowadays, much effort has been devoted to propose different algorithms, which perform the MPPT function by avoiding a direct measurement of solar irradiance, such as the perturb and observe [19], incremental conductance [20], and incremental resistance algorithms [21]. However, those methods may cause the operation to oscillate around the MPP and even may fail under rapidly changing irradiance conditions.

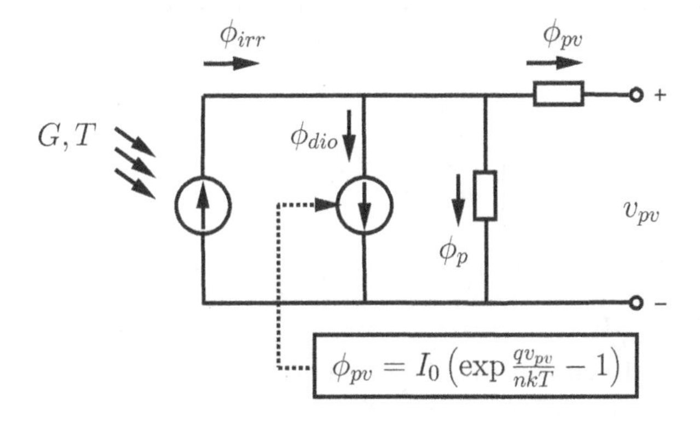

Figure 1.5 Circuital model for a single PV cell.

1.4.1 MODELLING OF UNCERTAIN PV POWER

Consider a single circuital model including all details of each cell, then the output current of the PV cell of Figure 1.5 may be expressed as [18]

$$\phi_{pv} = \phi_{irr} - \phi_{dio} - \phi_p, \tag{1.50}$$

where ϕ_{irr} denotes the photocurrent or irradiance current generated when the cell is exposed to sunlight, ϕ_{dio} is the current flowing through the antiparallel diode and induces the nonlinear characteristics of the PV cell, ϕ_p is a shunt current due to the shunt resistor branch. Substituting the relevant expressions for ϕ_{dio} and ϕ_p,

$$\phi_{pv} = \phi_{irr} - \phi_0 \left[\exp\left(\frac{q\left(v_{pv} + \phi_{pv} R_s\right)}{nkT} \right) - 1 \right] - \frac{v_{pv} + \phi_{pv} R_s}{R_p}, \tag{1.51}$$

where $q = 1.602 \times 10^{-19}$ C is the electron's electric charge, $k = 1.3806503 \times 10^{-23}$ J/K is the Boltzmann constant, T is the temperature of the cell, ϕ_0 is the diode saturation current or cell reverse saturation current, n is the ideality factor or the ideal constant of the diode, and R_s and ϕ_p represent the series and shunt resistance, respectively.

A PV power plant will usually contain a large number of cells in series, and the model of a single cell is generalized to an arbitrary number $N_s \times N_p$ of cells connected in series and parallel to form an array taking the final form [18]

$$\phi_{pv} = N_P \phi_{irr} - N_P \phi_0 \left[\exp\left(\frac{q\left(v_{pv} + \phi_{pv} \frac{N_s}{N_p} R_s\right)}{N_s nkT} \right) - 1 \right] - \frac{v_{pv} + \phi_{pv} \frac{N_s}{N_p} R_s}{\frac{N_s}{N_p} R_p}. \tag{1.52}$$

In the $\phi_{pv} - v_{pv}$ performance characteristic of a single PV cell and an array described by (1.51) and (1.52), respectively, the parameters ϕ_{irr}, ϕ_0, and R_p depend on the solar irradiance (G), the cell temperature (T), and the certain reference parameters $(G_{ref}, T_{ref}, \phi_{irr,ref}, \phi_{0,ref}$, and $R_{p,ref})$, as follows:

$$\begin{cases} \phi_{irr} = \left(k_1 + k_2(T)\right) G, \\ \phi_0 = \phi_{0,ref} \left(\frac{T}{T_{ref}} \right)^3 e^{k_3}, \\ R_p = k_4 G, \end{cases} \tag{1.53}$$

where $k_1 = \frac{\phi_{irr,ref}}{G_{ref}}, \phi_{irr,ref} \frac{G}{G_{ref}} \left(1 + \bar{\alpha}_T \left(T - T_{ref}\right)\right), k_2(T) = k_1 \bar{\alpha}_T \left(T - T_{ref}\right), k_3 = \frac{E_{g,ref}}{kT_{ref}} - \frac{E_g}{kT}, k_4 = \frac{R_{p,ref}}{G_{ref}}$.

According to the array power (1.25) and by taking the partial derivative of P_{pv} with respect to the PV voltage v_{pv}, we obtain

$$\frac{dP_{pv}}{dv_{pv}} = \phi_{pv} - \frac{d\phi_{pv}}{dv_{pv}} v_{pv}. \tag{1.54}$$

It follows from (1.50)-(1.54) that

$$\frac{d\phi_{pv}}{dv_{pv}} = -\frac{\frac{0.5qN_P\phi_{0,ref}\left(\frac{T}{T_{ref}}\right)^3 e^{k_3}e^{k_0}}{N_s nkT} + \frac{0.5N_p}{N_s k_4 G}}{1+\frac{0.5N_P\phi_{0,ref}\left(\frac{T}{T_{ref}}\right)^3 e^{k_3}qR_s}{N_p nkT}e^{k_0}}, \tag{1.55}$$

where $k_0 = \frac{q\left(v_{pv}+\phi_{pv}\frac{N_s}{N_p}R_s\right)}{N_s nkT}$.

To achieve the MPPT performance, we have $\frac{dP_{pv}}{dv_{pv}} = 0$. Here, by measuring the PV array current ϕ_{pv} and using the estimation G, we can then calculate the reference PV array voltage v_{pv}^*. When the condition $v_{pv} \to v_{pv}^*$ holds, the closed-loop PV control system achieves maximum power tracking control.

Based on the above description, the PV power system with unknown solar irradiation is given by

$$\begin{cases} \dot{v}_{pv} = \frac{1}{C_{pv}}\left(\phi_{pv} - \phi_L u\right), \\ \dot{\phi}_L = \frac{1}{L}R_0\left((\phi_0 - \phi_L) - R_L\phi_L - v_0\right) + \frac{1}{L}\left(V_D + v_{pv} - R_M\phi_L\right)u - \frac{V_D}{L}, \\ \dot{v}_0 = \frac{1}{C_0}\left(\phi_L - \phi_0\right), \\ 0\dot{G} = 0.5N_P\left(k_1 + k_2(T)\right)G^2 - G\phi_{pv} - \frac{0.5N_p v_{pv}}{k_4 N_s} \\ \quad - 0.5N_P\phi_{0,ref}\left(\frac{T}{T_{ref}}\right)^3 e^{k_3}\left[\exp(k_0) - 1\right]G. \end{cases} \tag{1.56}$$

Here, choose $z_1 = \frac{\phi_{pv}}{v_{pv}}, z_2 = \phi_L, z_3 = \frac{\phi_0}{\phi_L}, z_4 = \frac{V_D}{v_{pv}}$, $z_5 = v_{pv}$, $z_6 = T, z_7 = G$, $z_8 = \phi_{pv}, z_9 = \frac{\dot{v}_{pv}^*}{\phi_L}$, $z_{10} = \frac{v_{pv}^*}{\phi_L}$ as the fuzzy premise variables. Note that the proposed observer is to obtain the estimator of G, and then calculate the reference PV array voltage v_{pv}^*. Now, define $\zeta(t) = v_{pv} - v_{pv}^*$ and introduce the new state variable $\zeta(t)$, and define $x(t) = \begin{bmatrix} \zeta(t) & \phi_L & v_0 & G \end{bmatrix}^T$. Thus, it follows from (1.51) that the PV power nonlinear system is represented by

$$\begin{cases} E\dot{x}(t) = A(\mu)x(t) + B(\mu)u(t), \\ y(t) = Cx(t), \end{cases} \tag{1.57}$$

where $A(\mu) := \sum_{l=1}^{r} \mu_l A_l, B(\mu) := \sum_{l=1}^{r} \mu_l B_l$,

$$A_l = \begin{bmatrix} \frac{1}{C_{pv}} \mathscr{F}_1^l & -\mathscr{F}_9^l & 0 & 0 \\ -\frac{\mathscr{F}_4^l}{L} & \frac{R_0}{L}\mathscr{F}_3^l - \frac{R_0}{L} - \frac{R_0 R_L}{L} + \frac{\mathscr{F}_{10}^l}{L} & -\frac{R_0}{L} & 0 \\ 0 & \frac{1-\mathscr{F}_3^l}{C_0} & 0 & 0 \\ -\frac{0.5N_p}{k_4 N_s} & -\frac{0.5N_p \mathscr{F}_{10}^l}{k_4 N_s} & 0 & A_{l(1)} \end{bmatrix},$$

$$A_{l(1)} = 0.5N_P\left(k_1 + k_2\left(T\right)\right)\mathscr{F}_7^l - 0.5N_P\phi_{0,ref}\left(\frac{T}{T_{ref}}\right)^3 e^{k_3}\left[\exp\left(k_0\right) - 1\right] - \mathscr{F}_8^l,$$

$$E = \begin{bmatrix} 1 & 0 & 0 & 0 \\ 0 & 1 & 0 & 0 \\ 0 & 0 & 1 & 0 \\ 0 & 0 & 0 & 0 \end{bmatrix}, B_l = \begin{bmatrix} -\frac{\mathscr{F}_2^l}{C_{PV}} \\ \frac{1}{L}\left(V_D + \mathscr{F}_5^l - R_M \mathscr{F}_2^l\right) \\ 0 \\ 0 \end{bmatrix}, C = \begin{bmatrix} 1 & 0 & 0 & 0 \\ 0 & 1 & 0 & 0 \\ 0 & 0 & 1 & 0 \end{bmatrix},$$

$$\theta\left(\mathscr{F}_1^l, \mathscr{F}_5^l\right) = \frac{\phi_0 N_s R_{p,ref}}{G_{ref}} - \frac{N_s R_{p,ref} \mathscr{F}_1^l \mathscr{F}_5^l}{N_p G_{ref}} - \frac{\phi_0 N_s R_{p,ref}}{G_{ref}} \exp\left(\frac{q\left(\mathscr{F}_5^l + \mathscr{F}_1^l \mathscr{F}_5^l \frac{N_s}{N_p} R_s\right)}{N_s n k T}\right). \tag{1.58}$$

1.4.2 DESIGN OF OBSERVER-BASED CONTROLLER

In order to estimate G, a fuzzy state estimator is given by

Observer Rule $\mathscr{R}^l$: **IF** z_1 is $\mathscr{F}_1^l$ and z_2 is $\mathscr{F}_2^l$ and z_3 is $\mathscr{F}_3^l, \cdots$, and z_{10} is $\hat{\mathscr{F}}_{10}^l, \cdots$, and z_{10} is $\mathscr{F}_{10}^l$, **THEN**

$$\begin{cases} E\dot{\hat{x}}(t) = A_l \hat{x}(t) + B_l u(t) + L_l\left(y(t) - \hat{y}(t)\right), \\ \hat{y}(t) = C\hat{x}(t), l \in \mathscr{L} \end{cases} \tag{1.59}$$

where $\hat{x} \in \Re^{n_{\hat{x}}}$. If $n_{\hat{x}} < n_x$, the state estimator (1.59) becomes an observer with reducing dimensions. Otherwise, it is the one with full dimensions.

Similarly, the global T-S model is given by

$$\begin{cases} E\dot{\hat{x}}(t) = A(\mu, \hat{\mu})\hat{x}(t) + B(\mu)u(t) + L(\mu, \hat{\mu})\left(y(t) - \hat{y}(t)\right), \\ \hat{y}(t) = C\hat{x}(t). \end{cases} \tag{1.60}$$

Now, consider the following global fuzzy controller,

$$u(t) = K(\hat{\mu})\hat{x}(t), \tag{1.61}$$

where the notation $\hat{\mu}$ is induced by the estimated premise variable z_7. Here, without loss of generality, all premise variables on the controller are defined as $\hat{z}(t)$.

Define $e(t) = x(t) - \hat{x}(t)$, and it follows from (1.57)-(1.61) that

$$\bar{E}\dot{\bar{x}}(t) = \bar{A}(\mu, \hat{\mu})\bar{x}(t) + \bar{\omega}(t), \tag{1.62}$$

where

$$\bar{E}=\begin{bmatrix} E & 0 \\ 0 & E \end{bmatrix}, \bar{A}(\mu,\hat{\mu})=\begin{bmatrix} A(\mu)+B(\mu)K(\hat{\mu}) & -B(\mu)K(\hat{\mu}) \\ 0 & A(\mu)-L(\mu,\hat{\mu})C \end{bmatrix},$$

$$\bar{x}(t)=\begin{bmatrix} x(t) \\ e(t) \end{bmatrix}, \bar{\omega}(t)=\begin{bmatrix} 0 \\ (A(\mu)-A(\mu,\hat{\mu}))\hat{x}(t) \end{bmatrix}. \quad (1.63)$$

Given the closed-loop error system in (1.62), and for an $\mathscr{L}_2$-gain performance level $\gamma > 0$, the purpose of this section is to design a fuzzy observer-based controller in (1.60) and (1.61) such that the PV power system is asymptotically stable, and for any nonzero $\bar{\omega} \in \mathscr{L}_2\,[0\ \infty)$ the induced $\mathscr{L}_2$ norm of the operator from $\bar{\omega}$ to the voltage tracking synchronization ζ is less than γ

$$\int_0^\infty \zeta^T(s)\zeta(s)ds < \gamma^2 \int_0^\infty \bar{\omega}^T(s)\bar{\omega}(s)ds, \quad (1.64)$$

under zero initial conditions.

Based on the augmented closed-loop fuzzy control system in (1.62), the MPPT on the PV power system with unknown solar irradiation is proposed as below:

Theorem 1.5: Stability Analysis for Robust MPPT Control

Consider the PV power fuzzy system (1.57) using the fuzzy observer-based controller in the form of (1.60) and (1.61). For the matrix $\bar{P} \in \Re^{(n_x+n_{\hat{x}})\times(n_x+n_{\hat{x}})}$, the stability of maximum power generation of PV system is asymptotically achieved with the $\mathscr{H}_\infty$ performance index if the following inequalities hold:

$$\bar{E}^T\bar{P} = \bar{P}^T\bar{E} \geq 0, \quad (1.65)$$

$$\begin{bmatrix} \bar{P}^T\bar{A}(\mu,\hat{\mu})+\bar{A}^T(\mu,\hat{\mu})\bar{P}+F^TF & \bar{P}^T \\ \star & -\gamma^2 I \end{bmatrix} < 0, \quad (1.66)$$

where $F=\begin{bmatrix} 1 & 0 & 0 & 0 \end{bmatrix}$. ■

Proof. Consider $V(t) = \bar{x}^T(t)\bar{E}^T\bar{P}\hat{x}(t)$, where $\bar{E}^T\bar{P} = \bar{P}^T\bar{E} \geq 0$, and $\bar{P} \in \Re^{(n_x+n_{\hat{x}})\times(n_x+n_{\hat{x}})}$. It is well-known that the $\mathscr{H}_\infty$ performance can be verified if the following inequality holds,

$$\dot{V}(t)+\zeta^T(t)\zeta(t)-\bar{\gamma}^2\bar{\omega}^T(t)\bar{\omega}(t) < 0. \quad (1.67)$$

Thus, the proof is completed.

It is noted that the results on Theorem 1.5 are not LMIs. Here, a two-step processing is proposed. Firstly, define

$$\bar{P}=\begin{bmatrix} P_1 & 0 \\ 0 & P_2 \end{bmatrix}, \quad (1.68)$$

where $P_1 = \begin{bmatrix} P_{1(1)} & 0 \\ P_{1(2)} & P_{1(3)} \end{bmatrix}, P_2 = \begin{bmatrix} P_{2(1)} & 0 \\ P_{2(2)} & P_{2(3)} \end{bmatrix}$, $P_{1(1)} \in \Re^{n_x \times n_x}$ and $P_{2(1)} \in \Re^{n_{\hat{x}} \times n_{\hat{x}}}$ are the symmetric positive-definite matrix, $\{P_{1(1)}, P_{2(1)}, P_{1(2)}, P_{2(2)}\}$ are the matrices with suitable dimensions, $\{P_{1(3)}, P_{2(3)}\}$ are scalars. It is easy to see that

$$\begin{aligned} X &= \bar{P}^{-1} \\ &= \begin{bmatrix} X_1 & 0 \\ 0 & X_2 \end{bmatrix}, \end{aligned} \tag{1.69}$$

where $X_1 = \begin{bmatrix} X_{1(1)} & 0 \\ X_{1(2)} & X_{1(3)} \end{bmatrix}, X_2 = \begin{bmatrix} X_{2(1)} & 0 \\ X_{2(2)} & X_{2(3)} \end{bmatrix}$.

Submitting (1.68) into (1.66), it has

$$\begin{bmatrix} \bar{\Phi}(\mu, \hat{\mu}) + F^T F & \bar{P}^T \\ \star & \gamma^2 I \end{bmatrix} < 0, \tag{1.70}$$

where $\bar{\Phi}(\mu, \hat{\mu}) = \text{Sym}\left\{ \begin{bmatrix} P_1^T A(\mu) + P_1^T B(\mu) K(\hat{\mu}) & -P_1^T B(\mu) K(\hat{\mu}) \\ 0 & P_2^T A(\mu) - P_2^T L(\mu, \hat{\mu}) C \end{bmatrix} \right\}$.

By performing the congruence transformation to (1.70) by $\Gamma = \text{diag}\{ X^{-1} \quad I \}$, and using the Schur complement lemma, one gets

$$\begin{bmatrix} \Phi(\mu, \hat{\mu}) & I & X^{-T} F^T \\ \star & -\gamma^2 I & 0 \\ \star & \star & -I \end{bmatrix} < 0, \tag{1.71}$$

where $\Phi(\mu, \hat{\mu}) = \text{Sym}\left\{ \begin{bmatrix} A(\mu) X_1 + B(\mu) K(\hat{\mu}) X_1 & -B(\mu) K(\hat{\mu}) X_2 \\ 0 & A(\mu) X_2 - L(\mu, \hat{\mu}) C X_2 \end{bmatrix} \right\}$.

By extracting the fuzzy premise variables, one gets

$$\begin{aligned} \Phi(\mu, \hat{\mu}) &= \sum_{l=1}^{r} \sum_{s=1}^{r} \mu_l [z(t)] \hat{\mu}_s [\hat{z}(t)] \Phi_{ls} \\ &< 0, \end{aligned} \tag{1.72}$$

where $\Phi_{ls} = \text{Sym}\left\{ \begin{bmatrix} A_l X_1 + B_l K_s X_1 & -B_l K_s X_2 \\ 0 & A_l X_2 - L_s C X_2 \end{bmatrix} \right\}$.

It should be noted that the existing relaxation technique $\sum_{l=1}^{r} [\mu_l]^2 \Phi_{ll} + \sum_{l=1}^{r} \sum_{l<s\le r} \mu_l \mu_s \Phi_{ls} < 0$ is no longer applicable to the fuzzy controller synthesis, since $\mu_s \neq \hat{\mu}_s$. Similarly to the asynchronous relaxation technique in [22], it is assumed that $|\mu_l - \hat{\mu}_l| \le \delta_l, l \in \mathscr{L}$, where δ_l is a positive scalar. If $\Phi_{ls} + M_l \ge 0$, where M_l is a symmetric matrix, one gets

$$\begin{aligned} &\sum_{l=1}^{r} \sum_{s=1}^{r} \mu_l \hat{\mu}_s \Phi_{ls} \\ &= \sum_{l=1}^{r} \sum_{s=1}^{r} \mu_l \mu_s \Phi_{ls} + \sum_{l=1}^{r} \sum_{s=1}^{r} \mu_l (\hat{\mu}_s - \mu_s)(\Phi_{ls} + M_l) \\ &\le \sum_{l=1}^{r} \sum_{s=1}^{r} \mu_l \mu_s \left[\Phi_{ls} + \sum_{s=1}^{r} \delta_s (\Phi_{ls} + M_l) \right]. \end{aligned} \tag{1.73}$$

Therefore, upon defining $\Sigma_{ls} = \Phi_{ls} + \sum_{s=1}^{r} \delta_s (\Phi_{ls} + M_l)$, the existing relaxation technique from [23] can be applied to (1.73).

Now, assume that $|\mu_l - \hat{\mu}_l| \leq \delta_l$ and based on the results of (1.71) and (1.73), an algorithm to calculate the fuzzy controller and observer gains is proposed as below:

a) For the matrix $X_1 = \begin{bmatrix} X_{1(1)} & 0 \\ X_{1(2)} & X_{1(3)} \end{bmatrix}$, we solve the following inequality,

$$\Sigma_{ll} < 0, l \in \mathscr{L} \tag{1.74}$$

$$\Sigma_{ls} + \Sigma_{sl} < 0, 1 \leq l < s \leq r \tag{1.75}$$

where $\Sigma_{ls} = \Phi_{ls} + \sum_{s=1}^{r} \delta_s (\Phi_{ls} + M_l), \Phi_{ls} = A_l X_1 + B_l \bar{K}_s$, and then we obtain $\bar{K}_s$ and calculate $K_s = \bar{K}_s X_1^{-1}$.

b) Using the controller gain K_s, we solve the following inequality

$$\bar{\Sigma}_{ll} < 0, l \in \mathscr{L} \tag{1.76}$$

$$\bar{\Sigma}_{ls} + \bar{\Sigma}_{sl} < 0, 1 \leq l < s \leq r \tag{1.77}$$

where $\bar{\Sigma}_{ls} = \bar{\Phi}_{ls} + \sum_{s=1}^{r} \delta_s (\bar{\Phi}_{ls} + M_l), \bar{\Phi}_{ls} = \begin{bmatrix} \bar{\Phi}_{ls(1)} + F^T F & \bar{P}^T \\ \star & \gamma^2 I \end{bmatrix}$,

$\bar{\Phi}_{ls(1)} = \text{Sym}\left\{ \begin{bmatrix} P_1^T A_l + P_1^T B_l K_s & -P_1^T B_l K_s \\ 0 & P_2^T A_l - \bar{L}_s C \end{bmatrix} \right\}$. It obtains γ, P_1, P_2 and $\bar{L}_s$ and we calculate $L_s = P_2^{-T} \bar{L}_s$.

c) Using P_1 and P_2, we solve the following inequality

$$\tilde{\Sigma}_{ll} < 0, l \in \mathscr{L} \tag{1.78}$$

$$\tilde{\Sigma}_{ls} + \tilde{\Sigma}_{sl} < 0, 1 \leq l < s \leq r \tag{1.79}$$

where $\tilde{\Sigma}_{ls} = \bar{\Phi}_{ls} + \sum_{s=1}^{r} \delta_s (\bar{\Phi}_{ls} + M_l), \bar{\Phi}_{ls} = \begin{bmatrix} \bar{\Phi}_{ls(1)} + F^T F & \bar{P}^T \\ \star & \bar{\gamma}^2 I \end{bmatrix}$,

$\bar{\Phi}_{ls(1)} = \text{Sym}\left\{ \begin{bmatrix} P_1^T A_l + P_1^T B_l \tilde{K}_s & -P_1^T B_l \tilde{K}_s \\ 0 & P_2^T A_l - P_2^T \tilde{L}_s C \end{bmatrix} \right\}$. It obtains $\bar{\gamma}$ and $\tilde{K}_s$. If $\bar{\gamma} \leq \gamma$, use the controller gain $K_s = \tilde{K}_s$ and turn to the step b). If $\bar{\gamma} > \gamma$, output γ, K_s, L_s, and stop.

1.5 FINITE-TIME MPPT VIA SLIDING MODE CONTROL

This section will consider the finite-time stabilization via the sliding mode control for the PV power system with the MPPT. Without loss of generality, we only consider that the external disturbance appears in the output voltage v_{dc} over the time interval $[t_1, t_2]$, which is defined as below:

$$\mathbb{W}_{[t_1,t_2],\delta} \triangleq \left\{ \omega \in L_2[t_1,t_2] : \int_{t_1}^{t_2} \omega^2(s) ds \leq \delta \right\}, \tag{1.80}$$

where δ is a positive scalar.

Recall the PV power fuzzy system in (1.28) with external disturbances as below:

$$E\dot{x}(t) = A(\mu)x(t) + B(\mu)u(t) + D\omega(t), \tag{1.81}$$

where $A(\mu) := \sum_{l=1}^{r} \mu_l A_l$, $B(\mu) := \sum_{l=1}^{r} \mu_l B_l$, $D = \begin{bmatrix} 0 & 1 & 0 \end{bmatrix}^T$.

Before moving on, we extend the definition of the finite-time boundedness (FTB) in [24, 25] to the descriptor fuzzy system (1.81) as follows:

Definition 1.1. For a given time interval $[t_1, t_2]$, a symmetrical matrix $R > 0$, and two scalars c_1, c_2 subject to $0 < c_1 < c_2$, the descriptor fuzzy system (1.81) with $u(t) = 0$ is the FTB subject to $(c_1, c_2, [t_1, t_2], R, \mathrm{W}_{[t_1,t_2],\delta})$, if it satisfies

$$\begin{aligned} & x^T(t_1) E^T R E x(t_1) \leq c_1 \\ & \Longrightarrow x^T(t_2) E^T R E x(t_2) < c_2, \forall t \in [t_1, t_2], \end{aligned}$$

for all $\omega(t) \in \mathrm{W}_{[t_1,t_2],\delta}$.

This section aims at designing a fuzzy sliding mode controller (FSMC) such that the MPPT error will be bounded around zero in a finite time interval subject to $(c_1, c_2, [t_1, t_2], R, \mathrm{W}_{[t_1,t_2],\delta})$. First an FSMC law is designed, which ensures the state trajectories into the sliding surface in a finite-time T^* with $T^* \leq T$. And then, we calculate the scalar c^* satisfying $c_1 < c^* < c_2$, such that the resulting closed-loop system is the FTB subject to $(c_1, c^*, [0, T^*], R, \mathrm{W}_{[0,T^*],\delta})$.

1.5.1 DESIGN OF FSMC LAW FOR PV POWER WITH MPPT

Firstly, based on the descriptor fuzzy system (1.81), the integral-type sliding surface function is considered as below [26, 27]:

$$s(t) = GEx(t) - \int_0^t G[A(\mu) + B(\mu)K(\mu)]x(s)ds, \tag{1.82}$$

where the matrix G is chosen such that GB_l is the positive definite matrix.

In the following, based on the sliding surface function (1.82), we will design an FSMC law $u(t)$, which ensures the state trajectories of the fuzzy system (1.81) into the specified sliding surface $s(t) = 0$ in a finite-time duration.

Theorem 1.6: FSMC Law for PV Power System with MPPT

Consider the descriptor fuzzy system (1.81) representing the nonlinear PV system with the MPPT control problem. The reachability of the specified sliding surface (1.82) in the finite-time duration $[0, T^*]$ with $T^* \leq T$ can be ensured by the following FSMC law:

$$u(t) = u_b(t) + u_c(t), \tag{1.83}$$

with

$$u_b(t) = \sum_{l=1}^{r} \mu_l K_l x(t), u_c(t) = -\sum_{l=1}^{r} \mu_l [GB_l]^{-1} \rho(t) \operatorname{sgn}(s(t)), \tag{1.84}$$

where $K_l \in \Re^{1\times 3}$ denotes the fuzzy controller gains, $\rho(t) = \frac{\rho + \|GD\| \|\omega(t)\|}{\lambda_{\min}\left(GB_l [GB_p]^{-1}\right)}, \rho \geq \frac{1}{T} \|GEx(0)\|, (l,p) \in \mathscr{L}$, and sgn is a switching sign function defined as

$$\operatorname{sgn}(s(t)) = \begin{cases} -1, & \text{for } s(t) < 0, \\ 0, & \text{for } s(t) = 0, \\ 1, & \text{for } s(t) > 0. \end{cases} \tag{1.85}$$

■

Proof. It follows from (1.82)-(1.85) that

$$\begin{aligned} & s^T(t)\dot{s}(t) \\ & = -s^T(t)GB(\mu)\sum_{l=1}^{r} \mu_l [GB_l]^{-1} \rho(t) \operatorname{sgn}(s(t)) + s^T(t)GD\omega(t) \\ & \leq -\sum_{l=1}^{r}\sum_{p=1}^{r} \mu_l \mu_p \lambda_{\min}\left(GB_l [GB_p]^{-1}\right) \rho(t) \|s(t)\| + \|GD\| \|\omega(t)\| \|s(t)\| \\ & = -\rho \|s(t)\|. \end{aligned} \tag{1.86}$$

In addition, let us define

$$V_1(t) = \frac{1}{2} s^T(t) s(t). \tag{1.87}$$

We have

$$\begin{aligned} \dot{V}_1(t) & \leq -\rho \|s(t)\| \\ & = -\sqrt{2}\rho \sqrt{V_1(t)}. \end{aligned} \tag{1.88}$$

Based on [28], it yields

$$T^* \leq \frac{\sqrt{2}}{\rho} \sqrt{V_1(0)}. \tag{1.89}$$

Besides, it follows from (1.87) that

$$V_1(0) = \frac{1}{2} \|s(0)\|^2. \tag{1.90}$$

Substituting (1.90) into (1.89), one gets

$$T^* \leq \frac{1}{\rho} \|GEx(0)\|. \tag{1.91}$$

It follows from $\rho \geq \frac{1}{T}\|GEx(0)\|$ in (1.91) that

$$T^* \leq T, \tag{1.92}$$

which implies that the proposed FSMC law (1.83) can guarantee the state trajectories of the descriptor fuzzy system (1.81) into the specified sliding surface $s(t) = 0$ in a finite time T^* with $T^* \leq T$, thus completing this proof.

Note: It is noted that the proposed switching sign function $\mathrm{sgn}(\star)$ is discontinuous. The characteristic exhibits a high frequency oscillation, which is undesirable in practical applications. In order to eliminate chattering phenomena, an alternative approach is to employ the following switching function [29]:

$$\mathrm{sgn}\left(s(t)\right) = \begin{cases} -1, & \text{for } s(t) < -\rho, \\ \frac{1}{\rho}s, & \text{for } |s(t)| \leq \rho, \\ 1, & \text{for } s(t) > \rho. \end{cases}$$

It is easy to see that the proposed switching sign function becomes continuous and its value converges to the interval $[-\rho, \rho]$ instead of zero. In this case, the chattering conditions are eliminated.

1.5.2 REACHING PHASE IN FTB FOR PV POWER WITH FSMC LAW

By substituting the FSMC law (1.83) into (1.81), we obtain the resulting closed-loop control system as below:

$$\begin{aligned} E\dot{x}(t) = & \sum_{l=1}^{r}\sum_{p=1}^{r} \mu_l \mu_p \bar{A}_{lp} + D\omega(t) \\ & - \sum_{l=1}^{r}\sum_{p=1}^{r} \mu_l \mu_p B_l \left[GB_p\right]^{-1} \rho\left(t\right) \mathrm{sgn}\left(s(t)\right), \end{aligned} \tag{1.93}$$

where $\bar{A}_{lp} = A_l + B_l K_p$.

On the reaching phase within $[0, T^*]$, the sliding motion is generated outside of the sliding surface (1.82). By defining $\bar{\rho}(t) = \rho\left(t\right)\mathrm{sgn}(s(t))$, $\bar{\rho} = \frac{\rho}{\lambda_{\min}\left(GB_l\left[GB_p\right]^{-1}\right)}$, and $\varepsilon = \frac{\|GD\|}{\lambda_{\min}\left(GB_l\left[GB_p\right]^{-1}\right)}$, one gets

$$\begin{aligned} \bar{\rho}^2(t) &= \rho^2\left(t\right) \\ &= \left[\bar{\rho} + \varepsilon\left\|\omega(t)\right\|\right]^2 \\ &= \bar{\rho}^2 + 2\bar{\rho}\varepsilon\left\|\omega(t)\right\| + \varepsilon^2\left\|\omega(t)\right\|^2 \\ &\leq \left(1+\varepsilon^2\right)\bar{\rho}^2 + \left(1+\varepsilon^2\right)\left\|\omega(t)\right\|^2. \end{aligned} \tag{1.94}$$

Now, a sufficient condition for the FTB of closed-loop system (1.93) in the finite time interval $[0, T^*]$ is derived as below:

Theorem 1.7: Analysis on FTB for PV Power with FSMC Law

Consider the FSMC law (1.83). The resulting closed-loop PV control system in (1.93) is the FTB with respect to $(c_1, c^*, [0, T^*], R, W_{[0,T^*],\delta})$, if there exist the matrices $P = \begin{bmatrix} P_1 & 0 \\ P_2 & P_3 \end{bmatrix}$, $0 < P_1^T = P_1 \in \Re^{(n_x-1)\times(n_x-1)}$, $P_2 \in \Re^{1\times(n_x-1)}$ and P_3 that is a scalar, and the control gain $K_l \in \Re^{n_u \times n_x}$, and the positive scalars $\{c_1, c^*, \eta, \delta\}$, such that the following inequalities hold:

$$\Phi_{ll} < 0, 1 \le l \le r \tag{1.95}$$

$$\Phi_{lp} + \Phi_{pl} < 0, 1 \le l < p \le r \tag{1.96}$$

where

$$\Phi_{lp} = \begin{bmatrix} \Phi_{lp(1)} & P^T D & -P^T B_l [GB_p]^{-1} \\ \star & -\eta I & 0 \\ \star & \star & -\eta I \end{bmatrix},$$

$$\Phi_{lp(1)} = \text{Sym}\left\{A_l^T P + K_p^T B_l^T P\right\} - \eta E^T P. \tag{1.97}$$

Furthermore, the bounding is

$$\frac{\bar{\sigma}_{P1} c_1 + \left(\eta T^* \bar{\rho}^2 + \eta \delta\right)\left(1 + \varepsilon^2\right) + \eta \delta}{e^{-\eta T^*} \underline{\sigma}_{P1}} < c^*. \tag{1.98}$$

■

Proof. Consider the following Lyapunov functional

$$V_2(t) = x^T(t) E^T P x(t), \forall t \in [0, T^*]. \tag{1.99}$$

It is easy to see from Theorem 1.7 that $E^T P = P^T E \ge 0$.

Along the trajectory of system (1.93), one gets

$$\begin{aligned} \dot{V}_2(t) &= 2\left[E\dot{x}(t)\right]^T P x(t) \\ &= 2\left[\sum_{l=1}^{r}\sum_{p=1}^{r} \mu_l \mu_p \bar{A}_{lp} x(t)\right]^T P x(t) \\ &\quad - 2\left[\sum_{l=1}^{r}\sum_{p=1}^{r} \mu_l \mu_p B_l [GB_p]^{-1} \rho(t) \operatorname{sgn}(s(t))\right]^T P x(t) \\ &\quad + 2\left[D\omega(t)\right]^T P x(t). \end{aligned} \tag{1.100}$$

An auxiliary function is introduced as below:

$$J(t) = \dot{V}_2(t) - \eta V_2(t) - \eta \omega^2(t) - \eta \bar{\rho}^2(t), \tag{1.101}$$

where η is a positive scalar.

It follows from (1.100) and (1.101) that

$$\begin{aligned} J(t) = & 2\left[\sum_{l=1}^{r}\sum_{p=1}^{r}\mu_l\mu_p\bar{A}_{lp}x(t)\right]^T Px(t) \\ & -2\left[\sum_{l=1}^{r}\sum_{p=1}^{r}\mu_l\mu_p B_l\left[GB_p\right]^{-1}\bar{\rho}(t)\right]^T Px(t) \\ & +2\left[D\omega(t)\right]^T Px(t) - \eta x^T(t)E^T Px(t) - \eta\omega^2(t) - \eta\bar{\rho}^2(t) \\ = & \sum_{l=1}^{r}\sum_{p=1}^{r}\mu_l\mu_p\chi^T(t)\Phi_{lp}\chi(t), \end{aligned} \tag{1.102}$$

where $\chi(t) = \begin{bmatrix} x^T(t) & \omega^T(t) & \bar{\rho}^T(t) \end{bmatrix}^T$, and Φ_{lp} is defined in (1.97).

Because of (1.95) and (1.96) $J(t) < 0$, which implies that

$$\dot{V}_2(t) < \eta V_2(t) + \eta\omega^2(t) + \eta\bar{\rho}^2(t). \tag{1.103}$$

Multiplying both sides in (1.103) by $e^{-\eta t}$, and then integrating it from 0 to t, $t \in [0, T^*]$, it is easy to see that

$$\begin{aligned} e^{-\eta t}V_2(t) < & V_2(0) + \eta\int_0^t e^{-\eta s}\bar{\rho}^2(s)ds + \eta\int_0^t e^{-\eta s}\omega^2(s)ds \\ & \leq x^T(0)E^T Px(0) + \eta T^*\left(1+\varepsilon^2\right)\bar{\rho}^2 + \eta\left(1+\varepsilon^2\right)\delta + \eta\delta. \end{aligned} \tag{1.104}$$

On the other hand, it follows from (1.104) that

$$e^{-\eta t}V_2(t) \geq e^{-\eta T^*}x^T(t)E^T Px(t), \tag{1.105}$$

which implies that

$$\begin{aligned} e^{-\eta T^*}x^T(t)E^T Px(t) < & x^T(0)E^T Px(0) + \eta T^*\left(1+\varepsilon^2\right)\bar{\rho}^2 \\ & + \eta\left(1+\varepsilon^2\right)\delta + \eta\delta. \end{aligned} \tag{1.106}$$

Further, by specifying the matrix P as below:

$$P = \begin{bmatrix} P_1 & 0 \\ P_2 & P_3 \end{bmatrix}, \tag{1.107}$$

where $0 < P_1^T = P_1 \in \Re^{(n_x-1)\times(n_x-1)}$, $P_2 \in \Re^{1\times(n_x-1)}$ and P_3 is a scalar, it is easy to see that $E^T P = P^T E \geq 0$.

Now, we partition $x(t)$ as

$$x(t) = \begin{bmatrix} \bar{x}(t) \\ x_3(t) \end{bmatrix}, \tag{1.108}$$

where $\bar{x}(t) = \begin{bmatrix} x_1(t) \\ x_2(t) \end{bmatrix}$.

It follows from (1.104)-(1.108) that

$$e^{-\eta T^*}\bar{x}^T(t)P_1\bar{x}(t) < \bar{x}^T(0)P_1\bar{x}(0) + \eta T^*\left(1+\varepsilon^2\right)\bar{\rho}^2 + \eta\left(1+\varepsilon^2\right)\delta + \eta\delta. \tag{1.109}$$

Now, by introducing the matrix $0 < R_1^T = R_1 \in \Re^{n_x \times n_x}$, we further define

$$c_1 = \bar{x}^T(0)R_1\bar{x}(0),$$
$$\bar{\sigma}_{P1} = \lambda_{\max}\left(R_1^{-\frac{1}{2}}P_1R_1^{-\frac{1}{2}}\right), \underline{\sigma}_{P1} = \lambda_{\min}\left(R_1^{-\frac{1}{2}}P_1R_1^{-\frac{1}{2}}\right). \tag{1.110}$$

Based on the relationships (1.109) and (1.110), one gets

$$\bar{x}^T(t)R_1\bar{x}(t) < \frac{\bar{\sigma}_{P1}c_1 + \left(\eta T^*\bar{\rho}^2 + \eta\delta\right)\left(1+\varepsilon^2\right) + \eta\delta}{e^{-\eta T^*}\underline{\sigma}_{P1}}. \tag{1.111}$$

From Definition 1.1, the descriptor fuzzy system in (1.93) is FTB. This completes the proof.

Recall the fast dynamic subsystem in (1.88),

$$\sum_{l=1}^{r}\mu_l A_{l(1)}\bar{x}(t) - \sum_{l=1}^{r}\mu_l n_p \gamma I_{rs}\mathscr{F}_3^l x_3(t) = 0, \tag{1.112}$$

where $A_{l(1)} = \left[\begin{array}{cc} 1 - n_p\gamma I_{rs}\mathscr{F}_4^l & 0 \end{array}\right]$.

It follows from (1.110)-(1.112) that

$$\|\bar{x}(0)\| < \sqrt{\frac{c_1}{\lambda_{\min}(R_1)}},$$
$$\|\bar{x}(t)\| < \frac{\bar{\sigma}_{P1}c_1 + \left(\eta T^*\bar{\rho}^2 + \eta\delta\right)\left(1+\varepsilon^2\right) + \eta\delta}{e^{-\eta T^*}\underline{\sigma}_{P1}\lambda_{\min}(R_1)}, t \in [0, T^*]. \tag{1.113}$$

It follows from (1.112) and (1.113) that

$$\|x_3(0)\| < \left\|\frac{\sum_{l=1}^{r}\mu_l A_{l(1)}}{\sum_{l=1}^{r}\mu_l n_p\gamma I_{rs}\mathscr{F}_3^l}\right\|\sqrt{\frac{c_1}{\lambda_{\min}(R_1)}},$$
$$\|x_3(t)\| < \left\|\frac{\sum_{l=1}^{r}\mu_l A_{l(1)}}{\sum_{l=1}^{r}\mu_l n_p\gamma I_{rs}\mathscr{F}_3^l}\right\|c^*, t \in [0, T^*], \tag{1.114}$$

where c^* is defined in (1.98).

1.5.3 DESIGN PROCEDURE FOR MPPT ALGORITHM

The detailed calculating steps of the proposed MPPT algorithm for the PV system are summarized as below:

i) Use the descriptor system approach to represent the MPPT control problem of the PV system, as shown in (1.27);

ii) Use the T-S fuzzy model method to describe the nonlinear descriptor system as shown in (1.28);

iii) Choose a suitable matrix G, the time interval T, the controller gains K_l, the initial state $x(0)$, and construct the FSMC law as shown in Theorem 1.6;

iv) Based on the FSMC law, the reachability of the sliding surface (1.77) in the finite time T^* with $T^* \leq T$ is obtained;

v) Based on the the finite time T^*, we solve Theorem 1.7 to obtain the bounding c^*;

vi) Use (1.109) to calculate the boundary for the MPPT error ε_{pv}.

The controller gains K_l in (1.78) can be obtained by the following theorem.

Theorem 1.8: Design of MPPT Controller for PV Power

Recalling the FSMC law (1.78), the resulting closed-loop PV control system in (1.88) is asymptotically achieved if the following LMIs hold:

$$\Sigma_{ll} < 0, l \in \mathscr{L} \tag{1.115}$$

$$\Sigma_{ls} + \Sigma_{sl} < 0, 1 \leq l < s \leq r \tag{1.116}$$

where $\Sigma_{ls} = \text{Sym}(A_l X + B_l \bar{K}_s), X = \begin{bmatrix} X_{(1)} & 0 \\ X_{(2)} & X_{(3)} \end{bmatrix}, 0 < X_{(1)} = X_{(1)}^T \in \Re^{(n_x-1)\times(n_x-1)}$, $X_{(2)} \in \Re^{1\times(n_x-1)}$, $X_{(3)}$ is a scalar.

In that case, the controller gains can be calculated by

$$K_l = \bar{K}_l X^{-1}, l \in \mathscr{L} \tag{1.117}$$

■

1.6 SIMULATION STUDIES

1.6.1 SOLAR PV POWER WITH DC-DC BOOST CONVERTER

Consider a solar PV power with DC-DC boost converter as shown in Figure 1.2. The system parameters are chosen as below: $L = 0.0516\text{H}, C_0 = 0.0472\text{F}$. Now, we choose $\left\{\frac{v_{pv}}{\phi_{pv}}, \frac{\phi_0}{v_{dc}}, v_{dc}, \phi_{pv}\right\}$ as the fuzzy premise variables, and linearize the PV system around $\{26.39, 0.01, 19.00, 0.36\}$ and $\{29.29, 0.01, 6.00, 0.42\}$. Then, the succeeding system matrices of T-S fuzzy model can be obtained as below:

$$A_1 = \begin{bmatrix} 511.41 & -19.38 \\ 21.19 & -0.20 \end{bmatrix}, B_1 = \begin{bmatrix} 368.22 \\ -7.63 \end{bmatrix},$$
$$A_2 = \begin{bmatrix} 567.55 & -19.38 \\ 21.19 & -0.18 \end{bmatrix}, B_2 = \begin{bmatrix} 476.74 \\ -8.90 \end{bmatrix}.$$

Here, by applying Theorem 1.2, the fuzzy controller gains are given by

$$K_1 = \begin{bmatrix} -1.46 & 0.05 \end{bmatrix}, K_2 = \begin{bmatrix} -1.25 & 0.04 \end{bmatrix}.$$

As shown in Figure 1.6, the open loop system is unstable. Based on the above solutions, Figure 1.7 indicates that the state responses for the PV system converge to zero.

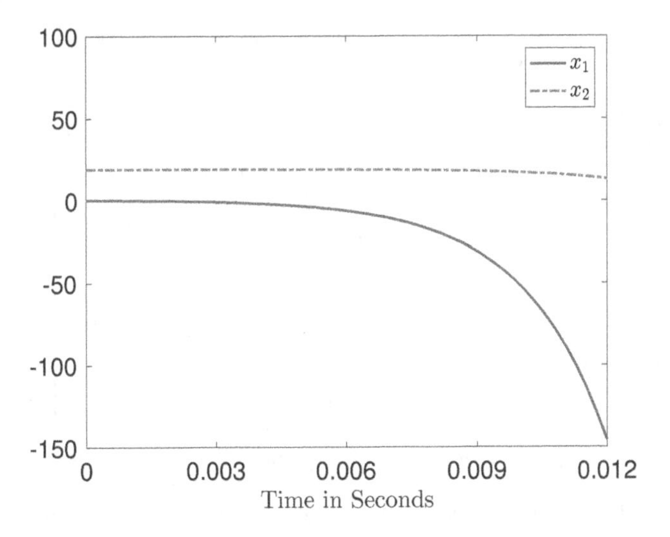

Figure 1.6 State responses of open-loop system with DC-DC boost converter.

1.6.2 SOLAR PV POWER WITH DC-DC BUCK CONVERTER

Consider the PV power fuzzy system with DC-DC buck converter in Figure 1.1. Here, the parameters of the PV system are chosen as below: $L = 0.0516\text{H}$, $C_0 = 0.0472\text{F}$, $n_s = 20, n_p = 5$, $\gamma = 0.4627$, $I_{rs} = 0.00015\text{A}$, $T = 313.15\text{K}$, $C_{pv} = 0.0101\text{F}$, $I_{ph} = 0.1\text{A}$, $R_0 = 1.1\Omega$, $R_L = 1.1\Omega$, $R_M = 0.85\Omega$, $V_D = 9.1\text{V}$. Now, we choose ϕ_{pv} as the fuzzy premise variable, and linearize the PV system around $\{0.36, 0.42\}$, and assume that $v_{pv} = 9.5\text{V}$, $v^*_{pv} = 10\text{V}$, $\dot{v}^*_{pv} = 0.01$. Then, the succeeding system matrices

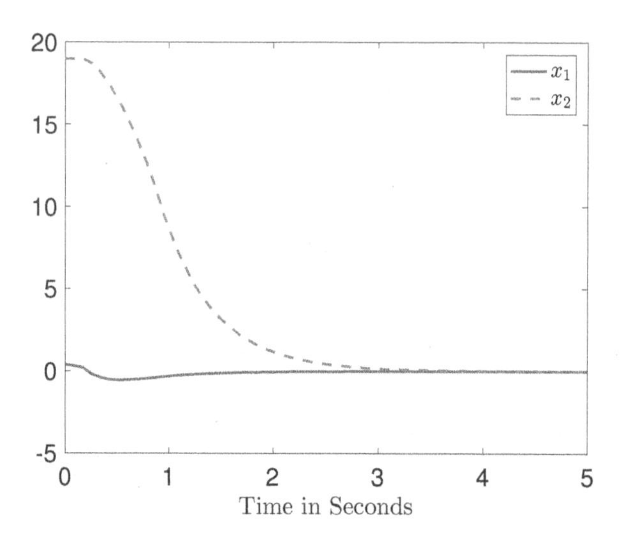

Figure 1.7 State responses of closed-loop control system with DC-DC boost converter.

of T-S fuzzy model can be obtained as below:

$$A_1 = \begin{bmatrix} 25.581 & -21.318 & 0 & -17.636 \\ -63.559 & 0 & 0 & 0 \\ 0 & 0 & -71.287 & -0.001 \\ 0 & 0 & -0.001 & -0.035468 \end{bmatrix}, B_1 = \begin{bmatrix} 356.35 \\ 0 \\ -24.752 \\ 0 \end{bmatrix},$$

$$A_2 = \begin{bmatrix} 25.581 & -21.318 & 0 & -17.636 \\ -63.559 & 0 & 0 & 0 \\ 0 & 0 & -83.168 & -0.001 \\ 0 & 0 & -0.001 & -0.035468 \end{bmatrix}, B_2 = \begin{bmatrix} 355.69 \\ 0 \\ -28.713 \\ 0 \end{bmatrix}.$$

Here, by applying Theorem 1.4, the fuzzy controller gains are given by

$$K_1 = \begin{bmatrix} -1.11 & 0.68 & 0.02 & 0.05 \end{bmatrix}, K_2 = \begin{bmatrix} -1.11 & 0.68 & 0.02 & 0.05 \end{bmatrix}.$$

As shown in Figure 1.8, the open loop system is unstable. Based on the above solutions, Figure 1.9 indicates that the state responses for the PV system converge to zero.

1.6.3 SOLAR PV POWER WITH MPPT CONTROL

Consider the PV power fuzzy system with DC-DC buck converter in (1.5). Here, the parameters of the PV system are chosen as below: $L = 0.0516\text{H}, C_0 = 0.0472\text{F}, n_s = 20, n_p = 5, \gamma = 0.4627, I_{rs} = 0.00015\text{A}, T = 313.15\text{K}, C_{pv} = 0.0101\text{F}$, $I_{ph} = 0.1\text{A}, R_0 = 1.1\Omega, R_L = 1.1\Omega, R_M = 0.85\Omega, V_D = 9.1\text{V}$. Now, we choose ϕ_{pv} as the fuzzy premise variable, and linearize the PV system around $\{0.36, 0.42\}$, and assume that $v_{pv} = 9.5\text{V}, v_{pv}^* = 10\text{V}, \dot{v}_{pv}^* = 0.01$. Then, the succeeding system matrices

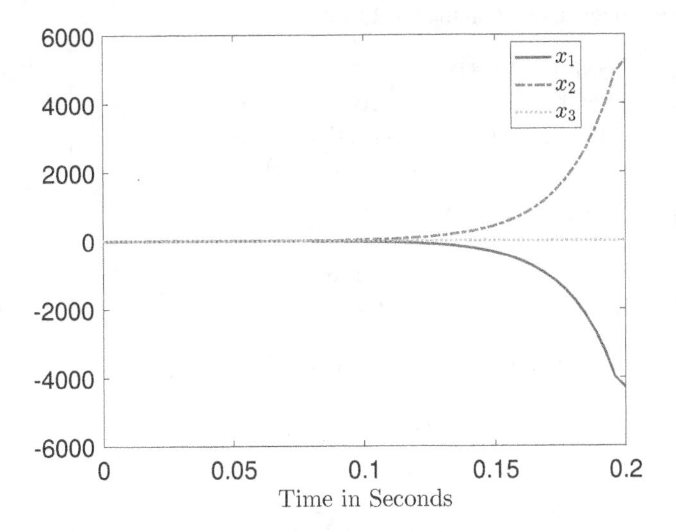

Figure 1.8 State responses of open-loop system with DC-DC buck converter.

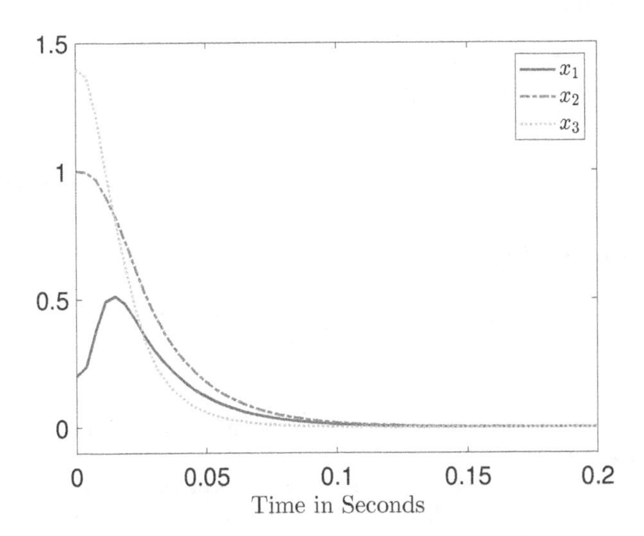

Figure 1.9 State responses of closed-loop control system with DC-DC buck converter.

of T-S fuzzy model can be obtained as below:

$$A_1 = \begin{bmatrix} 25.581 & -21.318 & 0 & -17.636 \\ -63.559 & 0 & 0 & 0 \\ 0 & 0 & -71.287 & -0.001 \\ 0 & 0 & -0.001 & -0.035468 \end{bmatrix}, B_1 = \begin{bmatrix} 356.35 \\ 0 \\ -24.752 \\ 0 \end{bmatrix},$$

$$A_2 = \begin{bmatrix} 25.581 & -21.318 & 0 & -17.636 \\ -63.559 & 0 & 0 & 0 \\ 0 & 0 & -83.168 & -0.001 \\ 0 & 0 & -0.001 & -0.035468 \end{bmatrix}, B_2 = \begin{bmatrix} 355.69 \\ 0 \\ -28.713 \\ 0 \end{bmatrix}.$$

Here, construct the FSMC law as shown in (1.83),

$$u(t) = u_b(t) + u_c(t),$$

where

$$u_b(t) = \sum_{l=1}^{r} \mu_l K_l x(t), u_c(t) = -\sum_{l=1}^{r} \mu_l \left[GB_l\right]^{-1} \rho(t) \operatorname{sgn}(s(t)),$$

with $K_1 = \begin{bmatrix} -1.11 & 0.68 & 0.02 & 0.05 \end{bmatrix}, K_2 = \begin{bmatrix} -1.11 & 0.68 & 0.02 & 0.05 \end{bmatrix}$, $G = \begin{bmatrix} 0.0330 & 0 & -0.0055 & 0 \end{bmatrix}, x(0) = \begin{bmatrix} 2 & 0.27 & 2 & 0 \end{bmatrix}^T, D = \begin{bmatrix} 0 & 0 & 1 & 0 \end{bmatrix}$, $\rho(t) = \frac{\rho + \|GD\| \|\omega(t)\|}{\lambda_{\min}\left(GB_l\left[GB_p\right]^{-1}\right)}, \rho \geq \frac{0.055}{3}, \omega(t) = 0.1\sin t, (l,p) \in \mathscr{L}$, and $\operatorname{sgn}(\star)$ is a switching sign function defined as

$$\operatorname{sgn}(s(t)) = \begin{cases} -1, & \text{for } s(t) < 0, \\ 0, & \text{for } s(t) = 0, \\ 1, & \text{for } s(t) > 0. \end{cases}$$

Given $R_1 = \operatorname{diag}\{1,1,1,1\}$. We use (1.83) to calculate $c_1 = 8.07, c^* = 98.85$.

1.7 REFERENCES

1. Li, X., Li, Y., and Seem, J. E. (2013). Maximum power point tracking for photovoltaic system using adaptive extremum seeking control. IEEE Transactions on Control Systems Technology, 21(6), 2315-2322.
2. Subudhi, B. and Pradhan, R. (2013). A comparative study on maximum power point tracking techniques for photovoltaic power systems. IEEE Transactions on Sustainable Energy, 4(1), 89-98.
3. Petrone, G., Spagnuolo, G., and Vitelli, M. (2012). An analog technique for distributed MPPT PV applications. IEEE Transactions on Industrial Electronics, 59(12), 4713-4722.
4. Zazo, H., Castillo, E. D., Jean Franois Reynaud, and Leyva, R. (2012). MPPT for photovoltaic modules via Newton-like extremum seeking control. Energies, 5(8), 2652-2666.
5. Ahmed, J. and Salam, Z. (2015). An improved perturb and observe (P&O) maximum power point tracking (MPPT) algorithm for higher efficiency. Applied Energy, 150, 97-108.

6. Beid S. and Doubabi S. (2014). DSP-based implementation of fuzzy output tracking control for a boost converter. IEEE Transactions on Industrial Electronics, 61(1): 196-209.
7. Ying, H. (2000). Fuzzy Control and Modeling: Analytical Foundations and Applications. New York: IEEE Press.
8. Zhong, Z., Lin, C. M., Shao, Z., and Xu, M. (2018). Decentralized event-triggered control for large-scale networked fuzzy systems. IEEE Transactions on Fuzzy Systems, 26(1): 29-45.
9. Chiu C. and Ouyang, Y. (2011). Robust maximum power tracking control of uncertain photovoltaic systems: A unified T-S fuzzy model-based approach. IEEE Transactions on Control Systems and Technology, 19(6): 1516-1526.
10. Ying, H. (1998). General SISO Takagi-Sugeno fuzzy systems with linear rule consequent are universal approximators. IEEE Transactions on Fuzzy Systems, 6(4), 582-587.
11. Mendel, J. M., Hagras, H., Tan, W. W., Melek, W. W., and Ying, H. (2014). Introduction To Type-2 Fuzzy Logic Control: Theory and Applications. New York: IEEE Press.
12. Chadli M., Karimi H., and Shi P. On stability and stabilization of singular uncertain Takagi-Sugeno fuzzy systems. Journal of the Franklin Institute, 2014, 351(3): 1453-1463.
13. Zolfaghari, M., Hosseinian, S., Fathi, S., Abedi, M., and Gharehpetian, G. (2018). A new power management scheme for parallel-connected PV systems in microgrids. IEEE Transactions on Sustainable Energy, doi: 10.1109/TSTE.2018.2799972.
14. Kim, S. (2018). Output voltage-tracking controller with performance recovery property for DC/DC boost converters. IEEE Transactions on Control Systems Technology, doi: 10.1109/TCST.2018.2806366
15. Mahmud, M., Hossain, M., Pota, H., and Roy, N. (2014). Robust nonlinear controller design for three-phase grid-connected photovoltaic systems under structured uncertainties. IEEE Transactions on Power Delivery, 2014, 29(3): 1221-1230.
16. Sangwongwanich, A., Abdelhakim, A., Yang, Y., and Zhou, K. (2018). Control of single-phase and three-phase DC/AC converters. In: Blaabjerg, F. (ed.) Control of Power Electronic Converters and Systems. London: Academic Press.
17. Wang, C., Li, X., Guo, L., and Li, Y. W. (2014). A nonlinear-disturbance-observer-based DC-bus voltage control for a hybrid AC/DC microgrid. IEEE Transactions on Power Electronics, 29(11), 6162-6177.
18. Carrasco, M., Mancilladavid, F., and Ortega, R. (2014). An estimator of solar irradiance in photovoltaic arrays with guaranteed stability properties. IEEE Transactions on Industrial Electronics, 2014, 61(7): 3359-3366.
19. Abdelsalam, A. K., Massoud, A. M., Ahmed, S., and Enjeti, P. N. (2011). High-performance adaptive perturb and observe MPPT technique for photovoltaic-based microgrids. IEEE Transactions on Power Electronics, 26(4), 1010-1021.
20. Garrigs, A., Blanes, J. M., Carrasco, J. A., and Ejea, J. B. (2007). Real time estimation of photovoltaic modules characteristics and its application to maximum power point operation. Renewable Energy, 32(6), 1059-1076.
21. Mei, Q., Shan, M., Liu, L., and Guerrero, J. M. (2011). A novel improved variable step-size incremental-resistance MPPT method for PV systems. IEEE Transactions on Industrial Electronics, 58(6), 2427-2434.
22. Zhang, D., Han, Q., and Jia, X. (2015). Network-based output tracking control for a class of T-S fuzzy systems that can not be stabilized by nondelayed output feedback controllers. IEEE Transactions on Cybernetics, 45(8): 1511-1524.

23. Zhong, Z. and Lin, C. (2017). Large-Scale Fuzzy Interconnected Control Systems Design and Analysis. Pennsylvania: IGI Global.
24. Basin, M., Yu, P., and Shtessel, Y. (2016). Finite- and fixed-time differentiators utilising HOSM techniques. IET Control Theory & Applications, 11(8), 1144-1152.
25. Song, J., Niu, Y., and Zou, Y. (2017). Finite-time stabilization via sliding mode control. IEEE Transactions on Automatic Control, 62(3), 1478-1483.
26. Basin, M. and Rodriguez-Ramirez, P. (2010). Sliding mode filter design for linear systems with unmeasured states. IEEE International Symposium on Intelligent Control, 58(8), 3616-3622.
27. Wu, L., Su, X., and Shi, P. (2012). Sliding mode control with bounded $\mathscr{L}_2$ gain performance of Markovian jump singular time-delay systems. Automatica, 48: 1929-1933.
28. Haddad W. and L'Afflitto, A. (2016). Finite-time stabilization and optimal feedback control. IEEE Transactions on Automatic Control, 61(4): 1069-1074.
29. Bartoszewicz, A. and Zuk, J. (2010). Sliding mode control: Basic concepts and current trends. IEEE International Symposium on Industrial Electronics, 3772-3777.
30. Muyeen, S. M. and Al-Durra, A. (2013). Modeling and control strategies of fuzzy logic controlled inverter system for grid interconnected variable speed wind generator. IEEE Systems Journal, 7(4), 817-824.

2 Fuzzy Modeling and Control of Wind Power

In recent years wind energy application, as an economic type of renewable energy, is rapidly growing. With increasing oil price, security threats, and environmental concerns, the portion of wind energy is expected to be 12% of total global energy by 2020. Wind turbine electricity generation depends on wind velocity and structure. The variable speed wind turbine, the most common type of wind conversion system, produces more power than a fixed speed turbine. The wind structure with the permanent magnet synchronous generator (PMSG) is an efficient configuration for variable speed systems. Its advantages include fewer repair requirements, DC excitation and high power-to-weight ratios. To connect the wind structure to the network, a full-scale converter is utilized providing a wide operation range for rotation speed and captured power [1].

In the existing literature, distinct control schemes are suggested for grid integrated operation of variable speed wind generators and most of those schemes are based on conventional proportional integral (PI) controllers. For reliable grid operation, the converters are controlled by using the direct feedback of torque and power in the cascaded control loops. The problem of reactive power control during grid faults is addressed using different energy storage devices by varying the number of control loop parameters. Most of these control schemes are based on voltage and stator flux linkage tracking control, using traditional PI controllers to access variable stator and rotor data. The challenging task in the above methods is in tuning the PI gains because utility plants are nonlinear and face uncertain operating conditions. The PI controller with fixed gains for a determined operating point provides an acceptable performance, but shows poor transient performance when the converter operation point varies continuously because of uncertain dynamics of the plant. Operating points of the grid interactive inverters vary with the natural conditions such as solar radiation or wind speed. Moreover grid specifications such as grid voltage, frequency and impedance might change during operation of an inverter. To address the operating point issues, many variations for PI have been proposed in the power electronics literature including the addition of a grid voltage feedforward path, multiple-state feedback and increasing the proportional gain. Generally, these variations can expand the PI controller bandwidth but, unfortunately, they also push the systems towards their stability boundary [2].

A wind power system with PMSG is shown in Figure 2.1. The use of a two-mass model for controller synthesis is motivated by the fact that the control laws derived from this model are more general and can be applied for wind turbines of different sizes. Particularly, these controllers are better adapted for high-flexibility wind turbines that cannot be properly modelled as one mass model. In fact, the two-mass model can report flexible modes in the drive train model that cannot be highlighted in a one-mass model.

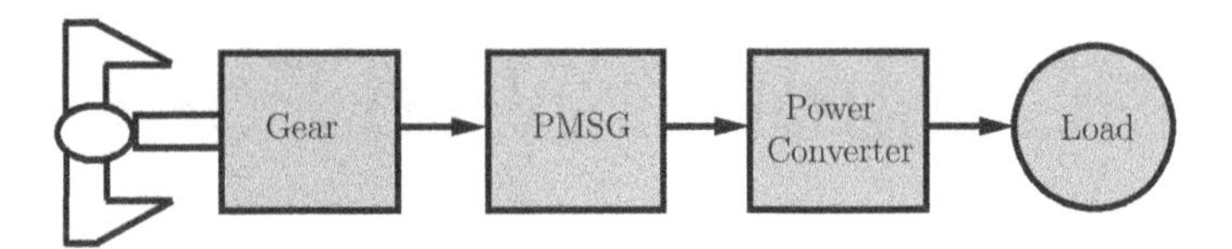

Figure 2.1 Wind turbine with PMSG.

In this chapter, an MPPT method is developed for stand-alone wind power generation systems via the T-S fuzzy-model-based approach. First, the nonlinear wind power system is represented in the T-S fuzzy model. Then, the MPPT problem of the considered wind power system is reformulated into the framework of descriptor systems. Meanwhile, we develop a fuzzy observer for state feedback control under partial state measurement. Furthermore, a fuzzy sliding mode controller is proposed in the framework of descriptor systems, such that the MPPT control performance will be achieved in a finite time interval. Finally, two numerical examples are provided to show the effectiveness of the proposed method.

2.1 MODELING OF WIND POWER

2.1.1 MODELING OF VARIABLE SPEED WIND POWER

Consider a two-mass model for the wind turbine with PMSG. The aerodynamic power captured by the rotor is given by [3]

$$P_a = \frac{1}{2}\rho\pi R^2 C_p(\lambda,\beta)v^3, \tag{2.1}$$

where ρ, R, C_p, β, and v, are air density, rotor radius, power coefficient, blade pitch angle, and wind speed, respectively. The tip speed ratio λ is defined as

$$\lambda = \frac{\omega_t R}{v}, \tag{2.2}$$

where ω_t is the rotor speed.

The aerodynamic torque is

$$T_a = \frac{1}{2\lambda}\rho\pi R^3 C_p(\lambda,\beta)v^2. \tag{2.3}$$

The rotor-side inertia J_r dynamics are given by the first order differential equation

$$J_r\dot{\omega}_t = T_a - T_{ls} - B_r\omega_t, \tag{2.4}$$

where J_r and B_r are rotor inertia and rotor external damping, respectively. The low-speed shaft torque T_{ls} acts as a braking torque on the rotor, and it is defined as

$$T_{ls} = K_{ls}(\theta_t - \theta_{ls}) + K_{ls}(\omega_t - \omega_{ls}), \tag{2.5}$$

where the superscript ls denotes the low-speed side on gear box. Correspondingly, it has

$$J_g \dot{\omega}_g = T_{hs} - T_{em} - B_g \omega_g, \tag{2.6}$$

where J_g and T_{em} are the generator inertia and electromagnetic torque, respectively.

Now, further defining the ratio of gearbox as n_g,

$$\begin{aligned} n_g &= \frac{T_{ls}}{T_{hs}} \\ &= \frac{\omega_g}{\omega_{ls}} \\ &= \frac{\theta_g}{\theta_{ls}}. \end{aligned} \tag{2.7}$$

According to the theory of the space vector, the stator voltage equations for PMSG can be represented in the rotating $d-q$ reference frame as follows [4, 5, 6]:

$$\begin{cases} u_d = \dot{\phi}_d - R_s \phi_d - \omega_g \psi_q, \\ u_q = \dot{\phi}_q - R_s \phi_q + \omega_g \psi_d, \\ T_{em} = \frac{3}{2}\frac{M}{2}\left(\psi_m \phi_q + \left(L_d - L_q\right) \phi_q \phi_d\right), \end{cases} \tag{2.8}$$

where u_d and u_q are the d-axis and q-axis stator terminal voltages, respectively; ϕ_d and ϕ_q are the d-axis and q-axis stator currents, respectively; R_s is the resistance of the stator windings; M is the number of poles; ω_g is the electrical angular velocity of the rotor; ψ_q and ψ_d are the d-axis and q-axis flux linkages of the PMSG, respectively, which are given by

$$\begin{cases} \psi_d = -L_d \phi_d + \psi_m, \\ \psi_q = -L_q \phi_q, \end{cases} \tag{2.9}$$

where L_d and L_q are the d-axis and q-axis inductances of the PMSG, respectively; and ψ_m is the flux linkage generated by the permanent magnets.

It follows from (2.1)-(2.9) that

$$\begin{cases} \dot{\omega}_t = k_1 T_a + k_2 \omega_t + k_3 \omega_g - k_4 T_{em}, \\ \dot{\omega}_g = k_5 T_a - k_6 \omega_t - k_7 \omega_g - k_8 T_{em}, \\ \dot{\phi}_d = R_s \phi_d - \omega_g L_q \phi_q + u_d, \\ \dot{\phi}_q = R_s \phi_q + \omega_g L_d \phi_d - \omega_g \psi_m + u_q, \\ 0 \dot{T}_{em} = \frac{3M}{4}\left(\psi_m \phi_q + \left(L_d - L_q\right) \phi_q \phi_d\right) - T_{em}, \end{cases} \tag{2.10}$$

where

$$k_1 = \frac{1}{J_r\left(1+\frac{K_{ls}}{J_r}\right)} + \frac{K_{ls}^2}{J_r^2 n_g^2 J_g\left(1+\frac{K_{ls}}{J_r}\right)^3\left(\frac{n_g^2 J_g + K_{ls}}{n_g^2 J_g} - \frac{K_{ls}^2}{n_g^2 J_r J_g\left(1+\frac{K_{ls}}{J_r}\right)}\right)},$$

$$k_2 = \frac{J_r K_{ls}^2\left(1+\frac{K_{ls}}{J_r}\right) - K_{ls}^3 + K_{ls}^2 B_r}{J_r^2 n_g^2 J_g\left(1+\frac{K_{ls}}{J_r}\right)^3\left(\frac{n_g^2 J_g + K_{ls}}{n_g^2 J_g} - \frac{K_{ls}^2}{n_g^2 J_r J_g\left(1+\frac{K_{ls}}{J_r}\right)}\right)} - \frac{K_{ls}+B_r}{J_r\left(1+\frac{K_{ls}}{J_r}\right)},$$

$$k_3 = \frac{K_{ls}}{n_g J_r\left(1+\frac{K_{ls}}{J_r}\right)} + \frac{\frac{K_{ls}^3}{n_g^2 J_r J_g\left(1+\frac{K_{ls}}{J_r}\right)} - \frac{K_{ls}^2}{n_g^2 J_g} - \frac{K_{ls} B_g}{J_g}}{n_g J_r\left(1+\frac{K_{ls}}{J_r}\right)^2\left(\frac{n_g^2 J_g + K_{ls}}{n_g^2 J_g} - \frac{K_{ls}^2}{n_g^2 J_r J_g\left(1+\frac{K_{ls}}{J_r}\right)}\right)},$$

$$k_4 = \frac{K_{ls}}{J_g n_g J_r\left(1+\frac{K_{ls}}{J_r}\right)^2\left(\frac{n_g^2 J_g + K_{ls}}{n_g^2 J_g} - \frac{K_{ls}^2}{n_g^2 J_r J_g\left(1+\frac{K_{ls}}{J_r}\right)}\right)},$$

$$k_5 = \frac{K_{ls}}{n_g J_g\left(J_r + K_{ls}\right)\left(1+\frac{K_{ls}}{n_g^2 J_g} - \frac{K_{ls}^2}{n_g^2 J_r J_g\left(1+\frac{K_{ls}}{J_r}\right)}\right)},$$

$$k_6 = \frac{K_{ls}^2 + K_{ls} B_r + K_{ls} J_r\left(1+\frac{K_{ls}}{J_r}\right)}{n_g J_r J_g\left(1+\frac{K_{ls}}{J_r}\right)\left(1+\frac{K_{ls}}{n_g^2 J_g} - \frac{K_{ls}^2}{n_g^2 J_r J_g\left(1+\frac{K_{ls}}{J_r}\right)}\right)},$$

$$k_7 = \frac{K_{ls}^2 + J_r\left(K_{ls} + n_g^2 B_g\right)\left(1+\frac{K_{ls}}{J_r}\right)}{n_g^2 J_g J_r\left(1+\frac{K_{ls}}{J_r}\right)\left(1+\frac{K_{ls}}{n_g^2 J_g} - \frac{K_{ls}^2}{n_g^2 J_r J_g\left(1+\frac{K_{ls}}{J_r}\right)}\right)},$$

$$k_8 = \frac{1}{J_g\left(1+\frac{K_{ls}}{n_g^2 J_g} - \frac{K_{ls}^2}{n_g^2 J_r J_g\left(1+\frac{K_{ls}}{J_r}\right)}\right)}. \tag{2.11}$$

We define $x(t) = \begin{bmatrix} \omega_t & \omega_g & \phi_d & \phi_q & T_{em} \end{bmatrix}^T$, $u(t) = \begin{bmatrix} u_d & u_q \end{bmatrix}^T$, and choose $z_1 = \phi_q, z_2 = \phi_d$, and $z_3 = \frac{C_p(\lambda,\beta)v^2}{\omega_t}$ as fuzzy premise variables. Thus, it follows from (2.10) and (2.11) that the wind energy conversion system with PMSG is represented by

Plant Rule $\mathscr{R}^l$: **IF** z_1 is $\mathscr{F}_1^l$ and z_2 is $\mathscr{F}_2^l$ and z_3 is $\mathscr{F}_3^l$, **THEN**

$$E\dot{x}(t) = A_l x(t) + B_l u(t), l \in \mathscr{L} := \{1, 2, \ldots, r\} \tag{2.12}$$

where $\mathscr{R}^l$ denotes the l-th fuzzy inference rule; r is the number of inference rules; $\mathscr{F}_\theta^l(\theta = 1,2,3)$ is the fuzzy set; $x(t) \in \Re^{n_x}$ and $u(t) \in \Re^{n_u}$ denote the system state and control input, respectively; $z(t) \triangleq [z_1, z_2, z_3]$ are the measurable variables; $\{A_l, B_l\}$ is the l-th local model as below:

$$A_l = \begin{bmatrix} k_2 + \frac{k_1}{2\lambda}\rho\pi R^3 \mathscr{F}_3^l & k_3 & 0 & 0 & -k_4 \\ \frac{k_5}{2\lambda}\rho\pi R^3 \mathscr{F}_3^l - k_6 & -k_7 & 0 & 0 & -k_8 \\ 0 & -L_q \mathscr{F}_1^l & R_s & 0 & 0 \\ 0 & L_d \mathscr{F}_2^l - \psi_m & 0 & R_s & 0 \\ 0 & 0 & (L_d - L_q)\mathscr{F}_1^l & \frac{3M}{4}\psi_m & -1 \end{bmatrix},$$

$$E = \begin{bmatrix} 1 & 0 & 0 & 0 & 0 \\ 0 & 1 & 0 & 0 & 0 \\ 0 & 0 & 1 & 0 & 0 \\ 0 & 0 & 0 & 1 & 0 \\ 0 & 0 & 0 & 0 & 0 \end{bmatrix}, B_l = \begin{bmatrix} 0 & 0 \\ 0 & 0 \\ 1 & 0 \\ 0 & 1 \\ 0 & 0 \end{bmatrix}. \tag{2.13}$$

Denote $\mathscr{F}^l := \prod_{\phi=1}^3 \mathscr{F}_\phi^l$ as the inferred fuzzy set, and $\mu_l[z(t)]$ as the normalized membership function, it yields

$$\mu_l[z(t)] := \frac{\prod_{\phi=1}^3 \mu_{\phi l}[z_\phi(t)]}{\sum_{\varsigma=1}^r \prod_{\phi=1}^3 \mu_\phi^\varsigma [z_\phi(t)]} \geq 0, \sum_{l=1}^r \mu_l[z(t)] = 1. \tag{2.14}$$

In the following, $\mu_l \triangleq \mu_l[z(t)]$ is denoted for brevity.

By fuzzy blending, the T-S fuzzy dynamic model is obtained by

$$E\dot{x}(t) = A(\mu)x(t) + B(\mu)u(t), \tag{2.15}$$

where $A(\mu) := \sum_{l=1}^r \mu_l A_l, B(\mu) := \sum_{l=1}^r \mu_l B_l$.

2.1.2 MODELING OF WIND POWER WITH DC LOAD

The power circuit with AC-DC converter is presented in Figure 2.2, where a three-phase fully controlled bridge comprising six switching devices is connected to the DC grid through smoothing inductor L. The DC side consists of a load R_L and a filter capacitor C. Table 2.1 gives the variables and their descriptions of the power converter system.

The mathematical formulation of the circuit system shown in Figure 2.2 is derived in the (a,b,c) coordinate frame as below:

$$\begin{cases} L\dot{\phi}_a = -r\phi_a - \frac{v_c}{3}(2u_a - u_b - u_c) + v_{ga}, \\ L\dot{\phi}_b = -r\phi_a - \frac{v_c}{3}(2u_a - u_b - u_c) + v_{ga}, \\ L\dot{\phi}_a = -r\phi_a - \frac{v_c}{3}(2u_a - u_b - u_c) + v_{ga}, \\ C\dot{v}_c = \phi_a u_a + \phi_b u_b + \phi_c u_c - \phi_0. \end{cases} \tag{2.16}$$

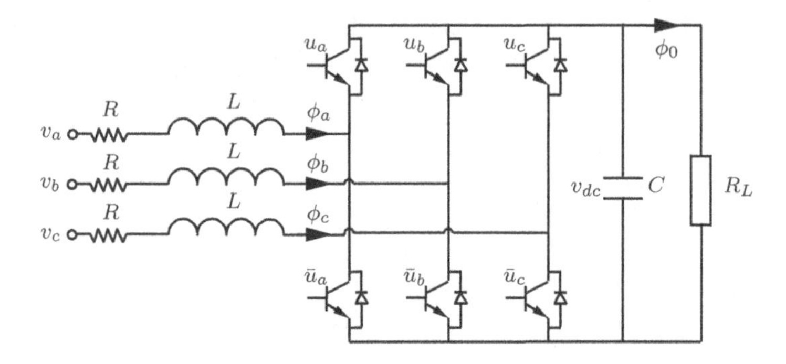

Figure 2.2 Power circuit of AC-DC converter.

Table 2.1
Parameters of AC-DC converter system.

Variable	Descriptions
C	DC-link capacitor
L	Smoothing inductor
r	Parasitic phase resistance
R_L	Load resistance
v_c	Output capacitor voltage
ω	Source grid voltage frequency
$\phi_{abc} = [\phi_a\ \phi_b\ \phi_c]^T$	Inductor current in (a,b,c) model
$u_{abc} = [u_a\ u_b\ u_c]^T$	Control inputs in (a,b,c) model
$v_{abc} = \left[v_{ga}\ v_{gb}\ v_{gc}\right]^T$	Grid voltage vector in (a,b,c) model

According to the power-invariant of the Clarke transformation,

$$\{\bullet\}_{\alpha\beta} = A\{\bullet\}_{abc}, A = \sqrt{\frac{2}{3}}\begin{bmatrix} 1 & -\frac{1}{2} & -\frac{1}{2} \\ 0 & \frac{\sqrt{3}}{2} & -\frac{\sqrt{3}}{2} \end{bmatrix}. \tag{2.17}$$

Then, the system (2.16) and (2.17) is given in a stationary (α,β) frame as below:

$$\begin{cases} L\dot{\phi}_{\alpha\beta} = v_{\alpha\beta} - \frac{v_{dc}}{2}u_{\alpha\beta}, \\ C\dot{v}_{dc} = \frac{1}{2}u_{\alpha\beta}^T\phi_{\alpha\beta} - \frac{v_{dc}}{R_L}, \end{cases} \tag{2.18}$$

where $\phi_{\alpha\beta}$ is vector of line currents; v_{dc} is output capacitor voltage; $v_{\alpha\beta}$ is vector of the source line voltages; $u_{\alpha\beta}$ is vector of control inputs.

Furthermore, it follows from instantaneous active P and reactive powers Q that

$$P = v_{\alpha\beta}^T\phi_{\alpha\beta}, Q = v_{\alpha\beta}^T J\phi_{\alpha\beta}, \tag{2.19}$$

where $J = \begin{bmatrix} 0 & -1 \\ 1 & 0 \end{bmatrix}$.

Using $\dot{v}_{\alpha\beta} = \omega J v_{\alpha\beta}$, and taking the time $\dot{v}_{\alpha\beta} = \omega J v_{\alpha\beta}$,

$$\begin{cases} \dot{P} = \omega v_{\alpha\beta}^T J^T \phi_{\alpha\beta} + \frac{1}{L} v_{\alpha\beta}^T v_{\alpha\beta} - v_{\alpha\beta}^T \frac{v_{dc}}{2L} u_{\alpha\beta}, \\ \dot{Q} = \omega v_{\alpha\beta}^T J^T J \phi_{\alpha\beta} + \frac{v_{\alpha\beta}^T J}{L} v_{\alpha\beta} - \frac{v_{\alpha\beta}^T J v_{dc}}{2L} u_{\alpha\beta}, \\ \dot{v}_{\alpha\beta} = \omega J v_{\alpha\beta}, \\ \dot{\phi}_{\alpha\beta} = \frac{1}{L} v_{\alpha\beta} - \frac{v_{dc}}{2L} u_{\alpha\beta}, \\ \dot{v}_{dc} = -\frac{1}{CR_L} v_{dc} + \frac{\phi_{\alpha\beta}^T}{2C} u_{\alpha\beta}. \end{cases} \tag{2.20}$$

Now, the control objective is to design a switching sequence of control vectors $u_{\alpha\beta}$ to regulate the output capacitor voltage v_{dc} to a desired constant value given by v_{dc}^*. Moreover, the system should achieve a near unity power factor (PF), meaning that the current vector $\phi_{\alpha\beta}$ should track a vector signal proportional to the line voltage $v_{\alpha\beta}$. The current tracking problem can be reinterpreted as a set point control problem if the outputs are chosen to be the active and reactive power P and Q. Thus, that is $P \to P^*, Q \to Q^*, v_{dc} \to v_{dc}^*$. Here, for the sake of simplicity we have assumed that the voltage source is balanced and free of harmonic distortion. We define $P_e = P - P^*, Q_e = Q - Q^*, v_e = v_{dc} - v_{dc}^*$, where v_{dc}^* is a given constant.

$$\dot{x}(t) = A(t)x(t) + B(t)u(t) + \omega(t), \tag{2.21}$$

where

$$x(t) = \begin{bmatrix} P_e & Q_e & v_{\alpha\beta}^T & \phi_{\alpha\beta}^T & v_e \end{bmatrix}^T,$$

$$A(t) = \begin{bmatrix} 0 & \omega L & \frac{v_{\alpha\beta}^T}{L} & 0 & 0 \\ 0 & -\omega L \frac{\|Jv_{\alpha\beta}\|^2}{\|v_{\alpha\beta}\|^2} & 0 & 0 & 0 \\ 0 & 0 & \omega J & 0 & 0 \\ 0 & 0 & \frac{1}{L} & 0 & 0 \\ 0 & 0 & 0 & 0 & -\frac{1}{CR_L} \end{bmatrix},$$

$$B(t) = \begin{bmatrix} -\frac{v_{\alpha\beta}^T (v_e + v_{dc}^*)}{2L} \\ -\frac{v_{\alpha\beta}^T J^T (v_e + v_{dc}^*)}{2L} \\ 0 \\ -\frac{v_{dc}}{2L} \\ \frac{\phi_{\alpha\beta}^T}{2C} \end{bmatrix}, \omega(t) = \begin{bmatrix} -\dot{P}^* + v_{\alpha\beta}^T \omega L \frac{v_{\alpha\beta}}{\|v_{\alpha\beta}\|^2} Q^* \\ -\dot{Q}^* - \omega L \frac{\|Jv_{\alpha\beta}\|^2}{\|v_{\alpha\beta}\|^2} Q^* \\ 0 \\ 0 \\ -\frac{1}{CR_L} v_{dc}^* \end{bmatrix}. \tag{2.22}$$

Here, one can choose $z_1 = v_{\alpha\beta}^T, z_2 = \frac{\|Jv_{\alpha\beta}\|^2}{\|v_{\alpha\beta}\|^2}$, $z_3 = v_{\alpha\beta}^T (v_e + v_{dc}^*)$, $z_4 = v_{\alpha\beta}^T J^T (v_e + v_{dc}^*)$, $z_5 = v_{dc}$, $z_6 = \omega$, and $z_7 = \phi_{\alpha\beta}^T$ as fuzzy premise variables. Thus, it follows from (2.21) that the wind power nonlinear system is represented by

Plant Rule $\mathscr{R}^l$: **IF** z_1 is $\mathscr{F}_1^l$ and z_2 is $\mathscr{F}_2^l$ and $\cdots$ and $\cdots$and z_7 is $\mathscr{F}_7^l$, **THEN**

$$\dot{x}(t) = A_l x(t) + B_l u(t) + \omega(t), l \in \mathscr{L} := \{1, 2, \ldots, r\} \tag{2.23}$$

where $\mathscr{R}^l$ denotes the l-th fuzzy inference rule; r is the number of inference rules; $\mathscr{F}_\theta^l(\theta = 1,2,3,4,5,6,7)$ is the fuzzy set; $x(t) \in \Re^{n_x}$ and $u(t) \in \Re^{n_u}$ denote the system state and control input, respectively; $z(t) \triangleq [z_1, z_2, z_3, z_4, z_5, z_6, z_7]$ are the measurable variables; $\{A_l, B_l\}$ is the l-th local model.

2.1.3 MODELING OF WIND POWER WITH AC LOAD

Nowadays, high-frequency power converters have been extensively used in many applications. The switching losses and electromagnetic interferences are major issues in these converters. Resonant converters can overcome these restrictions by using zero current and/or voltage switching [9]. AC-AC converters are required in order to vary the voltage across the load at fixed frequency.

Due to the large AC variations in currents and voltages of the resonant tank, the linearized model of the resonant converters has considerably large modeling error. This phenomenon makes the control of resonant converters more complex than pulse width modulation (PWM) converters. Due to sinusoidal approximation along with averaging and small signal approximation, the conventional controller design and stability analysis of resonant converters involve considerable error and limitations. Consequently, sophisticated control techniques with solid stability analysis are necessary. On the other hand, the accuracy of the small signal model and the designed controller diminish while the load resistance or the input voltage source have large variations. Therefore, it is essential to investigate the stability of the resonant converters using a nonlinear model [10].

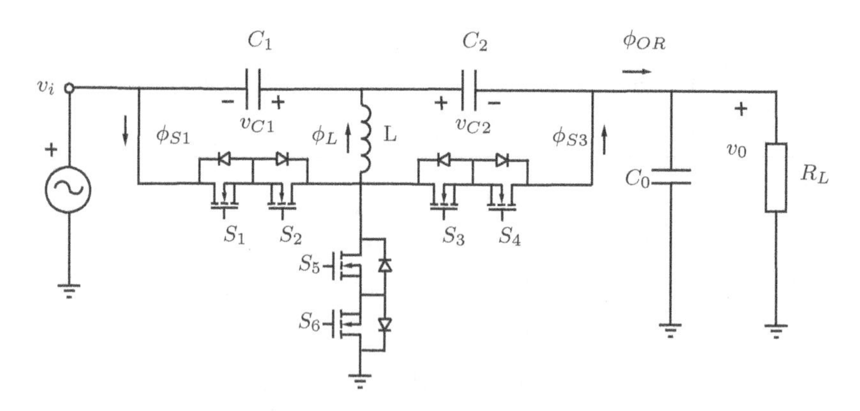

Figure 2.3 Power circuit of AC-AC resonant converter.

Figure 2.3 shows the power circuit of the AC-AC resonant converter. It consists of a third-order resonant circuit, three four-quadrant switches, and an output filter. The power flow between the AC source and the load is controlled by the proper selection of the operational modes of this converter: Energizing ($u = 1$) and deenergizing modes ($u = 0$). Each mode presents a discrete-time duration that coincides with an integer number of resonant periods. Moreover, in the resonant inductor the changes of the operational modes are always synchronized with the zero-current crossing point, which guarantees zero-current switching conditions. In energizing mode the AC source transfers energy to both the resonant circuit and the load when the converter operates. In this case, switches S1 and S2 are in ON-state during the first half-period of the energizing cycle; during the second half, switches S3 and S4 are in ON-state. The AC source does not transfer energy to the resonant circuit when the converter operates in deenergizing mode, and the energy stored in the resonant circuit is partially discharged onto the load. In this case, switches S5 and S6 are in ON-state during the first half-period of the deenergizing cycle, while in the second half, switches S3 and S4 are in ON-state.

Assuming that the line voltage v_i is generated by the wind generator with PMSG, and all the switches, inductors, and capacitors are ideal, the state-space model of the converter can be expressed as follows [9]:

$$\begin{cases} \dot{\phi}_L = \frac{1}{L}\left[v_a - v_{c2} - v_0\right], \\ C_1\dot{v}_{C1} + C_2\dot{v}_{C2} = \phi_L, \\ \dot{v}_0 = \frac{1}{C_0}\left[C_2\dot{v}_{C2} + \phi_{S3} - \frac{v_0}{R_L}\right], \end{cases} \tag{2.24}$$

where $v_a = uv_i clk + v_0(1 - clk), \phi_{S3} = (clk - 1)\phi_L$; v_i and f_i are line voltage and line frequency, respectively; L denotes the resonant tank inductance, R_L denotes the load resister; C_1 and C_2 are the resonant capacitances; v_0 and f_0 are output voltage and output frequency, respectively.

The value of the signal clk depends on the direction of the inductor current and the polarity of the input voltage, that is

$$clk = (1 + \operatorname{sgn}(\phi_L)\operatorname{sgn}(v_i))/2. \tag{2.25}$$

Based on the relationships (2.24) and (2.25),

$$\begin{cases} \dot{\phi}_L = -\frac{1}{L}v_{C2} - \frac{clk}{L}v_0 + \frac{1}{L}v_i clk u, \\ \dot{v}_{C2} = \frac{1}{C_2}\phi_L - \frac{C_1}{C_2}\dot{v}_{C1}, \\ \dot{v}_0 = \frac{clk}{C_0}\phi_L - \frac{C_1}{C_0}\dot{v}_{C1} - \frac{1}{R_L C_0}v_0. \end{cases} \tag{2.26}$$

Now, the control objective is to design a controller to regulate the output capacitor voltage v_0 to a desired constant value given by v_0^*. We define $e_0 = v_0 - v_0^*$ and $x(t) = \begin{bmatrix} \phi_L & v_{C2} & e_0 \end{bmatrix}^T$, and choose $z_1 = clk, z_2 = v_i$, and $z_3 = \frac{\dot{v}_{C1}}{v_{C2}}$ as fuzzy premise variables. Thus, it follows from (2.26) that the power circuit of AC-AC resonant converter is represented by

$$\dot{x}(t) = A(\mu)x(t) + B(\mu)u(t) + \omega(t), \tag{2.27}$$

where $A(\mu) := \sum_{l=1}^{r}\mu_l A_l, B(\mu) := \sum_{l=1}^{r}\mu_l B_l$,

$$A_l = \begin{bmatrix} 0 & -\frac{1}{L} & -\frac{1}{L}\mathscr{F}_1^l \\ \frac{1}{C_2} & -\frac{C_1}{C_2}\mathscr{F}_3^l & 0 \\ \frac{1}{C_0}\mathscr{F}_1^l & -\frac{C_1}{C_0}\mathscr{F}_3^l & -\frac{1}{R_L C_0} \end{bmatrix}, B_l = \begin{bmatrix} \frac{1}{L}\mathscr{F}_2^l\mathscr{F}_3^l \\ 0 \\ 0 \end{bmatrix}, \omega(t) = \begin{bmatrix} -\frac{clk}{L}v_0^* \\ 0 \\ -\frac{1}{RC_0}v_0^* \end{bmatrix}.$$

2.2 CONTROL OF WIND POWER WITH PMSG

2.2.1 STABILITY ANALYSIS OF WIND POWER

Consider a T-S fuzzy controller, which shares the same premise variables in (2.27) as follows:

Controller Rule $\mathscr{R}^l$: **IF** z_1 is $\mathscr{F}_1^l$ and z_2 is $\mathscr{F}_2^l$ and z_3 is $\mathscr{F}_3^l$, **THEN**

$$u(t) = K_l x(t), l \in \mathscr{L} \tag{2.28}$$

where $K_l \in \Re^{n_u \times n_x}$ are the controller gains to be designed.

Likewise, the total T-S fuzzy controller is given by

$$u(t) = K(\mu)x(t), \tag{2.29}$$

where $K(\mu) := \sum_{l=1}^{r} \mu_l K_l$.

It follows from (2.27) and (2.29) that the closed-loop wind power with PMSG is given by

$$\dot{x}(t) = (A(\mu) + B(\mu)K(\mu))x(t) + \omega(t). \tag{2.30}$$

Based on the closed-loop system in (2.31), the result on stability analysis is proposed as below:

Theorem 2.1: Stability Analysis of Wind Power

Consider the wind power with PMSG fuzzy system in (2.27) with $\omega(t) = 0$. A T-S fuzzy controller in (2.29) can be used to stabilize its closed-loop control system, if the following condition is satisfied:

$$P^T (A(\mu) + B(\mu)K(\mu)) + (A(\mu) + B(\mu)K(\mu))^T P < 0. \tag{2.31}$$

■

Proof. Similar to the proof in Theorem 1.3. Thus, it is deleted.

Note: Here, when considering a quadratic Lyapunov function $V(t) = x^T(t)Px(t)$, it is clear that if $P \equiv \sum_{l=1}^{r} \mu_l P_l$, the quadratic Lyapunov function turns into the fuzzy-basis-dependent Lyapunov function $V(t) = x^T(t)P(\mu)x(t)$. However, it requires that the time-derivative of μ_l is known a priori, which may be unpractical for the considered system.

2.2.2 DESIGN OF WIND POWER WITH MPPT CONTROL

The power coefficient curve $C_p(\lambda, \beta)$ has a unique maximum that corresponds to an optimal wind energy capture as below [3]:

$$C_p(\lambda^*, \beta^*) = C_p^*, \tag{2.32}$$

where

$$\lambda^* = \frac{\omega_t^* R}{v}. \tag{2.33}$$

Now, we define $e = \omega_t - \omega_t^*$. It follows from (2.20), (2.32) and (2.33) that

$$\begin{cases} \dot{e} = k_1 T_a + k_2 e + k_2 \omega_t^* + k_3 \omega_g - k_4 T_{em} - \lambda^* R \dot{v}, \\ \dot{\omega}_g = k_5 T_a - k_6 (e + \omega_t^*) - k_7 \omega_g - k_8 T_{em}, \\ \dot{\phi}_d = R_s \phi_d - \omega_g L_q \phi_q + u_d, \\ \dot{\phi}_q = R_s \phi_q + \omega_g L_d \phi_d - \omega_g \psi_m + u_q, \\ 0 \dot{T}_{em} = \frac{3M}{4} (\psi_m \phi_q + (L_d - L_q) \phi_q \phi_d) - T_{em}. \end{cases} \tag{2.34}$$

We define $x(t) = \begin{bmatrix} e & \omega_g & \phi_d & \phi_q & T_{em} \end{bmatrix}^T$, and choose $z_1 = \phi_q, z_2 = \phi_d$, $z_3 = \frac{C_p(\lambda,\beta)v^2}{e}$, and $z_4 = \frac{\dot{v}}{\omega_g}$ as fuzzy premise variables. Thus, it follows from (2.34) that the MPPT control problem of a wind power nonlinear system can be described by the following descriptor fuzzy system,

$$E\dot{x}(t) = A(\mu)x(t) + B(\mu)u(t), \tag{2.35}$$

where $A(\mu) := \sum_{l=1}^{r} \mu_l A_l, B(\mu) := \sum_{l=1}^{r} \mu_l B_l$,

$$A_l = \begin{bmatrix} k_2 + \frac{k_1}{2\lambda}\rho\pi R^3 \mathscr{F}_3^l & k_3 - \lambda^* R \mathscr{F}_4^l & 0 & 0 & -k_4 \\ \frac{k_5}{2\lambda}\rho\pi R^3 \mathscr{F}_3^l - k_6 & -k_7 & 0 & 0 & -k_8 \\ 0 & -L_q \mathscr{F}_1^l & R_s & 0 & 0 \\ 0 & L_d \mathscr{F}_2^l - \psi_m & 0 & R_s & 0 \\ 0 & 0 & (L_d - L_q)\mathscr{F}_1^l & \frac{3M}{4}\psi_m & -1 \end{bmatrix},$$

$$E = \begin{bmatrix} 1 & 0 & 0 & 0 & 0 \\ 0 & 1 & 0 & 0 & 0 \\ 0 & 0 & 1 & 0 & 0 \\ 0 & 0 & 0 & 1 & 0 \\ 0 & 0 & 0 & 0 & 0 \end{bmatrix}, B_l = \begin{bmatrix} 0 & 0 \\ 0 & 0 \\ 1 & 0 \\ 0 & 1 \\ 0 & 0 \end{bmatrix}. \tag{2.36}$$

Here, based on the descriptor fuzzy system in (2.35), the objective is to design a state-feedback fuzzy controller $u(t) = K(\mu)x(t)$, such that the closed-loop control system achieves the MPPT condition. The corresponding design result on the MPPT controller is summarized as below:

Theorem 2.2: Design of Wind Power with MPPT Control

Consider the wind power fuzzy system (2.35) using the T-S fuzzy controller laws in the form of (2.29). The MPPT of wind power generation is asymptotically achieved if the following LMIs hold:

$$\Sigma_{ll} < 0, l \in \mathscr{L} \tag{2.37}$$

$$\Sigma_{ls} + \Sigma_{sl} < 0, 1 \leq l < s \leq r \tag{2.38}$$

where $\Sigma_{ls} = \mathrm{Sym}(A_l X + B_l \bar{K}_s), X = \begin{bmatrix} X_{(1)} & 0 \\ X_{(2)} & X_{(3)} \end{bmatrix}, 0 < X_{(1)} = X_{(1)}^T \in \Re^{(n_x-1)\times(n_x-1)}$, $X_{(2)} \in \Re^{1\times(n_x-1)}$, $X_{(3)}$ is a scalar.

In that case, the controller gains can be calculated by

$$K_l = \bar{K}_l X^{-1}, l \in \mathscr{L}. \tag{2.39}$$

■

Proof. Similar to the proof in Theorem 1.4. Thus, it is deleted.

2.3 FINITE-TIME MPPT OF WIND POWER VIA SLIDING MODE CONTROL

It is noted that the wind turbine characteristics are provided by the manufacturers to obtain the MPPT performance. The accuracy of these wind turbine characteristic curves, however, is very sensitive to the operating environments. Furthermore, those operating environments, such as dirt, bugs, or ice on the blades, have an impact on the rotor speed. Therefore, it is meaningful to implement fast convergence and strict MPPT performance analysis on wind power systems under external disturbances.

Assume that no speed sensors are required in wind power system. Now, the MPPT problem of wind power generation in (2.35) is rewritten as

$$\begin{aligned} E\dot{x}(t) &= A(\mu)x(t) + B(\mu)u(t) + D\omega(t), \\ y(t) &= Cx(t), \end{aligned} \tag{2.40}$$

where $\omega(t)$ is external disturbance,

$$\begin{aligned} &A(\mu) := \sum_{l=1}^{r} \mu_l A_l, B(\mu) := \sum_{l=1}^{r} \mu_l B_l, \\ &C = \begin{bmatrix} 0 & 0 & 1 & 0 & 0 \\ 0 & 0 & 0 & 1 & 0 \end{bmatrix}, D = \begin{bmatrix} 1 & 1 & 0 & 0 & 0 \end{bmatrix}^T. \end{aligned} \tag{2.41}$$

Then, without loss of generality, the considered external disturbance is defined as below:

$$\mathbb{W}_{[t_1,t_2],\delta} \triangleq \left\{ \omega \in L_2[t_1,t_2] : \int_{t_1}^{t_2} \omega^2(s)ds \le \delta \right\}, \tag{2.42}$$

where δ is a positive scalar.

Before moving on, we extend the definition of the FTB in [11] to the descriptor fuzzy system (2.40) as follows:

Definition 2.1. For a given time interval $[t_1,t_2]$, a symmetrical matrix $R > 0$, and two scalars c_1, c_2 subject to $0 < c_1 < c_2$, the descriptor fuzzy system (2.42) with $u(t) = 0$ is the finite-time boundedness (FTB) subject to $(c_1, c_2, [t_1,t_2], R, \mathbb{W}_{[t_1,t_2],\delta})$, if it satisfies

$$x^T(t_1) E^T R E x(t_1) \le c_1 \Longrightarrow x^T(t_2) E^T R E x(t_2) < c_2, \forall t \in [t_1,t_2],$$

for all $\omega(t) \in \mathbb{W}_{[t_1,t_2],\delta}$.

The main aim here is to design a fuzzy sliding mode controller such that the MPPT control performance will be achieved in a finite time interval subject to $(c_1, c_2, [t_1, t_2], R, \mathrm{W}_{[t_1,t_2],\delta})$. Specifically, this part first designs an FSMC law, which ensures the state trajectories into the sliding surface in a finite-time T^* with $T^* \leq T$. And then, we calculate the bounding c^* satisfying $c_1 < c^* < c_2$, such that the resulting closed-loop system is the FTB subject to $(c_1, c^2, [0, T], R, \mathrm{W}_{[0,T],\delta})$.

2.3.1 DESIGN OF WIND POWER WITH FSMC LAW

Firstly, based on the descriptor fuzzy system (2.40), the integral-type sliding surface function is considered as below:

$$s(t) = GEx(t) - \int_0^t G[A(\mu) + B(\mu)K(\mu)C]x(s)ds, \tag{2.43}$$

where the matrix G is choosen such that GB_l is a positive definite matrix.

In the following, based on the sliding surface function (2.43), the objective is to design an FSMC law $u(t)$, which ensures the state trajectories of descriptor fuzzy system (2.40) into the specified sliding surface $s(t) = 0$ in a finite-time duration.

Theorem 2.3: Design of Wind Power with FSMC Law

Consider the descriptor fuzzy system (2.40) representing the nonlinear wind power system with MPPT control problem. The reachability of the specified sliding surface (2.43) in the finite-time duration $[0, T^*]$ with $T^* \leq T$ can be ensured by the following FSMC law:

$$u(t) = u_b(t) + u_c(t), \tag{2.44}$$

with

$$\begin{aligned} u_b(t) &= \sum_{l=1}^{r} \mu_l K_l C x(t), \\ u_c(t) &= -\sum_{l=1}^{r} \mu_l \left[GB_l\right]^{-1} \rho(t) \,\mathrm{sgn}\,(s(t)), \end{aligned} \tag{2.45}$$

where $K_l \in \Re^{2\times 5}$ denotes fuzzy controller gains, $\rho(t) = \frac{\rho + \|GD\| \|\omega(t)\|}{\lambda_{\min}\left(GB_l\left[GB_p\right]^{-1}\right)}, \rho \geq \frac{1}{T}\|GEx(0)\|, (l, p) \in \mathscr{L}$, and $\mathrm{sgn}(\star)$ is a switching sign function defined as

$$\mathrm{sgn}\,(s(t)) = \begin{cases} -1, & \text{for } s(t) < 0, \\ 0, & \text{for } s(t) = 0, \\ 1, & \text{for } s(t) > 0. \end{cases} \tag{2.46}$$

■

Proof. It follows from (2.43)-(2.46) that

$$\begin{aligned} s^T(t)\dot{s}(t) &= -s^T(t)GB(\mu)\sum_{l=1}^{r}\mu_l\left[GB_l\right]^{-1}\rho(t)\,\mathrm{sgn}(s(t)) + s^T(t)GD\omega(t) \\ &\leq -\sum_{l=1}^{r}\sum_{p=1}^{r}\mu_l\mu_p\lambda_{\min}\left(GB_l\left[GB_p\right]^{-1}\right)\rho(t)\left\|s(t)\right\| \\ &\quad + \left\|GD\right\|\left\|\omega(t)\right\|\left\|s(t)\right\| \\ &= -\rho\left\|s(t)\right\|. \end{aligned} \tag{2.47}$$

In addition, we define the following equation

$$V_1(t) = \frac{1}{2}s^T(t)s(t). \tag{2.48}$$

It has

$$\begin{aligned} \dot{V}_1(t) &\leq -\rho\left\|s(t)\right\| \\ &= -\sqrt{2}\rho\sqrt{V_1(t)}. \end{aligned} \tag{2.49}$$

Based on the proposal in [11], it yields

$$T^* \leq \frac{\sqrt{2}}{\rho}\sqrt{V_1(0)}. \tag{2.50}$$

Besides, it follows from (2.48) that

$$V_1(0) = \frac{1}{2}\left\|s(0)\right\|^2. \tag{2.51}$$

Substituting (2.51) into (2.50), one gets

$$T^* \leq \frac{1}{\rho}\left\|GEx(0)\right\|. \tag{2.52}$$

It follows from $\rho \geq \frac{1}{T}\left\|GEx(0)\right\|$ in (2.45) that

$$T^* \leq T, \tag{2.53}$$

which implies that the proposed FSMC law (2.45) can guarantee the state trajectories of the descriptor fuzzy system (2.40) into the specified sliding surface $s(t) = 0$ in a finite time T^* with $T^* \leq T$, thus completing this proof.

Note: It is noted that the proposed switching sign function sgn is discontinuous. The characteristic exhibits a high frequency oscillation, which is undesirable in practical applications. In order to eliminate chattering phenomena, an alternative approach is to employ the following switching function [12]:

$$\mathrm{sgn}(s(t)) = \begin{cases} -1, & \text{for } s(t) < -\rho, \\ \frac{1}{\rho}s, & \text{for } |s(t)| \leq \rho, \\ 1, & \text{for } s(t) > \rho. \end{cases}$$

It is easy to see that the proposed switching sign function becomes continuous and its value converges to the interval $[-\rho, \rho]$ instead of zero. In this case, the chattering conditions are eliminated.

2.3.2 REACHING PHASE IN FTB OF WIND POWER

By substituting the FSMC law (2.45) into (2.40), we obtain the resulting closed-loop control system as below:

$$E\dot{x}(t) = \sum_{l=1}^{r}\sum_{p=1}^{r} \mu_l\mu_p\bar{A}_{lp} + D\omega(t) - \sum_{l=1}^{r}\sum_{p=1}^{r} \mu_l\mu_p B_l\left[GB_p\right]^{-1}\rho(t)\,\mathrm{sgn}(s(t)), \quad (2.54)$$

where $\bar{A}_{lp} = A_l + B_l K_p C$.

On the reaching phase within $[0,T^*]$, the sliding motion is generated outside of the sliding surface (2.46). By defining $\bar{\rho}(t) = \rho(t)\mathrm{sgn}(s(t))$, $\bar{\rho} = \frac{\rho}{\lambda_{\min}\left(GB_l\left[GB_p\right]^{-1}\right)}$, and $\varepsilon = \frac{\|GD\|}{\lambda_{\min}\left(GB_l\left[GB_p\right]^{-1}\right)}$, one gets

$$\begin{aligned}\bar{\rho}^2(t) &= \rho^2(t) \\ &= \left[\bar{\rho} + \varepsilon\|\omega(t)\|\right]^2 \\ &= \bar{\rho}^2 + 2\bar{\rho}\varepsilon\|\omega(t)\| + \varepsilon^2\|\omega(t)\|^2 \\ &\leq \left(1+\varepsilon^2\right)\bar{\rho}^2 + \left(1+\varepsilon^2\right)\|\omega(t)\|^2. \end{aligned} \quad (2.55)$$

Now, a sufficient condition for the FTB of closed-loop system (2.43) in the finite time interval $[0,T^*]$ is derived as below:

Theorem 2.4: Reaching Phase in FTB

Consider the FSMC law (2.45), the resulting closed-loop wind power control system in (2.43) is FTB with respect to $(c_1, c^*, [0,T^*], R, W_{[0,T^*],\delta})$, if there exist matrices $P = \begin{bmatrix} P_1 & 0 \\ P_2 & P_3 \end{bmatrix}$, $0 < P_1^T = P_1 \in \Re^{(n_x-1)\times(n_x-1)}$, $P_2 \in \Re^{1\times(n_x-1)}$ and P_3 that is a scalar, and the control gain $K_l \in \Re^{n_u\times n_x}$, and positive scalars $\{c_1, c^*, \eta, \delta\}$, such that the following inequalities hold:

$$\Phi_{ll} < 0, 1 \leq l \leq r \quad (2.56)$$

$$\Phi_{lp} + \Phi_{pl} < 0, 1 \leq l < p \leq r \quad (2.57)$$

where

$$\Phi_{lp} = \begin{bmatrix} \Phi_{lp(1)} & P^T D & -P^T B_l\left[GB_p\right]^{-1} \\ \star & -\eta I & 0 \\ \star & \star & -\eta I \end{bmatrix},$$

$$\Phi_{lp(1)} = \mathrm{Sym}\left\{A_l^T P + C^T K_p^T B_l^T P\right\} - \eta E^T P. \quad (2.58)$$

Furthermore, the bounding is

$$\frac{\bar{\sigma}_{P1}c_1 + \left(\eta T^* \bar{\rho}^2 + \eta \delta\right)\left(1+\varepsilon^2\right) + \eta \delta}{e^{-\eta T^*}\underline{\sigma}_{P1}} < c^*. \tag{2.59}$$

■

Proof. Consider the following Lyapunov functional

$$V_2(t) = x^T(t) E^T P x(t), \forall t \in [0, T^*] \tag{2.60}$$

where $E^T P = P^T E \geq 0$.

Along the trajectory of system (2.54), one gets

$$\begin{aligned}
\dot{V}_2(t) &= 2\left[E\dot{x}(t)\right]^T P x(t) \\
&= 2\left[\sum_{l=1}^{r}\sum_{p=1}^{r} \mu_l \mu_p \bar{A}_{lp} x(t)\right]^T P x(t) \\
&\quad - 2\left[\sum_{l=1}^{r}\sum_{p=1}^{r} \mu_l \mu_p B_l \left[G B_p\right]^{-1} \rho(t) \operatorname{sgn}(s(t))\right]^T P x(t) \\
&\quad + 2\left[D\omega(t)\right]^T P x(t),
\end{aligned} \tag{2.61}$$

where $\bar{A}_{lp} = A_l + B_l K_p C$.

An auxiliary function $J(t)$ is introduced as below:

$$J(t) = \dot{V}_2(t) - \eta V_2(t) - \eta \omega^2(t) - \eta \bar{\rho}^2(t), \tag{2.62}$$

where η is a positive scalar.

It follows from (2.61) and (2.62) that

$$\begin{aligned}
J(t) &= 2\left[\sum_{l=1}^{r}\sum_{p=1}^{r} \mu_l \mu_p \bar{A}_{lp} x(t)\right]^T P x(t) \\
&\quad - 2\left[\sum_{l=1}^{r}\sum_{p=1}^{r} \mu_l \mu_p B_l \left[G B_p\right]^{-1} \bar{\rho}(t)\right]^T P x(t) + 2\left[D\omega(t)\right]^T P x(t) \\
&\quad - \eta x^T(t) E^T P x(t) - \eta \omega^2(t) - \eta \bar{\rho}^2(t) \\
&= \sum_{l=1}^{r}\sum_{p=1}^{r} \mu_l \mu_p \chi^T(t) \Phi_{lp} \chi(t),
\end{aligned} \tag{2.63}$$

where $\chi(t) = \begin{bmatrix} x^T(t) & \omega^T(t) & \bar{\rho}^T(t) \end{bmatrix}^T$, and Φ_{lp} is defined in (2.57).

Because of (2.62) and (2.63), $J(t) < 0$ implies that

$$\dot{V}_2(t) < \eta V_2(t) + \eta \omega^2(t) + \eta \bar{\rho}^2(t). \tag{2.64}$$

Multiplying both sides in (2.64) by $e^{-\eta t}$, and then integrating it from 0 to t, $t \in [0, T^*]$. It is easy to see that

$$\begin{aligned} e^{-\eta T^*} V_2(t) < {} & V_2(0) + \eta \int_0^{T^*} e^{-\eta s} \bar{\rho}^2(s) ds \\ & + \eta \int_0^{T^*} e^{-\eta s} \omega^2(s) ds \\ & \leq x^T(0) E^T P x(0) + \eta T^* \left(1+\varepsilon^2\right) \bar{\rho}^2 \\ & + \eta \left(1+\varepsilon^2\right) \delta + \eta \delta. \end{aligned} \tag{2.65}$$

On the other hand, it follows from (2.60) that

$$e^{-\eta T^*} V_2(t) \geq e^{-\eta T^*} x^T(t) E^T P x(t), \tag{2.66}$$

which implies that

$$\begin{aligned} e^{-\eta T^*} x^T(t) E^T P x(t) < {} & x^T(0) E^T P x(0) + \eta T^* \left(1+\varepsilon^2\right) \bar{\rho}^2 \\ & + \eta \left(1+\varepsilon^2\right) \delta + \eta \delta. \end{aligned} \tag{2.67}$$

Further, by specifying the matrix P as below:

$$P = \begin{bmatrix} P_1 & 0 \\ P_2 & P_3 \end{bmatrix}, \tag{2.68}$$

where $0 < P_1^T = P_1 \in \Re^{(n_x-1)\times(n_x-1)}$, $P_2 \in \Re^{1\times(n_x-1)}$ and P_3 is a scalar, it is easy to see that $E^T P = P^T E \geq 0$.

Now, we partition $x(t)$ as

$$x(t) = \begin{bmatrix} \bar{x}(t) \\ x_5(t) \end{bmatrix}, \tag{2.69}$$

where $\bar{x}(t) \in \Re^{(n_x-1)}$.

It follows from (2.66) and (2.67) that

$$\begin{aligned} e^{-\eta T^*} \bar{x}^T(t) P_1 \bar{x}(t) < {} & \bar{x}^T(0) P_1 \bar{x}(0) + \eta T^* \left(1+\varepsilon^2\right) \bar{\rho}^2 \\ & + \eta \left(1+\varepsilon^2\right) \delta + \eta \delta. \end{aligned} \tag{2.70}$$

Then, by introducing the matrix $0 < R_1^T = R_1 \in \Re^{(n_x-1)\times(n_x-1)}$, and further defining

$$\begin{aligned} & c_1 = \bar{x}^T(0) R_1 \bar{x}(0), \\ & \bar{\sigma}_{P1} = \lambda_{\max}\left(R_1^{-\frac{1}{2}} P_1 R_1^{-\frac{1}{2}}\right), \underline{\sigma}_{P1} = \lambda_{\min}\left(R_1^{-\frac{1}{2}} P_1 R_1^{-\frac{1}{2}}\right). \end{aligned} \tag{2.71}$$

Based on the relationships (2.70) and (2.71), one gets

$$\begin{aligned} x^T(t)E^T PEx(t) &= \bar{x}^T(t)R_1\bar{x}(t) \\ &< \frac{\bar{\sigma}_{P1}c_1 + (\eta T^*\bar{\rho}^2 + \eta\delta)(1+\varepsilon^2) + \eta\delta}{e^{-\eta T^*}\underline{\sigma}_{P1}} \\ &< c^*. \end{aligned} \tag{2.72}$$

From the Definition 2.1, the descriptor fuzzy control system in (2.54) is the FTB. This completes the proof.

Recall the tracking speed error $\varepsilon = \omega_t - \omega_t^*$ in (2.34). In order to calculate the bounding ε at the finite time convergence, we specify the matrix P_1 as

$$P_1 = \begin{bmatrix} P_{1(1)} & 0 \\ 0 & P_{1(2)} \end{bmatrix}, \tag{2.73}$$

where $P_{1(1)}$ is a scalar, and $0 < P_{1(2)}^T = P_{1(2)} \in \Re^{(n_x-1)\times(n_x-1)}$, and $P_{1(1)} >> \|P_{1(2)}\|$.

Based on the Definition 2.1, it is easy to see that

$$\varepsilon(0) < \frac{c_1}{\lambda_{\min}(P_1)}, \tag{2.74}$$

$$\varepsilon(t) < \frac{c_1 + \eta T^*(1+\varepsilon^2)\bar{\rho}^2 + \eta(1+\varepsilon^2)\delta + \eta\delta}{e^{-\eta T^*}\lambda_{\min}(P_1)}. \tag{2.75}$$

2.3.3 DESIGN PROCEDURE FOR MPPT ALGORITHM

The detailed calculating steps of the proposed MPPT algorithm for the wind power system are summarized as below:

i) Use the descriptor system approach to represent the MPPT control problem of the wind power system, as shown in (2.34);

ii) Use the T-S fuzzy model method to describe the nonlinear descriptor system as shown in (2.35);

iii) Choose a suitable matrix G, the time interval T, the controller gains K_l, the initial state $x(0)$, and construct the FSMC law as shown in (2.44);

iv) Based on the FSMC law, the reachability of the sliding surface (2.52) in finite time T^* with $T^* \le T$ is obtained.

v) Based on the the finite time T^*, we solve Theorem 2.4 to obtain the bounding c^*;

vi) Use (2.74) and (2.75) to calculate the bounding for the tracking speed error ε.

2.4 SIMULATION STUDIES

2.4.1 MPPT CONTROL OF WIND POWER WITH PMSG

Consider the PMSG with MPPT in (2.35). Here, the parameters of the PMSG system are chosen as below: $B_r = 0.2736$Nm/rad/s, $B_g = 0.002$Nm/rad/s, $R = 0.2165\Omega, R_s = 2.875\Omega, L_q = L_D = 0.0085H, M = 4, \Psi = 0.175, n_g = 43.165$,

$J_r = 32.5\text{kgm}^2, J_g = 0.0034kgm^2, K_{ls} = 26.91\text{Nm/rad/s}, \rho = 1.25kg/m^3$, $c_{p1} = 21\text{F}, c_{p2} = 125.22\text{F}, c_{p3} = 9.7792\text{F}, c_{p4} = 0.0068\text{F}$. Now, we choose $\{v, \omega_t, \omega_t^*, \phi_d, \phi_q\}$ as the fuzzy premise variable, and linearize the PMSG system around $\{17, 6, 6.28, 1, 1.5\}$ and $\{20, 7, 7.39, 1.5, 1\}$, and assume that $\lambda^* = 8, , \dot{v} = 0.1$. Then, the succeeding system matrices of T-S fuzzy model can be obtained as below:

$$A_1 = \begin{bmatrix} -0.4691 & -0.0181 & 0 & 0 & 0.027 \\ 0.0992 & 0.0025 & 0 & 0 & 4.7034 \\ 0 & -0.0128 & 2.875 & 0 & 0 \\ 0 & -0.1665 & 0 & 2.875 & 0 \\ 0 & 0 & 0 & 0.525 & -1 \end{bmatrix}, B_1 = \begin{bmatrix} 0 & 0 \\ 0 & 0 \\ 1 & 0 \\ 0 & 1 \\ 0 & 0 \end{bmatrix},$$

$$A_2 = \begin{bmatrix} -0.4691 & -0.0140 & 0 & 0 & 0.027 \\ 0.0992 & 0.0025 & 0 & 0 & 4.7034 \\ 0 & -0.0085 & 2.875 & 0 & 0 \\ 0 & -0.1623 & 0 & 2.875 & 0 \\ 0 & 0 & 0 & 0.525 & -1 \end{bmatrix}, B_2 = \begin{bmatrix} 0 & 0 \\ 0 & 0 \\ 1 & 0 \\ 0 & 1 \\ 0 & 0 \end{bmatrix}.$$

Here, by applying Theorem 2.2, the fuzzy controller gains are given by

$$K_1 = \begin{bmatrix} 0.0006 & 0.0152 & -3.2501 & 0.0126 & 0.0206 \\ -0.8398 & -7.5040 & 0.0566 & -15.4730 & -21.3070 \end{bmatrix},$$

$$K_2 = \begin{bmatrix} 0.0006 & 0.0098 & -3.2501 & 0.0113 & 0.0206 \\ -0.8699 & -8.0326 & 0.0617 & -16.0430 & -22.1140 \end{bmatrix}.$$

As shown in Figure 2.4, the open loop system is unstable. Based on the above solutions, Figure 2.5 indicates that the state responses for the PMSG system converge to zero.

2.4.2 FTB OF SMC OF WIND POWER WITH PMSG

Consider the PMSG with MPPT in (2.35). Here, the parameters of the PMSG system are chosen as below: $B_r = 0.2736\text{Nm/rad/s}, B_g = 0.002\text{Nm/rad/s}, R = 0.2165\Omega, R_s = 2.875\Omega, L_q = L_D = 0.0085\text{H}, M = 4, \Psi = 0.175, n_g = 43.165, J_r = 32.5\text{kgm}^2, J_g = 0.0034kgm^2, K_{ls} = 26.91\text{Nm/rad/s}, \rho = 1.25kg/m^3, c_{p1} = 21\text{F}, c_{p2} = 125.22\text{F}, c_{p3} = 9.7792\text{F}, c_{p4} = 0.0068\text{F}$. Now, we choose $\{v, \omega_t, \omega_t^*, \phi_d, \phi_q\}$ as the fuzzy premise variable, and linearize the PV system around $\{17, 6, 6.28, 1, 1.5\}$ and $\{20, 7, 7.39, 1.5, 1\}$, and assume that $\lambda^* = 8, , \dot{v} = 0.1$. Then, the succeeding system

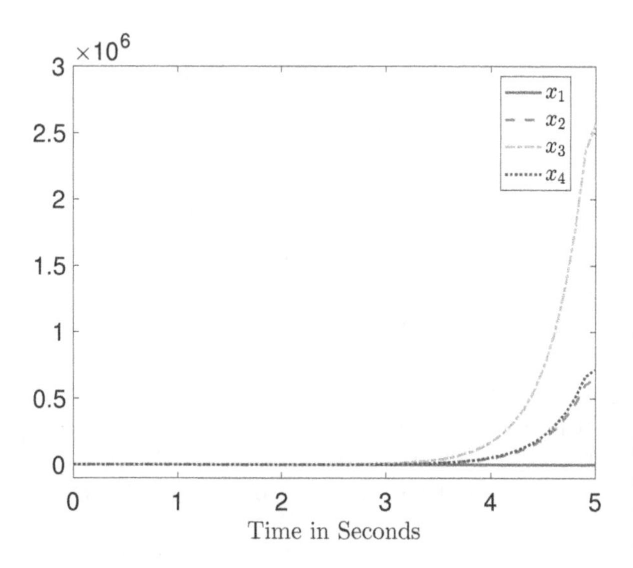

Figure 2.4 State responses of open-loop system.

matrices of T-S fuzzy model can be obtained as below:

$$A_1 = \begin{bmatrix} -0.4691 & -0.0181 & 0 & 0 & 0.027 \\ 0.0992 & 0.0025 & 0 & 0 & 4.7034 \\ 0 & -0.0128 & 2.875 & 0 & 0 \\ 0 & -0.1665 & 0 & 2.875 & 0 \\ 0 & 0 & 0 & 0.525 & -1 \end{bmatrix}, B_1 = \begin{bmatrix} 0 & 0 \\ 0 & 0 \\ 1 & 0 \\ 0 & 1 \\ 0 & 0 \end{bmatrix},$$

$$A_2 = \begin{bmatrix} -0.4691 & -0.0140 & 0 & 0 & 0.027 \\ 0.0992 & 0.0025 & 0 & 0 & 4.7034 \\ 0 & -0.0085 & 2.875 & 0 & 0 \\ 0 & -0.1623 & 0 & 2.875 & 0 \\ 0 & 0 & 0 & 0.525 & -1 \end{bmatrix}, B_2 = \begin{bmatrix} 0 & 0 \\ 0 & 0 \\ 1 & 0 \\ 0 & 1 \\ 0 & 0 \end{bmatrix}.$$

Here, by applying Theorem 2.2, the fuzzy controller gains are given by

$$K_1 = \begin{bmatrix} 0.0006 & 0.0152 & -3.2501 & 0.0126 & 0.0206 \\ -0.8398 & -7.5040 & 0.0566 & -15.4730 & -21.3070 \end{bmatrix},$$

$$K_2 = \begin{bmatrix} 0.0006 & 0.0098 & -3.2501 & 0.0113 & 0.0206 \\ -0.8699 & -8.0326 & 0.0617 & -16.0430 & -22.1140 \end{bmatrix}.$$

Now, we construct the FSMC law as shown in (2.44),

$$u(t) = u_b(t) + u_c(t),$$

where

$$u_b(t) = \sum_{l=1}^{r} \mu_l K_l C x(t), u_c(t) = -\sum_{l=1}^{r} \mu_l \left[GB_l\right]^{-1} \rho(t) \operatorname{sgn}(s(t)),$$

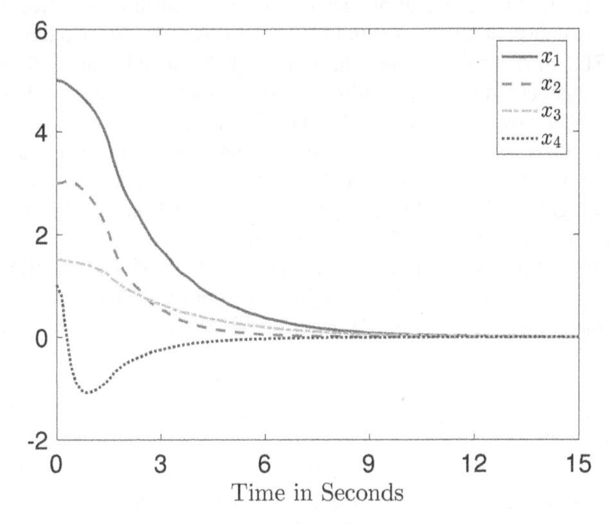

0

Figure 2.5 State responses of closed-loop control system.

with $x(0) = \begin{bmatrix} 1 \\ 1 \\ 1 \\ 1 \\ 0 \end{bmatrix}, D = \begin{bmatrix} 1 \\ 1 \\ 0 \\ 0 \\ 0 \end{bmatrix}, C = \begin{bmatrix} 0 & 0 & 0 & 0 & 0 \\ 0 & 0 & 0 & 0 & 0 \\ 0 & 0 & 1 & 0 & 0 \\ 0 & 0 & 0 & 1 & 0 \\ 0 & 0 & 0 & 0 & 0 \end{bmatrix}$,
$G = \begin{bmatrix} 0 & 0 & 6.11 & 0 & 0 \\ 0 & 0 & 0 & 6.11 & 0 \end{bmatrix}, \rho(t) = \frac{\rho + \|GD\|\|\omega(t)\|}{\lambda_{\min}\left(GB_l\left[GB_p\right]^{-1}\right)}, \rho \geq 2.88, \omega(t) = 0.1\sin t$, $(l,p) \in \mathscr{L}, T = 3$, and $\mathrm{sgn}(\star)$ is a switching sign function defined as

$$\mathrm{sgn}(s(t)) = \begin{cases} -1, & \text{for } s(t) < 0, \\ 0, & \text{for } s(t) = 0, \\ 1, & \text{for } s(t) > 0. \end{cases}$$

Given $R_1 = \mathrm{diag}\{1,1,1,1,1\}$. By using (2.44) to calculate $c_1 = 4, c^* < 974.1, e < 1090.64, e^* = 2395.6$.

2.5 REFERENCES

1. Shariatpanah, H., Fadaeinedjad, R., and Rashidinejad, M. (2013). A new model for PMSG-based wind turbine with yaw control. IEEE Transactions on Energy Conversion, 28(4), 929-937.
2. Raju, S. K. and Pillai, G. N. (2015). Design and implementation of type-2 fuzzy logic controller for DFIG-based wind energy systems in distribution networks. IEEE Transactions on Sustainable Energy, 7(1), 1-9.
3. Boukhezzar, B. and Siguerdidjane, H. (2010). Comparison between linear and nonlinear control strategies for variable speed wind turbines. Control Engineering Practice, 18(12), 1357-1368.

4. Li, S. and Li, J. (2017). Output predictor-based active disturbance rejection control for a wind energy conversion system with PMSG. IEEE Access, 5(99), 5205-5214.
5. Wei, C., Zhang, Z., Qiao, W., and Qu, L. (2016). An adaptive network-based reinforcement learning method for mppt control of PMSG wind energy conversion systems. IEEE Transactions on Power Electronics, 31(11), 7837-7848.
6. Hosseinzadeh, M. and Salmasi, F. R. (2015). Robust optimal power management system for a hybrid AC/DC micro-grid. IEEE Transactions on Sustainable Energy, 6(3), 675-687.
7. Liu, X., Wang, P., and Loh, P. C. (2011). A hybrid AC/DC microgrid and its coordination control. IEEE Transactions on Smart Grid, 2(2), 278-286.
8. Liu, J., Yin, Y., Luo, W., Vazquez, S., Franquelo, L. G., and Wu, L. (2018). Sliding mode control of a three-phase AC/DC voltage source converter under unknown load conditions: industry applications. IEEE Transactions on Systems Man & Cybernetics Systems, 48(10), 1771-1780.
9. Vicuna, L. G. G. D., Castilla, M., Miret, J., Matas, J., and Guerrero, J. M. (2009). Sliding-mode control for a single-phase AC/AC quantum resonant converter. IEEE Transactions on Industrial Electronics, 56(9), 3496-3504.
10. Molla-Ahmadian, H., Tahami, F., Karimpour, A., and Pariz, N. (2015). Hybrid control of DC-DC series resonant converters: the direct piecewise affine approach. Power Electronics IEEE Transactions on, 30(3), 1714-1723.
11. Song, J., Niu, Y., and Zou, Y. (2017). Finite-time stabilization via sliding mode control. IEEE Transactions on Automatic Control, 62(3), 1478-1483.
12. Bartoszewicz, A. and ZUk, J. (2010). Sliding mode control-basic concepts and current trends. IEEE International Symposium on Industrial Electronics, 3772-3777.

3 Fuzzy Modeling and Control Energy Storage Systems

Energy storage enables large-scale integration of distributed renewable energy sources. The benefits of storage can be appreciated, because system reliability cannot be guaranteed if renewable energy sources lack adequate storage facilities [1]. A storage unit is required to maintain the power balance between power generation and demand, especially in the power electronic-based microgrids or those based on photovoltaic generators with low inertia. The microgrids can substantially benefit from the availability of energy storages, generation, transmission, distribution, and consumption. For example, storage can eliminate or delay expansion of the transmission infrastructure or generation capacity. Storage can be combined with nondispatchable energy resources such as wind and solar generators to turn them into dispatchable power. On the consumers' side, storage can be employed for peak-shaving by storing the locally generated energy until it is needed.

Energy storage systems also play an important role in the power performance. For all the energy storage devices, batteries and supercapacitors are the most appropriate. However, the energy storage systems with only batteries or supercapacitors have significant limitations, and cannot meet the requirements of energy density and power density simultaneously. Generally, batteries have high energy density but their power density needs to be improved. Specifically, some modern batteries exhibit high power density but they may be unsuitable because of their size or cost. The supercapacitor has high power density, and its life cycle is about 100 times higher than that of battery. However, its drawback is that its energy density is lower. For these reasons, hybrid energy storage systems combining batteries and supercapacitors are perfect solutions for energy storage systems because of their high energy and power densities in electric vehicles (EVs) [2]. Despite its benefits, energy storage has not been fully utilized. In addition to cost, other limiting factors are lacks of appropriate control and management strategies. Future research is needed to investigate and develop control methodologies for the different energy storage technologies, and their applications.

The objective of this chapter is the development of a model simplification framework for the lead-acid batteries, the lithium-ion (Li-ion) batteries, and the supercapacitors. The original PDE-based battery model is first reformulated in a switching fuzzy-based model to precisely characterize its charge and discharge operations. Then, a fuzzy state of charge (SOC) estimation approach for the lead-acid batteries, the Li-ion batteries, and the supercapacitors has been proposed. Finally, a numerical example is provided to show the effectiveness of the proposed method.

3.1 MODELING AND CONTROL OF LEAD-ACID BATTERIES

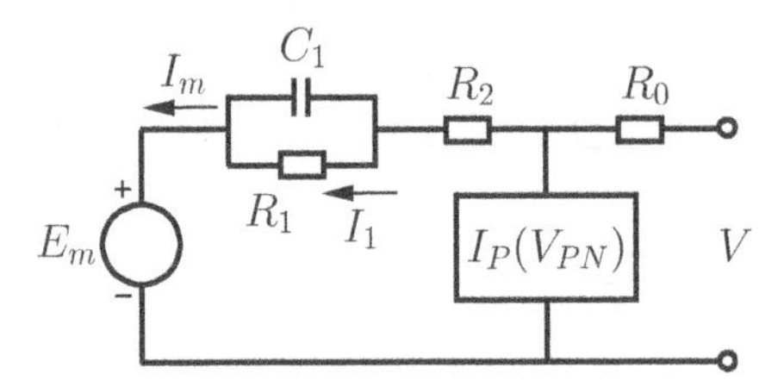

Figure 3.1 Lead-acid battery equivalent network.

Conveniently, the dynamics behavior of electrochemical batteries can be modeled based on equivalent electric networks. Although these networks contain elements that are nonlinear and depend on battery state-of-charge and electrolyte temperature, they are very useful for the electrical engineers, since they allow engineers to think in terms of electric quantities, instead of internal battery electrochemical reactions. The third-order model proposed has an accuracy satisfactory for the majority of uses; for particular situations more sophisticated models can be derived from the general model structure proposed as shown in Figure 3.1. The third-order model has proven testified to be a good compromise between complexity and precision [3].

3.1.1 MODELING OF LEAD-ACID BATTERIES

Assume that state variables are the current I_1, the extracted charge Q_e, and electrolyte temperature θ. Therefore, the dynamic equations of the model are given by [3]:

$$\begin{cases} \frac{dI_1}{dt} = \frac{1}{\tau_1}\left(I_m - I_1\right), \\ \frac{dQ_e}{dt} = -I_m, \\ \frac{d\theta}{dt} = \frac{1}{C_\theta}\left[P_s - \frac{(\theta-\theta_a)}{R_\theta}\right], \end{cases} \tag{3.1}$$

where $\tau_1 = R_1C_1$, C_θ and R_θ is thermal capacitor and resistance between the battery and its environment, θ_a is the temperature of the environment surrounding the battery, and P_s is the thermal source generated internally in the battery.

The assumed equations for E_m, R_0, R_1, R_2 are

$$\begin{cases} E_m = E_{m0} - K_E\left(273+\theta\right)\left(1-SOC\right), \\ R_0 = R_{00}\left[1+A_0\left(1-SOC\right)\right], \\ R_1 = -R_{10}\ln\left(DOC\right), \\ R_2 = R_{20}\frac{\exp[A_{21}(1-SOC)]}{1+\exp(A_{22}I_m/I^*)}, \end{cases} \tag{3.2}$$

where $E_{m0}, K_E, R_{00}, A_0, R_{10}, R_{20}, A_{21}, A_{22}$ are constant for a particular battery. SOC is an indicator of how full is a battery with reference to the maximum capacity the battery is able to deliver at the given temperature θ, DOC is an indicator of how full is the battery with reference to the actual discharge regime, and the current to be utilized in the expression of the capacity $C(I,\theta)$ is $I_{avg} = I_1$, SOC is state-of-charge, that is

$$SOC = 1 - Q_e/C(0,\theta) \\ = 1 - Q_e/(K_C C(I^*)), \tag{3.3}$$

where I^* is a reference current, and K_C is an empirical coefficient for a given I^*. DOC is depth-of-charge, that is

$$DOC = 1 - Q_e/C(I_{avg},\theta). \tag{3.4}$$

The behavior of the parasitic branch is actually strongly nonlinear. The following equation can be used to match the Tafel gassing-current relationship as below [3]:

$$I_p = V_{PN} G_{p0} \exp\left(V_{PN}/V_{p0} + A_p\left(1 - \theta/\theta_f\right)\right), \tag{3.5}$$

where θ_f is the electrolyte freezing temperature that depends mainly on the electrolyte specific gravity, and can normally be assumed as equal to $-40^{o}C$; The parameters G_{p0}, V_{p0}, and A_p are constant for a particular battery.

3.1.2 CHARGE MODELING

Assume that V is the terminal voltage and I is the input current (positive in discharging and negative in charging). Now, we define $z_1 = \frac{1}{\tau_1}$, $z_2 = \exp\left(V_{PN}/V_{p0} + A_p\left(1 - \theta/\theta_f\right)\right)$, $z_3 = R_1$, $z_4 = R_2$, $z_5 = R_0$. It follows from (3.1)-(3.5) that [3]

$$\begin{cases} \frac{dI_1}{dt} = -z_1 I_1 - z_1 z_2 G_{p0} V + (z_1 z_2 z_5 G_{p0} + z_1) I, \\ \frac{dQ_e}{dt} = G_{p0} z_2 V - (1 + z_2 z_5 G_{p0}) I, \\ \frac{d\theta}{dt} = -\frac{\theta}{C_\theta R_\theta} + \frac{\theta_a}{C_\theta R_\theta} + \frac{1}{C_\theta} P_s, \\ 0\frac{dV}{dt} = z_3 I_1 - (z_2 z_4 G_{p0} + 1) V + (z_5 + z_6 + G_{p0} z_2 z_5 z_6) I + E_m, \end{cases} \tag{3.6}$$

where $\tau_1 = R_1 C_1$, θ_a is the temperature of the environment surrounding the battery, and P_s is the thermal source generated internally in the battery.

Further define $x(t) = \begin{bmatrix} I_1 & Q_e & \theta & V \end{bmatrix}^T$, and it follows from (3.6) that

$$\begin{cases} E\dot{x}(t) = A(t)x(t) + B(t)u(t) + \omega(t), \\ y(t) = Cx(t), \end{cases} \tag{3.7}$$

where $E = \text{diag}\{1,1,1,0\}$, $u(t) = I, \omega(t) = \begin{bmatrix} 0 & 0 & \frac{\theta_a}{C_\theta R_\theta} + \frac{1}{C_\theta}P_s & E_m \end{bmatrix}^T$, and

$$A(t) = \begin{bmatrix} -z_1 & 0 & 0 & -z_1 z_2 G_{p0} \\ 0 & 0 & 0 & G_{p0} z_2 \\ 0 & 0 & -\frac{1}{C_\theta R_\theta} & 0 \\ z_3 & 0 & 0 & -(z_2 z_4 G_{p0} + 1) \end{bmatrix},$$

$$B(t) = \begin{bmatrix} z_1 z_2 z_5 G_{p0} + z_1 \\ -(1 + z_2 z_5 G_{p0}) \\ 0 \\ z_5 + z_6 + G_{p0} z_2 z_5 z_6 \end{bmatrix}, C = \begin{bmatrix} 0 & 0 & 1 & 0 \\ 0 & 0 & 0 & 1 \end{bmatrix}. \tag{3.8}$$

Now, choose $z_1 - z_5$ as fuzzy premise variables, similar to the processing on (1.7), the nonlinear system in (3.7) can be represented by the following fuzzy system,

$$\begin{cases} E\dot{x}(t) = A(\mu)x(t) + B(\mu)u(t) + \omega(t), \\ y(t) = Cx(t). \end{cases} \tag{3.9}$$

Note: It is important to note that since during discharge $R_2 \cong 0$ and $I_p \cong 0$, if only the discharge behavior is to be simulated, the resistor R_2 and the whole parasitic branch can be omitted from the model. The discharge modeling can be given by the next subsection.

Given the lead-acid battery fuzzy model in (3.9), and for an $\mathscr{L}_2$-gain performance level $\gamma > 0$, the purpose of this section is to design a fuzzy output-feedback controller $u(t) = K(\mu)y(t)$ such that the closed-loop fuzzy control system is asymptotically stable, and for any nonzero $\omega \in \mathscr{L}_2\,[0\ \infty)$ the induced $\mathscr{L}_2$ norm of the operator from ω to the extracted charge Q_e is less than γ

$$\int_0^\infty Q_e^2(s)ds < \gamma^2 \int_0^\infty \omega^T(s)\omega(s)ds, \tag{3.10}$$

under zero initial conditions.

Based on the augmented closed-loop fuzzy control system in (3.9), the result on $\mathscr{H}_\infty$ performance analysis is proposed as below:

Theorem 3.1: $\mathscr{H}_\infty$ Performance Analysis for Charge Modeling

Consider the lead-acid charge modeling in (3.9) using the output feedback controller $u(t) = K(\mu)Cy(t)$. Then, the stability of closed-loop charging control system is asymptotically achieved with $\mathscr{H}_\infty$ performance index if the following inequalities hold:

$$E^T P = P^T E \geq 0, \tag{3.11}$$

$$\begin{bmatrix} P^T A(\mu) + A^T(\mu)P + F^T F & P^T \\ \star & -\gamma^2 I \end{bmatrix} < 0, \tag{3.12}$$

where $\bar{A}(\mu) = A(\mu) + B(\mu)K(\mu)C$, $F = \begin{bmatrix} 0 & 1 & 0 & 0 \end{bmatrix}$. ■

Proof. Consider $V(t) = x^T(t)E^TPx(t)$. It is well-known that $\mathscr{H}_\infty$ performance can be verified if the following inequality holds,

$$\dot{V}(t) + Q_e^2 - \gamma^2\omega^T(t)\omega(t) < 0. \tag{3.13}$$

Thus, the proof is completed.

3.1.3 DISCHARGE MODELING

During discharge, $R_2 \cong 0$ and $I_p \cong 0$. Thus, the dynamic equations of the discharging model are given by [3]

$$\begin{cases} \frac{dI_1}{dt} = -\frac{1}{\tau_1}I_1 + \frac{1}{\tau_1}I, \\ \frac{dQ_e}{dt} = -I, \\ \frac{d\theta}{dt} = -\frac{\theta}{C_\theta R_\theta} + \frac{\theta_a}{C_\theta R_\theta} + \frac{1}{C_\theta}P_s, \\ 0\frac{dV}{dt} = I_1R_1 - V + IR_0 + E_m. \end{cases} \tag{3.14}$$

Now, define $z_1 = \frac{1}{\tau_1}, z_2 = R_0, z_3 = R_1, x(t) = \begin{bmatrix} I_1 & Q_e & \theta & V \end{bmatrix}^T$, and based on Thevenin's theorem,

$$\begin{cases} E\dot{x}(t) = A(t)x(t) + B(t)u(t) + \omega(t), \\ y(t) = Cx(t), \end{cases} \tag{3.15}$$

where $E = \text{diag}\{1,1,1,0\}$, $u(t) = I, \omega(t) = \begin{bmatrix} 0 & 0 & \frac{\theta_a}{C_\theta R_\theta} + \frac{1}{C_\theta}P_s & E_m \end{bmatrix}^T$, and

$$A(t) = \begin{bmatrix} -z_1 & 0 & 0 & 0 \\ 0 & 0 & 0 & 0 \\ 0 & 0 & -\frac{1}{C_\theta R_\theta} & 0 \\ z_3 & 0 & 0 & -1 \end{bmatrix}, B(t) = \begin{bmatrix} z_1 \\ -1 \\ 0 \\ R_0 \end{bmatrix}, C = \begin{bmatrix} 0 & 0 & 1 & 0 \\ 0 & 0 & 0 & 1 \end{bmatrix}.$$

Choose $z_1 - z_3$ as fuzzy premise variables, similar to the fuzzy processing on (1.7). The nonlinear system in (3.15) can be represented by the following fuzzy system,

$$\begin{cases} E\dot{x}(t) = A(\mu)x(t) + B(\mu)u(t) + \omega(t), \\ y(t) = Cx(t), \end{cases} \tag{3.16}$$

where $A(\mu) := \sum_{l=1}^{r} \mu_l A_l, B(\mu) := \sum_{l=1}^{r} \mu_l B_l, A_l$ and B_l the l-th local model.

Note that, based on the fuzzy system (3.16), the result on $\mathscr{H}_\infty$ performance analysis is similar to Theorem 3.1, thus it is omitted.

3.1.4 SWITCHING CHARGE AND DISCHARGE OPERATIONS

Consider the fast switching between charge and discharge. Those operations can be represented by the following system under arbitrary switching,

$$\begin{cases} E\dot{x}(t) = A_i(t)x(t) + B_i(t)u(t) + \omega(t), \\ y(t) = Cx(t), i = \{1,2\} \end{cases} \tag{3.17}$$

where $x(t) = \begin{bmatrix} I_1 & Q_e & \theta & V \end{bmatrix}^T, u(t) = I, \omega(t) = \begin{bmatrix} 0 & 0 & \frac{\theta_a}{C_\theta R_\theta} + \frac{1}{C_\theta} P_s & E_m \end{bmatrix}^T$,

$$A_1(t) = \begin{bmatrix} -z_1 & 0 & 0 & -z_1 z_2 G_{p0} \\ 0 & 0 & 0 & G_{p0} z_2 \\ 0 & 0 & -\frac{1}{C_\theta R_\theta} & 0 \\ z_3 & 0 & 0 & -(z_2 z_4 G_{p0} + 1) \end{bmatrix},$$

$$B_1(t) = \begin{bmatrix} z_1 z_2 z_5 G_{p0} + z_1 \\ -(1 + z_2 z_5 G_{p0}) \\ 0 \\ z_5 + z_6 + G_{p0} z_2 z_5 z_6 \end{bmatrix}, C = \begin{bmatrix} 0 & 0 & 1 & 0 \\ 0 & 0 & 0 & 1 \end{bmatrix},$$

$$A_2(t) = \begin{bmatrix} -z_1 & 0 & 0 & 0 \\ 0 & 0 & 0 & 0 \\ 0 & 0 & -\frac{1}{C_\theta R_\theta} & 0 \\ z_3 & 0 & 0 & -1 \end{bmatrix}, B_2(t) = \begin{bmatrix} z_1 \\ -1 \\ 0 \\ R_0 \end{bmatrix}. \tag{3.18}$$

Similar to the fuzzy processing on (1.7), the nonlinear switching system in (3.17) can be represented by the following switching fuzzy system,

$$\begin{cases} E\dot{x}(t) = A_i(\mu_i)x(t) + B_i(\mu_i)u(t) + \omega(t), \\ y(t) = Cx(t), i = \{1,2\} \end{cases} \tag{3.19}$$

where $A_i(\mu_i) := \sum_{l=1}^{r} \mu_l A_{il}, B_i(\mu_i) := \sum_{l=1}^{r} \mu_l B_{il}, A_{il}$ and B_{il} the l-th local model of the i-th switching subsystem.

Note that in lead-acid batteries, the charge model (3.9) and the discharge model (3.16) are described in the uniform framework of switching systems as shown in (3.19). Based on the switching fuzzy system, the result on $\mathscr{H}_\infty$ performance analysis is summarized as below.

Theorem 3.2: $\mathscr{H}_\infty$ Performance Analysis for Charge and Discharge Switching

Consider the lead-acid switching fuzzy system in (3.19) using the output feedback controller $u(t) = K_i(\mu_i)Cy(t)$. Then, the stability of a closed-loop swtching control system is asymptotically achieved with $\mathscr{H}_\infty$ performance index if the following inequalities hold:

$$E^T P = P^T E \geq 0, \tag{3.20}$$

$$\begin{bmatrix} P^T A_i(\mu_i) + A_i^T(\mu_i)P + F^T F & P^T \\ \star & -\gamma^2 I \end{bmatrix} < 0, \tag{3.21}$$

where $\bar{A}_i(\mu_i) = A_i(\mu_i) + B_i(\mu_i)K_i(\mu_i)C$, $F = \begin{bmatrix} 0 & 1 & 0 & 0 \end{bmatrix}$. ■

Proof. Consider $V(t) = x_i^T(t)E^T P x_i(t)$. It is well-known that $\mathscr{H}_\infty$ performance can be verified if the following inequality holds,

$$\dot{V}_i(t) + Q_e^2 - \gamma^2 \omega^T(t)\omega(t) < 0, i = \{1,2\}. \tag{3.22}$$

Thus, the proof is completed.

3.1.5 SOC ESTIMATION OF SWITCHING OPERATIONS

Based on the charge and discharge systems on lead-acid batteries (3.19), the objective is to estimate the SOC by using the following fuzzy descriptor observer,

Plant Rule $\mathscr{R}_i^s$: **IF** z_{i1} is $\hat{\mathscr{F}}_{i1}^s$ and z_{i2} is $\hat{\mathscr{F}}_{i2}^s, \cdots$, and z_{ig} is $\hat{\mathscr{F}}_{ig}^s$, **THEN**

$$\begin{cases} E\dot{\hat{x}}(t) = A_{is}\hat{x}(t) + B_{is}u(t) + L_{is}(y(t) - \hat{y}(t)), \\ \hat{y}(t) = C\hat{x}(t), \end{cases} \tag{3.23}$$

where $\hat{x} \in \mathfrak{R}^{n_{\hat{x}}}$, and L_{is} is the observer gain to be determined.

Note that lead-acid batteries are operated in the charge mode for $i = 1, g = 5$ and the discharge mode for $i = 2, g = 3$. The global fuzzy observer is given by

$$\begin{cases} E\dot{\hat{x}}(t) = A_i(\hat{\mu}_i)\hat{x}(t) + B_i(\hat{\mu}_i)u(t) + L_i(\hat{\mu}_i)(y(t) - \hat{y}(t)), \\ \hat{y}(t) = C\hat{x}(t), \end{cases} \tag{3.24}$$

where the notation $\hat{\mu}_i$ is induced by the estimated premise variable z_{ig}.

Now, consider the following global fuzzy controller,

$$u(t) = K_i(\hat{\mu}_i)\hat{x}(t). \tag{3.25}$$

Define $e(t) = x(t) - \hat{x}(t)$, and it follows from (3.19), (3.23)-(3.25) that

$$\bar{E}\dot{\bar{x}}(t) = \bar{A}_i(\mu_i, \hat{\mu}_i)\bar{x}(t) + \bar{\omega}(t), \tag{3.26}$$

where

$$\bar{E} = \begin{bmatrix} E & 0 \\ 0 & E \end{bmatrix}, \bar{A}_i(\mu_i, \hat{\mu}_i) = \begin{bmatrix} A_i(\mu_i) + B_i(\mu_i)K_i(\hat{\mu}_i) & -B_i(\mu_i)K_i(\hat{\mu}_i) \\ 0 & A_i(\mu_i) - L_i(\mu_i, \hat{\mu}_i)C \end{bmatrix},$$
$$\bar{x}(t) = \begin{bmatrix} x(t) \\ e(t) \end{bmatrix}, \bar{\omega}(t) = \begin{bmatrix} \omega(t) \\ (A_i(\mu_i) - A_i(\hat{\mu}_i))\hat{x}(t) \end{bmatrix}. \tag{3.27}$$

Given the closed-loop error system in (3.26), and for a $\mathcal{L}_2$-gain performance level $\gamma > 0$, the purpose of this section is to design a fuzzy observer-based controller in (3.24) and (3.25) such that the error system of lead-acid battery is asymptotically stable, and for any nonzero $\bar{\omega} \in \mathcal{L}_2\,[0\ \infty)$ the induced $\mathcal{L}_2$ norm of the operator from $\bar{\omega}$ to the extracted charge Q_e is less than γ

$$\int_0^{\infty} Q_e^2(s)\,ds < \gamma^2 \int_0^{\infty} \bar{\omega}^T(s)\bar{\omega}(s)ds, \tag{3.28}$$

under zero initial conditions.

Based on the closed-loop error system in (3.26), the $\mathcal{H}_\infty$ performance analysis on state of charge (SOC) estimation of lead-acid batteries is proposed as below:

Theorem 3.3: $\mathcal{H}_\infty$ Performance Analysis for SOC Estimation

Consider the lead-acid battery switching system (3.19) using the fuzzy observer-based controller in the form of (3.24) and (3.25). For matrix $\bar{P} \in \Re^{(n_x+n_{\tilde{x}})\times(n_x+n_{\tilde{x}})}$, the error system of lead-acid battery is asymptotically achieved with $\mathcal{H}_\infty$ performance index if the following inequalities hold:

$$\bar{E}^T\bar{P} = \bar{P}^T\bar{E} \geq 0, \tag{3.29}$$

$$\begin{bmatrix} \bar{P}^T\bar{A}_i(\mu_i,\hat{\mu}_i)+\bar{A}_i^T(\mu_i,\hat{\mu}_i)\bar{P}+\bar{F}^T\bar{F} & \bar{P}^T \\ \star & -\gamma^2 I \end{bmatrix} < 0, \tag{3.30}$$

where $\bar{F} = \begin{bmatrix} 0 & 1 & 0 & 0 & 0 & 0 & 0 & 0 \end{bmatrix}$. ■

Proof. Consider a common Lyapunov function $V(t) = \bar{x}^T(t)\bar{E}^T\bar{P}\bar{x}(t)$, where $\bar{E}^T\bar{P} = \bar{P}^T\bar{E} \geq 0$, and $\bar{P} \in \Re^{(n_x+n_{\tilde{x}})\times(n_x+n_{\tilde{x}})}$. It is well-known that $\mathcal{H}_\infty$ performance can be verified if the following inequality holds,

$$\dot{V}(t) + Q_e^2 - \gamma^2\bar{\omega}^T(t)\bar{\omega}(t) < 0, i = \{1,2\}. \tag{3.31}$$

Thus, the proof is completed.

It is noted that the results on Theorem 3.3 are not LMIs. Here, a two-step processing proposed (1.68)-(1.73) can be used to obtain the observer and controller gains.

3.2 MODELING AND CONTROL OF LI-ION BATTERIES

A Li-ion battery cell is composed of the negative electrode, the positive electrode, the separator, and the electrolyte [4]. The separator keeps the electrodes apart and the electrodes and separator are immersed in the electrolyte. The electrolyte is an ionic solution that can exchange lithium with the electrodes and provides electrical insulation. Each electrode has a certain potential due to the electrochemical couple formed by its material and the lithium dissolved in the electrolyte. Electrons are attracted by this difference of potential but cannot be exchanged from one electrode to the other within the battery unlike lithium ions. An electron that belongs to an electrode can go through an external electrical circuit to reach the other electrode if this circuit exists. Then, lithium is removed from its source electrode and another is inserted in its electrode of destination. This conserves the charges' equilibrium in the electrodes and in the electrolyte at any time. From a macroscopic point of view, this flow of electrons between the electrodes through an external circuit corresponds to the current.

In order to estimate the battery SOC under different circumstances accurately, a proper model that describes the battery characteristics is indispensable. The most commonly used lithium-ion batteries include the electrochemical mechanism models [5], the neural network models [6, 7], and the equivalent circuit models [8]. The electrochemical mechanism models are often utilized for the battery mechanism analysis and the electrode and electrolyte material selections. However, since parameters in a battery model are related to the battery's structure, dimensions, and materials, the electrochemical mechanism models are too complicated for SOC estimations. The neural network models can be utilized to simulate the characteristics of the battery, as the neural network has the ability of learning nonlinear dynamics. Therefore, complicated system dynamics that involve battery states and inputs can be modeled by neural networks without knowing the complicated internal battery mechanisms. Nonetheless, neural network models need a large amount of training data. The equivalent circuit models utilize the circuit elements to simulate the battery's characteristics, and have the advantages of low computational complexity and good flexibility in battery materials and sizes.

3.2.1 LI-ION BATTERIES BASED ON SINGLE PARTICLE MODEL (SPM)

To simplify the electrochemical model, we assume that the behavior of a single particle of the average size of the particles in the electrode represents the behavior of the whole electrode as shown in [5].

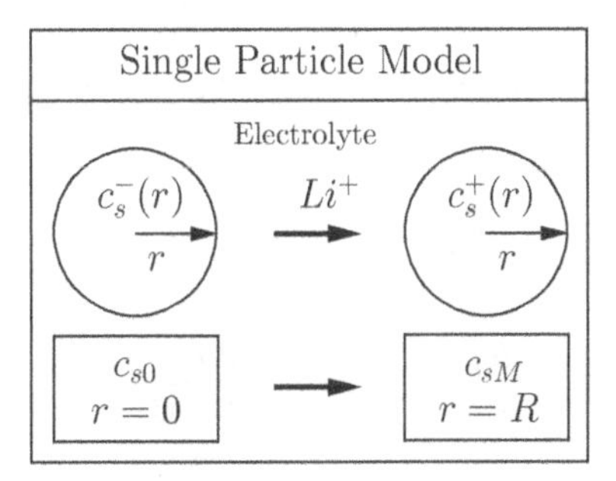

Figure 3.2 Schematic of the SPM.

The electrodes in the SPM electrochemical model are approximated as spherical particles. The SPM is a simplified model adopted for this brief. It is depicted in Figure 3.2 [9] and the nomenclature is listed in Table 3.1.

Table 3.1
Parameters of Li-ion batteries.

Symbol	Definition and Unit
A	Current collector area (m^2)
$\pm$	Positive/negative electrode
$a_s^{\pm}$	Specific surface area(m^2/m^3)
c_e	Electrolyte phase Li-ion concentration (mol/m^3)
$c_s^{\pm}$	Solid phase Li-ion concentration (mol/m^3)
$c_{s,e}^{\pm}$	Solid phase Li-ion concentration at surface (mol/m^3)
$c_{s,\max}^{\pm}$	Solid phase Li-ion saturation concentration (mol/m^3)
$D_s^{\pm}$	Effective diffusion coefficient in solid phase (m^2/s)
F	Faraday's constant (C/mol)
$L^{\pm}$	Length of the electrodes (m)
r	Radial coordinate (cm)
$R^{\pm}$	Radius of solid active particle (m)
$\bar{R}$	Universal gas constant (J/mol-K)
R_f	Contact film resistance (Ω)
T	Temperature (K)
$U^{\pm}$	Open circuit potential
$a^{\pm}$	Charge transfer coefficient
$\varepsilon_s^{\pm}$	Active material volume fraction (dimensionless)

The concentration of solid lithium with respect to r, the radial coordinate, is driven by the solid diffusion equation, which is described by the partial differential equation (PDE) [9]:

$$\frac{dc_s^{\pm}}{dt} = \frac{D_s^{\pm}}{r^2}\frac{d}{dr}\left[r^2\frac{dc_s^{\pm}}{dr}\right], \tag{3.32}$$

where c_s is the lithium concentration in the solid phase, along with two boundary conditions,

$$\begin{aligned} \left.\frac{dc_s^{\pm}}{dr}\right|_{r=0} &= 0, \\ \left.\frac{dc_s^{\pm}}{dr}\right|_{r=R^{\pm}} &= \frac{\pm I}{a_s^{\pm}FD_s^{\pm}AL^{\pm}}, \end{aligned} \tag{3.33}$$

where $c_s^{\pm}$ s is the Li-ion concentration of the positive and negative electrodes and I is the charge/discharge current. The specific surface area can be computed as $a_s^{\pm} = 3\varepsilon_s^{\pm}/R_s^{\pm}$. The output voltage map is derived using the Butler–Volmer kinetics, electrical potential, and electrode thermodynamic properties and is given by,

$$V = \frac{\bar{R}T}{\varepsilon^+ F}\sinh^{-1}\left(\frac{I}{2a_s^+ AL^+ i_0^+}\right) - \frac{\bar{R}T}{\varepsilon^- F}\sinh^{-1}\left(\frac{I}{2a_s^- AL^- i_0^-}\right)$$
$$+U^+\left(c_{s,e}^+\right) - U^-\left(c_{s,e}^-\right) - R_f I, \quad (3.34)$$

where $c_{s,e}^+$ and $c_{s,e}^-$ are the surface concentrations of the positive and negative electrodes, respectively; U^+ and U^- are the open circuit potentials of the positive and negative electrodes, respectively; $i_0^\pm$ are the exchange current densities given by

$$i_0^+ = k_0^+ \sqrt{\lambda_e c_{s,e}^+ \left(c_{s,\max}^+ - c_{s,e}^+\right)},$$
$$i_0^+ = k_0^+ \sqrt{\lambda_e c_{s,e}^+ \left(c_{s,\max}^+ - c_{s,e}^+\right)}. \quad (3.35)$$

The spatial domain is discretized into $(M+1)$ nodes, where $\{c_{s0}, c_{s1}, \cdots, c_{sM}\}$ are the Li-ion concentration states at the nodes, which leads to the following ordinary differential equations:

$$\begin{cases} \frac{dc_{s0}}{dt} = -3\varepsilon c_{s0} + 3\varepsilon c_{s1}, \\ \frac{dc_{sm}}{dt} = \left(1-\frac{1}{m}\right)\varepsilon c_{s(m-1)} - 2\varepsilon c_{sm} + \left(1+\frac{1}{m}\right)\varepsilon c_{s(m+1)}, \\ \frac{dc_{sM}}{dt} = \left(1-\frac{1}{M}\right)\varepsilon c_{s(M-1)} - \left(1-\frac{1}{M}\right)\varepsilon c_{sM} - \left(1+\frac{1}{M}\right) bI, \end{cases} \quad (3.36)$$

with $m = 0, 1, \ldots, (M-1)$, discretization step $\Delta = R/M, \varepsilon = D_s^-/\Delta^2$, and $b = 1/a_s F\Delta AL^-$. It is noted that the bulk SOC information can be computed using some combination of the states in (3.36), whereas c_{sM} indicates the surface SOC.

The volume-averaged normalized bulk SOC can be computed from (3.32)-(3.36) using the following formula:

$$\text{SOC}_{\text{Bulk}} = \frac{1}{4\pi R^3 c_{s,\max}^-}\int_0^R 4\pi r^2 c_s^-(r,t)\,dr. \quad (3.37)$$

This formula can also be used on the discretized SPM to compute the bulk SOC of the cell as:

$$\text{SOC}_{\text{Bulk}} = \{(4\Delta^3)/(4R^3 c_{s,\max}^-)\}\sum_{j=1}^{M} j^2 c_{sj}. \quad (3.38)$$

The output voltage equation can be formed from (3.34) by substituting $c_{s,e}^- = c_{sM}$ and $c_{s,e}^+ = k_1 c_{sM} + k_2$, where k_1 and k_2 are constants in the algebraic relationship between the positive and negative electrode Li-ion concentrations. Then, the output voltage expression can be given by

$$V = \frac{\bar{R}T}{\varepsilon^+ F}\sinh^{-1}\left(\frac{I}{2a_s^+ AL^+ i_0^+}\right) - \frac{\bar{R}T}{\varepsilon^- F}\sinh^{-1}\left(\frac{I}{2a_s^- AL^- i_0^-}\right)$$
$$+U^+\left(k_1 c_{sM} + k_2\right) - U^-\left(c_{sM}\right) - R_f I, \quad (3.39)$$

where $i_0^{\pm}$ are the exchange current densities given by

$$i_0^+ = k_0^+ \sqrt{c_e\left(k_1 c_{sM} + k_2\right)\left(c_{s,\max}^+ - \left(k_1 c_{sM} + k_2\right)\right)},$$
$$i_0^+ = k_0^- \sqrt{c_e c_{sM}\left(c_{s,\max}^- - c_{sM}\right)}. \tag{3.40}$$

It follows from (3.36)-(3.40), the state-space model of the Li-ion cell can be written compactly in the following form:

$$\dot{x}(t) = Ax(t) + Bu(t),$$
$$y(t) = h(x(t), u(t)), \tag{3.41}$$

where the system states $x(t) = [c_{s0}, c_{s1}, \cdots, c_{sM}]^T$, the input current $u(t) = I$, output voltage $y(t) = V$, the matrices A and B can be derived from (3.36), and the output function h is formed by (3.39).

Choose $z_1 = \frac{\bar{R}T}{\varepsilon^+ FI}\sinh^{-1}\left(\frac{I}{2a_s^+ AL^+ i_0^+}\right) - \frac{\bar{R}T}{\varepsilon^- FI}\sinh^{-1}\left(\frac{I}{2a_s^- AL^- i_0^-}\right) - R_f$, and $z_2 = \frac{U^+(k_1 c_{sM} + k_2)}{c_{sM}} - \frac{U^-(c_{sM})}{c_{sM}}$ as fuzzy premise variables, then the nonlinear output measurement y can be represented by

$$y(t) = C(\mu)x(t) + D(\mu)u(t), \tag{3.42}$$

where $C(\mu) := \sum_{l=1}^{r} \mu_l C_l, D(\mu) := \sum_{l=1}^{r} \mu_l D_l, C_l$ and D_l the l-th local model.

3.2.2 LI-ION BATTERIES BASED ON CIRCUIT MODEL

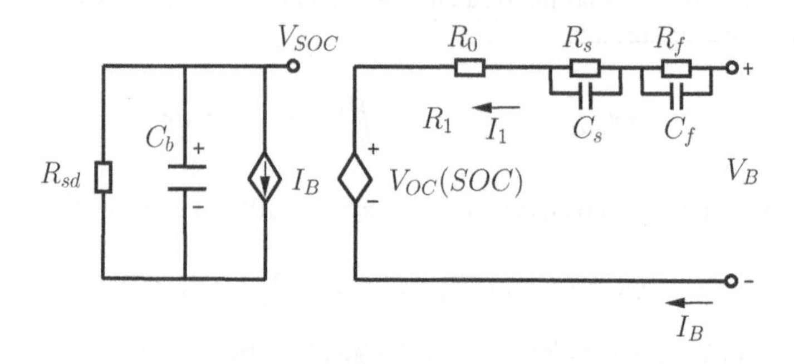

Figure 3.3 Schematic of circuit model for lithium-ion battery.

In the framework of circuit models, the Li-ion battery can be represented by the equivalent circuit model with two interrelated subcircuits, which influence each other through a voltage-controlled voltage source and a current-controlled current source, as shown in Figure 3.3. In Figure 3.3, the left subcircuit is utilized to simulate the SOC and remaining runtime of the battery; C_b is used to represent the full charge stored in the battery and R_{sd} denotes the self-discharge resistor; V_{SOC} denotes the SOC of the battery quantitatively. In Figure 3.3, the right subcircuit represents the transient response and $V-I$ curve of the battery, the resistor R_0 is used to characterize the charge and discharge energy losses of the battery, the R_C networks (R_s, C_s) and (R_f, C_f) are utilized to present the short-term and long-term transient responses of the battery, respectively, R_f and C_f are larger than R_s and C_s, the resistances and capacitances in the circuit are the functions of the battery's SOC, the terminal voltage of the battery is represented by V_B. The current of the battery is represented by I_B, which is positive/negative when the battery is in the discharging/charging mode. The nonlinear mapping from the battery's SOC to the OC is represented by [7]

$$V_{OC} = f(V_{SOC}), \tag{3.43}$$

where $f(\cdot)$ is a nonlinear function with its first-order derivative being assumed to exist.

In practice, in order to simplify the battery model, C_b can be considered as the battery's nominal capacity and R_{sd} can be simplified as a large constant resistor with the temperature of the battery varying within a small range. Since the effects of the change rates of the capacitances C_f and C_s on the battery's $V-I$ characteristics can be neglected, the dynamics of the voltages across the capacitors C_b, C_s, and C_f denoted as V_{SOC}, V_s, and V_f are expressed as

$$\begin{cases} \dot{V}_{SOC} = -\frac{1}{R_{sd}C_b}V_{SOC} - \frac{1}{C_b}I_B, \\ \dot{V}_s = -\frac{1}{R_s(SOC)C_s(SOC)}V_s + \frac{1}{C_s(SOC)}I_B, \\ \dot{V}_f = -\frac{1}{R_f(SOC)C_f(SOC)}V_f + \frac{1}{C_f(SOC)}I_B, \end{cases} \tag{3.44}$$

where SOC denotes the battery's state of charge. The terminal voltage V_B is obtained by

$$V_B = V_{OC} - R_0(SOC)I_B - V_f - V_s. \tag{3.45}$$

Based on (3.44) and (3.45),

$$\begin{cases} \dot{V}_{SOC} = -\frac{1}{R_{sd}C_b}V_{SOC} - \frac{1}{C_b}I_B, \\ \dot{V}_s = -\frac{1}{R_s(SOC)C_s(SOC)}V_s + \frac{1}{C_s(SOC)}I_B, \\ \dot{V}_f = -\frac{1}{R_f(SOC)C_f(SOC)}V_f + \frac{1}{C_f(SOC)}I_B, \\ 0\dot{V}_B = -V_B + V_{OC} - R_0(SOC)I_B - V_f - V_s. \end{cases} \tag{3.46}$$

Define $x(t) = \begin{bmatrix} V_{SOC} & V_s & V_f & V_B \end{bmatrix}^T$, $u(t) = I_B$. The Li-ion battery system in (3.45) can be represented by the following nonlinear model,

$$\begin{cases} E\dot{x}(t) = A(t)x(t) + B(t)u(t) + \omega(t), \\ y(t) = Cx(t), \end{cases} \tag{3.47}$$

where $E = \mathrm{diag}\{1,1,1,0\}$, $\omega(t)$ denotes the disturbance, $z_1 = R_s(SOC), z_2 = C_s(SOC), z_3 = R_f(SOC), z_4 = C_f(SOC)$, $z_5 = R_0(SOC)$, and

$$A(t) = \begin{bmatrix} -\frac{1}{R_{sd}C_b} & 0 & 0 & 0 \\ 0 & -\frac{1}{z_1 z_2} & 0 & 0 \\ 0 & 0 & -\frac{1}{z_3 z_4} & 0 \\ 0 & -1 & -1 & -1 \end{bmatrix}, B(t) = \begin{bmatrix} -\frac{1}{C_b} \\ \frac{1}{z_2} \\ \frac{1}{z_4} \\ -z_5 \end{bmatrix},$$

$$\omega(t) = \begin{bmatrix} 0 \\ 0 \\ 0 \\ 0 \\ f(V_{SOC}) \end{bmatrix}, C = \begin{bmatrix} 0 & 0 & 0 & 1 \end{bmatrix}. \tag{3.48}$$

Choose $z_1 - z_5$ as the fuzzy premise variables. The nonlinear systems in (3.47) can be represented by the following fuzzy model,

$$\begin{cases} E\dot{x}(t) = A(\mu)x(t) + B(\mu)u(t) + \omega(t), \\ y(t) = Cx(t), \end{cases} \tag{3.49}$$

where $A(\mu) := \sum_{l=1}^{r} \mu_l A_l, B(\mu) := \sum_{l=1}^{r} \mu_l B_l, A_l$ and B_l the l-th local model.

3.2.3 STABILITY ANALYSIS OF SOC ESTIMATION SYSTEM

Based on the particle model of a Li-ion battery in (3.41) and (3.42), the objective is to estimate SOC by using the following fuzzy observer,

Observer Rule $\mathscr{R}^s$: IF z_1 is $\hat{\mathscr{F}}_1^s$ and z_2 is $\hat{\mathscr{F}}_2^s$, **THEN**

$$\begin{cases} \dot{\hat{x}}(t) = A\hat{x}(t) + Bu(t) + L_s(y(t) - \hat{y}(t)), \\ \hat{y}(t) = C_s\hat{x}(t) + D_s u(t), s \in \mathscr{L} \end{cases} \tag{3.50}$$

where $\hat{x} \in \Re^{n_{\hat{x}}}$, and L_s is observer gain to be determined.

The global fuzzy observer is given by,

$$\begin{cases} \dot{\hat{x}}(t) = A\hat{x}(t) + Bu(t) + L(\hat{\mu})(y(t) - \hat{y}(t)), \\ \hat{y}(t) = C(\hat{\mu})\hat{x}(t) + D(\hat{\mu})u(t), \end{cases} \tag{3.51}$$

where the notation $\hat{\mu}$ is induced by the estimated premise variable, and $L(\hat{\mu}) = \sum_{s=1}^{r} \hat{\mu}_s L_s, C(\hat{\mu}) = \sum_{s=1}^{r} \hat{\mu}_s C_s, D(\hat{\mu}) = \sum_{s=1}^{r} \hat{\mu}_s D_s$.

Now, consider the following global fuzzy controller,

$$u(t) = K(\hat{\mu})\hat{x}(t), \tag{3.52}$$

where $K(\hat{\mu}) = \sum_{s=1}^{r} \hat{\mu}_s K_s$, and K_s is control gain to be determined.

Define $e(t) = x(t) - \hat{x}(t)$, and it follows from (3.41), (3.42), (3.51) and (3.52) that

$$\dot{\bar{x}}(t) = \bar{A}(\mu, \hat{\mu})\bar{x}(t), \tag{3.53}$$

where

$$\bar{A}(\mu,\hat{\mu}) = \begin{bmatrix} A+BK(\hat{\mu}) & -BK(\hat{\mu}) \\ L(\hat{\mu})C(\hat{\mu},\mu)+L(\hat{\mu})D(\hat{\mu},\mu)K(\hat{\mu}) & A-L(\hat{\mu})C(\hat{\mu})-L(\hat{\mu})D(\hat{\mu},\mu)K(\hat{\mu}) \end{bmatrix},$$

$$\bar{x}(t) = \begin{bmatrix} x(t) \\ e(t) \end{bmatrix}, C(\hat{\mu},\mu) = C(\hat{\mu}) - C(\mu), D(\hat{\mu},\mu) = D(\hat{\mu}) - D(\mu). \quad (3.54)$$

Based on the closed-loop error system in (3.53), the stability analysis on the SOC estimation of Li-ion batteries is proposed as below:

Theorem 3.4: Stability Analysis Based on Li-Ion Battery's SPM

Consider the Li-ion battery's particle model (3.41) and (3.42) using the fuzzy observer-based controller in the form of (3.51) and (3.52). For matrix $P \in \Re^{(n_x+n_{\hat{x}})\times(n_x+n_{\hat{x}})}$, the error system of Li-ion battery in (3.53) is asymptotically achieved if the following inequalities hold:

$$P^T\bar{A}(\mu,\hat{\mu}) + \bar{A}^T(\mu,\hat{\mu})P < 0, \quad (3.55)$$

where $\bar{A}(\mu,\hat{\mu})$ is defined in (3.54). ■

Proof. This proof is similar to Theorem 1.1, thus is completed.

Based on the circuit mode of Li-ion battery in (3.49), the objective is to estimate the SOC by using the following fuzzy observer,

Observer Rule $\mathscr{R}^s$: **IF** z_1 is $\hat{\mathscr{F}}_1^s$ and z_2 is $\hat{\mathscr{F}}_2^s$, ..., and z_5 is $\hat{\mathscr{F}}_5^s$, **THEN**

$$\begin{cases} E\dot{\hat{x}}(t) = A_s\hat{x}(t) + B_su(t) + L_s(y(t) - \hat{y}(t)), \\ \hat{y}(t) = C\hat{x}(t), s \in \mathscr{L} \end{cases} \quad (3.56)$$

where $\hat{x} \in \Re^{n_{\hat{x}}}$, and L_s is observer gain to be determined.

The global fuzzy observer is given by

$$\begin{cases} E\dot{\hat{x}}(t) = A(\hat{\mu})\hat{x}(t) + B(\hat{\mu})u(t) + L(\hat{\mu})(y(t) - \hat{y}(t)), \\ \hat{y}(t) = C\hat{x}(t), \end{cases} \quad (3.57)$$

where the notation $\hat{\mu}$ is induced by the estimated premise variable, and $L(\hat{\mu}) = \sum_{s=1}^r \hat{\mu}_sL_s, C(\hat{\mu}) = \sum_{s=1}^r \hat{\mu}_sC_s, D(\hat{\mu}) = \sum_{s=1}^r \hat{\mu}_sD_s$.

Now, consider the following global fuzzy controller,

$$u(t) = K(\hat{\mu})\hat{x}(t). \quad (3.58)$$

Define $e(t) = x(t) - \hat{x}(t)$, and it follows from (3.49), (3.57), and (3.58) that

$$\bar{E}\dot{\bar{x}}(t) = \bar{A}(\mu,\hat{\mu})\bar{x}(t) + \bar{\omega}(t), \quad (3.59)$$

where

$$\bar{E} = \begin{bmatrix} E & 0 \\ 0 & E \end{bmatrix}, \bar{A}(\mu,\hat{\mu}) = \begin{bmatrix} A(\mu)+B(\mu)K(\hat{\mu}) & -B(\mu)K(\hat{\mu}) \\ 0 & A(\mu)-L(\hat{\mu})C \end{bmatrix},$$
$$\bar{x}(t) = \begin{bmatrix} x(t) \\ e(t) \end{bmatrix}, \bar{\omega}(t) = \begin{bmatrix} \omega(t) \\ (A(\mu)-A(\hat{\mu}))\hat{x}(t)+(B(\mu)-B(\mu))u(t)+\bar{\omega}(t) \end{bmatrix}. \tag{3.60}$$

Given the closed-loop error system in (3.59), and for an $\mathscr{L}_2$-gain performance level $\gamma > 0$, the purpose of this section is to design a fuzzy observer-based controller in (3.57) and (3.58) such that the error system of Li-ion battery is asymptotically stable, and for any nonzero $\bar{\omega} \in \mathscr{L}_2[0\ \infty)$ the induced $\mathscr{L}_2$ norm of the operator from $\bar{\omega}$ to the extracted charge V_{SOC} is less than γ

$$\int_0^{\infty} V_{SOC}^2(s)\,ds < \gamma^2 \int_0^{\infty} \bar{\omega}^T(s)\bar{\omega}(s)ds, \tag{3.61}$$

under zero initial conditions.

Based on the closed-loop error system in (3.59), the $\mathscr{H}_\infty$ performance analysis on SOC estimation of Li-ion batteries is proposed as below:

Theorem 3.5: $\mathscr{H}_\infty$ Performance Analysis for Circuit Model for Li-Ion Battery

Consider the circuit mode of a Li-ion battery (3.49) using the fuzzy observer-based controller in the form of (3.57) and (3.58). For matrix $\bar{P} \in \Re^{(n_x+n_{\bar{x}})\times(n_x+n_{\bar{x}})}$, the error system of Li-ion battery is asymptotically achieved with $\mathscr{H}_\infty$ performance index if the following inequalities hold:

$$\bar{E}^T\bar{P} = \bar{P}^T\bar{E} \geq 0, \tag{3.62}$$

$$\begin{bmatrix} \bar{P}^T\bar{A}(\mu,\hat{\mu})+\bar{A}^T(\mu,\hat{\mu})\bar{P}+F^TF & \bar{P}^T \\ \star & -\gamma^2 I \end{bmatrix} < 0, \tag{3.63}$$

where $F = \begin{bmatrix} 1 & 0 & 0 & 0 & 0 & 0 & 0 & 0 \end{bmatrix}$, and $\bar{A}(\mu,\hat{\mu})$ is defined (3.60). ■

Proof: This proof is similar to Theorem 3.3, thus is completed.

Note: It is noted that the results on Theorem 3.3 are not LMIs. Here, a two-step processing proposed (1.63)-(1.68) can be used to obtain the observer and controller gains.

3.2.4 DESIGN OF OBSERVER-BASED FUZZY CONTROLLER

It is noted that Theorem 3.4 is not an LMI-based result. Here, a two-step processing is proposed. Firstly, define

$$X = \begin{bmatrix} X_1 & 0 \\ 0 & X_2 \end{bmatrix}, X = P^{-1}, \tag{3.64}$$

where $\{X_1, X_2\} \in \Re^{n_x \times n_x}$ are symmetric positive-definite matrices. By performing the congruence transformation to (3.64) by $\Gamma = \text{diag}\{X, X\}$, one has

$$\Phi(\hat{\mu}, \mu) < 0, \tag{3.65}$$

where

$$\begin{aligned}
&\Phi(\hat{\mu}, \mu) = \text{Sym}\left\{\begin{bmatrix} A + BK(\hat{\mu})X_1 & -BK(\hat{\mu})X_2 \\ \Phi_1(\hat{\mu}, \mu) & \Phi_2(\hat{\mu}, \mu) \end{bmatrix}\right\}, \\
&\Phi_1(\hat{\mu}, \mu) = L(\hat{\mu})C(\hat{\mu}, \mu)X_1 + L(\hat{\mu})D(\hat{\mu}, \mu)K(\hat{\mu})X_1, \\
&\Phi_2(\hat{\mu}, \mu) = AX_2 - L(\hat{\mu})C(\hat{\mu})X_2 - L(\hat{\mu})D(\hat{\mu}, \mu)K(\hat{\mu})X_2.
\end{aligned} \tag{3.66}$$

By extracting the fuzzy premise variables, one gets

$$\begin{aligned}
\Phi(\hat{\mu}, \mu) &= \sum_{f=1}^{r} \sum_{g=1}^{r} \sum_{s=1}^{r} \sum_{l=1}^{r} \hat{\mu}_f \hat{\mu}_g \hat{\mu}_s \mu_l \Phi_{fgsl} \\
&< 0,
\end{aligned} \tag{3.67}$$

where

$$\begin{aligned}
&\Phi_{fgls} = \text{Sym}\left\{\begin{bmatrix} A + BK_s X_1 & -BK_s X_2 \\ \Phi_{(1)fgls} & \Phi_{(2)fgls} \end{bmatrix}\right\}, \\
&\Phi_{(1)fgls} = L_f (C_g - C_l) X_1 + L_f (D_g - D_l) K_s X_1, \\
&\Phi_{(2)fgls} = AX_2 - L_f (C_g - C_l) X_2 - L_f (D_g - D_l) K_s X_2.
\end{aligned} \tag{3.68}$$

Now, define

$$P = \begin{bmatrix} P_1 & 0 \\ 0 & P_2 \end{bmatrix}, \tag{3.69}$$

where $\{P_1, P_2\} \in \Re^{n_x \times n_x}$ are symmetric positive-definite matrices.

Submitting (3.69) into (3.55), we have

$$\bar{\Phi}(\hat{\mu}, \mu) < 0, \tag{3.70}$$

where

$$\begin{aligned}
&\bar{\Phi}(\hat{\mu}, \mu) = \text{Sym}\left\{\begin{bmatrix} P_1 A + P_1 BK(\hat{\mu}) & -P_1 BK(\hat{\mu}) \\ \bar{\Phi}_1(\hat{\mu}, \mu) & \bar{\Phi}_2(\hat{\mu}, \mu) \end{bmatrix}\right\}, \\
&\bar{\Phi}_1(\hat{\mu}, \mu) = P_2 L(\hat{\mu})C(\hat{\mu}, \mu) + P_2 L(\hat{\mu})D(\hat{\mu}, \mu)K(\hat{\mu}), \\
&\bar{\Phi}_2(\hat{\mu}, \mu) = P_2 A - P_2 L(\hat{\mu})C(\hat{\mu}) - P_2 L(\hat{\mu})D(\hat{\mu}, \mu)K(\hat{\mu}).
\end{aligned} \tag{3.71}$$

Define $\bar{L}_s = P_2 L_s$, and it is easy to see that the following inequality implies (3.70),

$$\begin{aligned}
\bar{\Phi}(\hat{\mu}, \mu) &\leq \sum_{s=1}^{r} \sum_{l=1}^{r} \hat{\mu}_s \mu_l \bar{\Phi}_{sssl} \\
&< 0,
\end{aligned} \tag{3.72}$$

where

$$\bar{\Phi}_{sssl} = \text{Sym}\left\{\left[\begin{array}{cc} P_1A+P_1BK_s & -P_1BK_s \\ \bar{\Phi}_{(1)sssl} & \bar{\Phi}_{(2)sssl} \end{array}\right]\right\},$$
$$\bar{\Phi}_{(1)sssl} = \bar{L}_s\left(C_s-C_l\right)+\bar{L}_s\left(D_s-D_l\right)K_s,$$
$$\bar{\Phi}_{(2)sssl} = P_2A-\bar{L}_s\left(C_s-C_l\right)-\bar{L}_s\left(D_s-D_l\right)K_s. \tag{3.73}$$

Based on the result of (3.67) and (3.72), an algorithm to calculate the fuzzy controller and observer gains is proposed as below:

a) We solve the following inequality,

$$\Sigma_s < 0, s \in \mathscr{L} \tag{3.74}$$

where $\Sigma_s = A+B\bar{K}_s$, and then we obtain $\bar{K}_s$ and calculate $K_s = \bar{K}_s X_1^{-1}$.

b) Using the controller gain K_s, we solve the following inequality

$$\tilde{\Phi}_{llll} < 0, l \in \mathscr{L} \tag{3.75}$$
$$\tilde{\Phi}_{sssl}+\tilde{\Phi}_{llls} < 0, 1 \leq l < s \leq r \tag{3.76}$$

where $\tilde{\Phi}_{sssl} = \bar{\Phi}_{sssl}+\sum_{s=1}^{r}\delta_s\left(\bar{\Phi}_{sssl}+M_l\right), \bar{\Phi}_{sssl} = \text{Sym}\left\{\left[\begin{array}{cc} P_1A+P_1BK_s & -P_1BK_s \\ \bar{\Phi}_{(1)sssl} & \bar{\Phi}_{(2)sssl} \end{array}\right]\right\}$. It obtains $\bar{L}_s$ and we calculate $L_s = P_2^{-1}\bar{L}_s$.

It is also noted that Theorem 3.5 is not an LMI-based result. Here, a two-step processing is proposed. Firstly, define

$$\bar{X} = \bar{P}^{-1}. \tag{3.77}$$

We perform the congruence transformation to (3.63) by diag$\{\bar{X}, I\}$, and specify the matrix $\bar{X}$ as

$$\bar{X} = \text{diag}\left\{\bar{X}_{(1)}, \bar{X}_{(2)}\right\}, \tag{3.78}$$

where $\bar{X}_{(1)} = \left[\begin{array}{cc} X_{(1)1} & 0 \\ X_{(1)2} & X_{(1)3} \end{array}\right], \bar{X}_{(2)} = \left[\begin{array}{cc} X_{(2)1} & 0 \\ X_{(2)2} & X_{(2)3} \end{array}\right]$, $\{X_{(1)1}, X_{(1)2}, X_{(1)3}, X_{(2)1}, X_{(2)2}, X_{(2)3}\} \in \Re^{n_x \times n_x}$, and $\left\{X_{(1)1}, X_{(2)1}\right\}$ are symmetric positive-definite matrices.

Based on the procedure (3.77) and the relation (3.78), we have

$$\Phi(\hat{\mu}, \mu) < 0, \tag{3.79}$$

where $\Phi(\hat{\mu}, \mu) = \text{Sym}\left(A(\mu)X_1+B(\mu)K(\hat{\mu})\bar{X}_{(1)}\right)$.

By extracting the fuzzy premise variables, and define $\bar{K}_s = K_s X_{(1)}$, we have

$$\begin{aligned}\Phi(\hat{\mu}, \mu) &= \sum_{s=1}^{r}\sum_{l=1}^{r}\hat{\mu}_s\mu_l\Phi_{sl} \\ &< 0,\end{aligned} \tag{3.80}$$

where $\Phi_{sl} = \text{Sym}(A_l X_1+B_l\bar{K}_s)$.

We specify

$$\bar{P} = \text{diag}\left\{\bar{P}_{(1)}, \bar{P}_{(2)}\right\}, \tag{3.81}$$

where $\bar{P}_{(1)} = \begin{bmatrix} P_{(1)1} & 0 \\ P_{(1)2} & P_{(1)3} \end{bmatrix}, \bar{P}_{(2)} = \begin{bmatrix} P_{(2)1} & 0 \\ P_{(2)2} & P_{(2)3} \end{bmatrix}$, $\{P_{(1)1}, P_{(1)2}, P_{(1)3}, P_{(2)1}, P_{(2)2}, P_{(2)3}\} \in \Re^{n_x \times n_x}$, and $\left\{P_{(1)1}, P_{(2)1}\right\}$ are symmetric positive-definite matrices.

By submitting (3.81) into (3.63), we have

$$\bar{\Phi}(\hat{\mu}, \mu) < 0, \tag{3.82}$$

where

$$\bar{\Phi}(\hat{\mu}, \mu) = \text{Sym}\left(\begin{bmatrix} \bar{P}_{(1)}A(\mu) + \bar{P}_{(1)}B(\mu)K(\hat{\mu}) & -\bar{P}_{(1)}B(\mu)K(\hat{\mu}) \\ 0 & \bar{P}_{(2)}A(\mu) - \bar{P}_{(2)}L(\hat{\mu})C \end{bmatrix}\right). \tag{3.83}$$

Define $\bar{L}_s = \bar{P}_{(2)}L_s$. By extracting the fuzzy premise variables, we get

$$\begin{aligned} \bar{\Phi}(\hat{\mu}, \mu) &= \textstyle\sum_{s=1}^{r}\sum_{l=1}^{r} \hat{\mu}_s \mu_l \bar{\Phi}_{sl} \\ &< 0, \end{aligned} \tag{3.84}$$

where

$$\bar{\Phi}_{sl} = \begin{bmatrix} \bar{P}_{(1)}A_l + \bar{P}_{(1)}B_l K_s & -\bar{P}_{(1)}B_l K_s \\ 0 & \bar{P}_{(2)}A_l - \bar{L}_s C \end{bmatrix}. \tag{3.85}$$

Based on the result of (3.80) and (3.84), an algorithm to calculate the fuzzy controller and observer gains is proposed as below:

a) We solve the following inequality,

$$\bar{\Sigma}_{ll} < 0, l \in \mathscr{L} \tag{3.86}$$

$$\bar{\Sigma}_{sl} + \bar{\Sigma}_{ls} < 0, 1 \leq l < s \leq r \tag{3.87}$$

where $\bar{\Sigma}_{ls} = \Sigma_{ls} + \sum_{s=1}^{r} \delta_s \left(\Sigma_{ls} + M_l\right), \Sigma_{ls} = \text{Sym}\left(A_l \bar{X}_{(1)} + B_l \bar{K}_s\right)$, and then we obtain $\bar{K}_s$ and calculate $K_s = \bar{K}_s \bar{X}_{(1)}^{-1}$.

b) Using the controller gain K_s, and we solve the following inequality

$$\tilde{\Phi}_{ll} < 0, l \in \mathscr{L} \tag{3.88}$$

$$\tilde{\Phi}_{sl} + \tilde{\Phi}_{ls} < 0, 1 \leq l < s \leq r \tag{3.89}$$

where $\tilde{\Phi}_{sl} = \bar{\Phi}_{sl} + \sum_{s=1}^{r} \delta_s \left(\bar{\Phi}_{sl} + M_l\right), \bar{\Phi}_{sl} = \begin{bmatrix} \bar{P}_{(1)}A_l + \bar{P}_{(1)}B_l K_s & -\bar{P}_{(1)}B_l K_s \\ 0 & \bar{P}_{(2)}A_l - \bar{L}_s C \end{bmatrix}$.

It obtains $\bar{L}_s$ and we calculate $L_s = \bar{P}_{(2)}^{-1}\bar{L}_s$.

3.3 MODELING OF SUPERCAPACITORS

It is noted that charging a capacitor and similarly a supercapacitor, from zero charge to full charge, with an CV source results in 50% energy loss, irrespective of the internal and line resistances [10]. Its circuit is shown in Figure 3.4, and the differential

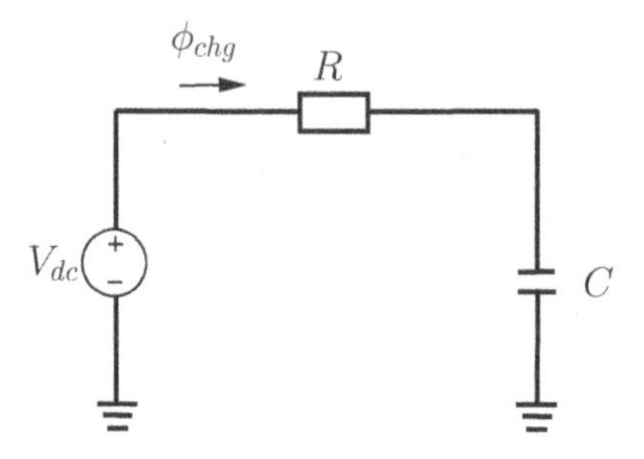

Figure 3.4 Schematic of supercapacitor.

equation is given by

$$R\dot{q} + \frac{q}{C} = V_{dc}, \tag{3.90}$$

where C is the nominal capacitance, R is the total internal resistance, V_{dc} is the charging voltage, and q denotes the stored charge. If V_{dc} remains constant over time, the solution to the above-mentioned differential equation from a zero initial charge condition can be obtained as

$$q = CV_{dc}\left[1 - e^{-t/RC}\right], \tag{3.91}$$

and the charging current of CV $\phi_{chg} = \phi_{cv}$ is given by

$$I_{cv} = \frac{V_{dc}}{R}e^{-t/RC}. \tag{3.92}$$

3.4 SIMULATION STUDIES

Consider the circuit mode of a Li-ion battery (3.49) using the fuzzy observer-based controller in the form of (3.57) and (3.58). Here, the parameters of the Li-ion battery are chosen as below: $C_b = 0.1296\text{F}, R_{sd} = 1 \times 10^9\Omega, R_s = 0.5 \times 10^{-3}\Omega, C_s = 0.8 \times 10^4\text{F}, R_f = 0.0015\Omega, R_0 = 0.008\Omega$. Now, we assume that $SOC = (40\%,\ 80\%)$, and choose $C_s(SOC)$ as the fuzzy premise variable, and linearize the PV system around $\{0.24 \times 10^5, 0.25 \times 10^5\}$. Then, the succeeding system matrices of T-S fuzzy model can be obtained as below:

$$A_1 = \begin{bmatrix} -7.716\times10^{-15} & 0 & 0 & 0 \\ 0 & -0.25 & 0 & 0 \\ 0 & 0 & -0.0027778 & 0 \\ 0 & -1 & -1 & -1 \end{bmatrix},$$

$$A_2 = \begin{bmatrix} -7.716\times10^{-15} & 0 & 0 & 0 \\ 0 & -0.25 & 0 & 0 \\ 0 & 0 & -0.0027778 & 0 \\ 0 & -1 & -1 & -1 \end{bmatrix},$$

$$B_1 = \begin{bmatrix} -7.72\times10^{-6} \\ 0.000125 \\ 4.1667\times10^{-6} \\ -0.008 \end{bmatrix}, B_2 = \begin{bmatrix} -7.72\times10^{-6} \\ 0.000125 \\ 4.1667\times10^{-6} \\ -0.008 \end{bmatrix}.$$

Here, by applying the algorithm with (3.86)-(3.89), the fuzzy controller gains and observer gains are given by

$$K_1 = \begin{bmatrix} -0.0082608 & 0.078412 & 2.9997\times10^{6} & -0.32745 \end{bmatrix},$$

$$K_2 = \begin{bmatrix} -0.0082613 & 0.077869 & 2.9997\times10^{6} & -0.32721 \end{bmatrix},$$

$$L_1 = \begin{bmatrix} 0 \\ -11.924 \\ -14.588 \\ -3.2521 \end{bmatrix}, L_2 = \begin{bmatrix} 0 \\ 12.837 \\ 17.607 \\ 4.0573 \end{bmatrix}, \gamma = 4.032.$$

3.5 REFERENCES

1. Olivares, D. E., Mehrizi-Sani, A., Etemadi, A. H., Canizares, C. A., Iravani, R., Kazerani, M., et al. (2014). Trends in microgrid control. IEEE Transactions on Smart Grid, 5(4), 1905-1919.
2. Wang, B., Xu, J., Wai, R. J., and Cao, B. (2017). Adaptive sliding-mode with hysteresis control strategy for simple multimode hybrid energy storage system in electric vehicles. IEEE Transactions on Industrial Electronics, 64(2), 1404-1414.
3. Massimo, C. (2000). New dynamical models of lead-acid batteries. IEEE Transactions on Power Systems, 15(4), 1184-1900.
4. Blondel, P., Postoyan, R., Raël, S., Benjamin, S., and Desprez, P. (2018). Nonlinear circle-criterion observer design for an electrochemical battery model. IEEE Transactions on Control Systems Technology, doi: 10.1109/TCST.2017.2782787.
5. Moura, S. J., Argomedo, F. B., Klein, R., Mirtabatabaei, A., and Krstic, M. (2015). Battery state estimation for a single particle model with electrolyte dynamics. IEEE Transactions on Control Systems Technology, 25(2), 453-468.
6. Lin, F. J., Huang, M. S., Yeh, P. Y., Tsai, H. C., and Kuan, C. H. (2012). DSP-based probabilistic fuzzy neural network control for Li-ion battery charger. IEEE Transactions on Power Electronics, 27(8), 3782-3794.
7. Chen, J., Ouyang, Q., Xu, C., and Su, H. (2018). Neural network-based state of charge observer design for lithium-ion batteries. IEEE Transactions on Control Systems Technology, doi:10.1109/TCST.2017.2664726.

8. Zhang, C., Zhang, Y., and Li, Y. (2015). A novel battery state-of-health estimation method for hybrid electric vehicles. IEEE/ASME Transactions on Mechatronics, 20(5), 2604-2612.
9. Dey, S., Ayalew, B., and Pisu, P. (2015). Nonlinear robust observers for state-of-charge estimation of lithium-ion cells based on a reduced electrochemical model. IEEE Transactions on Control Systems Technology, 23(5), 1935-1942.
10. Parvini, Y., Vahidi, A., and Fayazi, S. (2018). Heuristic versus optimal charging of supercapacitors, lithium-ion, and lead-acid batteries: An efficiency point of view. IEEE Transactions on Control Systems Technology, 26(1), 167-180.

Part II

Coordinated Fuzzy Control for Microgrids

Preview

For a microgrid, many renewable energy sources are interconnected and distributed in different geographical areas. Microgrids using multiple-converters are represented as a class of interconnected subsystems with nonlinear dynamics. The stability of a single subsystem cannot guarantee the stability of an entire microgrid. These conditions, such as interconnections, distributions, and nonlinear dynamics increase many technical and operational control difficulties. In that case, coordinated control should be implemented in order to achieve a global stability for microgrids. Based on the means of communication between the interface converters, it can be realized either by using decentralized, centralized, or distributed control.

In order to guarantee stability of microgrids, effective control strategies should be developed. From the communication perspective, overall control of microgrids can be divided into the following three categories as below:

1) Centralized control: Data from distributed units are collected in a centralized aggregator, processed and feedback commands are sent back to them via communication networks.

2) Decentralized control: Communication networks among distributed units do not exist and power lines are used as the only channel of local communication.

3) Distributed control: Communication networks exist, but are implemented between units and coordinated control strategies.

The basic configurations of these control structures are depicted in the three chapters in this section. The chapters also provide detailed overviews of the significant features of local and coordinated control strategies.

In this part, coordinated controls for multi-photovoltaic systems with DC-AC loads are considered. Some results of stability analyses and centralized, decentralized, and distributed controller designs are derived in terms of LMIs. This section considers only multi-photovoltaic systems with DC-AC loads. However, those derived results can be easily extended to other generator systems.

4 Centralized Fuzzy Control

Centralized controllers gather system-wide data and require extensive communication networks to issue commands. All components of microgrids are commanded by communications from a single central controller [1, 2]. It is important to note that the controller receives all sensor data available and determines all input signals of the plant. In other words, all information is assumed to be available for a single unit that designs and applies the controller to the plant [3].

This chapter covers the tracking problems of voltage synchronization of multi-photovoltaic and multi-wind systems. All generator subsystems act as one group toward a common synchronization goal and are implemented through communications networks. This chapter examines a network-based controller utilizing sampled data measurement and a controller using time-triggered zero-order hold (ZOH). It also provides a numerical example is provided to show the effectiveness of the proposed method.

4.1 MODELING OF MULTI-PV GENERATORS

4.1.1 MODELING OF MULTI-PVS WITH DC LOAD

Recall the solar PV generator with DC-DC buck converter as below [4]:

$$\begin{cases} \dot{v}_{PV} = \frac{1}{C_{PV}}\left(\phi_{PV} - \phi_L u\right), \\ \dot{\phi}_L = \frac{1}{L}\left(R_0\left(\phi_0 - \phi_L\right) - R_L\phi_L - v_0\right) + \frac{1}{L}\left(V_D + v_{PV} - R_M\phi_L\right)u - \frac{V_D}{L}, \\ \dot{v}_0 = \frac{1}{C_0}\left(\phi_L - \phi_0\right), \end{cases} \tag{4.1}$$

where v_{PV}, ϕ_L, and v_0 are the voltage of the PV array, the current of the inductance L, and the voltage of the capacitance C_0, respectively; R_0, R_L, and R_M are the internal resistances of the capacitance C_0, the inductance L, and the power MOSFET, respectively; V_D is the forward voltage of the power diode; ϕ_0 is the measurable load current.

Consider an interconnected multi-PV generator with DC load as shown in Figure 4.1. Based on the Thevenin's theorem, it has

$$v_{0(i)} = \phi_{0(i)}R_{load} + \cdots + \phi_{0(N)}R_{load}, \tag{4.2}$$

where the subscript i denotes the i-th subystem, $i \in \mathscr{N} := \{1, 2, \ldots, N\}$, N denotes the number of subystems, R_{load} is the load resistance, $\phi_{0(i)}$ is the line current of the i-th subystem.

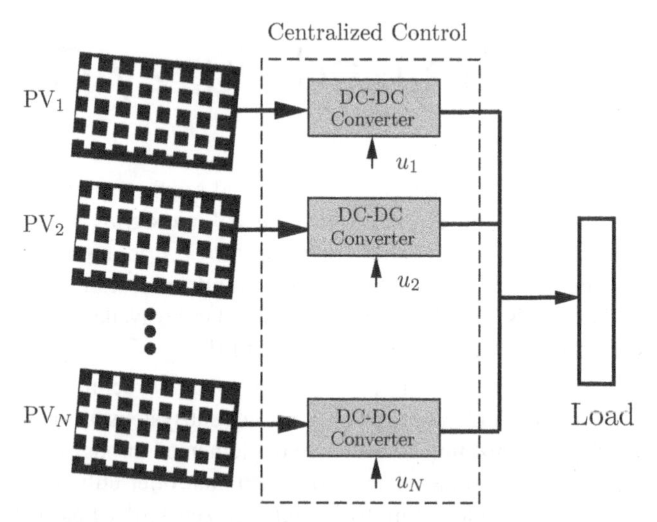

Figure 4.1 Centralized control for multi-PV generator with DC load.

Based on the relations (4.1) and (4.2), it has

$$\begin{cases} \dot{v}_{PV(i)} = \frac{1}{C_{PV(i)}}\phi_{PV(i)} - \frac{\phi_{L(i)}}{C_{PV(i)}}u_{(i)}, \\ \dot{\phi}_{L(i)} = -\left(\frac{R_{0(i)}}{L_{(i)}} + \frac{R_{L(i)}}{L_{(i)}}\right)\phi_{L(i)} + \left(\frac{R_{0(i)}}{L_{(i)}R_{load}} - \frac{1}{L_{(i)}}\right)v_{0(i)} - \sum_{j=1,j\neq i}^{N}\frac{R_{0(i)}}{L_{(i)}}\phi_{0(j)} \\ \quad +\frac{1}{L_{(i)}}\left(V_{D(i)} + v_{PV(i)} - R_{M(i)}\phi_{L(i)}\right)u_{(i)} - \frac{V_{D(i)}}{L_{(i)}}, \\ \dot{v}_{0(i)} = \frac{1}{C_{0(i)}}\phi_{L(i)} - \frac{1}{R_{load}C_{0(i)}}v_{0(i)} + \sum_{j=1,j\neq i}^{N}\frac{1}{C_{0(i)}}\phi_{0(j)}. \end{cases} \tag{4.3}$$

Now, consider the tracking problem of voltage synchronization for multi-photovoltaic systems with DC loads. Assume the reference voltage is v_0^*, and define $e_{0(i)} = v_{0(i)} - v_0^*$, and

$$x_i(t) = \begin{bmatrix} v_{PV(i)} & \phi_{L(i)} & e_{0(i)} \end{bmatrix}^T,$$

$$\omega_i(t) = \begin{bmatrix} 0 & \left(\frac{R_{0(i)}}{L_{(i)}R_{load}} - \frac{1}{L_{(i)}}\right)v_0^* - \frac{V_{D(i)}}{L_{(i)}} & -\frac{v_0^*}{R_{load}C_{0(i)}} \end{bmatrix}^T,$$

$$A_{ii}(t) = \begin{bmatrix} \frac{1}{C_{PV(i)}}\frac{\phi_{PV(i)}}{v_{PV(i)}} & 0 & 0 \\ 0 & -\frac{(R_{0(i)}+R_{L(i)})}{L_{(i)}} & \frac{R_{0(i)}-R_{load}}{L_{(i)}R_{load}} \\ 0 & \frac{1}{C_{0(i)}} & -\frac{1}{R_{load}C_{0(i)}} \end{bmatrix},$$

$$B_{ii}(t) = \begin{bmatrix} -\frac{\phi_{L(i)}}{C_{PV(i)}} \\ \frac{1}{L_{(i)}}\left(V_{D(i)} + v_{PV(i)} - R_{M(i)}\phi_{L(i)}\right) \\ 0 \end{bmatrix}, A_{ij}(t) = \begin{bmatrix} 0 & 0 & 0 \\ 0 & -\frac{R_{0(i)}}{L_{(i)}}\frac{\phi_{0(j)}}{\phi_{L(j)}} & 0 \\ 0 & \frac{1}{C_{0(i)}}\frac{\phi_{0(j)}}{\phi_{L(j)}} & 0 \end{bmatrix}. \tag{4.4}$$

Then, by transforming (4.3) into the state-space framework,

$$\dot{x}(t) = A(t)x(t) + B(t)u(t) + \omega(t), \tag{4.5}$$

where

$$\begin{aligned} x(t) &= \begin{bmatrix} x_1^T(t) & x_2^T(t) & \cdots & x_N^T(t) \end{bmatrix}^T, \\ A(t) &= \begin{bmatrix} A_{11}(t) & A_{12}(t) & \cdots & A_{1N}(t) \\ A_{21}(t) & A_{22}(t) & \cdots & A_{2N}(t) \\ \vdots & \vdots & \ddots & \vdots \\ A_{N1}(t) & A_{N2}(t) & \cdots & A_{NN}(t) \end{bmatrix}, \omega(t) = \begin{bmatrix} \omega_1(t) \\ \omega_2(t) \\ \vdots \\ \omega_N(t) \end{bmatrix}, \\ B(t) &= \begin{bmatrix} B_1(t) & 0 & \cdots & 0 \\ 0 & B_2(t) & \cdots & 0 \\ \vdots & \vdots & \ddots & \vdots \\ 0 & 0 & \cdots & B_N(t) \end{bmatrix}, u(t) = \begin{bmatrix} u_1(t) \\ u_2(t) \\ \vdots \\ u_N(t) \end{bmatrix}. \end{aligned} \tag{4.6}$$

Choose $z_{i1}(t) = \frac{\phi_{PV(i)}}{v_{PV(i)}}, z_{i2}(t) = \phi_{L(i)}, z_{i3}(t) = v_{PV(i)}$, and $z_{i4}(t) = \frac{\dot{\phi}_{0(i)}}{\phi_{L(i)}}$ as the fuzzy premise variables. Thus, the PV power nonlinear system in (4.5) can be represented by the following fuzzy system,

Plant Rule $\mathscr{R}^l$: **IF** $z_{11}(t)$ is $\mathscr{F}_{11}^l$ and $\cdots$ and $z_{N1}(t)$ is $\mathscr{F}_{N1}^l$, $z_{12}(t)$ is $\mathscr{F}_{12}^l$ and $\cdots$ and $z_{N2}(t)$ is $\mathscr{F}_{N2}^l$, $\cdots$, $z_{14}(t)$ is $\mathscr{F}_{14}^l$ and $\cdots$ and $z_{N4}(t)$ is $\mathscr{F}_{N4}^l$, **THEN**

$$\dot{x}(t) = A_l x(t) + B_l u(t) + \omega(t), l \in \mathscr{L} := \{1, 2, \ldots, r\} \tag{4.7}$$

where $\mathscr{R}^l$ denotes the l-th fuzzy inference rule; r is the number of inference rules; $z(t) \triangleq [z_{11}, \cdots, z_{N1}, z_{12}, \cdots, z_{N2}, z_{13}, \cdots, z_{N3}, z_{14}, \cdots, z_{N4}]$ are the measurable variables; $\{A_l, B_l\}$ is the l-th local model.

Denote as $\mathscr{F}^l := \prod_{\phi=1}^{4N} \mathscr{F}_\phi^l$ the inferred fuzzy set, and $\mu_l[z(t)]$ as the normalized membership function, it yields

$$\mu_l[z(t)] := \frac{\prod_{\phi=1}^{g} \mu_{l\phi}[z_\phi(t)]}{\sum_{\varsigma=1}^{r} \prod_{\phi=1}^{g} \mu_{\varsigma\phi}[z_\phi(t)]} \geq 0, \sum_{l=1}^{r} \mu_l[z(t)] = 1. \tag{4.8}$$

We denote $\mu_l := \mu_l[z(t)]$ for simplicity.

After fuzzy blending, the global T-S fuzzy dynamic model is given by

$$\begin{cases} \dot{x}(t) = A(\mu)x(t) + B(\mu)u(t) + \omega(t), \\ y(t) = Cx(t), \end{cases} \tag{4.9}$$

where $y(t)$ denotes the voltage error, $C = \underbrace{\begin{bmatrix} 0 & 0 & 1 & 0 & 0 & 1 & \cdots & 0 & 0 & 1 \end{bmatrix}}_{N}$,

and $A(\mu) := \sum_{l=1}^{r} \mu_l A_l, B(\mu) := \sum_{l=1}^{r} \mu_l B_l$.

Given the tracking error system in (4.9), and for an $\mathscr{L}_2$-gain performance level $\gamma > 0$, the purpose of this section is to design a controller such that the closed-loop fuzzy control system is asymptotically stable, and for any nonzero $\omega \in \mathscr{L}_2[0\ \infty)$ the induced $\mathscr{L}_2$ norm of the operator from ω to the tracking voltage error $y(t)$ is less than γ

$$\int_0^\infty y^T(s)y(s)ds < \gamma^2 \int_0^\infty \omega^T(s)\omega(s)ds, \tag{4.10}$$

under zero initial conditions.

Note: The premise variable $z(t)$ includes the dynamics of all subsystems, which gives rise to the computational complexity. In other words, the number of matrix inequalities to be solved is too large. Consequently, a new solution for fuzzy controller design involving the fewest possible number of LMIs is worth exploring. A detailed discussion on this issue is given in Chapter 5.

4.1.2 MODELING OF MULTI-PHOTOVOLTAIC SYSTEM WITH AC LOAD

We recall the PV power system with DC-AC converter in (1.13) as below:

$$\begin{cases} \dot{v}_{PV} = \frac{1}{C_{pv}}(\phi_{PV} - \frac{1.5u_d}{v_{pv}}\phi_d), \\ \dot{\phi}_d = -\frac{R_1}{L_1}\phi_d - \omega\phi_q + \frac{1}{L_1}e_d, \\ \dot{\phi}_q = \omega\phi_d - \frac{R_1}{L_1}\phi_q + \frac{1}{L_1}e_q, \end{cases} \tag{4.11}$$

where u_d is the d-axis component of grid voltage; ω is the fundamental angular frequency; L_1 is the filter inductance; R_1 denotes the equivalent resistances describing the system loss; ϕ_d and ϕ_q are the active and reactive components of the grid side current, respectively; v_{PV} and ϕ_{PV} are the PV voltage and current, respectively. Neglecting the conversion loss of the converters, the active power pg transferred between the DC subgrid and the AC grid can be expressed by

$$\begin{aligned} P &= v_{PV}\phi_{dc} \\ &= 1.5u_d\phi_d. \end{aligned} \tag{4.12}$$

Here, consider an interconnected PV system with AC load as shown in Figure 4.2, where the three-phase DC-AC converter is controlled as a current source in order to track a certain current reference in the synchronous dq-reference frame. The d- and q-axis current references (ϕ_d^* and ϕ_q^*) can be calculated directly from the desired active and reactive power as below [5]:

$$\phi_d^* = \frac{2P^*}{3v_d}, \tag{4.13}$$

$$\phi_q^* = -\frac{2Q^*}{3v_d}. \tag{4.14}$$

Assuming that the load is a linear resistance R and $R_{line(i)}$ denotes the resistance at the i-th power line. Based on Kirchhoff's voltage law (KVL),

$$u_{d(i)} = \phi_{d(i)}R_{line(i)} + R_{load}\left[\sum_{i=1}^{N}\phi_{d(i)}\right]. \tag{4.15}$$

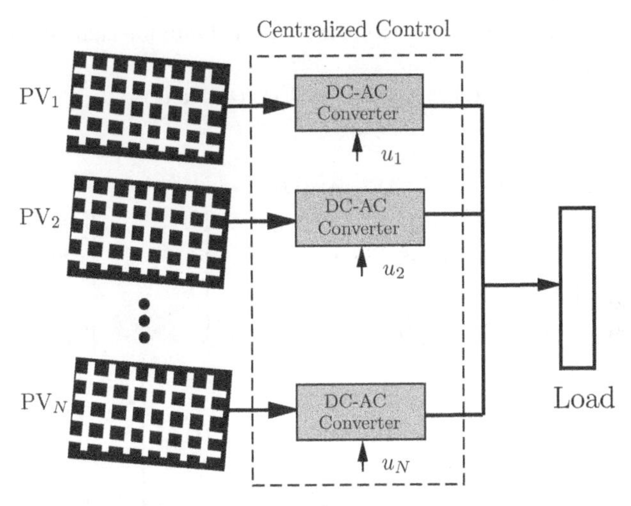

Figure 4.2 Interconnected PV generator with AC load.

Define $\phi_{d(i)} = e_{d(i)} + \phi_d^*, \phi_{q(i)} = e_{q(i)} + \phi_q^*$, and it follows from (4.11)-(4.15) that

$$\begin{cases} \dot{v}_{PV(i)} = \frac{1}{C_{pv(i)}} \phi_{PV(i)} - \left(1.5R_{line(i)} + R_{load}\right) \frac{\phi_{d(i)}}{v_{pv(i)}} \left(e_{d(i)} + \phi_d^*\right) \\ \quad - \sum_{j=1, j \neq i}^{N} \frac{R_{load}\phi_{d(i)}}{v_{pv(i)}} \left(e_{d(j)} + \phi_d^*\right), \\ \dot{e}_{d(i)} = -\frac{R_{1(i)}}{L_{1(i)}} \left(e_{d(i)} + \phi_d^*\right) - \omega_{(i)} \left(e_{q(i)} + \phi_q^*\right) + \frac{1}{L_{1(i)}} e_{d(i)} - \dot{\phi}_d^*, \\ \dot{e}_{q(i)} = \omega_{(i)} \left(e_{d(i)} + \phi_d^*\right) - \frac{R_{1(i)}}{L_{1(i)}} \left(e_{q(i)} + \phi_q^*\right) + \frac{1}{L_{1(i)}} e_{q(i)} - \dot{\phi}_q^*. \end{cases} \tag{4.16}$$

Now, further define

$$x_i(t) = \begin{bmatrix} v_{PV(i)} & e_{d(i)} & e_{q(i)} \end{bmatrix}^T,$$

$$A_{ii}(t) = \begin{bmatrix} \frac{1}{C_{PV(i)}} \frac{\phi_{PV(i)}}{v_{PV(i)}} & -\left(1.5R_{line(i)} + R_{load}\right) \frac{\phi_{d(i)}}{v_{pv(i)}} & 0 \\ 0 & -\frac{R_{1(i)}}{L_{1(i)}} & -\omega_{(i)} \\ 0 & \omega_{(i)} & -\frac{R_{1(i)}}{L_{1(i)}} \end{bmatrix},$$

$$B_i(t) = \begin{bmatrix} 0 & 0 \\ \frac{1}{L_{1(i)}} & 0 \\ 0 & \frac{1}{L_{1(i)}} \end{bmatrix}, u_i(t) = \begin{bmatrix} e_{d(i)} \\ e_{q(i)} \end{bmatrix}, A_{ij}(t) = \begin{bmatrix} 0 & -\frac{R_{load}\phi_{d(i)}}{v_{pv(i)}} & 0 \\ 0 & 0 & 0 \\ 0 & 0 & 0 \end{bmatrix},$$

$$\omega_i(t) = \begin{bmatrix} -\left(1.5R_{line(i)} + R_{load}\right) \frac{\phi_{d(i)}}{v_{pv(i)}} \phi_d^* - (N-1) \frac{R_{load}\phi_{d(i)}}{v_{pv(i)}} \phi_d^* \\ -\frac{R_{1(i)}}{L_{1(i)}} \phi_d^* - \omega_{(i)} \phi_q^* - \dot{\phi}_d^* \\ \omega_{(i)} \phi_d^* - \frac{R_{1(i)}}{L_{1(i)}} \phi_q^* - \dot{\phi}_q^* \end{bmatrix}. \tag{4.17}$$

Then, the nonlinear system in (4.16) is transformed into the state-space representation as below:

$$\dot{x}(t) = A(t)x(t) + B(t)u(t) + \omega(t), \tag{4.18}$$

where

$$x(t) = \begin{bmatrix} x_1^T(t) & x_2^T(t) & \cdots & x_N^T(t) \end{bmatrix}^T,$$

$$A(t) = \begin{bmatrix} A_{11}(t) & A_{12}(t) & \cdots & A_{1N}(t) \\ A_{21}(t) & A_{22}(t) & \cdots & A_{2N}(t) \\ \vdots & \vdots & \ddots & \vdots \\ A_{N1}(t) & A_{N2}(t) & \cdots & A_{NN}(t) \end{bmatrix}, \omega(t) = \begin{bmatrix} \omega_1(t) \\ \omega_2(t) \\ \vdots \\ \omega_N(t) \end{bmatrix},$$

$$B(t) = \begin{bmatrix} B_1(t) & 0 & \cdots & 0 \\ 0 & B_2(t) & \cdots & 0 \\ \vdots & \vdots & \ddots & \vdots \\ 0 & 0 & \cdots & B_N(t) \end{bmatrix}, u(t) = \begin{bmatrix} u_1(t) \\ u_2(t) \\ \vdots \\ u_N(t) \end{bmatrix}. \tag{4.19}$$

Choose $z_{i1}(t) = \frac{\phi_{PV(i)}}{v_{PV(i)}}$, $z_{i2}(t) = \frac{\phi_{d(i)}}{v_{pv(i)}}$, and $z_{i3}(t) = \omega_i$ as the fuzzy premise variables. Thus, it follows from (4.18) that the PV power nonlinear system is represented by

Plant Rule $\mathscr{R}^l$: **IF** $z_{11}(t)$ is $\mathscr{F}_{11}^l$ and $\cdots$ and $z_{N1}(t)$ is $\mathscr{F}_{N1}^l$, $z_{12}(t)$ is $\mathscr{F}_{12}^l$ and $\cdots$ and $z_{N2}(t)$ is $\mathscr{F}_{N2}^l$, $z_{13}(t)$ is $\mathscr{F}_{13}^l$ and $\cdots$ and $z_{N3}(t)$ is $\mathscr{F}_{N3}^l$, **THEN**

$$\dot{x}(t) = A_l x(t) + B_l u(t), l \in \mathscr{L} := \{1, 2, \ldots, r\} \tag{4.20}$$

where $\mathscr{R}^l$ denotes the l-th fuzzy inference rule; r is the number of inference rules; $z(t) \triangleq [z_{11}, \cdots, z_{N1}, z_{12}, \cdots, z_{N2}, , z_{13}, \cdots, z_{N3}]$ are the measurable variables; $\{A_l, B_l\}$ is the l-th local model.

Denoting as $\mathscr{F}^l := \prod_{\phi=1}^{3N} \mathscr{F}_\phi^l$ the inferred fuzzy set, and $\mu_l[z(t)]$ as the normalized membership function yields

$$\mu_l[z(t)] := \frac{\prod_{\phi=1}^{g} \mu_{l\phi}[z_\phi(t)]}{\sum_{\varsigma=1}^{r} \prod_{\phi=1}^{g} \mu_{\varsigma\phi}[z_\phi(t)]} \geq 0, \sum_{l=1}^{r} \mu_l[z(t)] = 1. \tag{4.21}$$

We denote $\mu_l := \mu_l[z(t)]$ for simplicity.

After fuzzy blending, the global T-S fuzzy dynamic model is given by

$$\begin{cases} \dot{x}(t) = A(\mu)x(t) + B(\mu)u(t) + \omega(t), \\ y(t) = Cx(t), \end{cases} \tag{4.22}$$

where $y(t)$ denotes the voltage error, $C = \underbrace{\begin{bmatrix} 0 & 1 & 1 & 0 & 1 & 1 & \cdots & 0 & 1 & 1 \end{bmatrix}}_{N}$,

and $A(\mu) := \sum_{l=1}^{r} \mu_l A_l, B(\mu) := \sum_{l=1}^{r} \mu_l B_l$.

Given the tracking error system in (4.22), and for a $\mathscr{L}_2$-gain performance level $\gamma > 0$, the purpose of this section is to design a controller such that the closed-loop fuzzy control system is asymptotically stable, and for any nonzero $\omega \in \mathscr{L}_2\,[0\ \infty)$ the induced $\mathscr{L}_2$ norm of the operator from ω to the tracking error $y(t)$ is less than γ

$$\int_0^\infty y^T(s)y(s)ds < \gamma^2 \int_0^\infty \omega^T(s)\omega(s)ds, \tag{4.23}$$

under zero initial conditions.

4.2 MODELING OF MULTI-MACHINE WIND GENERATORS

4.2.1 MODELING OF MULTI-WIND SYSTEMS WITH DC LOADS

We recall the wind generator with DC load in (2.20) as below:

$$\begin{cases} \dot{P} = \omega v_{\alpha\beta}^T J^T \phi_{\alpha\beta} + \frac{1}{L} v_{\alpha\beta}^T v_{\alpha\beta} - v_{\alpha\beta}^T \frac{v_{dc}}{2L} u_{\alpha\beta}, \\ \dot{Q} = \omega v_{\alpha\beta}^T J^T J \phi_{\alpha\beta} + \frac{v_{\alpha\beta}^T J}{L} v_{\alpha\beta} - \frac{v_{\alpha\beta}^T J v_{dc}}{2L} u_{\alpha\beta}, \\ \dot{v}_{\alpha\beta} = \omega J v_{\alpha\beta}, \\ \dot{\phi}_{\alpha\beta} = \frac{1}{L} v_{\alpha\beta} - \frac{v_{dc}}{2L} u_{\alpha\beta}, \\ \dot{v}_{dc} = -\frac{1}{CR_L} v_{dc} + \frac{\phi_{\alpha\beta}^T}{2C} u_{\alpha\beta}, \end{cases} \tag{4.24}$$

where P and Q are the instantaneous active and reactive powers, respectively; $\phi_{\alpha\beta}$ is vector of line currents respectively; v_{dc} is output capacitor voltage; $v_{\alpha\beta}$ is vector of the source line voltages; $u_{\alpha\beta}$ is vector of control inputs; $J = \begin{bmatrix} 0 & -1 \\ 1 & 0 \end{bmatrix}$.

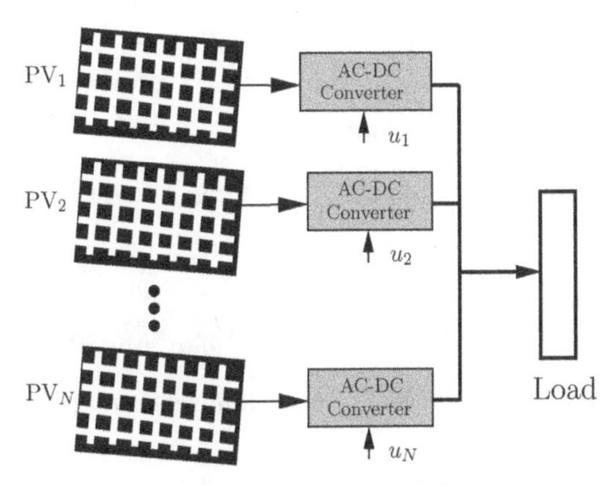

Figure 4.3 Interconnected multi-wind generator with DC load.

Consider an interconnected multi-wind generator with DC load as shown in Figure 4.3. Based on the Thevenin's theorem,

$$v_{dc(i)} = \phi_{0(i)}R_L + \cdots + \phi_{0(N)}R_L, \tag{4.25}$$

where the subscript i denotes the i-th subsystem, $i \in \mathcal{N} := \{1,2,\ldots,N\}$, N denotes the number of subsystems, R_{load} is the load resistance, $\phi_{0(i)}$ is the line current of the i-th subsystem. Based on the relationships on (4.24) and (4.25), it has

$$\begin{cases} \dot{P}_{(i)} = \omega_{(i)} v^T_{\alpha\beta(i)} J^T_{(i)} \phi_{\alpha\beta(i)} + \frac{1}{L_{(i)}} v^T_{\alpha\beta(i)} v_{\alpha\beta(i)} - v^T_{\alpha\beta(i)} \frac{v_{dc(i)}}{2L_{(i)}} u_{\alpha\beta(i)}, \\ \dot{Q}_{(i)} = \omega_{(i)} v^T_{\alpha\beta(i)} J^T_{(i)} J_{(i)} \phi_{\alpha\beta(i)} + \frac{v^T_{\alpha\beta(i)} J_{(i)}}{L_{(i)}} v_{\alpha\beta(i)} - \frac{v^T_{\alpha\beta(i)} J_{(i)} v_{dc(i)}}{2L_{(i)}} u_{\alpha\beta(i)}, \\ \dot{v}_{\alpha\beta(i)} = \omega_{(i)} J_{(i)} v_{\alpha\beta(i)}, \\ \dot{\phi}_{\alpha\beta(i)} = \frac{1}{L_{(i)}} v_{\alpha\beta(i)} - \frac{v_{dc(i)}}{2L_{(i)}} u_{\alpha\beta(i)}, \\ \dot{v}_{dc(i)} = -\frac{v_{dc(i)}}{C_{(i)}R_L} + \sum_{j=1, j\neq i}^{N} \frac{1}{C_{(i)}} \phi_{0(j)} + \frac{\phi^T_{\alpha\beta(i)}}{2C_{(i)}} u_{\alpha\beta(i)}. \end{cases} \tag{4.26}$$

Now, the control objective is to design a centralized control vector $u_{\alpha\beta}$ to regulate the output voltage $v_{dc(i)}$ to a desired constant value v^*_{dc}. Moreover, the outputs are chosen to be the active and reactive power P_i and Q_i subject to $P_{(i)} \to P^*, Q_{(i)} \to Q^*, v_{dc(i)} \to v^*_{dc}$. Here, for the sake of simplicity we have assumed that the voltage source is balanced and free of harmonic distortion. Define $P_{e(i)} = P_{(i)} - P^*, Q_{e(i)} = Q_{(i)} - Q^*, v_{e(i)} = v_{dc(i)} - v^*_{dc}$, where v^*_{dc} is a given constant.

$$\begin{cases} \dot{P}_{e(i)} = \omega_{(i)} v^T_{\alpha\beta(i)} J^T_{(i)} \phi_{\alpha\beta(i)} + \frac{1}{L_{(i)}} v^T_{\alpha\beta(i)} v_{\alpha\beta(i)} - v^T_{\alpha\beta(i)} \frac{v_{dc(i)}}{2L_{(i)}} u_{\alpha\beta(i)}, \\ \dot{Q}_{e(i)} = \omega_{(i)} v^T_{\alpha\beta(i)} J^T_{(i)} J_{(i)} \phi_{\alpha\beta(i)} + \frac{v^T_{\alpha\beta(i)} J_{(i)}}{L_{(i)}} v_{\alpha\beta(i)} - \frac{v^T_{\alpha\beta(i)} J_{(i)} v_{dc(i)}}{2L_{(i)}} u_{\alpha\beta(i)}, \\ \dot{v}_{\alpha\beta(i)} = \omega_{(i)} J_{(i)} v_{\alpha\beta(i)}, \\ \dot{\phi}_{\alpha\beta(i)} = \frac{1}{L_{(i)}} v_{\alpha\beta(i)} - \frac{v_{dc(i)}}{2L_{(i)}} u_{\alpha\beta(i)}, \\ \dot{v}_{e(i)} = -\frac{v_{e(i)} + v^*_{dc}}{C_{(i)}R_L} + \sum_{j=1, j\neq i}^{N} \frac{1}{C_{(i)}} \frac{\phi_{0(j)}}{v_{e(j)}} v_{e(j)} + \frac{\phi^T_{\alpha\beta(i)}}{2C_{(i)}} u_{\alpha\beta(i)}. \end{cases} \tag{4.27}$$

Define $z_{i1} = v^T_{\alpha\beta(i)}, z_{i2} = \frac{\omega_{(i)} \left\| J_{(i)} v_{\alpha\beta(i)} \right\|^2}{\left\| v_{\alpha\beta(i)} \right\|^2}, z_{i3} = v^T_{\alpha\beta(i)} v_{dc(i)}, z_{i4} = v^T_{\alpha\beta(i)} J^T_{(i)} v_{dc(i)}, z_{i5} = \omega_{(i)}, z_{i6} = v_{dc(i)}, z_{i7} = \frac{\phi_{0(i)}}{v_{e(i)}}, z_{i8} = \phi^T_{\alpha\beta(i)}$, and it follows from (4.27) that

$$\dot{x}_i(t) = A_{ii}(t)x_i(t) + \sum_{j=1, j\neq i}^{N} A_{ij}(t)x_j(t) + B_i(t)u_i(t) + \omega_i(t), \tag{4.28}$$

where $x_i(t) = \begin{bmatrix} P_{e(i)} & Q_{e(i)} & v^T_{\alpha\beta(i)} & \phi^T_{\alpha\beta(i)} & v_{e(i)} \end{bmatrix}^T$, and

$$A_{ii}(t) = \begin{bmatrix} 0 & L_{(i)}z_{i5} & \frac{z_{i1}}{L_{(i)}} & 0 & 0 \\ 0 & -L_{(i)}z_{i2} & 0 & 0 & 0 \\ 0 & 0 & L_{(i)}z_{i5} & 0 & 0 \\ 0 & 0 & \frac{1}{L_{(i)}} & 0 & 0 \\ 0 & 0 & 0 & 0 & -\frac{1}{C_{(i)}R_{L(i)}} \end{bmatrix}, A_{ij}(t) = \begin{bmatrix} 0 & 0 & 0 & 0 & 0 \\ 0 & 0 & 0 & 0 & 0 \\ 0 & 0 & 0 & 0 & 0 \\ 0 & 0 & 0 & 0 & 0 \\ 0 & 0 & 0 & 0 & \frac{z_{j7}}{C_{(i)}} \end{bmatrix},$$

$$B_i(t) = \begin{bmatrix} -\frac{z_{i3}}{2L_{(i)}} \\ -\frac{z_{i4}}{2L_{(i)}} \\ 0 \\ -\frac{z_{i6}}{2L_{(i)}} \\ \frac{z_{i8}}{2C_{(i)}} \end{bmatrix}, \omega_i(t) = \begin{bmatrix} -\dot{P}^* + v^T_{\alpha\beta(i)}\omega_{(i)}L_{(i)}\frac{v_{\alpha\beta(i)}}{\left\|v_{\alpha\beta(i)}\right\|^2}Q^* \\ -\dot{Q}^* - \omega_{(i)}L_{(i)}\frac{\left\|J_{(i)}v_{\alpha\beta(i)}\right\|^2}{\left\|v_{\alpha\beta(i)}\right\|^2}Q^* \\ 0 \\ 0 \\ -\frac{1}{C_{(i)}R_{L(i)}}v^*_{dc} \end{bmatrix}. \tag{4.29}$$

Here, one can choose $z_{i1} - z_{i8}$ as fuzzy premise variables. Then, the nonlinear system in (4.27) is transformed into the state-space representation as below:

$$\dot{x}(t) = A(\mu)x(t) + B(\mu)u(t) + \omega(t), \tag{4.30}$$

where

$$x(t) = \begin{bmatrix} x_1^T(t) & x_2^T(t) & \cdots & x_N^T(t) \end{bmatrix}^T,$$
$$A(\mu) = \begin{bmatrix} A_{11}(\mu) & A_{12}(\mu) & \cdots & A_{1N}(\mu) \\ A_{21}(\mu) & A_{22}(\mu) & \cdots & A_{2N}(\mu) \\ \vdots & \vdots & \ddots & \vdots \\ A_{N1}(\mu) & A_{N2}(\mu) & \cdots & A_{NN}(\mu) \end{bmatrix}, \omega(t) = \begin{bmatrix} \omega_1(t) \\ \omega_2(t) \\ \vdots \\ \omega_N(t) \end{bmatrix},$$
$$B(t) = \begin{bmatrix} B_1(\mu) & 0 & \cdots & 0 \\ 0 & B_2(\mu) & \cdots & 0 \\ \vdots & \vdots & \ddots & \vdots \\ 0 & 0 & \cdots & B_N(\mu) \end{bmatrix}, u(t) = \begin{bmatrix} u_1(t) \\ u_2(t) \\ \vdots \\ u_N(t) \end{bmatrix}. \tag{4.31}$$

4.2.2 MODELING OF MULTI-WIND GENERATOR WITH AC LOAD

Recall the wind generator model in (2.26) as below:

$$\begin{cases} \dot{\phi}_L = -\frac{1}{L}v_{C2} - \frac{clk}{L}v_0 + \frac{1}{L}v_i clku, \\ \dot{v}_{C2} = \frac{1}{C_2}\phi_L - \frac{C_1}{C_2}\dot{v}_{C1}, \\ \dot{v}_0 = \frac{clk}{C_0}\phi_L - \frac{C_1}{C_0}\dot{v}_{C1} - \frac{1}{R_L C_0}v_0. \end{cases} \tag{4.32}$$

We now consider interconnected multi-machine wind generator with AC load as shown in Figure 4.4. Based on the Thevenin's theorem, it has

$$v_{0(i)} = \phi_{0(i)}R_L + \cdots + \phi_{0(N)}R_L, \tag{4.33}$$

where the subscript i denotes the i-th subsystem, $i \in \mathcal{N} := \{1,2,\ldots,N\}$, N denotes the number of subsystems, R_L is the load resistance, $\phi_{0(i)}$ is the line current of the i-th subsystem.

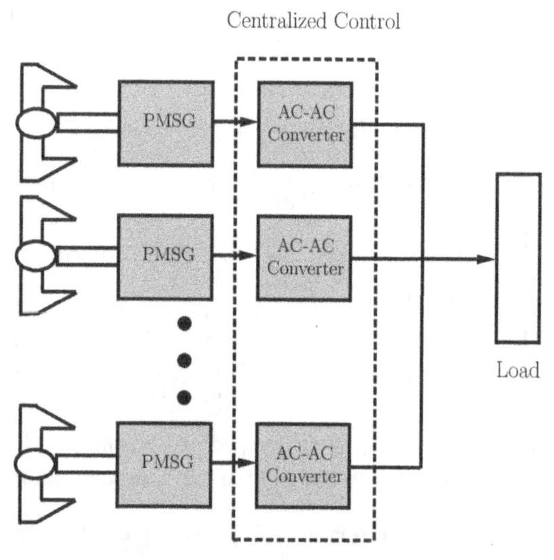

Figure 4.4 Interconnected multi-wind generator with AC load.

Submitting (4.33) into (4.32) yields

$$\begin{cases} \dot{\phi}_{L(i)} = -\frac{1}{L}v_{C2(i)} - \frac{z_{i1}(t)R_L}{L_{(i)}}\sum_{j=1}^{N} z_{j2}(t)\,\phi_{L(j)} + \frac{1}{L_{(i)}}z_{i2}(t)\,u_i, \\ \dot{v}_{C2(i)} = \frac{1}{C_{2(i)}}\phi_{L(i)} - \frac{C_{1(i)}z_{4i}(t)}{C_{2(i)}}v_{C2(i)}, \\ \dot{v}_{0(i)} = \frac{z_{i1}(t)}{C_{0(i)}}\phi_{L(i)} - \frac{C_{1(i)}z_{4i}(t)}{C_{0(i)}}v_{C2(i)} - \frac{v_{0(i)}}{R_L C_{0(i)}}, \end{cases} \tag{4.34}$$

where $z_{i1}(t) = clk_{(i)}, z_{i2}(t) = v_{i(i)}clk_{(i)}, z_{i3}(t) = \frac{\phi_{0(i)}}{\phi_{L(i)}}$, and $z_{4i}(t) = \frac{\dot{v}_{C1(i)}}{v_{C2(i)}}$.

Now, further define $e_{0(i)} = v_{0(i)} - v_{ref}$, where v_{ref} is reference voltage, and

$$x_i(t) = \begin{bmatrix} \phi_{L(i)} & v_{C2(i)} & e_{0(i)} \end{bmatrix}^T, A_{ii}(t) = \begin{bmatrix} 0 & -\frac{1}{L} & 0 \\ \frac{1}{C_{2(i)}} & 0 & -\frac{C_{1(i)}z_{4i}(t)}{C_{2(i)}} \\ \frac{z_{i1}(t)}{C_{0(i)}} & -\frac{C_{1(i)}z_{4i}(t)}{C_{0(i)}} & -\frac{1}{R_L C_{0(i)}} \end{bmatrix},$$

$$A_{ij}(t) = \begin{bmatrix} 0 & \frac{z_{i1}(t)R_L}{L_{(i)}}z_{j2}(t) & 0 \\ 0 & 0 & 0 \\ 0 & 0 & 0 \end{bmatrix}, B_i(t) = \begin{bmatrix} \frac{z_{i2}(t)}{L_{(i)}} \\ 0 \\ 0 \end{bmatrix}, \omega_i(t) = \begin{bmatrix} 0 \\ -\frac{C_{1(i)}\dot{v}_{C1(i)}v_{ref}}{C_{2(i)}v_{C2(i)}} \\ -\frac{v_{ref}}{R_L C_{0(i)}} \end{bmatrix}. \tag{4.35}$$

Then, the nonlinear system in (4.34) is transformed into the state-space representation as below:

$$\dot{x}(t) = A(t)x(t) + B(t)u(t) + \omega(t), \tag{4.36}$$

where

$$x(t) = \begin{bmatrix} x_1^T(t) & x_2^T(t) & \cdots & x_N^T(t) \end{bmatrix}^T,$$

$$A(t) = \begin{bmatrix} A_{11}(t) & A_{12}(t) & \cdots & A_{1N}(t) \\ A_{21}(t) & A_{22}(t) & \cdots & A_{2N}(t) \\ \vdots & \vdots & \ddots & \vdots \\ A_{N1}(t) & A_{N2}(t) & \cdots & A_{NN}(t) \end{bmatrix}, u(t) = \begin{bmatrix} u_1(t) \\ u_2(t) \\ \vdots \\ u_N(t) \end{bmatrix},$$

$$B(t) = \begin{bmatrix} B_1(t) & 0 & \cdots & 0 \\ 0 & B_2(t) & \cdots & 0 \\ \vdots & \vdots & \ddots & \vdots \\ 0 & 0 & \cdots & B_N(t) \end{bmatrix}, \omega(t) = \begin{bmatrix} \omega_1(t) \\ \omega_2(t) \\ \vdots \\ \omega_N(t) \end{bmatrix}. \tag{4.37}$$

Choose $z_{i1}(t) - z_{4i}(t)$ as fuzzy premise variables. Thus, the wind power nonlinear system in (4.36) can be represented by the following fuzzy system,

Plant Rule $\mathscr{R}^l$: **IF** $z_{11}(t)$ is $\mathscr{F}_{11}^l$ and $\cdots$ and $z_{N1}(t)$ is $\mathscr{F}_{N1}^l$, $z_{12}(t)$ is $\mathscr{F}_{12}^l$ and $\cdots$ and $z_{N2}(t)$ is $\mathscr{F}_{N2}^l$, $\cdots$, $z_{14}(t)$ is $\mathscr{F}_{14}^l$ and $\cdots$ and $z_{N4}(t)$ is $\mathscr{F}_{N4}^l$, **THEN**

$$\dot{x}(t) = A_l x(t) + B_l u(t) + \omega(t), l \in \mathscr{L} := \{1, 2, \ldots, r\} \tag{4.38}$$

where $\mathscr{R}^l$ denotes the l-th fuzzy inference rule; r is the number of inference rules; $z(t) \triangleq [z_{11}, \cdots, z_{N1}, z_{12}, \cdots, z_{N2}, z_{13}, \cdots, z_{N3}, z_{14}, \cdots, z_{N4}]$ are the measurable variables; $\{A_l, B_l\}$ is the l-th local model.

By fuzzy blending, it has

$$\dot{x}(t) = A(\mu)x(t) + B(\mu)u(t) + \omega(t), \tag{4.39}$$

where $A(\mu) := \sum_{l=1}^{r} \mu_l A_l, B(\mu) := \sum_{l=1}^{r} \mu_l B_l$.

4.3 CENTRALIZED CONTROL OF TRACKING SYNCHRONIZATION

4.3.1 CENTRALIZED FUZZY CONTROL

Assume that the premise variables between the fuzzy system and controller are synchronous and the following centralized fuzzy controller is given by

$$u(t) = K(\mu)x(t), \tag{4.40}$$

where $K(\mu) := \sum_{l=1}^{r} \mu_l [z(t)] K_l, K_l \in \Re^{Nn_{ui} \times Nn_{xi}}$ are controller gains to be determined.

In the case, the closed-loop fuzzy control system is given by

$$\dot{x}(t) = (A(\mu) + B(\mu)K(\mu))x(t). \tag{4.41}$$

Based on the closed-loop fuzzy control system in (4.41), the result on stability analysis is proposed as below:

Theorem 4.1: Stability Analysis

For a multi-PV power system with DC load in (4.5) with $\omega(t) \equiv 0$, and a centralized T-S fuzzy controller in the form of (4.40), the asymptotic stability of the closed-loop fuzzy control system is achieved, if the following condition is satisfied:

$$\text{Sym}\left(P\left(A(\mu)+B(\mu)K(\mu)\right)\right) < 0, \tag{4.42}$$

where $\text{Sym}(\star) = (\star)+(\star)^T$, $0 < P = P^T \in \Re^{Nn_{xi}\times Nn_{xi}}$, $K(\mu) \in \Re^{Nn_{ui}\times Nn_{xi}}$. ■

Proof. Consider the following Lyapunov functional:

$$V(t) = x^T(t)Px(t), \tag{4.43}$$

where $0 < P = P^T \in \Re^{Nn_{xi}\times Nn_{xi}}$.

By differentiating $V(t)$ shown in (4.43) with respect to time and using the closed-loop control system in (4.41), it yields

$$\dot{V}(t) = \text{Sym}\left(P\left(A(\mu)+B(\mu)K(\mu)\right)\right). \tag{4.44}$$

The inequality in (4.42) is obtained directly.

Note: A quadratic Lyapunov function $V(t) = x^T(t)Px(t)$ is considered in (4.43). It is clear that if $P \equiv \sum_{l=1}^{r} \mu_l P_l$, (4.43) turns into the fuzzy-basis-dependent Lyapunov function $V(t) = x^T(t)P(\mu)x(t)$. However, it requires that the time-derivative of μ_l is a priori, which may be unpractical for the considered system.

Note: In Theorem 4.1, we just consider the multi-PV power with DC load in (4.5). However, the obtained result can be extended easily to other systems, such as the multi-PV power with AC load in (4.18), multi-wind generators with DC loads (4.30) and multi-wind generators with AC loads (4.36).

4.3.2 DESIGN OF STABILIZATION CONTROLLER

Based on Theorem 4.1, the fuzzy controller gains can be calculated as below:

Theorem 4.2: Design of Centralized Controller

For a multi-PV power system with $\omega(t) \equiv 0$, and a centralized T-S fuzzy controller in the form of (4.40), the asymptotic stability of the closed-loop fuzzy control system is achieved if the following LMIs are satisfied:

$$\Sigma_{ll} < 0, l \in \mathscr{L} \tag{4.45}$$

$$\Sigma_{ls} + \Sigma_{sl} < 0, 1 \leq l < s \leq r \tag{4.46}$$

where $\Sigma_{ls} = \text{Sym}(A_l X + B_l \bar{K}_s)$, $0 < X = X^T \in \Re^{n_x \times n_x}$, $\bar{K}_s \in \Re^{n_u \times n_x}$.

In that case, the controller gains can be calculated by

$$K_l = \bar{K}_l X^{-1}, l \in \mathscr{L}. \tag{4.47}$$

■

Proof. By performing the congruence transformation to (4.44) by $X = P^{-1}$, and defining $\bar{K}_l = K_l X$, then by extracting the fuzzy premise variable, the design result on centralized controller can be directly obtained. Thus, the proof is completed.

4.3.3 CENTRALIZED SAMPLED-DATA CONTROLLER WITH EVENT-TRIGGERED ZOH

Before moving on, the following assumptions are firstly required [7].

Assumption 4.1. The sampler in each subsystem is synchronous clock-driven. Let h denote the upper bound of sampling intervals, we have

$$t_{k+1} - t_k \le h, k \in \mathbb{N} \tag{4.48}$$

where $h > 0$.

Assumption 4.2. The zero-order-hold (ZOH) is event-driven, and it uses the latest sampled-data signals and holds them until the next transmitted data are received.

Define $\rho^s(t) = t - t_k$, it has

$$0 \le \rho^s(t) < h. \tag{4.49}$$

It is noted that in the context of digital control systems, both $z(t)$ and $x(t)$ are involved in the sampled-data measurement. Now, without loss of generality, we further assume that both $z(t)$ and $x(t)$ are updated at the same time. Then, a centralized state-feedback fuzzy controller can be given by

Plant Rule $\mathscr{R}_i^s$: **IF** $z_{11}(t_k)$ is $\mathscr{F}_{11}^s$ and $\cdots$ and $z_{N1}(t_k)$ is $\mathscr{F}_{N1}^s$, $z_{12}(t_k)$ is $\mathscr{F}_{12}^s$ and $\cdots$ and $z_{N2}(t_k)$ is $\mathscr{F}_{N2}^s$, $\cdots$, $z_{14}(t_k)$ is $\mathscr{F}_{14}^s$ and $\cdots$ and $z_{N4}(t_k)$ is $\mathscr{F}_{N4}^s$, **THEN**

$$u(t) = K_s x(t_k), s \in \mathscr{L}, t \in [t_k, t_{k+1}) \tag{4.50}$$

where $K_s \in \Re^{Nn_{ui} \times Nn_{xi}}, s \in \mathscr{L}, i \in \mathscr{N}$ are controller gains to be determined; $z(t_k) := [z_{11}(t_k), z_{12}(t_k), z_{13}(t_k), z_{14}(t); \cdots; z_{i1}(t_k), z_{i2}(t_k), z_{i3}(t_k), z_{i4}(t_k); \cdots; z_{N1}(t_k), z_{N2}(t_k), z_{N3}(t_k), z_{N4}(t_k)]$; $z(t_k)$ and $x(t_k)$ denote the updating signals in the fuzzy controller.

Similarly, the overall centralized state-feedback fuzzy controller is

$$u(t) = K(\hat{\mu}) x(t_k), t \in [t_k, t_{k+1}) \tag{4.51}$$

where

$$\begin{cases} K(\hat{\mu}) := \sum_{s=1}^{r} \hat{\mu}_s[z(t_k)] K_s, \ \sum_{s=1}^{r} \hat{\mu}_s[z(t_k)] = 1, \\ \hat{\mu}_s[z(t_k)] := \dfrac{\Pi_{\phi=1}^{g} \hat{\mu}_{s\phi}[z_\phi(t_k)]}{\Sigma_{\varsigma=1}^{r} \Pi_{\phi=1}^{g} \hat{\mu}_{\varsigma\phi}[z_\phi(t_k)]} \ge 0. \end{cases} \tag{4.52}$$

In the following, we will denote $\hat{\mu}_s := \hat{\mu}_s[z(t_k)]$ for brevity.

We further define

$$x(v) = x(t_k) - x(t). \tag{4.53}$$

Submitting (4.53) into the fuzzy controller (4.51) and combined with the fuzzy system (4.5), the closed-loop fuzzy control system is

$$\dot{x}(t) = (A(\mu) + B(\mu)K(\hat{\mu}))x(t) + B(\mu)K(\hat{\mu})x(v). \tag{4.54}$$

Now, we introduce the following Lyapunov function:

$$\begin{aligned} V(t) = {} & x^T(t)Px(t) + h^2 \int_{t_k}^{t} \dot{x}^T(\alpha)Q\dot{x}(\alpha)d\alpha \\ & - \frac{\pi^2}{4}\int_{t_k}^{t} [x(\alpha) - x(t_k)]^T Q[x(\alpha) - x(t_k)]d\alpha, \end{aligned} \tag{4.55}$$

where $\{P, Q\} \in \Re^{Nn_{xi} \times Nn_{xi}}$ are positive definite symmetric matrices.

By using Wirtinger's inequality in [2], it can be known that $V(t) > 0$. Based on the new model in (4.54) and the Lyapunov function in (4.55), a sufficient condition for existing a centralized sampled-data controller can be given as below:

Lemma 4.1: Stability Analysis of Centralized Sampled-Data Control

The closed-loop fuzzy system in (4.54) using a centralized sampled-data fuzzy controller (4.51), is asymptotically stable, if there exist the symmetric positive definite matrices $\{P, W\} \in \Re^{Nn_{xi} \times Nn_{xi}}$, $K(\hat{\mu}) \in \Re^{Nn_{ui} \times Nn_{xi}}$, and the positive scalars h, such that the following matrix inequalities hold:

$$\Theta + \mathrm{Sym}(\mathbb{G}\mathbb{A}(\mu, \hat{\mu})) < 0, \tag{4.56}$$

where

$$\Theta = \begin{bmatrix} h^2Q & P & 0 \\ \star & 0 & 0 \\ \star & \star & -\frac{\pi^2}{4}Q \end{bmatrix}, \mathbb{A}(\mu, \hat{\mu}) = \begin{bmatrix} -I & A(\mu) + B(\mu)K(\hat{\mu}) & B(\mu)K(\hat{\mu}) \end{bmatrix}. \tag{4.57}$$

■

Proof. By taking the time derivative of $V(t)$ in (4.55), we have

$$\dot{V}(t) = \sum_{i=1}^{N} \{2x^T(t)P\dot{x}(t) + h^2\dot{x}^T(t)Q\dot{x}(t) - \frac{\pi^2}{4}x^T(v)Qx(v)\}. \tag{4.58}$$

Define the matrix $\mathbb{G} \in \Re^{3Nn_{xi} \times Nn_{xi}}$ and $\chi(t) = \begin{bmatrix} \dot{x}^T(t) & x^T(t) & x^T(v) \end{bmatrix}^T$, and it follows from (4.54) that

$$0 = 2\chi^T(t)\mathbb{G}\mathbb{A}(\mu,\hat{\mu})\chi(t), \tag{4.59}$$

where $\mathbb{A}(\mu,\hat{\mu}) = \begin{bmatrix} -I & A(\mu)+B(\mu)K(\hat{\mu}) & B(\mu)K(\hat{\mu}) \end{bmatrix}$.

It follows from (4.58)-(4.59) that the result on (4.56) can be obtained directly.

It is noted that the result on (4.56) is not an LMI-based result. It is also noted that when the asynchronized information of μ_l and $\hat{\mu}_l$ is unknown, the designed result generally leads to the linear controller instead of the fuzzy one [9]. From a practical perspective, obtaining a priori knowledge of μ_l and $\hat{\mu}_l$ is possible. Thus, we assume that the asynchronized condition is subject to

$$\underline{\rho}_l \leq \frac{\hat{\mu}_l}{\mu_l} \leq \bar{\rho}_l, \tag{4.60}$$

where $\underline{\rho}_l$ and $\bar{\rho}_l$ are positive scalars.

It follows from that the design result on the centralized sampled-data fuzzy controller can be summarized as below:

Theorem 4.3: Design of Centralized Sampled-Data Fuzzy Control Using Asynchronized Method

The closed-loop fuzzy system in (4.54) using a centralized sampled-data fuzzy controller (4.51), is asymptotically stable, if there exist the symmetric positive definite matrices $\bar{P}, \in \Re^{Nn_{xi} \times Nn_{xi}}$, and matrices $G \in \Re^{Nn_{xi} \times Nn_{xi}}$, $M_{ls} = M_{sl}^T \in \Re^{4Nn_{xi} \times 4Nn_{xi}}$, $\bar{K}_s \in \Re^{Nn_{ui} \times Nn_{xi}}$, and the positive scalars $\{h, \underline{\rho}_l, \bar{\rho}_l\}$, such that for all $(l,s) \in \mathscr{L}$, the following LMIs hold:

$$\bar{\rho}_l \Sigma_{ll} + M_{ll} < 0, \tag{4.61}$$

$$\underline{\rho}_l \Sigma_{ll} + M_{ll} < 0, \tag{4.62}$$

$$\bar{\rho}_s \Sigma_{ls} + \bar{\rho}_l \Sigma_{sl} + M_{ls} + M_{sl} < 0, \tag{4.63}$$

$$\underline{\rho}_s \Sigma_{ls} + \underline{\rho}_l \Sigma_{sl} + M_{ls} + M_{sl} < 0, \tag{4.64}$$

$$\underline{\rho}_s \Sigma_{ls} + \bar{\rho}_l \Sigma_{sl} + M_{ls} + M_{sl} < 0, \tag{4.65}$$

$$\bar{\rho}_s \Sigma_{ls} + \underline{\rho}_l \Sigma_{sl} + M_{ls} + M_{sl} < 0, \tag{4.66}$$

$$\begin{bmatrix} M_{11} & \cdots & M_{1r} \\ \vdots & \ddots & \vdots \\ M_{r1} & \cdots & M_{rr} \end{bmatrix} > 0, \tag{4.67}$$

where

$$\Sigma_{ls} = \bar{\Theta} + \mathrm{Sym}\left(\bar{\mathbb{I}}\tilde{\mathbb{A}}_{ls}\right), \bar{\Theta} = \begin{bmatrix} h^2\bar{Q} & \bar{P} & 0 \\ \star & 0 & 0 \\ \star & \star & -\frac{\pi^2}{4}\bar{Q} \end{bmatrix},$$

$$\bar{\mathbb{I}} = \begin{bmatrix} I \\ I \\ 0 \end{bmatrix}, \tilde{\mathbb{A}}_{ls} = \begin{bmatrix} -G & A_l G + B_l \bar{K}_s & B_l \bar{K}_s \end{bmatrix}. \quad (4.68)$$

In that case, the proposed sampled-data fuzzy controller gains can be calculated by

$$K_s = \bar{K}_s G^{-1}, s \in \mathscr{L}. \quad (4.69)$$

■

Proof. It follows from (4.61) that

$$h^2\bar{Q} - \mathrm{Sym}\{G\} < 0, \quad (4.70)$$

which implies that G is a nonsingular matrix.

We further define

$$\mathbb{G} = \begin{bmatrix} G^{-1} & G^{-1} & 0 \end{bmatrix}^T, \Gamma_1 := \mathrm{diag}\{ G \quad G \quad G \}, \bar{P} = G^T P G, \bar{Q} = G^T Q G. \quad (4.71)$$

By performing a congruence transformation by Γ_1 to (4.56), and extracting the fuzzy membership functions,

$$\sum_{l=1}^{r}\sum_{s=1}^{r} \mu_l \hat{\mu}_s \Sigma_{ls} < 0, \quad (4.72)$$

where

$$\Sigma_{ls} = \bar{\Theta} + \mathrm{Sym}\left(\bar{\mathbb{I}}\bar{\mathbb{A}}_{ls}\right), \bar{\Theta} = \begin{bmatrix} h^2\bar{Q} & \bar{P} & 0 \\ \star & 0 & 0 \\ \star & \star & -\frac{\pi^2}{4}\bar{Q} \end{bmatrix},$$

$$\bar{\mathbb{I}} = \begin{bmatrix} I \\ I \\ 0 \end{bmatrix}, \bar{\mathbb{A}}_{ls} = \begin{bmatrix} -G & A_l G + B_l K_s G & B_l K_s G \end{bmatrix}. \quad (4.73)$$

Then, define $\bar{K}_s = K_s G$, and by using the asynchronous method proposed in [16], the inequalities (4.61)-(4.67) can be obtained, thus completing his proof.

It is worth pointing output that the number of LMIs on Theorem 4.3 is large. We should also note that the existing relaxation technique $\sum_{l=1}^{r}[\mu_l]^2 \Sigma_{ll} + \sum_{l=1}^{r}\sum_{l<s\leq r}^{r} \mu_l \mu_s [\Sigma_{sl} + \Sigma_{ls}] < 0$ is no longer applicable to fuzzy controller synthesis, since $\mu_s \neq \hat{\mu}_s$. Similarly to the relaxation technique in [10], it is assumed that

$$|\mu_l - \hat{\mu}_l| \leq \delta_l, l \in \mathscr{L}, \quad (4.74)$$

where δ_l is a positive scalar. If $\Sigma_{ls} + M_l \geq 0$, where M_l is a symmetric matrix, one obtains

$$\begin{aligned}\sum_{l=1}^{r}\sum_{s=1}^{r}\mu_l\hat{\mu}_s\Sigma_{ls} &= \sum_{l=1}^{r}\sum_{s=1}^{r}\mu_l\mu_s\Sigma_{ls} + \sum_{l=1}^{r}\sum_{s=1}^{r}\mu_l\left(\hat{\mu}_s - \mu_s\right)\left(\Sigma_{ls} + M_l\right) \\ &\leq \sum_{l=1}^{r}\sum_{s=1}^{r}\mu_l\mu_s\left[\Sigma_{ls} + \sum_{f=1}^{r}\delta_{if}\left(\Sigma_{fl} + M_l\right)\right]. \end{aligned} \quad (4.75)$$

Therefore, upon defining $\Sigma_{ls} = \Phi_{ls} + \sum_{s=1}^{r}\delta_s\left(\Phi_{ls} + M_l\right)$, the existing relaxation technique from [7] can be applied to (4.75).

Based on the assumption in (4.74), and the result of (4.72), the design result on fuzzy control is proposed as below:

Theorem 4.4: Design of Centralized Sampled-Data Fuzzy Control Using Synchronized Method

The closed-loop fuzzy system in (4.54) using a centralized sampled-data fuzzy controller (4.51), is asymptotically stable, if there exist the symmetric positive definite matrices $\bar{P} \in \Re^{Nn_{xi}\times Nn_{xi}}$, and matrices $G \in \Re^{Nn_{xi}\times Nn_{xi}}$, $M_l = M_l^T \in \Re^{4Nn_{xi}\times 4Nn_{xi}}$, $\bar{K}_s \in \Re^{Nn_{ui}\times Nn_{xi}}$, and the positive scalars $\{h, \delta_l\}$, such that for all $(l,s) \in \mathscr{L}$, the following LMIs hold:

$$\Sigma_{ll} < 0, l \in \mathscr{L} \quad (4.76)$$

$$\Sigma_{ls} + \Sigma_{sl} < 0, 1 \leq l < s \leq r \quad (4.77)$$

where

$$\begin{aligned}&\Sigma_{ls} = \Phi_{ls} + \sum_{f=1}^{r}\delta_f\left(\Phi_{lf} + M_l\right), \Phi_{ls} = \bar{\Theta} + \mathrm{Sym}\left(\bar{\mathbb{I}}\tilde{\mathbb{A}}_{ls}\right), \\ &\bar{\Theta} = \begin{bmatrix} h^2\bar{Q} & \bar{P} & 0 \\ \star & 0 & 0 \\ \star & \star & -\frac{\pi^2}{4}\bar{Q}\end{bmatrix}, \bar{\mathbb{I}} = \begin{bmatrix} I \\ I \\ 0\end{bmatrix}, \tilde{\mathbb{A}}_{ls} = \begin{bmatrix} -G & A_lG + B_l\bar{K}_s & B_l\bar{K}_s\end{bmatrix}.\end{aligned} \quad (4.78)$$

In that case, the centralized controller gains can be calculated by

$$K_l = \bar{K}_l G^{-1}, l \in \mathscr{L}. \quad (4.79)$$

■

Note that when the asynchronized information of μ_l and $\hat{\mu}_l$ is unknown, the design result on a centralized sampled-data linear controller can be directly derived as below:

Theorem 4.5: Centralized Sampled-Data Linear Controller Design

The closed-loop fuzzy system in (4.54) using a centralized sampled-data linear controller $u(t) = Kx(t_k), t \in [t_k, t_{k+1})$, is asymptotically stable, if there exist the symmetric positive definite matrices $\bar{P} \in \Re^{Nn_{xi} \times Nn_{xi}}$, and matrices $G \in \Re^{Nn_{xi} \times Nn_{xi}}$, $\bar{K} \in \Re^{Nn_{ui} \times Nn_{xi}}$, and the positive scalars h, such that for all $l \in \mathscr{L}$, the following LMIs hold:

$$\bar{\Theta} + \mathrm{Sym}\left(\bar{\mathbb{I}}\tilde{\mathbb{A}}_l\right) < 0, \tag{4.80}$$

where

$$\bar{\Theta} = \begin{bmatrix} h^2\bar{Q} & \bar{P} & 0 \\ \star & 0 & 0 \\ \star & \star & -\frac{\pi^2}{4}\bar{Q} \end{bmatrix}, \bar{\mathbb{I}} = \begin{bmatrix} I \\ I \\ 0 \end{bmatrix}, \tilde{\mathbb{A}}_l = \begin{bmatrix} -G & A_lG + B_l\bar{K} & B_l\bar{K} \end{bmatrix}. \tag{4.81}$$

In that case, the proposed linear controller gains can be calculated by

$$K = \bar{K}G^{-1}. \tag{4.82}$$

■

4.3.4 CENTRALIZED SAMPLED-DATA CONTROLLER DESIGN WITH TIME-TRIGGERED ZOH

This subsection will focus on the centralized sampled-data controller design for multi-PV fuzzy system in (4.9). It is noted that the obtained results can be easily expanded to the other fuzzy systems in (4.22), (4.30), and (4.39). Before moving on, the assumptions are made as below [7]:

Assumption 4.3. Sensors are time-driven, and satisfy the set,

$$\underline{s} \le s^{k+1} - s^k \le \bar{s}, k \in \mathbb{N}. \tag{4.83}$$

Assumption 4.4. ZOHs are time-driven, and satisfy the set,

$$\underline{z} \le z^{k+1} - z^k \le \bar{z}, k \in \mathbb{N}. \tag{4.84}$$

Note: The work in [7] considered even-based zero order holds in the sense that control signals are applied as soon as new data becomes available. However, that case does not always happen in engineering applications when actuators have low updated rates. For example, according to the work in [8], the refresh rates in actuators such as electric cylinders, solenoids, shape memory alloys, and electroactive polymers are 10 Hz or less. Those actuators do not immediately refresh the control signal and induce time-delays, which can affect stability and performance of the closed-loop system. Thus, here the time-driven ZOHs are considered in (4.41).

Generally, ZOHs receive new data from controllers as soon as sensors send new sampling data when considering event-driven ZOHs. Based on Assumptions 4.1

and 4.2, it is easy to see that, the proposed ZOHs are time-driven and are not updated immediately. Here, we define the time elapsed as $\rho^z(t)$ [8],

$$0 \le \rho^z(t) < \bar{z}. \tag{4.85}$$

Thus, it is easy to see that

$$\begin{aligned}\rho^{zs}(t) &= \rho^z(t) + t - s^k \\ &= \rho^z(t) + \rho^s(t), t \in [z^k, z^{k+1})\end{aligned} \tag{4.86}$$

where $0 \le \rho^s(t) < \bar{s}$, $0 \le \rho^{zs}(t) < \bar{z} + \bar{s}$.

Now, consider the following centralized sampled-data fuzzy controller:

Controller Rule $\mathscr{R}^f$: **IF** $z_{11}(z^k)$ is $\mathscr{F}_{11}^f$ and $\cdots$ and $z_{N1}(z^k)$ is $\mathscr{F}_{N1}^f$, $z_{12}(z^k)$ is $\mathscr{F}_{12}^f$ and $\cdots$ and $z_{N2}(z^k)$ is $\mathscr{F}_{N2}^f$, $\cdots$, $z_{14}(z^k)$ is $\mathscr{F}_{14}^f$ and $\cdots$ and $z_{N4}(z^k)$ is $\mathscr{F}_{N4}^f$, **THEN**

$$u(t) = K_f x(z^k), t \in [z^k, z^{k+1}), \tag{4.87}$$

where $K_f \in \Re^{n_u \times n_x}$ are controller gains to be determined.

Similarly, the overall fuzzy controller can be represented by

$$u(t) = K(\hat{\mu}) x(z^k), t \in [z^k, z^{k+1}) \tag{4.88}$$

where $K(\hat{\mu}) := \sum_{f=1}^{r} \hat{\mu}_f K_f$.

Here, we further define

$$x(v) = x(z^k) - x(t). \tag{4.89}$$

Submitting (4.89) into the fuzzy controller (4.88) and combined with the fuzzy system (4.9), the closed-loop fuzzy control system is

$$\dot{x}(t) = (A(\mu) + B(\mu)K(\hat{\mu}))x(t) + B(\mu)K(\hat{\mu})x(v). \tag{4.90}$$

Note: The premise variable $z(t)$ undergoes time-driven sensors and ZOHs, and is implemented by the proposed fuzzy controller in (4.88). Hence the premise variable spaces in asynchronous form between $z(t)$ and $z(z^k)$ are more practical.

This subsection will first present a sufficient condition to guarantee the asymptotic stability of the closed-loop control system. Then, by using some linearization techniques of matrix inequality, the centralized sampled-data controller design result is derived in terms of LMIs.

Now, we introduce the following LKF:

$$\begin{aligned}V(t) &= x^T(t) P x(t) + [\bar{z} + \bar{s}]^2 \int_{z^k}^{t} \dot{x}^T(\alpha) Q \dot{x}(\alpha) d\alpha \\ &- \frac{\pi^2}{4} \int_{z^k}^{t} \left[x(\alpha) - x\left(z^k\right)\right]^T Q \left[x(\alpha) - x\left(z^k\right)\right] d\alpha,\end{aligned} \tag{4.91}$$

where $\{P, Q\} \in \Re^{Nn_{xi} \times Nn_{xi}}$ are positive definite symmetric matrices.

By using Wirtinger's inequality [2], we can see that $V(t) > 0$. Based on the new model in (4.90) and the LKF in (4.91), a sufficient condition for existing a centralized sampled-data controller can be given as below:

Lemma 4.2: Stability Analysis

The closed-loop fuzzy system in (4.90) using a centralized sampled-data fuzzy controller (4.88), is asymptotically stable, if there exist positive definite symmetric matrices $\{P, Q\} \in \Re^{Nn_{xi} \times Nn_{xi}}$, $K(\hat{\mu}) \in \Re^{Nn_{ui} \times Nn_{xi}}$, and the positive scalars $\bar{z}, \bar{s}$, such that the following matrix inequalities hold:

$$\begin{bmatrix} \mathrm{Sym}\{P(A(\mu)+B(\mu)K(\hat{\mu}))\} & \star & \star \\ (A(\mu)+B(\mu)K(\hat{\mu}))^T P & -\frac{\pi^2}{4}Q & \star \\ [\bar{z}+\bar{s}]\, Q(A(\mu)+B(\mu)K(\hat{\mu})) & QB(\mu)K(\hat{\mu}) & -Q \end{bmatrix} < 0. \tag{4.92}$$

■

Proof. By taking the time derivative of $V(t)$ in (4.91), we have

$$\begin{aligned} \dot{V}(t) &= 2x^T(t)P\dot{x}(t) + [\bar{z}+\bar{s}]^2 \dot{x}^T(t)Q\dot{x}(t) - \frac{\pi^2}{4}x^T(v)Qx(v) \\ &= \bar{x}^T(t)\Theta(\mu,\hat{\mu})\bar{x}(t), \end{aligned} \tag{4.93}$$

where

$$\begin{aligned} \bar{x}(t) &= \begin{bmatrix} x(t) \\ x(v) \end{bmatrix}, \Theta(\mu,\hat{\mu}) = \Theta_1(\mu,\hat{\mu}) + \Theta_2(\mu,\hat{\mu}), \\ \Theta_1(\mu,\hat{\mu}) &= \begin{bmatrix} \mathrm{Sym}\{P(A(\mu)+B(\mu)K(\hat{\mu}))\} & P(A(\mu)+B(\mu)K(\hat{\mu})) \\ \star & -\frac{\pi^2}{4}Q \end{bmatrix}, \\ \Theta_2(\mu,\hat{\mu}) &= [\bar{z}+\bar{s}]^2 (\star) Q \begin{bmatrix} (A(\mu)+B(\mu)K(\hat{\mu})) & B(\mu)K(\hat{\mu}) \end{bmatrix}. \end{aligned} \tag{4.94}$$

It is easy to see that the following inequality implies $\dot{V}(t) < 0$,

$$\Theta(\mu,\hat{\mu}) < 0. \tag{4.95}$$

Now, by using Schur complement lemma, the inequality in (4.92) is directly obtained.

It is noted that the obtained result on (4.92) is nonlinear. In order to transform it into linear form, it follows from (4.90) that

$$0 = 2\left[\dot{x}^T(t)P + x^T(t)\varepsilon P\right]\left[-\dot{x}(t) + (A(\mu)+B(\mu)K(\hat{\mu}))x(t) + B(\mu)K(\hat{\mu})x(v)\right], \tag{4.96}$$

where ε is a positive scalar.

It follows from (4.93) and (4.96) that

$$\begin{bmatrix} [\bar{z}+\bar{s}]^2 Q - 2P & (1-\varepsilon)P + PA(\mu) + PB(\mu)K(\hat{\mu}) & PB(\mu)K(\hat{\mu}) \\ \star & \text{Sym}\{\varepsilon PA(\mu) + \varepsilon PB(\mu)K(\hat{\mu})\} & \varepsilon PB(\mu)K(\hat{\mu}) \\ \star & \star & -\frac{\pi^2}{4}Q \end{bmatrix} < 0. \quad (4.97)$$

By performing the congruence transformation to (4.97) by $\Gamma = \text{diag}\{X,X,X\}$, where $X = P^{-1}$, and define $\bar{Q} = XQX, \bar{K}(\hat{\mu}) = K(\hat{\mu})X$,

$$\begin{bmatrix} [\bar{z}+\bar{s}]^2 \bar{Q} - 2X & (1-\varepsilon)X + A(\mu)X + B(\mu)\bar{K}(\hat{\mu}) & B(\mu)\bar{K}(\hat{\mu}) \\ \star & \text{Sym}\{\varepsilon A(\mu)X + \varepsilon B(\mu)\bar{K}(\hat{\mu})\} & \varepsilon B(\mu)\bar{K}(\hat{\mu}) \\ \star & \star & -\frac{\pi^2}{4}\bar{Q} \end{bmatrix} < 0. \quad (4.98)$$

Note that when the asynchronized information of μ_l and $\hat{\mu}_l$ is unknown, the designed result generally leads to the linear controller instead of the fuzzy one [9]. From a practical perspective, obtaining a priori knowledge of μ_l and $\hat{\mu}_l$ is possible. Thus, we assume that the asynchronized condition is subject to

$$\underline{\rho}_l \leq \frac{\hat{\mu}_l}{\mu_l} \leq \bar{\rho}_l, \quad (4.99)$$

where $\underline{\rho}_l$ and $\bar{\rho}_l$ are positive scalars.

It follows (4.92) and (4.99) that the design result on the centralized sampled-data fuzzy controller can be summarized as below:

Theorem 4.6: Design of Centralized Fuzzy Controller Using Asynchronized Method

Consider the multi-photovoltaic system with DC load in (4.5). A centralized sampled-data fuzzy controller in the form (4.88) can guarantee that the resulting closed-loop fuzzy control system with the assumption (4.99) is asymptotically stable, if there exist the symmetric positive definite matrices $\{\bar{P}, \bar{Q}\} \in \Re^{Nn_{xi} \times Nn_{xi}}$, and matrices $M_{ls} = M_{sl}^T \in \Re^{3Nn_{xi} \times 3Nn_{xi}}$, $\bar{K}_s \in \Re^{Nn_{ui} \times Nn_{xi}}$, and positive scalars $\left\{\bar{s}, \bar{z}, \bar{\rho}_l, \underline{\rho}_l, \varepsilon\right\}$, such that for all $(l,s) \in \mathscr{L}$, the following LMIs hold:

$$\bar{\rho}_l \Sigma_{ll} + M_{ll} < 0, \tag{4.100}$$

$$\underline{\rho}_l \Sigma_{ll} + M_{ll} < 0, \tag{4.101}$$

$$\bar{\rho}_s \Sigma_{ls} + \bar{\rho}_l \Sigma_{sl} + M_{ls} + M_{sl} < 0, \tag{4.102}$$

$$\underline{\rho}_s \Sigma_{ls} + \underline{\rho}_l \Sigma_{sl} + M_{ls} + M_{sl} < 0, \tag{4.103}$$

$$\underline{\rho}_s \Sigma_{ls} + \bar{\rho}_l \Sigma_{sl} + M_{ls} + M_{sl} < 0, \tag{4.104}$$

$$\bar{\rho}_s \Sigma_{ls} + \underline{\rho}_l \Sigma_{sl} + M_{ls} + M_{sl} < 0, \tag{4.105}$$

$$\begin{bmatrix} M_{11} & \cdots & M_{1r} \\ \vdots & \ddots & \vdots \\ M_{r1} & \cdots & M_{rr} \end{bmatrix} > 0, \tag{4.106}$$

where

$$\Sigma_{ls} = \begin{bmatrix} [\bar{z}+\bar{s}]^2 \bar{Q} - 2X & (1-\varepsilon)X + A_l X + B_l \bar{K}_s & B_l \bar{K}_s \\ \star & \mathrm{Sym}\{\varepsilon A_l X + \varepsilon B_l \bar{K}_s\} & \varepsilon B_l \bar{K}_s \\ \star & \star & -\frac{\pi^2}{4}\bar{Q} \end{bmatrix}. \tag{4.107}$$

In that case, the proposed fuzzy controller gains can be calculated by

$$K_s = \bar{K}_s X^{-1}, s \in \mathscr{L}. \tag{4.108}$$

■

Note that the number of LMIs in Theorem 4.6 is large. It is also noted that the existing relaxation technique $\sum_{l=1}^{r} [\mu_l]^2 \Sigma_{ll} + \sum_{l=1}^{r} \sum_{l<s\le r}^{r} \mu_l \mu_s \Sigma_{ls} < 0$ is no longer applicable to fuzzy controller synthesis, since $\mu_s \neq \hat{\mu}_s$. Similarly to the asynchronous relaxation technique in [10], it is assumed that

$$|\mu_l - \hat{\mu}_l| \le \delta_l, l \in \mathscr{L}, \tag{4.109}$$

where δ_l is a positive scalar. If $\Sigma_{ls} + M_l \ge 0$, where M_l is a symmetric matrix, we have

$$\begin{aligned} &\sum_{l=1}^{r} \sum_{s=1}^{r} \mu_l \hat{\mu}_s \Sigma_{ls} \\ &= \sum_{l=1}^{r} \sum_{s=1}^{r} \mu_l \mu_s \Sigma_{ls} + \sum_{l=1}^{r} \sum_{s=1}^{r} \mu_l (\hat{\mu}_s - \mu_s)(\Sigma_{ls} + M_l) \\ &\le \sum_{l=1}^{r} \sum_{s=1}^{r} \mu_l \mu_s \left[\Sigma_{ls} + \sum_{s=1}^{r} \delta_s (\Sigma_{ls} + M_l)\right]. \end{aligned} \tag{4.110}$$

Therefore, upon defining $\Sigma_{ls} = \Phi_{ls} + \sum_{s=1}^{r} \delta_s (\Phi_{ls} + M_l)$, the existing relaxation technique from [7] can be applied to (4.110).

Based on the assumption in (4.109), a small number of LMIs to the design of a centralized sampled-data fuzzy controller is derived as below.

Theorem 4.7: Design of Centralized Fuzzy Controller Using Synchronized Method

Consider the multi-photovoltaic system with DC load in (4.5). A centralized sampled-data fuzzy controller in the form (4.88) can guarantee that the resulting closed-loop fuzzy control system with the assumption (4.109) is asymptotically stable, if there exist the symmetric positive definite matrices $\{X, \bar{Q}\} \in \Re^{Nn_{xi} \times Nn_{xi}}$, and matrices $M_l = M_l^T \in \Re^{3Nn_{xi} \times 3Nn_{xi}}$, $\bar{K}_s \in \Re^{Nn_u \times Nn_x}$, and positive scalars $\{\bar{s}, \bar{z}, \delta_l, \varepsilon\}$, such that for all $(l,s) \in \mathscr{L}$, the following LMIs hold:

$$\Sigma_{ll} < 0, l \in \mathscr{L} \tag{4.111}$$

$$\Sigma_{ls} + \Sigma_{sl} < 0, 1 \leq l < s \leq r \tag{4.112}$$

where

$$\Sigma_{ls} = \Phi_{ls} + \sum\nolimits_{f=1}^{r} \delta_f \left(\Phi_{lf} + M_l\right),$$
$$\Phi_{ls} = \begin{bmatrix} [\bar{z}+\bar{s}]^2 \bar{Q} - 2X & (1-\varepsilon)X + A_l X + B_l \bar{K}_s & B_l \bar{K}_s \\ \star & \text{Sym}\{\varepsilon A_l X + \varepsilon B_l \bar{K}_s\} & \varepsilon B_l \bar{K}_s \\ \star & \star & -\frac{\pi^2}{4}\bar{Q} \end{bmatrix}. \tag{4.113}$$

In that case, the controller gains can be calculated by

$$K_l = \bar{K}_l X^{-1}, l \in \mathscr{L}. \tag{4.114}$$

■

Note that when the asynchronized information of μ_l and $\hat{\mu}_l$ is unknown, the design result on a centralized sampled-data linear controller can be directly derived as below:

Theorem 4.8: Design of Centralized Sampled-Data Linear Controller

Consider the multi-photovoltaic system with DC load in (4.5). A centralized sampled-data fuzzy controller in the form (4.88) can guarantee that the resulting closed-loop fuzzy control system with unknown asynchronized information is asymptotically stable, if there exist the symmetric positive definite matrices $\{X, \bar{Q}\} \in \Re^{Nn_{xi} \times Nn_{xi}}$, and matrix $\bar{K} \in \Re^{Nn_{ui} \times Nn_{xi}}$, and positive scalars $\{\bar{s}, \bar{z}, \varepsilon\}$, such that for all $l \in \mathscr{L}$, the following LMIs hold:

$$\begin{bmatrix} [\bar{z}+\bar{s}]^2 \bar{Q} - 2X & (1-\varepsilon)X + A_l X + B_l \bar{K} & B_l \bar{K} \\ \star & \text{Sym}\{\varepsilon A_l X + \varepsilon B_l \bar{K}\} & \varepsilon B_l \bar{K} \\ \star & \star & -\frac{\pi^2}{4}\bar{Q} \end{bmatrix} < 0, \tag{4.115}$$

In that case, the proposed linear controller gains can be calculated by

$$K = \bar{K}X^{-1}. \tag{4.116}$$

■

When considering external disturbances, $\mathscr{H}_\infty$ performance can be guaranteed by the following theorem,

---

Theorem 4.9: $\mathscr{H}_\infty$ Fuzzy Controller Design Using Synchronized Method

Consider the multi-photovoltaic system with DC load in (4.5). A centralized sampled-data fuzzy controller in the form (4.88) can guarantee that the resulting closed-loop fuzzy control system with the assumption (4.109) is asymptotically stable, if there exist the symmetric positive definite matrices $\{X, \bar{Q}\} \in \Re^{Nn_{xi} \times Nn_{xi}}$, and matrices $M_l = M_l^T \in \Re^{3Nn_{xi} \times 3Nn_{xi}}$, $\bar{K}_s \in \Re^{Nn_{ui} \times Nn_{xi}}$, and positive scalars $\{\bar{s}, \bar{z}, \delta_l, \varepsilon\}$, such that for all $(l,s) \in \mathscr{L}$, the following LMIs hold:

$$\Sigma_{ll} < 0, l \in \mathscr{L} \tag{4.117}$$

$$\Sigma_{ls} + \Sigma_{sl} < 0, 1 \leq l < s \leq r \tag{4.118}$$

where

$$\Sigma_{ls} = \Phi_{ls} + \sum_{f=1}^{r} \delta_f \left(\Phi_{lf} + M_l\right),$$

$$\Phi_{ls} = \begin{bmatrix} [\bar{z}+\bar{s}]^2 \bar{Q} - 2X & (1-\varepsilon)X + A_l X + B_l \bar{K}_s & B_l \bar{K}_s & 0 \\ \star & \text{Sym}\{\varepsilon A_l X + \varepsilon B_l \bar{K}_s\} & \varepsilon B_l \bar{K}_s & X^T C^T \\ \star & \star & -\frac{\pi^2}{4}\bar{Q} & 0 \\ \star & \star & \star & -\gamma^2 I \end{bmatrix}. \tag{4.119}$$

In that case, the controller gains can be calculated by

$$K_l = \bar{K}_l X^{-1}, l \in \mathscr{L}. \tag{4.120}$$

■

4.3.5 CENTRALIZED SAMPLED-DATE CONTROL WITH TIME DELAY

This section studies the networked control problem for microgrids using a centralized framework. Each component in the considered microgrid is represented by a T-S model and exchanges its information through a digital channel with time delays.

Before moving on, the assumption is first given as below [8]:

Assumption 4.5. The sampled signals at the instant t_k are executed in a centralized controller inducing a constant time delay

$$\tau \geq 0. \tag{4.121}$$

It follows from (4.83), (4.84) and (4.121) that

$$\tau \leq \rho^{zs}(t) < \bar{z}+\bar{s}+\tau, t \in [z^k, z^{k+1}). \tag{4.122}$$

Here, the following fuzzy controller is proposed,

$$u(t) = K(\hat{\mu})x(z^k), t \in [z^k, z^{k+1}) \tag{4.123}$$

where $K(\hat{\mu}) := \sum_{f=1}^{r} \hat{\mu}_f K_f$.

Further define

$$x(v) = x(z^k) - x(t-\tau). \tag{4.124}$$

Submitting (4.124) into the fuzzy controller (4.123) and combined with the fuzzy system (4.9), the closed-loop fuzzy control system is

$$\dot{x}(t) = A(\mu)x(t) + B(\mu)K(\hat{\mu})x(t-\tau) + B(\mu)K(\hat{\mu})x(v) + \omega(t). \tag{4.125}$$

Now, we introduce the following LKF by utilizing Wirtinger's inequality [15]:

$$V(t) = V_1(t) + V_2(t), t \in [t_k, t_{k+1}) \tag{4.126}$$

with

$$\begin{cases} V_1(t) = x^T(t)Px(t) + \int_{t-\tau}^{t} x^T(\alpha)Qx(\alpha)d\alpha + \tau\int_{-\tau}^{0}\int_{t+\beta}^{t} \dot{x}^T(\alpha)Z\dot{x}(\alpha)d\alpha d\beta, \\ V_2(t) = (\bar{z}+\bar{s}-\tau)^2 \int_{t_k-\tau}^{t} \dot{x}^T(\alpha)W\dot{x}(\alpha)d\alpha \\ \quad - \frac{\pi^2}{4}\int_{t_k-\tau}^{t-\tau} [x(\alpha)-x(t_k-\tau)]^T W [x(\alpha)-x(t_k-\tau)]d\alpha, \end{cases} \tag{4.127}$$

where $\{P, Q, Z, W\} \in \Re^{Nn_x \times Nn_x}$ are symmetric matrices, and $P > 0, Z > 0, W > 0$.

Inspired by [11], we do not require that the matrix Q in (4.127) is necessarily positive definite. To ensure the positive property of $V(t)$, we give the following lemma:

Lemma 4.3: Stability Analysis with Novel Lyapunov-Krasovskii Functional

Consider the Lyapunov–Krasovskii functional (LKF) in (4.126), then $V(t) \geq \varepsilon\|x(t)\|^2$, where $\varepsilon > 0$, $x(t) = \begin{bmatrix} x_1^T(t) & x_2^T(t) & \cdots & x_N^T(t) \end{bmatrix}^T$, if there exist the symmetric positive definite matrices $\{P, Z, W\} \in \Re^{Nn_{xi} \times Nn_{xi}}$, and symmetric matrix $Q \in \Re^{Nn_{xi} \times Nn_{xi}}$, such that the following inequalities hold:

$$\begin{bmatrix} \frac{1}{\tau}P + Z & -Z \\ \star & Q+Z \end{bmatrix} > 0. \tag{4.128}$$

■

Proof. Firstly, by using Jensen's inequality [12],

$$\begin{aligned}
&\tau \int_{-\tau}^{0} \int_{t+\beta}^{t} x^T(\alpha) Z x(\alpha) d\alpha d\beta \\
&\geq \tau \int_{-\tau}^{0} \frac{-1}{\beta} \left[\int_{t+\beta}^{t} \dot{x}^T(\alpha) d\alpha \right] Z \left[\int_{t+\beta}^{t} \dot{x}(\alpha) d\alpha \right] d\beta \\
&= \tau \int_{-\tau}^{0} \frac{-1}{\beta} [x(t) - x(t+\beta)]^T Z [x(t) - x(t+\beta)] d\beta \\
&= \tau \int_{0}^{\tau} \frac{1}{\beta} [x(t) - x(t-\beta)]^T Z [x(t) - x(t-\beta)] d\beta \\
&\geq \int_{0}^{\tau} [x(t) - x(t-\beta)]^T Z [x(t) - x(t-\beta)] d\beta \\
&= \int_{t-\tau}^{t} [x(t) - x(\alpha)]^T Z [x(t) - x(\alpha)] d\alpha.
\end{aligned} \tag{4.129}$$

It follows from (4.127) and (4.129) that

$$\begin{aligned}
V_1(t) =& x^T(t) P x(t) + \int_{t-\tau}^{t} x^T(\alpha) Q x(\alpha) d\alpha + \tau \int_{-\tau}^{0} \int_{t+\beta}^{t} \dot{x}^T(\alpha) Z \dot{x}(\alpha) d\alpha d\beta \\
\geq & \int_{t-\tau}^{t} \begin{bmatrix} x(t) \\ x(\alpha) \end{bmatrix}^T \begin{bmatrix} \frac{1}{\tau} P + Z & -Z \\ \star & Q + Z \end{bmatrix} \begin{bmatrix} x(t) \\ x(\alpha) \end{bmatrix} d\alpha.
\end{aligned} \tag{4.130}$$

For $V_2(t)$ given in (4.127), it has $x(\alpha) - x(t_k - \tau) = 0$ when $\alpha = t_k - \tau$. With the help of Wirtinger's inequality [15], it is easy to see that $V_2(t) \geq 0$. Therefore, there always exists a positive scalar ε such that the inequality $V(t) \geq \varepsilon \|x(t)\|^2$ holds if the inequality in (4.128) holds, thus completing this proof.

Based on the LKF in (4.126), a sufficient condition for the stability of the closed-loop fuzzy control system in (4.125) is given by the following theorem.

Theorem 4.10: $\mathscr{H}_\infty$ Performance Analysis of Centralized Fuzzy Control

Consider the multi-photovoltaic system with DC load in (4.5), and a fuzzy controller in the form of (4.123), the closed-loop fuzzy control system in (4.125) is asymptotically stable with $\mathscr{H}_\infty$ performance index γ, if there exist the symmetric positive definite matrices $\{P, Z, W\} \in \Re^{Nn_{xi} \times Nn_{xi}}$, and symmetric matrix $Q \in \Re^{Nn_{xi} \times Nn_{xi}}$, and positive scalars $\{\bar{z}, \tau, \bar{s}, \varepsilon\}$, such that the following matrix inequalities hold:

$$\begin{bmatrix} \frac{1}{\tau}P+Z & -Z \\ \star & Q+Z \end{bmatrix} > 0, \tag{4.131}$$

$$\begin{bmatrix} \Theta_1 & (1-\varepsilon)P+PA(\mu) & PB(\mu)K(\hat{\mu}) & PB(\mu)K(\hat{\mu}) & P \\ \star & \Theta_2 & \varepsilon PB(\mu)K(\hat{\mu})+Z & \varepsilon PB(\mu)K(\hat{\mu}) & \varepsilon P \\ \star & \star & -Q-Z & 0 & 0 \\ \star & \star & \star & -\frac{\pi^2}{4}W & 0 \\ \star & \star & \star & \star & -\gamma^2 I \end{bmatrix} < 0, \tag{4.132}$$

where $\Theta_1 = [\bar{z}+\bar{s}]^2 W - 2P + \tau^2 Z$, and $\Theta_2 = \mathrm{Sym}(\varepsilon PA(\mu)) + Q - Z$. ■

Proof. By taking the time derivative of $V(t)$, one has

$$\begin{aligned} \dot{V}_1(t) \leq & \, 2x^T(t)P\dot{x}(t) + x^T(t)Qx(t) - x^T(t-\tau)Qx(t-\tau) \\ & + \tau^2\dot{x}^T(t)Z\dot{x}(t) - \tau\int_{t-\tau}^{t}\dot{x}^T(\alpha)Z\dot{x}(\alpha)d\alpha, \end{aligned} \tag{4.133}$$

$$\dot{V}_2(t) \leq (\bar{z}+\bar{s})^2\dot{x}^T(t)W\dot{x}(t) - \frac{\pi^2}{4}x^T(v)Wx(v). \tag{4.134}$$

Based on Jensen's inequality [12],

$$\begin{aligned} -\tau\int_{t-\tau}^{t}\dot{x}^T(\alpha)Z\dot{x}(\alpha)d\alpha & \leq -\left[\int_{t-\tau}^{t}\dot{x}(\alpha)d\alpha\right]^T Z\left[\int_{t-\tau}^{t}\dot{x}(\alpha)d\alpha\right] \\ & = -(x(t)-x(t-\tau))^T Z(x(t)-x(t-\tau)). \end{aligned} \tag{4.135}$$

Consider the following performance index,

$$J = \dot{V}(t) + y^T(t)y(t) - \gamma^2\omega^T(t)\omega(t). \tag{4.136}$$

It is well-known that $J < 0$ implies the closed-loop control system is asymptotically stable with $\mathscr{H}_\infty$ performance.

Define

$$\begin{aligned} \chi(t) &= \begin{bmatrix} x^T(t) & x^T(t-\tau) & x^T(v) & \omega^T(t) \end{bmatrix}^T, \\ \mathbb{A}(\mu,\hat{\mu}) &= \begin{bmatrix} A(\mu) & B(\mu)K(\hat{\mu}) & B(\mu)K(\hat{\mu}) & I \end{bmatrix}. \end{aligned} \tag{4.137}$$

It is follows from (4.133)-(4.137) that

$$\begin{aligned} J &\leq 2x^T(t)P\mathbb{A}(\mu,\hat{\mu})\chi(t) + x^T(t)Qx(t) - x^T(t-\tau)Qx(t-\tau) \\ &+ \tau^2\chi^T(t)\mathbb{A}^T(\mu,\hat{\mu})Z\mathbb{A}(\mu,\hat{\mu})\chi(t) - (x(t)-x(t-\tau))^T Z(x(t)-x(t-\tau)) \\ &+ (\bar{z}+\bar{s})^2\chi^T(t)\mathbb{A}^T(\mu,\hat{\mu})W\mathbb{A}(\mu,\hat{\mu})\chi(t) - \frac{\pi^2}{4}x^T(v)Wx(v) \\ &+ y^T(t)y(t) - \gamma^2\omega^T(t)\omega(t) \\ &= \chi^T(t)\Theta(\mu,\hat{\mu})\chi(t) \end{aligned} \tag{4.138}$$

where

$$\begin{aligned} \Theta(\mu,\hat{\mu}) &= \begin{bmatrix} \Theta_1(\mu) & Z+PB(\mu)K(\hat{\mu}) & PB(\mu)K(\hat{\mu}) & P \\ \star & -Q-Z & 0 & 0 \\ \star & \star & -\frac{\pi^2}{4}W & 0 \\ \star & \star & \star & -\gamma^2 I \end{bmatrix} \\ &+ \tau^2\mathbb{A}^T(\mu,\hat{\mu})Z\mathbb{A}(\mu,\hat{\mu}) + (\bar{z}+\bar{s})^2\mathbb{A}^T(\mu,\hat{\mu})W\mathbb{A}(\mu,\hat{\mu}), \\ \Theta_1(\mu) &= \mathrm{Sym}(PA(\mu)) + Q - Z + C^TC. \end{aligned} \tag{4.139}$$

By using Schur complement lemma to $\Theta(\mu,\hat{\mu}) < 0$, the result on (4.132) can be directly obtained. Thus, the proof is completed.

It is noted that the obtained result on (4.132) is nonlinear. In order to transform it into linear form, it follows from (4.125) that

$$\begin{aligned} 0 &= 2\left[\dot{x}^T(t)P + x^T(t)\varepsilon P\right] \\ &\times \left[-\dot{x}(t) + A(\mu)x(t) + B(\mu)K(\hat{\mu})x(t-\tau) + B(\mu)K(\hat{\mu})x(v) + \omega(t)\right], \end{aligned} \tag{4.140}$$

where ε is a positive scalar.

It follows from (4.133), (4.134), (4.136), and (4.140) that

$$\begin{bmatrix} \Theta_1 & (1-\varepsilon)P + PA(\mu) & PB(\mu)K(\hat{\mu}) & PB(\mu)K(\hat{\mu}) & P \\ \star & \mathrm{Sym}(\varepsilon PA(\mu)) + Q - Z + C^TC & \varepsilon PB(\mu)K(\hat{\mu}) + Z & \varepsilon PB(\mu)K(\hat{\mu}) & \varepsilon P \\ \star & \star & -Q-Z & 0 & 0 \\ \star & \star & \star & -\frac{\pi^2}{4}W & 0 \\ \star & \star & \star & \star & -\gamma^2 I \end{bmatrix} < 0, \tag{4.141}$$

where $\Theta_1 = [\bar{z}+\bar{s}]^2 W - 2P + \tau^2 Z$.

By using the Schur complement lemma, and performing the congruence transformation to (4.141) by $\Gamma = \mathrm{diag}\{X,X,X,X,I,I\}$, where $X = P^{-1}$, and define $\bar{W} =$

$XWX, \bar{Z} = XZX, \bar{Q} = XQX, \bar{K}(\hat{\mu}) = K(\hat{\mu})X,$

$$\begin{bmatrix} \bar{\Theta}_1 & (1-\varepsilon)X + A(\mu)X & B(\mu)\bar{K}(\hat{\mu}) & B(\mu)\bar{K}(\hat{\mu}) & I & 0 \\ \star & \mathrm{Sym}(\varepsilon A(\mu)X) + \bar{Q} - \bar{Z} & \varepsilon B(\mu)\bar{K}(\hat{\mu}) + \bar{Z} & \varepsilon B(\mu)\bar{K}(\hat{\mu}) & \varepsilon I & XC^T \\ \star & \star & -\bar{Q} - \bar{Z} & 0 & 0 & 0 \\ \star & \star & \star & -\frac{\pi^2}{4}\bar{W} & 0 & 0 \\ \star & \star & \star & \star & -\gamma^2 I & 0 \\ \star & \star & \star & \star & \star & -I \end{bmatrix} < 0, \tag{4.142}$$

where $\bar{\Theta}_1 = [\bar{z} + \bar{s}]^2 \bar{W} - 2X + \tau^2 \bar{Z}$.

It is noted that the information of μ_l and $\hat{\mu}_l$ is asynchronized. Here, assume that the asynchronized information is known, and it is subject to [16]

$$\underline{\rho}_l \le \frac{\hat{\mu}_l}{\mu_l} \le \bar{\rho}_l, \tag{4.143}$$

where $\underline{\rho}_l$ and $\bar{\rho}_l$ are positive scalars.

It follows from (4.142) and (4.143) and the asynchronized method proposed in [16] that the design result on the centralized sampled-data fuzzy controller can be summarized as below:

Theorem 4.11: Design of $\mathscr{H}_\infty$ Centralized Fuzzy Controller Using Asynchronized Method

Consider the multi-photovoltaic system with DC load in (4.5). A centralized sampled-data fuzzy controller in the form (4.123) can guarantee that the closed-loop fuzzy control system (4.125) is asymptotically stable with $\mathscr{H}_\infty$ performance γ, if there exist the symmetric positive definite matrices $\{P, \bar{Z}, \bar{W}, X\} \in \Re^{Nn_{xi} \times Nn_{xi}}$, and matrices $M_{ls} = M_{sl}^T \in \Re^{N(4n_{xi}+n_{\omega i}) \times N(4n_{xi}+n_{\omega i})}$, $\bar{Q} = \bar{Q}^T \in \Re^{Nn_{xi} \times Nn_{xi}}$, $\bar{K}_s \in \Re^{Nn_{ui} \times Nn_{xi}}$, and positive scalars $\left\{\bar{s}, \bar{z}, \tau, \bar{\rho}_l, \underline{\rho}_l\right\}$, such that for all $(l, s) \in \mathscr{L}$, the following LMIs hold:

$$\begin{bmatrix} \frac{1}{\tau}X+\bar{Z} & -\bar{Z} \\ \star & \bar{Q}+\bar{Z} \end{bmatrix} > 0, \tag{4.144}$$

$$\bar{\rho}_l \Sigma_{ll} + M_{ll} < 0, \tag{4.145}$$

$$\underline{\rho}_l \Sigma_{ll} + M_{ll} < 0, \tag{4.146}$$

$$\bar{\rho}_s \Sigma_{ls} + \bar{\rho}_l \Sigma_{sl} + M_{ls} + M_{sl} < 0, \tag{4.147}$$

$$\underline{\rho}_s \Sigma_{ls} + \underline{\rho}_l \Sigma_{sl} + M_{ls} + M_{sl} < 0, \tag{4.148}$$

$$\underline{\rho}_s \Sigma_{ls} + \bar{\rho}_l \Sigma_{sl} + M_{ls} + M_{sl} < 0, \tag{4.149}$$

$$\bar{\rho}_s \Sigma_{ls} + \underline{\rho}_l \Sigma_{sl} + M_{ls} + M_{sl} < 0, \tag{4.150}$$

$$\begin{bmatrix} M_{11} & \cdots & M_{1r} \\ \vdots & \ddots & \vdots \\ M_{r1} & \cdots & M_{rr} \end{bmatrix} > 0, \tag{4.151}$$

where

$$\Sigma_{ls} = \begin{bmatrix} \bar{\Theta}_1 & (1-\varepsilon)X + A_l X & B_l \bar{K}_s & B_l \bar{K}_s & I & 0 \\ \star & \mathrm{Sym}(\varepsilon A_l X) + \bar{Q} - \bar{Z} & \varepsilon B_l \bar{K}_s + \bar{Z} & \varepsilon B_l \bar{K}_s & \varepsilon I & XC^T \\ \star & \star & -\bar{Q} - \bar{Z} & 0 & 0 & 0 \\ \star & \star & \star & -\frac{\pi^2}{4}\bar{W} & 0 & 0 \\ \star & \star & \star & \star & -\gamma^2 I & 0 \\ \star & \star & \star & \star & \star & -I \end{bmatrix},$$

$$\bar{\Theta}_1 = [\bar{z}+\bar{s}]^2 \bar{W} - 2X + \tau^2 \bar{Z}. \tag{4.152}$$

In that case, the proposed fuzzy controller gains can be calculated by

$$K_s = \bar{K}_s X^{-1}, s \in \mathscr{L}. \tag{4.153}$$

■

Note that the number of LMIs on Theorem 4.11 is large. It is also noted that the existing relaxation technique $\sum_{l=1}^{r} [\mu_l]^2 \Sigma_{ll} + \sum_{l=1}^{r} \sum_{l<s\leq r}^{r} \mu_l \mu_s \Sigma_{ls} < 0$ is no longer applicable to fuzzy controller synthesis, since $\mu_s \neq \hat{\mu}_s$. Similarly to the asynchronous relaxation technique in [10], it is assumed that

$$|\mu_l - \hat{\mu}_l| \leq \delta_l, l \in \mathscr{L}, \tag{4.154}$$

where δ_l is a positive scalar. If $\Sigma_{ls} + M_l \geq 0$, where M_l is a symmetric matrix, one obtains

$$\begin{aligned}
&\sum_{l=1}^{r}\sum_{s=1}^{r} \mu_l \hat{\mu}_s \Sigma_{ls} \\
&= \sum_{l=1}^{r}\sum_{s=1}^{r} \mu_l \mu_s \Sigma_{ls} + \sum_{l=1}^{r}\sum_{s=1}^{r} \mu_l \left(\hat{\mu}_s - \mu_s\right)\left(\Sigma_{ls} + M_l\right) \\
&\leq \sum_{l=1}^{r}\sum_{s=1}^{r} \mu_l \mu_s \left[\Sigma_{ls} + \sum_{f=1}^{r} \delta_f \left(\Sigma_{lf} + M_l\right)\right].
\end{aligned} \tag{4.155}$$

Therefore, upon defining $\Sigma_{ls} = \Phi_{ls} + \sum_{s=1}^{r} \delta_s \left(\Phi_{ls} + M_l\right)$, the existing relaxation technique from [7] can be applied to (4.155).

Based on the assumption in (4.154), and the result of (4.152), (4.155), a relaxation algorithm to design asynchronous fuzzy controller is proposed as below:

Theorem 4.12: Design of $\mathscr{H}_\infty$ Centralized Fuzzy Controller Using Synchronized Method

Consider the multi-photovoltaic system with DC load in (4.5). A centralized T-S fuzzy controller in (4.123) can be used to stabilize its closed-loop control system with $\mathscr{H}_\infty$ performance γ, if there exist the symmetric positive definite matrices $\{P, \bar{Z}, \bar{W}, X\} \in \mathfrak{R}^{Nn_{xi} \times Nn_{xi}}$, and matrices $M_l = M_l^T \in \mathfrak{R}^{N(4n_{xi}+n_{\omega i}) \times N(4n_{xi}+n_{\omega i})}$, $\bar{Q} = \bar{Q}^T \in \mathfrak{R}^{Nn_{xi} \times Nn_{xi}}$, $\bar{K}_s \in \mathfrak{R}^{Nn_{ui} \times Nn_{xi}}$, and positive scalars $\{\bar{s}, \bar{z}, \tau, \delta_l\}$, such that for all $(l, s) \in \mathscr{L}$, the following LMIs hold:

$$\begin{bmatrix} \frac{1}{\tau}X + \bar{Z} & -\bar{Z} \\ \star & \bar{Q} + \bar{Z} \end{bmatrix} > 0, \tag{4.156}$$

$$\Sigma_{ll} < 0, l \in \mathscr{L} \tag{4.157}$$

$$\Sigma_{ls} + \Sigma_{sl} < 0, 1 \leq l < s \leq r \tag{4.158}$$

where

$$\Sigma_{ls} = \Phi_{ls} + \sum_{s=1}^{r} \delta_s \left(\Phi_{ls} + M_l\right),$$

$$\Phi_{ls} = \begin{bmatrix} \bar{\Theta}_1 & (1-\varepsilon)X + A_l X & B_l \bar{K}_s & B_l \bar{K}_s & I & 0 \\ \star & \mathrm{Sym}(\varepsilon A_l X) + \bar{Q} - \bar{Z} & \varepsilon B_l \bar{K}_s + \bar{Z} & \varepsilon B_l \bar{K}_s & \varepsilon I & XC^T \\ \star & \star & -\bar{Q} - \bar{Z} & 0 & 0 & 0 \\ \star & \star & \star & -\frac{\pi^2}{4}\bar{W} & 0 & 0 \\ \star & \star & \star & \star & -\gamma^2 I & 0 \\ \star & \star & \star & \star & \star & -I \end{bmatrix}.$$

$$\bar{\Theta}_1 = [\bar{z} + \bar{s}]^2 \bar{W} - 2X + \tau^2 \bar{Z}. \tag{4.159}$$

In that case, the controller gains can be calculated by

$$K_l = \bar{K}_l X^{-1}, l \in \mathscr{L}. \tag{4.160}$$

■

Note that when the asynchronized information of μ_l and $\hat{\mu}_l$ is unknown, the design result on a centralized sampled-data linear controller can be directly derived as below:

Theorem 4.13: Design of $\mathscr{H}_\infty$ Centralized Sampled-Data Linear Controller

Consider the multi-photovoltaic system with DC load in (4.5). A centralized T-S fuzzy controller in (4.123) can be used to stabilize its closed-loop control system with $\mathscr{H}_\infty$ performance γ, if there exist the symmetric positive definite matrices $\{P, \bar{Z}, \bar{W}, X\} \in \Re^{Nn_{xi} \times Nn_{xi}}$, and matrices $\bar{Q} = \bar{Q}^T \in \Re^{Nn_{xi} \times Nn_{xi}}$, $\bar{K} \in \Re^{Nn_{ui} \times Nn_{xi}}$, and positive scalars $\{\bar{s}, \bar{z}, \tau, \varepsilon\}$, such that for all $l \in \mathscr{L}$, the following LMIs hold:

$$\begin{bmatrix} \frac{1}{\tau}X + \bar{Z} & -\bar{Z} \\ \star & \bar{Q} + \bar{Z} \end{bmatrix} > 0, \quad (4.161)$$

$$\begin{bmatrix} \bar{\Theta}_1 & (1-\varepsilon)X + A_l X & B_l \bar{K} & B_l \bar{K} & I & 0 \\ \star & \mathrm{Sym}(\varepsilon A_l X) + \bar{Q} - \bar{Z} & \varepsilon B_l \bar{K} + \bar{Z} & \varepsilon B_l \bar{K} & \varepsilon I & XC^T \\ \star & \star & -\bar{Q} - \bar{Z} & 0 & 0 & 0 \\ \star & \star & \star & -\frac{\pi^2}{4}\bar{W} & 0 & 0 \\ \star & \star & \star & \star & -\gamma^2 I & 0 \\ \star & \star & \star & \star & \star & -I \end{bmatrix} < 0, \quad (4.162)$$

where $\bar{\Theta}_1 = [\bar{z} + \bar{s}]^2 \bar{W} - 2X + \tau^2 \bar{Z}$.

In that case, the proposed linear controller gains can be calculated by

$$K = \bar{K} X^{-1}. \quad (4.163)$$

■

4.4 SIMULATION STUDIES

Consider the multi-PV power with DC load in (4.9) using the centralized sampled-data fuzzy controller (4.51). Here, the parameters of the multi-PV power are chosen as below: $L_{(1)} = 0.0516\text{H}, L_{(2)} = 0.0513\text{H}, C_{0(1)} = 0.0472\text{F}, C_{0(2)} = 0.0471\text{F}, C_{PV(1)} = 0.0101\text{F}, C_{PV(2)} = 0.0102\text{F}, R_{0(1)} = 1.1\Omega, R_{0(2)} = 1.2\Omega, R_{L(1)} = 1.8\Omega, R_{L(2)} = 1.3\Omega, R_{M(1)} = 0.85\Omega, R_{M(2)} = 0.86\Omega, R_{Load} = 2.5\Omega, V_{D(1)} = 9.1\text{V}, V_{D(2)} = 9.1\text{V}$.Now, we choose $z_{i1}(t) = \frac{\phi_{PV(i)}}{v_{PV(i)}}, z_{i2}(t) = \phi_{L(i)}, z_{i3}(t) = v_{PV(i)}$, and $z_{i4}(t) = \frac{\phi_{0(i)}}{\phi_{L(i)}}$ as fuzzy premise variables, and linearize the first PV system around $\{0.037895, 0.25, 9.5, 4\}$ and $\{0.029268, 0.25, 12.3, 4\}$, and linearize the second PV system around $\{0.037895, 0.25, 9.5, 4\}$ and $\{0.0288, 0.25, 12.3, 4\}$. Then, the succeeding system matrices of T-S fuzzy model can be obtained as below:

$$A_1 = \begin{bmatrix} 3.752 & 0 & 0 & 0 & 0 & 0 \\ 0 & -56.202 & -10.853 & 0 & -85.271 & 0 \\ 0 & 21.186 & -8.4746 & 0 & -84.746 & 0 \\ 0 & 0 & 0 & 3.7152 & 0 & 0 \\ 0 & -93.567 & 0 & 0 & -48.733 & -10.136 \\ 0 & -84.926 & 0 & 0 & 21.231 & -8.4926 \end{bmatrix},$$

$$A_2 = \begin{bmatrix} 2.898 & 0 & 0 & 0 & 0 & 0 \\ 0 & -56.202 & -10.853 & 0 & -85.271 & 0 \\ 0 & 21.186 & -8.4746 & 0 & -84.746 & 0 \\ 0 & 0 & 0 & 2.8235 & 0 & 0 \\ 0 & -93.567 & 0 & 0 & -48.733 & -10.136 \\ 0 & -84.926 & 0 & 0 & 21.231 & -8.4926 \end{bmatrix},$$

$$B_1 = \begin{bmatrix} -24.752 & 0 \\ 356.35 & 0 \\ 0 & 0 \\ 0 & -24.51 \\ 0 & 358.38 \\ 0 & 0 \end{bmatrix}, B_2 = \begin{bmatrix} -24.752 & 0 \\ 410.61 & 0 \\ 0 & 0 \\ 0 & -24.51 \\ 0 & 416.86 \\ 0 & 0 \end{bmatrix}.$$

Here, by applying Theorem 4.2, the fuzzy controller gains are given by

$$K_1 = \begin{bmatrix} 0.001434 & 0.098330 & -0.010608 & 0.002173 & 0.134170 & 0.052377 \\ 0.002243 & 0.147090 & 0.054282 & 0.001591 & 0.088660 & 0.000519 \end{bmatrix},$$

$$K_2 = \begin{bmatrix} 0.001434 & 0.094640 & -0.010608 & 0.002173 & 0.134170 & 0.052377 \\ 0.002243 & 0.147090 & 0.054282 & 0.001591 & 0.085240 & 0.000519 \end{bmatrix}.$$

4.5 REFERENCES

1. Dragicevic, T., Lu, X., Vasquez, J. C., and Guerrero, J. M. (2016). DC microgrids. Part I: A review of control strategies and stabilization techniques. IEEE Transactions on Power Electronics, 31(7), 4876-4891.
2. Zhong, Z.X., Lin, C. M., Shao, Z. H., and Xu, M. (2018). Decentralized event-triggered control for large-scale networked fuzzy systems. IEEE Transactions on Fuzzy Systems, 26(1): 29-50.
3. Lin, W. and Bitar, E. (2018). Decentralized stochastic control of distributed energy resources. IEEE Transactions on Power Systems, 33(1): 888-900.
4. Chiu C. and Ouyang, Y. (2011). Robust maximum power tracking control of uncertain photovoltaic systems: A unified T-S fuzzy model-based approach. IEEE Transactions on Control Systems and Technology, 19(6): 1516-1526.
5. Sangwongwanich, A., Abdelhakim, A., Yang, Y., and Zhou, K. (2018). Control of Power Electronic Converters and Systems, 153-173.
6. Bidram, A., Davoudi, A., Lewis, F. L., and Nasirian, V. (2017). Cooperative synchronization in distributed microgrid control. Advances in Industrial Control. Berlin: Springer-Verlag.

7. Zhong, Z.X. and Lin C.M. (2017). Large-Scale Fuzzy Interconnected Control Systems Design and Analysis, Pennsylvania: IGI Global.
8. Moarref, M. and Rodrigues, L. (2014). Stability and stabilization of linear sampled-data systems with multi-rate samplers and time driven zero order holds. Automatica, 50(10), 2685-2691.
9. Lam, H. K. and Narimani, M. (2009). Stability analysis and performance design for fuzzy-model-based control system under imperfect premise matching. IEEE Transactions on Fuzzy Systems, 17(4), 949-961.
10. Zhang, D., Han, Q. L., and Jia, X. (2017). Network-based output tracking control for a class of T-S fuzzy systems that can not be stabilized by nondelayed output feedback controllers. IEEE Transactions on Cybernetics, 45(8), 1511-1524.
11. Xu, S., Lam, J., Zhang, B., and Zou, Y. (2015). New insight into delay-dependent stability of time-delay systems. International Journal of Robust and Nonlinear Control, 25(7), 961-970.
12. Gu, K. (2002). An integral inequality in the stability problem of time-delay systems. Proceedings of 2000 IEEE Conference on Decision and Control, Vol 3, pp. 2805-2810.
13. Heemels, W. P. M. H. and Donkers, M. C. F. (2013). Model-based periodic event-triggered control for linear systems. Automatica, 49(3), 698-711.
14. Fridman, E. (2010). A refined input delay approach to sampled-data control . Automatica, 46(2), 421-427.
15. Liu K. and Fridman E. (2012). Wirtinger's inequality and Lyapunov-based sampled-data stabilization. Automatica, 48(1), 102-108.
16. Arino, C. and Sala, A. (2008). Extensions to stability analysis of fuzzy control systems subject to uncertain grades of membership. IEEE Transactions on Systems, Man and Cyberetics, Part B, 38(2), 558-563.

5 Decentralized Fuzzy Control

Nowadays, the smart grid is expected to operate under an updated philosophy with a significant increase in the level of monitoring, communication, and control and coordination. The use of renewable resources in the smart grid is distinct by its distributed nature as opposed to the large centralized power plants in the current grid. The centralized scheme will not be able to operate under the significantly increasing computational burdens. Therefore noncentralized techniques are better suited to provide the required functionality. Noncentralized techniques, in turn, can be either decentralized or distributed [1, 2]. It can be concluded that decentralized technique relies only on local information and does not need any components other than interface converters. Therefore, it is easy to implement decentralized control methods in decentralized systems [3].

In this chapter, we consider the problems of tracking voltage synchronization in multiple pV and multiple wind generator systems. All generator systems must act as one group to achieve synchronization through local information exchanges. Thus we examine sampled-data measurement and time-triggered ZOH and provide a numerical example.

5.1 MODELING OF MULTI-PV GENERATORS

5.1.1 MODELING OF MULTI-PV POWER WITH DC LOAD

Recall a multi-photovoltaic system with DC load in (4.3) as below:

$$\begin{cases} \dot{v}_{PV(i)} = \frac{1}{C_{PV(i)}}\phi_{PV(i)} - \frac{\phi_{L(i)}}{C_{PV(i)}}u_{(i)}, \\ \dot{\phi}_{L(i)} = -\left(\frac{R_{0(i)}}{L_{(i)}} + \frac{R_{L(i)}}{L_{(i)}}\right)\phi_{L(i)} + \left(\frac{R_{0(i)}}{L_{(i)}R_{load}} - \frac{1}{L_{(i)}}\right)v_{0(i)} - \sum_{j=1,j\neq i}^{N}\frac{R_{0(i)}}{L_{(i)}}\phi_{0(j)} \\ \quad + \frac{1}{L_{(i)}}\left(V_{D(i)} + v_{PV(i)} - R_{M(i)}\phi_{L(i)}\right)u_{(i)} - \frac{V_{D(i)}}{L_{(i)}}, \\ \dot{v}_{0(i)} = \frac{1}{C_{0(i)}}\phi_{L(i)} - \frac{1}{R_{load}C_{0(i)}}v_{0(i)} + \sum_{j=1,j\neq i}^{N}\frac{1}{C_{0(i)}}\phi_{0(j)}. \end{cases} \tag{5.1}$$

Define $e_{0(i)} = v_{0(i)} - v_0^*$, and $x_i(t) = \left[\begin{array}{ccc} v_{PV(i)} & \phi_{L(i)} & e_{0(i)} \end{array}\right]^T$, and choose $z_{i1}(t) = \frac{\phi_{PV(i)}}{v_{PV(i)}}, z_{i2}(t) = \phi_{L(i)}, z_{i3}(t) = v_{PV(i)}$, and $z_{i4}(t) = \frac{\phi_{0(i)}}{\phi_{L(i)}}$ as fuzzy premise variables. Thus, it follows from (5.1) that the multi-photovoltaic system with DC load is represented by

Plant Rule $\mathscr{R}_i^l$: **IF** $z_{i1}(t)$ is $\mathscr{F}_{i1}^l$, $z_{i2}(t)$ is $\mathscr{F}_{i2}^l$, $z_{i3}(t)$ is $\mathscr{F}_{i3}^l$, $z_{i4}(t)$ is $\mathscr{F}_{i4}^l$, **THEN**

$$\dot{x}_i(t) = A_{iil}x_i(t) + B_{il}u_i(t) + \sum_{\substack{j=1 \\ j\neq i}}^{N} A_{ijl}x_j(t) + \omega_i(t), l \in \mathscr{L}_i := \{1, 2, \ldots, r_i\} \tag{5.2}$$

where $\mathscr{R}_i^l$ denotes the l-th fuzzy inference rule; r_i is the number of inference rules; $z_i(t) \triangleq [z_{i1}, z_{i2}, z_{i3}, z_{i4}]$ are the measurable variables; $\{A_{iil}, B_{il}, A_{ijl}\}$ is the l-th local model.

Define $\mu_{il}[z_i(t)]$ as the normalized membership function of the inferred fuzzy set $\mathscr{F}_i^l := \prod_{\phi=1}^g \mathscr{F}_{i\phi}^l$ and

$$\mu_{il}[z_i(t)] := \frac{\prod_{\phi=1}^g \mu_{il\phi}[z_{i\phi}(t)]}{\sum_{\varsigma=1}^{r_i} \prod_{\phi=1}^g \mu_{i\varsigma\phi}[z_{i\phi}(t)]} \geq 0, \sum_{l=1}^{r_i} \mu_{il}[z_i(t)] = 1. \tag{5.3}$$

In the following, we will denote $\mu_{il} := \mu_{il}[z_i(t)]$ for brevity.

By fuzzy blending, the global T-S fuzzy dynamic model can be obtained as follows:

$$\dot{x}_i(t) = A_{ii}(\mu_i)x_i(t) + B_i(\mu_i)u_i(t) + \sum_{\substack{j=1 \\ j\neq i}}^{N} A_{ij}(\mu_i)x_j(t) + \omega_i(t), \tag{5.4}$$

where $A_{ii}(\mu_i) := \sum_{l=1}^{r_i} \mu_{il}A_{iil}, B_i(\mu_i) := \sum_{l=1}^{r_i} \mu_{il}B_{il}, A_{ij}(\mu_i) := \sum_{l=1}^{r_i} \mu_{il}A_{ijl}$, and

$$A_{iil} = \begin{bmatrix} \frac{1}{C_{PV(i)}}F_{i1}^l & 0 & 0 \\ 0 & -\frac{(R_{0(i)}+R_{L(i)})}{L_{(i)}} - \frac{V_{D(i)}}{L_{(i)}F_{i2}^l} & \frac{R_{0(i)}-R_{line(i)}-R_{load}}{L_{(i)}} \\ 0 & \frac{1}{C_{0(i)}(R_{line(i)}+R_{load})} & -\frac{1}{C_{0(i)}(R_{line(i)}+R_{load})} \end{bmatrix},$$

$$B_{il} = \begin{bmatrix} -\frac{F_{i2}^l}{C_{PV(i)}} \\ \frac{1}{L_{(i)}}\left(V_{D(i)} + F_{i3}^l - R_{M(i)}F_{i2}^l\right) \\ 0 \end{bmatrix}, A_{ijl} = \begin{bmatrix} 0 & 0 & 0 \\ 0 & 0 & -\frac{R_{load}}{L_{(i)}} \\ 0 & 0 & -F_{i4}^l\frac{R_{load}}{(R_{line(i)}+R_{load})} \end{bmatrix}. \tag{5.5}$$

Note: In Theorem 4.1 the centralized fuzzy controller is designed for multi-photovoltaic systems with DC loads as shown in (4.108). It is noted that the considered system is represented by an augmented form in high-dimension cases and the criteria in Theorem 4.1 are not suitable for computation since the system decomposition or its eigenvalues are needed, and are not easy to obtain.

5.1.2 MODELING OF MULTI-PV GENERATORS WITH AC LOAD

Recall a multi-photovoltaic system with AC load in (4.16) as below:

$$\begin{cases} \dot{v}_{PV(i)} = \frac{1}{C_{pv(i)}}\phi_{PV(i)} - \left(1.5R_{line(i)} + R_{load}\right)\frac{\phi_{d(i)}}{v_{pv(i)}}\left(e_{d(i)} + \phi_d^*\right) \\ \quad - \sum_{j=1, j\neq i}^{N} \frac{R_{load}\phi_{d(i)}}{v_{pv(i)}}\left(e_{d(j)} + \phi_d^*\right), \\ \dot{e}_{d(i)} = -\frac{R_{1(i)}}{L_{1(i)}}\left(e_{d(i)} + \phi_d^*\right) - \omega_{(i)}\left(e_{q(i)} + \phi_q^*\right) + \frac{1}{L_{1(i)}}e_{d(i)} - \dot{\phi}_d^*, \\ \dot{e}_{q(i)} = \omega_{(i)}\left(e_{d(i)} + \phi_d^*\right) - \frac{R_{1(i)}}{L_{1(i)}}\left(e_{q(i)} + \phi_q^*\right) + \frac{1}{L_{1(i)}}e_{q(i)} - \dot{\phi}_q^*. \end{cases} \tag{5.6}$$

Define $x_i(t) = \begin{bmatrix} v_{PV(i)} & e_{d(i)} & e_{q(i)} \end{bmatrix}^T$, and

$$A_{ii}(t) = \begin{bmatrix} \frac{1}{C_{PV(i)}}\frac{\phi_{PV(i)}}{v_{PV(i)}} & -\left(1.5R_{line(i)} + R_{load}\right)\frac{\phi_{d(i)}}{v_{pv(i)}} & 0 \\ 0 & -\frac{R_{1(i)}}{L_{1(i)}} & -\omega_{(i)} \\ 0 & \omega_{(i)} & -\frac{R_{1(i)}}{L_{1(i)}} \end{bmatrix},$$

$$B_i(t) = \begin{bmatrix} 0 & 0 \\ \frac{1}{L_{1(i)}} & 0 \\ 0 & \frac{1}{L_{1(i)}} \end{bmatrix}, u_i(t) = \begin{bmatrix} e_{d(i)} \\ e_{q(i)} \end{bmatrix}, A_{ij}(t) = \begin{bmatrix} 0 & -\frac{R_{load}\phi_{d(i)}}{v_{pv(i)}} & 0 \\ 0 & 0 & 0 \\ 0 & 0 & 0 \end{bmatrix},$$

$$\omega_i(t) = \begin{bmatrix} -\left(1.5R_{line(i)} + R_{load}\right)\frac{\phi_{d(i)}}{v_{pv(i)}}\phi_d^* - (N-1)\frac{R_{load}\phi_{d(i)}}{v_{pv(i)}}\phi_d^* \\ -\frac{R_{1(i)}}{L_{1(i)}}\phi_d^* - \omega_{(i)}\phi_q^* - \dot{\phi}_d^* \\ \omega_{(i)}\phi_d^* - \frac{R_{1(i)}}{L_{1(i)}}\phi_q^* - \dot{\phi}_q^* \end{bmatrix}. \quad (5.7)$$

Choose $z_{i1}(t) = \frac{\phi_{PV(i)}}{v_{PV(i)}}$, $z_{i2}(t) = \frac{\phi_{d(i)}}{v_{pv(i)}}$, and $z_{i3}(t) = \omega_i$ as fuzzy premise variables. Thus, it follows from (5.6) that the nonlinear PV power system is represented by

Plant Rule $\mathscr{R}_i^l$: **IF** $z_{i1}(t)$ is $\mathscr{F}_{i1}^l$, $z_{i2}(t)$ is $\mathscr{F}_{i2}^l$, $z_{i3}(t)$ is $\mathscr{F}_{i3}^l$, **THEN**

$$\dot{x}_i(t) = A_{iil}x_i(t) + B_{il}u_i(t) + \sum_{\substack{j=1 \\ j\neq i}}^{N} A_{ijl}x_j(t) + \omega_i(t), l \in \mathscr{L}_i := \{1, 2, \ldots, r\} \quad (5.8)$$

where $\mathscr{R}_i^l$ denotes the l-th fuzzy inference rule; r is the number of inference rules; $z(t) \triangleq [z_{i1}, z_{i2}, z_{i3}]$ are the measurable variables; $\{A_{iil}, B_{il}\}$ is the l-th local model.

By fuzzy blending, the global T-S fuzzy dynamic model can be obtained as follows:

$$\dot{x}_i(t) = A_{ii}(\mu_i)x_i(t) + B_i(\mu_i)u_i(t) + \sum_{j=1 j\neq i}^{N} A_{ij}(\mu_i)x_j(t) + \omega_i(t), \quad (5.9)$$

where $A_{ii}(\mu_i) := \sum_{l=1}^{r_i} \mu_{il}A_{iil}, B_i(\mu_i) := \sum_{l=1}^{r_i} \mu_{il}B_{il}, A_{ij}(\mu_i) := \sum_{l=1}^{r_i} \mu_{il}A_{ijl}$.

5.2 MODELING OF MULTI-MACHINE WIND GENERATOR

5.2.1 MODELING OF MULTI-MACHINE WIND WITH DC LOAD

Recall the multi-machine wind generator with DC load in (4.27) as below:

$$\begin{cases} \dot{P}_{e(i)} = \omega_{(i)}v_{\alpha\beta(i)}^T J_{(i)}^T \phi_{\alpha\beta(i)} + \frac{1}{L_{(i)}}v_{\alpha\beta(i)}^T v_{\alpha\beta(i)} - v_{\alpha\beta(i)}^T \frac{v_{dc(i)}}{2L_{(i)}}u_{\alpha\beta(i)}, \\ \dot{Q}_{e(i)} = \omega_{(i)}v_{\alpha\beta(i)}^T J_{(i)}^T J_{(i)}\phi_{\alpha\beta(i)} + \frac{v_{\alpha\beta(i)}^T J_{(i)}}{L_{(i)}}v_{\alpha\beta(i)} - \frac{v_{\alpha\beta(i)}^T J_{(i)}v_{dc(i)}}{2L_{(i)}}u_{\alpha\beta(i)}, \\ \dot{v}_{\alpha\beta(i)} = \omega_{(i)}J_{(i)}v_{\alpha\beta(i)}, \\ \dot{\phi}_{\alpha\beta(i)} = \frac{1}{L_{(i)}}v_{\alpha\beta(i)} - \frac{v_{dc(i)}}{2L_{(i)}}u_{\alpha\beta(i)}, \\ \dot{v}_{e(i)} = -\frac{v_{e(i)}+v_{dc}^*}{C_{(i)}R_L} + \sum_{j=1,j\neq i}^{N}\frac{1}{C_{(i)}}\frac{\phi_{0(j)}}{v_{e(j)}}v_{e(j)} + \frac{\phi_{\alpha\beta(i)}^T}{2C_{(i)}}u_{\alpha\beta(i)}. \end{cases} \quad (5.10)$$

Define $z_{i1} = v_{\alpha\beta(i)}^T, z_{i2} = \frac{\omega_{(i)}\left\|J_{(i)}v_{\alpha\beta(i)}\right\|^2}{\left\|v_{\alpha\beta(i)}\right\|^2}$, $z_{i3} = v_{\alpha\beta(i)}^T v_{dc(i)}, z_{i4} = v_{\alpha\beta(i)}^T J_{(i)}^T v_{dc(i)}$, $z_{i5} = \omega_{(i)}$, $z_{i6} = v_{dc(i)}, z_{i7} = \frac{\phi_{0(i)}}{v_{e(i)}}, z_{i8} = \phi_{\alpha\beta(i)}^T$, and it follows from (5.10) that

$$\dot{x}_i(t) = A_{ii}(t)x_i(t) + \sum_{j=1, j\neq i}^{N} A_{ij}(t)x_j(t) + B_i(t)u_i(t) + \omega_i(t), \tag{5.11}$$

where $x_i(t) = \begin{bmatrix} P_{e(i)} & Q_{e(i)} & v_{\alpha\beta(i)}^T & \phi_{\alpha\beta(i)}^T & v_{e(i)} \end{bmatrix}^T$, and

$$A_{ii}(t) = \begin{bmatrix} 0 & L_{(i)}z_{i5} & \frac{z_{i1}}{L_{(i)}} & 0 & 0 \\ 0 & -L_{(i)}z_{i2} & 0 & 0 & 0 \\ 0 & 0 & L_{(i)}z_{i5} & 0 & 0 \\ 0 & 0 & \frac{1}{L_{(i)}} & 0 & 0 \\ 0 & 0 & 0 & 0 & -\frac{1}{C_{(i)}R_{L(i)}} \end{bmatrix}, A_{ij}(t) = \begin{bmatrix} 0 & 0 & 0 & 0 & 0 \\ 0 & 0 & 0 & 0 & 0 \\ 0 & 0 & 0 & 0 & 0 \\ 0 & 0 & 0 & 0 & 0 \\ 0 & 0 & 0 & 0 & \frac{z_{j7}}{C_{(i)}} \end{bmatrix},$$

$$B_i(t) = \begin{bmatrix} -\frac{z_{i3}}{2L_{(i)}} \\ -\frac{z_{i4}}{2L_{(i)}} \\ 0 \\ -\frac{z_{i6}}{2L_{(i)}} \\ \frac{z_{i8}}{2C_{(i)}} \end{bmatrix}, \omega_i(t) = \begin{bmatrix} -\dot{P}^* + v_{\alpha\beta(i)}^T \omega_{(i)} L_{(i)} \frac{v_{\alpha\beta(i)}}{\left\|v_{\alpha\beta(i)}\right\|^2} Q^* \\ -\dot{Q}^* - \omega_{(i)} L_{(i)} \frac{\left\|J_{(i)}v_{\alpha\beta(i)}\right\|^2}{\left\|v_{\alpha\beta(i)}\right\|^2} Q^* \\ 0 \\ 0 \\ -\frac{1}{C_{(i)}R_{L(i)}} v_{dc}^* \end{bmatrix}. \tag{5.12}$$

Here, one can choose $z_{i1} - z_{i8}$ as fuzzy premise variables. Then, the nonlinear system in (5.11) is transformed into the following T-S fuzzy system,

$$\dot{x}_i(t) = A_{ii}(\mu_i)x_i(t) + B_i(\mu_i)u_i(t) + \sum_{j=1 j\neq i}^{N} A_{ij}(\mu_i)x_j(t) + \omega_i(t), \tag{5.13}$$

where $A_{ii}(\mu_i) := \sum_{l=1}^{r_i} \mu_{il}A_{iil}, B_i(\mu_i) := \sum_{l=1}^{r_i} \mu_{il}B_{il}, A_{ij}(\mu_i) := \sum_{l=1}^{r_i} \mu_{il}A_{ijl}$.

5.2.2 MODELING OF MULTI-MACHINE WIND GENERATOR WITH AC LOAD

Recall the multi-machine wind generator with AC load in (4.34) as below:

$$\begin{cases} \dot{\phi}_{L(i)} = -\frac{1}{L}v_{C2(i)} - \frac{z_{i1}(t)R_L}{L_{(i)}} \sum_{j=1}^{N} z_{j2}(t)\phi_{L(j)} + \frac{1}{L_{(i)}} z_{i2}(t)u_i, \\ \dot{v}_{C2(i)} = \frac{1}{C_{2(i)}}\phi_{L(i)} - \frac{C_{1(i)}z_{4i}(t)}{C_{2(i)}}v_{C2(i)}, \\ \dot{v}_{O(i)} = \frac{z_{i1}(t)}{C_{O(i)}}\phi_{L(i)} - \frac{C_{1(i)}z_{4i}(t)}{C_{O(i)}}v_{C2(i)} - \frac{v_{O(i)}}{R_L C_{O(i)}}, \end{cases} \tag{5.14}$$

where $z_{i1}(t) = clk_{(i)}, z_{i2}(t) = v_{i(i)}clk_{(i)}, z_{i3}(t) = \frac{\phi_{O(i)}}{\phi_{L(i)}}$, and $z_{4i}(t) = \frac{\dot{v}_{C1(i)}}{v_{C2(i)}}$.

Now, further define $e_{0(i)} = v_{0(i)} - v_{ref}$, where v_{ref} is reference voltage, and

$$x_i(t) = \begin{bmatrix} \phi_{L(i)} & v_{C2(i)} & e_{0(i)} \end{bmatrix}^T, A_{ii}(t) = \begin{bmatrix} 0 & -\frac{1}{L} & 0 \\ \frac{1}{C_{2(i)}} & 0 & -\frac{C_{1(i)}z_{4i}(t)}{C_{2(i)}} \\ \frac{z_{i1}(t)}{C_{0(i)}} & -\frac{C_{1(i)}z_{4i}(t)}{C_{0(i)}} & -\frac{1}{R_L C_{0(i)}} \end{bmatrix},$$

$$A_{ij}(t) = \begin{bmatrix} 0 & \frac{z_{i1}(t)R_L}{L_{(i)}} z_{j2}(t) & 0 \\ 0 & 0 & 0 \\ 0 & 0 & 0 \end{bmatrix}, B_i(t) = \begin{bmatrix} \frac{z_{i2}(t)}{L_{(i)}} \\ 0 \\ 0 \end{bmatrix}, \omega_i(t) = \begin{bmatrix} 0 \\ -\frac{C_{1(i)}\dot{v}_{C1(i)}v_{ref}}{C_{2(i)}v_{C2(i)}} \\ -\frac{v_{ref}}{R_L C_{0(i)}} \end{bmatrix}. \tag{5.15}$$

Then, the nonlinear system in (5.14) is transformed into the state-space representation as below:

$$\dot{x}_i(t) = A_{ii}(t)x_i(t) + \sum_{j=1, j\neq i}^{N} A_{ij}(t)x_j(t) + B_i(t)u_i(t) + \omega_i(t). \tag{5.16}$$

Choose $z_{i1}(t) - z_{4i}(t)$ as fuzzy premise variables. Thus, the nonlinear wind power system in (5.14) can be represented by the following fuzzy system,

Plant Rule $\mathscr{R}_i^l$: **IF** $z_{11}(t)$ is $\mathscr{F}_{11}^l$ and $\cdots$ and $z_{N1}(t)$ is $\mathscr{F}_{N1}^l$, $z_{12}(t)$ is $\mathscr{F}_{12}^l$ and $\cdots$ and $z_{N2}(t)$ is $\mathscr{F}_{N2}^l$, $\cdots$, $z_{14}(t)$ is $\mathscr{F}_{14}^l$ and $\cdots$ and $z_{N4}(t)$ is $\mathscr{F}_{N4}^l$, **THEN**

$$\dot{x}_i(t) = A_{iil}x_i(t) + B_{il}u_i(t) + \sum_{\substack{j=1 \\ j\neq i}}^{N} A_{ijl}x_j(t) + \omega_i(t), l \in \mathscr{L}_i := \{1, 2, \ldots, r\} \tag{5.17}$$

where $\mathscr{R}_i^l$ denotes the l-th fuzzy inference rule; r is the number of inference rules; $z(t) \triangleq [z_{11}, \cdots, z_{N1}, z_{12}, \cdots, z_{N2}, z_{13}, \cdots, z_{N3}, z_{14}, \cdots, z_{N4}]$ are the measurable variables; $\{A_l, B_l\}$ is the l-th local model.

By fuzzy blending, the global T-S fuzzy dynamic model can be obtained as follows:

$$\dot{x}_i(t) = A_{ii}(\mu_i)x_i(t) + B_i(\mu_i)u_i(t) + \sum_{j=1 j\neq i}^{N} A_{ij}(\mu_i)x_j(t) + \omega_i(t), \tag{5.18}$$

where $A_{ii}(\mu_i) := \sum_{l=1}^{r_i} \mu_{il}A_{iil}, B_i(\mu_i) := \sum_{l=1}^{r_i} \mu_{il}B_{il}, A_{ij}(\mu_i) := \sum_{l=1}^{r_i} \mu_{il}A_{ijl}$.

5.3 DECENTRALIZED CONTROL OF TRACKING SYNCHRONIZATION

5.3.1 DECENTRALIZED FUZZY CONTROL

Before moving on, the following lemma is used to obtain the main results [4].

Lemma 5.1: Relaxation Technique on Fuzzy Rule

Consider the interconnected matrix A_{ijl} in the system (5.4), (5.9), (5.13), and (5.18). For the symmetric positive definite matrix $W_i \in \Re^{n_{xi} \times n_{xi}}$, the following inequality holds:

$$\sum_{i=1}^{N} \sum_{\substack{j=1 \\ j \neq i}}^{N} A_{ij}(\mu_i) W_i A_{ij}^T(\mu_i) \leq \sum_{i=1}^{N} \sum_{\substack{j=1 \\ j \neq i}}^{N} \sum_{l=1}^{r_i} \mu_{il} A_{ijl} W_i A_{ijl}^T. \tag{5.19}$$

■

Proof. Note that for $(i, j) \in N, j \neq i, l \in \mathscr{L}_i$

$$\left[A_{ijl} - A_{ijf}\right] W_i \left[A_{ijl} - A_{ijf}\right]^T \geq 0, \tag{5.20}$$

which implies that

$$A_{ijl} W_i A_{ijl}^T + A_{ijf} W_i A_{ijf}^T \geq A_{ijl} W_i A_{ijf}^T + A_{ijf} W_i A_{ijl}^T. \tag{5.21}$$

By taking the relations in (5.20) and (5.21),

$$\begin{aligned}
\sum_{i=1}^{N} \sum_{\substack{j=1 \\ j \neq i}}^{N} A_{ij}(\mu_i) W_i A_{ij}^T(\mu_i) &= \sum_{i=1}^{N} \sum_{\substack{j=1 \\ j \neq i}}^{N} \sum_{l=1}^{r_i} \sum_{f=1}^{r_i} \mu_{il} \mu_{if} A_{ijl} W_i A_{ijf}^T \\
&= \frac{1}{2} \sum_{i=1}^{N} \sum_{\substack{j=1 \\ j \neq i}}^{N} \sum_{l=1}^{r_i} \sum_{f=1}^{r_i} \mu_{il} \mu_{if} \left[A_{ijl} W_i A_{ijf}^T + A_{ijf} W_i A_{ijl}^T\right] \\
&\leq \frac{1}{2} \sum_{i=1}^{N} \sum_{\substack{j=1 \\ j \neq i}}^{N} \sum_{l=1}^{r_i} \sum_{f=1}^{r_i} \mu_{il} \mu_{if} \left[A_{ijl} W_i A_{ijl}^T + A_{ijf} W_i A_{ijf}^T\right] \\
&= \frac{1}{2} \sum_{i=1}^{N} \sum_{\substack{j=1 \\ j \neq i}}^{N} \sum_{l=1}^{r_i} \mu_{il} A_{ijl} W_i A_{ijl}^T + \frac{1}{2} \sum_{i=1}^{N} \sum_{\substack{j=1 \\ j \neq i}}^{N} \sum_{f=1}^{r_i} \mu_{is} A_{ijf} W_i A_{ijf}^T \\
&= \sum_{i=1}^{N} \sum_{\substack{j=1 \\ j \neq i}}^{N} \sum_{l=1}^{r_i} \mu_{il} A_{ijl} W_i A_{ijl}^T.
\end{aligned} \tag{5.22}$$

This completes the proof.

Assuming the premise variables of the fuzzy system and controller are synchronous, a decentralized fuzzy controller is given by

$$u_i(t) = K_i(\mu_i) x_i(t), \tag{5.23}$$

where $K_i(\mu_i) := \sum_{l=1}^{r_i} \mu_{il} K_{il}$.

In the case, the i-th closed-loop fuzzy control system is given by

$$\dot{x}_i(t) = \bar{A}_i(\mu_i)x_i(t) + \sum_{\substack{j=1 \\ j\neq i}}^{N} A_{ij}(\mu_i)x_j(t), \tag{5.24}$$

where $\bar{A}_i(\mu_i) = A_i(\mu_i) + B_i(\mu_i)K_i(\mu_i)$.

Note: A cluster of controllers is called the decentralized control as shown in (5.23). They are mutually independent, can be designed to execute the overall control task.

Here, a sufficient condition for the stability of the closed-loop fuzzy control system in (5.24) is given by the following theorem.

Theorem 5.1: Stability Analysis of Decentralized Fuzzy Control

The closed-loop fuzzy system in (5.24) using a decentralized fuzzy controller (5.23), is asymptotically stable, if there exist the symmetric positive definite matrices $\{P_i, W_i\} \in \Re^{n_{xi}\times n_{xi}}$, such that the following matrix inequalities hold:

$$\text{Sym}\left(P_i\bar{A}_i(\mu_i)\right) + \sum_{\substack{j=1 \\ j\neq i}}^{N} P_iA_{ij}(\mu_i)W_iA_{ij}^T(\mu_i)P_i + \sum_{\substack{j=1 \\ j\neq i}}^{N} W_j^{-1} < 0. \tag{5.25}$$

■

Proof. Consider the following Lyapunov functional,

$$\begin{aligned} V(t) &= \sum_{i=1}^{N} V_i(t) \\ &= \sum_{i=1}^{N} x_i^T(t)P_ix_i(t), \end{aligned} \tag{5.26}$$

where $P_i \in \Re^{n_{xi}\times n_{xi}}$ is the symmetric positive definite matrix.

By taking the time derivative of $V(t)$ along the trajectory of the system in (5.24),

$$\begin{aligned} \dot{V}(t) &= 2\sum_{i=1}^{N} x_i^T(t)P_i\left[\bar{A}_i(\mu_i)x_i(t) + \sum_{\substack{j=1 \\ j\neq i}}^{N} A_{ij}(\mu_i)x_j(t)\right] \\ &= \sum_{i=1}^{N} x_i^T(t)\left\{\text{Sym}\left(P_i\bar{A}_i(\mu_i)\right)\right\}x_i(t) + 2\sum_{i=1}^{N}\sum_{\substack{j=1 \\ j\neq i}}^{N} x_i^T(t)P_iA_{ij}(\mu_i)x_j(t). \end{aligned} \tag{5.27}$$

Note that

$$2\bar{x}^T\bar{y} \leq \bar{x}^T M^{-1}\bar{x} + \bar{y}^T M\bar{y}, \tag{5.28}$$

where $\bar{x}, \bar{y} \in \Re^n$ and symmetric matrix $M > 0$.

Define $0 < W_i = W_i^T \in \Re^{n_{xi} \times n_{xi}}$, and by using the relation of (5.28), we have

$$\begin{aligned} &2\sum_{i=1}^{N}\sum_{\substack{j=1\\ j\neq i}}^{N} x_i^T(t)P_iA_{ij}(\mu_i)x_j(t) \\ &\leq \sum_{i=1}^{N}\sum_{\substack{j=1\\ j\neq i}}^{N} x_i^T(t)P_iA_{ij}(\mu_i)W_iA_{ij}^T(\mu_i)P_ix_i(t) + \sum_{i=1}^{N}\sum_{\substack{j=1\\ j\neq i}}^{N} x_j^T(t)W_i^{-1}x_j(t). \end{aligned} \tag{5.29}$$

It is easy to obtain the stability result on (5.25). Thus, the proof is completed.

It is noted that the result on (5.25) is not an LMI-based result. In the following, by using some linearization techniques of matrix inequality, the decentralized controller design result is derived in terms of LMIs.

Theorem 5.2: Design of Decentralized Fuzzy Controller

The closed-loop fuzzy system in (5.24) using a decentralized fuzzy controller (5.23), is asymptotically stable, if there exist the symmetric positive definite matrices $\{X_i, W_i, W_0\} \in \Re^{n_{xi} \times n_{xi}}$, $W_i \geq W_0$, and matrix $\bar{K}_{is} \in \Re^{n_{ui} \times n_{xi}}$, such that for all $i \in \mathscr{N}$ the following LMIs hold:

$$\Sigma_{ll} < 0, l \in \mathscr{L}_i \tag{5.30}$$

$$\Sigma_{ls} + \Sigma_{sl} < 0, 1 \leq l < s \leq r \tag{5.31}$$

where

$$\Sigma_{ls} = \begin{bmatrix} \text{Sym}(A_{il}X_i + B_{il}\bar{K}_{is}) + \sum\limits_{\substack{j=1\\ j\neq i}}^{N} A_{ijl}W_iA_{ijl}^T & X_i \\ \star & -\frac{1}{(N-1)}W_0 \end{bmatrix}. \tag{5.32}$$

In that case, the decentralized controller gains can be calculated by

$$K_{is} = \bar{K}_{is}X_i^{-1}. \tag{5.33}$$

■

Proof. It follows from $W_0 \leq W_i, i \in N$ that

$$W_0^{-1} \geq W_i^{-1}, \tag{5.34}$$

which implies $(N-1)W_0^{-1} \geq \sum_{j=1, j\neq i}^{N} W_j^{-1}$.

By using Schur complement lemma, it is easy to see that the following inequality implies (5.25),

$$\begin{bmatrix} \mathrm{Sym}\left(P_i\bar{A}_i(\mu_i)\right) + \sum_{j=1, j\neq i}^{N} P_iA_{ij}(\mu_i)W_iA_{ij}^T(\mu_i)P_i & I \\ \star & -\frac{1}{(N-1)}W_0 \end{bmatrix} < 0. \quad (5.35)$$

Define $X_i = P_i^{-1}$, and by performing the congruence transformation to (5.36) by $\Gamma_i = \mathrm{diag}\{X_i, I\}$, and extracting the fuzzy premise variables, the results from (5.30) and (5.31) are easy to obtain, thus completing this proof.

5.3.2 DECENTRALIZED SAMPLED-DATA CONTROL WITH EVENT-DRIVEN ZOH

Before moving on, the following assumptions are firstly required [4, 5].

Assumption 5.1. The sampler in each subsystem is clock-driven. Let h_i denote the upper bound of sampling intervals,

$$t_{k+1}^i - t_k^i \leq h_i, k \in \mathbb{N} \quad (5.36)$$

where $h_i > 0$.

Assumption 5.2. The zero-order-hold (ZOH) is event-driven, and it uses the latest sampled-data signals and holds them until the next transmitted data are received.

Define $\rho_i^s(t) = t - t_k^i$, it has

$$0 \leq \rho_i^s(t) < h_i. \quad (5.37)$$

It is noted that in the context of digital control systems, both $z_i(t)$ and $x_i(t)$ are involved in the sampled-data measurement. Now, without loss of generality, we further assume that both $z_i(t)$ and $x_i(t)$ are packed, transmitted, and updated at the same time. Then, a decentralized state-feedback fuzzy controller can be given by

Controller Rule $\mathscr{R}_i^s$: **IF** $z_{i1}(t_k^i)$ is $\mathscr{F}_{i1}^s$ and $z_{i2}(t_k^i)$ is $\mathscr{F}_{i2}^s$ and $\cdots$ and $z_{ig}(t_k^i)$ is $\mathscr{F}_{ig}^s$, **THEN**

$$u_i(t) = K_{is}x_i(t_k^i), t \in [t_k^i, t_{k+1}^i) \quad (5.38)$$

where $K_{is} \in \Re^{n_{ui} \times n_{xi}}, s \in \mathscr{L}_i, i \in \mathscr{N}$ are controller gains to be determined; $z_i(t_k^i) := [z_{i1}(t_k^i), z_{i2}(t_k^i), \ldots, z_{iG}(t_k^i)]$; $z_i(t_k^i)$ and $x_i(t_k^i)$ denote the updating signals in the fuzzy controller.

Similarly, the overall decentralized state-feedback fuzzy controller is

$$u_i(t) = K_i(\hat{\mu}_i)x_i(t_k^i), t \in [t_k^i, t_{k+1}^i) \quad (5.39)$$

where

$$\begin{cases} K_i(\hat{\mu}_i) := \sum_{s=1}^{r_i} \hat{\mu}_{is}\left[z_i(t_k^i)\right] K_{is}, \sum_{s=1}^{r_i} \hat{\mu}_{is}\left[z_i(t_k^i)\right] = 1, \\ \hat{\mu}_{is}\left[z_i(t_k^i)\right] := \frac{\prod_{\phi=1}^{g} \hat{\mu}_{is\phi}\left[z_{i\phi}(t_k^i)\right]}{\sum_{\varsigma=1}^{r_i}\prod_{\phi=1}^{g} \hat{\mu}_{i\varsigma\phi}\left[z_{i\phi}(t_k^i)\right]} \geq 0. \end{cases} \tag{5.40}$$

In the following, we will denote $\hat{\mu}_{is} := \hat{\mu}_{is}\left[z_i(t_k^i)\right]$ for brevity.

Note: t_k^i is relative to the clock on the i-th subsystem. In other words, the sampled-data clocks can be different among all subsystems.

We further define

$$x_i(v) = x_i(t_k^i) - x_i(t). \tag{5.41}$$

Submitting (5.41) into the fuzzy controller (5.39) and combining it with the fuzzy system (5.4), the closed-loop fuzzy control system is

$$\dot{x}_i(t) = (A_{ii}(\mu_i) + B_i(\mu_i)K_i(\hat{\mu}_i))\,x_i(t) + B_i(\mu_i)K_i(\hat{\mu}_i)x_i(v) + \sum_{\substack{j=1 \\ j\neq i}}^{N} A_{ij}(\mu_i)x_j(t). \tag{5.42}$$

Now, we introduce the following Lyapunov function:

$$\begin{aligned} V(t) &= \sum_{i=1}^{N} V_i(t) \\ &= \sum_{i=1}^{N} x_i^T(t) P_i x_i(t) + \sum_{i=1}^{N} h_i^2 \int_{t_k^i}^{t} \dot{x}_i^T(\alpha) Q_i \dot{x}_i(\alpha)\, d\alpha \\ &\quad - \sum_{i=1}^{N} \frac{\pi^2}{4} \int_{t_k^i}^{t} \left[x_i(\alpha) - x_i(t_k^i)\right]^T Q_i \left[x_i(\alpha) - x_i(t_k^i)\right] d\alpha, \end{aligned} \tag{5.43}$$

where $\{P_i, Q_i\} \in \Re^{n_{xi}\times n_{xi}}$ are positive definite symmetric matrices.

By using Wirtinger's inequality in [4], it can be known that $V(t) > 0$. Based on the new model in (5.42) and the Lyapunov function in (5.43), a sufficient condition for a decentralized sampled-data controller can be given as below:

Lemma 5.2: Stability Analysis of Decentralized Sampled-Data Control

The closed-loop fuzzy system in (5.42) using a decentralized sampled-data fuzzy controller (5.39), is asymptotically stable, if there exist the symmetric positive definite matrices $\{P_i, W_i, Q_i\} \in \Re^{n_{xi}\times n_{xi}}$, $K_i(\hat{\mu}_i) \in \Re^{n_{ui}\times n_{xi}}$, and the positive scalars h_i, such that the following matrix inequalities hold:

$$\Theta_i + \mathrm{Sym}\left(\mathbb{G}_i \mathbb{A}_i(\mu_i, \hat{\mu}_i)\right) + \sum_{\substack{j=1 \\ j\neq i}}^{N} \mathbb{G}_i A_{ij}(\mu_i) W_i A_{ij}^T(\mu_i) \mathbb{G}_i^T < 0, \tag{5.44}$$

where

$$\Theta_i = \begin{bmatrix} h_i^2 Q_i & P_i & 0 \\ \star & \sum_{\substack{j=1 \\ j\neq i}}^{N} W_j^{-1} & 0 \\ \star & \star & -\frac{\pi^2}{4} Q_i \end{bmatrix},$$

$$\mathbb{A}_i(\mu_i, \hat{\mu}_i) = \begin{bmatrix} -I & A_i(\mu_i) + B_i(\mu_i) K_i(\hat{\mu}_i) & B_i(\mu_i) K_i(\hat{\mu}_i) \end{bmatrix} \tag{5.45}$$

■

Proof. By taking the time derivative of $V(t)$ in (5.43), one has

$$\dot{V}(t) = \sum_{i=1}^{N} \{2x_i^T(t) P_i \dot{x}_i(t) + h_i^2 \dot{x}_i^T(t) Q_i \dot{x}_i(t) - \frac{\pi^2}{4} x_i^T(v) Q_i x_i(v)\}. \tag{5.46}$$

Define the matrix $\mathbb{G}_i \in \Re^{3n_{xi} \times n_{xi}}$ and $\chi_i(t) = \begin{bmatrix} \dot{x}_i^T(t) & x_i^T(t) & x_i^T(v) \end{bmatrix}^T$, and it follows from (5.42) that

$$0 = \sum_{i=1}^{N} 2\chi_i^T(t) \mathbb{G}_i \mathbb{A}_i(\mu_i, \hat{\mu}_i) \chi_i(t) + \sum_{i=1}^{N} 2\chi_i^T(t) \mathbb{G}_i \sum_{\substack{j=1 \\ j\neq i}}^{N} A_{ij}(\mu_i) x_j(t), \tag{5.47}$$

where $\mathbb{A}_i(\mu_i, \hat{\mu}_i) = \begin{bmatrix} -I & A_i(\mu_i) + B_i(\mu_i) K_i(\hat{\mu}_i) & B_i(\mu_i) K_i(\hat{\mu}_i) \end{bmatrix}$.

By introducing matrix $0 < W_i = W_i^T \in \Re^{n_{xi} \times n_{xi}}$, and using the relation of (5.28),

$$\begin{aligned} &\sum_{i=1}^{N} 2\chi_i^T(t) \mathbb{G}_i \sum_{\substack{j=1 \\ j\neq i}}^{N} A_{ij}(\mu_i) x_j(t) \\ &\leq \sum_{i=1}^{N} \sum_{\substack{j=1 \\ j\neq i}}^{N} \chi_i^T(t) \mathbb{G}_i A_{ij}(\mu_i) W_i A_{ij}^T(\mu_i) \mathbb{G}_i^T \chi_i(t) + \sum_{i=1}^{N} \sum_{\substack{j=1 \\ j\neq i}}^{N} x_j^T(t) W_i^{-1} x_j(t) \\ &= \sum_{i=1}^{N} \sum_{\substack{j=1 \\ j\neq i}}^{N} \chi_i^T(t) \mathbb{G}_i A_{ij}(\mu_i) W_i A_{ij}^T(\mu_i) \mathbb{G}_i^T \chi_i(t) + \sum_{i=1}^{N} \sum_{\substack{j=1 \\ j\neq i}}^{N} x_i^T(t) W_j^{-1} x_i(t). \end{aligned} \tag{5.48}$$

It follows from (5.46)-(5.48) that the result on (5.44) can be obtained directly.

It is noted that the result on (5.44) is not an LMI-based result. It is also noted that when the asynchronized information of μ_{il} and $\hat{\mu}_{il}$ is unknown, the designed result generally leads to the linear controller instead of the fuzzy one [6]. From practical perspective, obtaining a priori knowledge of μ_{il} and $\hat{\mu}_{il}$ is possible. Thus, we assume that the asynchronized condition is subject to

$$\underline{\rho}_{il} \leq \frac{\hat{\mu}_{il}}{\mu_{il}} \leq \bar{\rho}_{il}, \tag{5.49}$$

where $\underline{\rho}_{il}$ and $\bar{\rho}_{il}$ are positive scalars.

It follows from (5.44) and (5.49) that the design result on the decentralized sampled-data fuzzy controller can be summarized as below:

Theorem 5.3: Design of Decentralized Sampled-Data Fuzzy Control Using Asynchronized Method

The closed-loop fuzzy system in (5.42) using a decentralized sampled-data fuzzy controller (5.39), is asymptotically stable, if there exist the symmetric positive definite matrices $\{\bar{P}_i, W_i, W_0, \bar{Q}_i\} \in \Re^{n_{xi} \times n_{xi}}$, $W_0 \le W_i$, and matrices $G_i \in \Re^{n_{xi} \times n_{xi}}$, $M_{ils} = M_{isl}^T \in \Re^{4n_{xi} \times 4n_{xi}}$, $\bar{K}_{is} \in \Re^{n_{ui} \times n_{xi}}$, and the positive scalars $\{h_i, \underline{\rho}_{il}, \bar{\rho}_{il}\}$, such that for all $(l,s) \in \mathscr{L}_i$, the following LMIs hold:

$$\bar{\rho}_{il}\Sigma_{ill} + M_{ill} < 0, \tag{5.50}$$

$$\underline{\rho}_{il}\Sigma_{ill} + M_{ill} < 0, \tag{5.51}$$

$$\bar{\rho}_{is}\Sigma_{ils} + \bar{\rho}_{il}\Sigma_{isl} + M_{ils} + M_{isl} < 0, \tag{5.52}$$

$$\underline{\rho}_{is}\Sigma_{ils} + \underline{\rho}_{il}\Sigma_{isl} + M_{ils} + M_{isl} < 0, \tag{5.53}$$

$$\underline{\rho}_{is}\Sigma_{ils} + \bar{\rho}_{il}\Sigma_{isl} + M_{ils} + M_{isl} < 0, \tag{5.54}$$

$$\bar{\rho}_{is}\Sigma_{ils} + \underline{\rho}_{il}\Sigma_{isl} + M_{ils} + M_{isl} < 0, \tag{5.55}$$

$$\begin{bmatrix} M_{i11} & \cdots & M_{i1r} \\ \vdots & \ddots & \vdots \\ M_{ir1} & \cdots & M_{irr} \end{bmatrix} > 0, \tag{5.56}$$

where

$$\Sigma_{ils} = \begin{bmatrix} \bar{\Theta}_i + \text{Sym}\left(\bar{\mathbb{I}}_i \tilde{\mathbb{A}}_{ils}\right) + \sum_{\substack{j=1 \\ j \neq i}}^{N} \bar{\mathbb{I}}_i A_{ijl} W_i A_{ijl}^T \bar{\mathbb{I}}_i^T & \bar{G}_i \\ \star & -\frac{1}{(N-1)} W_0 \end{bmatrix},$$

$$\bar{\Theta}_i = \begin{bmatrix} h_i^2 \bar{Q}_i & \bar{P}_i & 0 \\ \star & 0 & 0 \\ \star & \star & -\frac{\pi^2}{4}\bar{Q}_i \end{bmatrix}, \bar{G}_i = \begin{bmatrix} 0 \\ G_i^T \\ 0 \end{bmatrix}, \bar{\mathbb{I}}_i = \begin{bmatrix} I \\ I \\ 0 \end{bmatrix},$$

$$\tilde{\mathbb{A}}_{ils} = \begin{bmatrix} -G_i & A_{il}G_i + B_{il}\bar{K}_{is} & B_{il}\bar{K}_{is} \end{bmatrix}. \tag{5.57}$$

In that case, the proposed sampled-data fuzzy controller gains can be calculated by

$$K_{is} = \bar{K}_{is} G_i^{-1}, s \in \mathscr{L}_i. \tag{5.58}$$

■

Proof. Define $W_0 \le W_i, i \in \mathscr{N}$,

$$W_0^{-1} \ge W_i^{-1}, \tag{5.59}$$

which implies $(N-1)W_0^{-1} \geq \sum_{j=1, j\neq i}^{N} W_j^{-1}$.

It is easy to see that the following inequality holds, which implies (5.44),

$$\begin{bmatrix} \bar{\Theta}_i + \mathrm{Sym}\left(\mathbb{G}_i \mathbb{A}_i(\mu_i, \hat{\mu}_i)\right) + \sum_{j=1, j\neq i}^{N} \mathbb{G}_i A_{ij}(\mu_i) W_i A_{ij}^T(\mu_i) \mathbb{G}_i^T & \mathbb{I} \\ \star & -\frac{1}{(N-1)} W_0 \end{bmatrix} < 0, \tag{5.60}$$

where $\mathbb{G}_i$ is defined in (5.47), and

$$\bar{\Theta}_i = \begin{bmatrix} h_i^2 Q_i & P_i & 0 \\ \star & 0 & 0 \\ \star & \star & -\frac{\pi^2}{4} Q_i \end{bmatrix}, \mathbb{I} = \begin{bmatrix} 0 \\ I \\ 0 \end{bmatrix}. \tag{5.61}$$

It follows from (5.60) and (5.61) that

$$h_i^2 Q_i - \mathrm{Sym}\{G_i\} < 0, i \in \mathscr{N} \tag{5.62}$$

which implies that $G_i, i \in \mathscr{N}$ are nonsingular matrices.

We further define

$$\begin{cases} \mathbb{G}_i = \begin{bmatrix} G_i^{-1} & G_i^{-1} & 0 \end{bmatrix}^T, \Gamma_1 := \mathrm{diag}\{ G_i \quad G_i \quad G_i \quad I \}, \\ \bar{P}_i = G_i^T P_i G_i, \bar{Q}_i = G_i^T Q_i G_i. \end{cases} \tag{5.63}$$

By substituting (5.63) into (5.60), and performing a congruence transformation by Γ_1, and extracting the fuzzy membership functions, we have

$$\sum_{l=1}^{r_i} \sum_{f=1}^{r_i} \sum_{s=1}^{r_i} \mu_{il} \mu_{if} \hat{\mu}_{is} \Sigma_{ilfs} < 0, \tag{5.64}$$

where

$$\Sigma_{ilfs} = \begin{bmatrix} \bar{\Theta}_i + \mathrm{Sym}\left(\bar{\mathbb{I}}_i \bar{\mathbb{A}}_{ils}\right) + \sum_{j=1, j\neq i}^{N} \bar{\mathbb{I}}_i A_{ijl} W_i A_{ijf}^T \bar{\mathbb{I}}_i^T & \bar{G}_i \\ \star & -\frac{1}{(N-1)} W_0 \end{bmatrix},$$

$$\bar{\Theta}_i = \begin{bmatrix} h_i^2 \bar{Q}_i & \bar{P}_i & 0 \\ \star & 0 & 0 \\ \star & \star & -\frac{\pi^2}{4} \bar{Q}_i \end{bmatrix}, \bar{G}_i = \begin{bmatrix} 0 \\ G_i^T \\ 0 \end{bmatrix}, \bar{\mathbb{I}}_i = \begin{bmatrix} I \\ I \\ 0 \end{bmatrix},$$

$$\bar{\mathbb{A}}_{ils} = \begin{bmatrix} -G_i & A_{il} G_i + B_{il} K_{is} G_i & B_{il} K_{is} G_i \end{bmatrix}. \tag{5.65}$$

Then, by defining $\bar{K}_{is} = K_{is} G_i$, and by using Lemma 5.1, the following inequality implies (5.64),

$$\sum_{l=1}^{r_i} \sum_{s=1}^{r_i} \mu_{il} \hat{\mu}_{is} \Sigma_{ils} < 0, \tag{5.66}$$

where Σ_{ils} is defined in (5.57).

By taking the relation in (5.49) and using the asynchronous method proposed in [7], the inequality in (5.66) holds if the inequalities (5.50)-(5.56) hold, thus completing this proof.

It is worth nothing that the number of LMIs on Theorem 5.4 is large. It is also noted that the existing relaxation technique $\sum_{l=1}^{r_i}[\mu_{il}]^2\Sigma_{ill}+\sum_{l=1}^{r_i}\sum_{l<s\leq r_i}^{r_i}\mu_{il}\mu_{is}\Sigma_{ls}<0$ is no longer applicable to fuzzy controller synthesis, since $\mu_{is}\neq\hat{\mu}_{is}$. Similarly to the relaxation technique in [8], it is assumed that

$$|\mu_{il}-\hat{\mu}_{il}|\leq\delta_{il}, l\in\mathscr{L}_i, \tag{5.67}$$

where δ_{il} is a positive scalar. If $\Sigma_{ils}+M_{il}\geq 0$, where M_{il} is a symmetric matrix, we have

$$\begin{aligned}
&\sum_{l=1}^{r_i}\sum_{s=1}^{r_i}\mu_{il}\hat{\mu}_{is}\Sigma_{ils}\\
&=\sum_{l=1}^{r_i}\sum_{s=1}^{r_i}\mu_{il}\mu_{is}\Sigma_{ils}+\sum_{l=1}^{r_i}\sum_{s=1}^{r_i}\mu_{il}\left(\hat{\mu}_{is}-\mu_{is}\right)\left(\Sigma_{ils}+M_{il}\right)\\
&\leq\sum_{l=1}^{r_i}\sum_{s=1}^{r_i}\mu_{il}\mu_{is}\left[\Sigma_{ils}+\sum_{f=1}^{r_i}\delta_{if}\left(\Sigma_{ilf}+M_{il}\right)\right].
\end{aligned} \tag{5.68}$$

Therefore, upon defining $\Sigma_{ils}=\Phi_{ils}+\sum_{s=1}^{r_i}\delta_{is}\left(\Phi_{ils}+M_{il}\right)$, the existing relaxation technique from [9] can be applied to (5.68).

Based on the assumption in (5.67), and the result of (5.66) and (5.68), the design result on fuzzy control is proposed as below:

Theorem 5.4: Design of Decentralized Sampled-Data Fuzzy Control Using Synchronized Method

The closed-loop fuzzy system in (5.42) using a decentralized sampled-data fuzzy controller (5.39), is asymptotically stable, if there exist the symmetric positive definite matrices $\{\bar{P}_i, W_i, W_0, \bar{Q}_i\}\in\Re^{n_{xi}\times n_{xi}}$, $W_0\leq W_i$, and matrices $G_i\in\Re^{n_{xi}\times n_{xi}}$, $M_{il}=M_{il}^T\in\Re^{4n_{xi}\times 4n_{xi}}$, $\bar{K}_{is}\in\Re^{n_{ui}\times n_{xi}}$, and the positive scalars $\{h_i,\delta_{il}\}$, such that for all $(l,s)\in\mathscr{L}_i$, the following LMIs hold:

$$\Sigma_{ill}<0, l\in\mathscr{L}_i \tag{5.69}$$

$$\Sigma_{ils}+\Sigma_{isl}<0, 1\leq l<s\leq r_i \tag{5.70}$$

where

$$\Sigma_{ils} = \Phi_{ils} + \sum_{f=1}^{r_i} \delta_{if}\left(\Phi_{ilf} + M_{il}\right),$$

$$\Phi_{ils} = \begin{bmatrix} \bar{\Theta}_i + \mathrm{Sym}\left(\bar{\mathbb{I}}_i \tilde{\mathbb{A}}_{ils}\right) + \sum\limits_{\substack{j=1 \\ j\neq i}}^{N} \bar{\mathbb{I}}_i A_{ijl} W_i A_{ijl}^T \bar{\mathbb{I}}_i^T & \bar{G}_i \\ \star & -\frac{1}{(N-1)} W_0 \end{bmatrix},$$

$$\bar{\Theta}_i = \begin{bmatrix} h_i^2 \bar{Q}_i & \bar{P}_i & 0 \\ \star & 0 & 0 \\ \star & \star & -\frac{\pi^2}{4}\bar{Q}_i \end{bmatrix}, \bar{G}_i = \begin{bmatrix} 0 \\ G_i^T \\ 0 \end{bmatrix}, \bar{\mathbb{I}}_i = \begin{bmatrix} I \\ I \\ 0 \end{bmatrix},$$

$$\tilde{\mathbb{A}}_{ils} = \begin{bmatrix} -G_i & A_{il} G_i + B_{il} \bar{K}_{is} & B_{il} \bar{K}_{is} \end{bmatrix}. \tag{5.71}$$

In that case, the controller gains can be calculated by

$$K_{il} = \bar{K}_{il} G_i^{-1}. \tag{5.72}$$

■

Note that when the asynchronized information of μ_{il} and $\hat{\mu}_{il}$ is unknown, the design result on a decentralized sampled-data linear controller can be directly derived as below:

Theorem 5.5: Decentralized Sampled-Data Linear Controller Design

The closed-loop fuzzy system in (5.42) using a decentralized sampled-data linear controller $u_i(t) = K_i x_i(t_k^i), t \in [t_k^i, t_{k+1}^i)$, is asymptotically stable, if there exist the symmetric positive definite matrices $\{\bar{P}_i, W_i, W_0, \bar{Q}_i\} \in \Re^{n_{xi} \times n_{xi}}$, $W_0 \leq W_i$, and matrices $G_i \in \Re^{n_{xi} \times n_{xi}}$, $\bar{K}_i \in \Re^{n_{ui} \times n_{xi}}$, and the positive scalars h_i, such that for all $l \in \mathscr{L}_i$, the following LMIs hold:

$$\begin{bmatrix} \bar{\Theta}_i + \mathrm{Sym}\left(\bar{\mathbb{I}}_i \tilde{\mathbb{A}}_{il}\right) + \sum\limits_{\substack{j=1 \\ j\neq i}}^{N} \bar{\mathbb{I}}_i A_{ijl} W_i A_{ijl}^T \bar{\mathbb{I}}_i^T & \bar{G}_i \\ \star & -\frac{1}{(N-1)} W_0 \end{bmatrix} < 0, \tag{5.73}$$

where

$$\bar{\Theta}_i = \begin{bmatrix} h_i^2 \bar{Q}_i & \bar{P}_i & 0 \\ \star & 0 & 0 \\ \star & \star & -\frac{\pi^2}{4}\bar{Q}_i \end{bmatrix}, \bar{G}_i = \begin{bmatrix} 0 \\ G_i^T \\ 0 \end{bmatrix}, \bar{\mathbb{I}}_i = \begin{bmatrix} I \\ I \\ 0 \end{bmatrix},$$

$$\tilde{\mathbb{A}}_{il} = \begin{bmatrix} -G_i & A_{il} G_i + B_{il} \bar{K}_i & B_{il} \bar{K}_i \end{bmatrix}. \tag{5.74}$$

In that case, the proposed linear controller gains can be calculated by

$$K_i = \bar{K}_i G_i^{-1}. \tag{5.75}$$

■

5.3.3 DECENTRALIZED SAMPLED-DATA CONTROL WITH TIME-DRIVEN ZOH

Before moving on, consider the following assumptions [10]:

Assumption 5.3. Sensors are time-driven, and satisfy the set,

$$\underline{s}_i \leq s_i^{k+1} - s_i^k \leq \bar{s}_i, k \in \mathbb{N}_i. \tag{5.76}$$

Assumption 5.4. ZOHs are time-driven, and satisfy the set,

$$\underline{z}_i \leq z_i^{k+1} - z_i^k \leq \bar{z}_i, k \in \mathbb{N}_i. \tag{5.77}$$

Note: The work in [9] considered even-based ZOHs in the sense that control signals are applied as soon as new data becomes available. However, that does not always happen in engineering applications when actuators have low updated rates. For example, according to the work in [10], the refresh rates in actuators such as electric cylinders, solenoids, shape memory alloys, and electroactive polymers are 10 Hz or less. Those actuators do not immediately refresh the control signals and induce time-delays, which can affect stability and performance of the closed-loop system. Thus, this section considers the time-driven ZOHs in (5.77).

Generally, event-driven ZOHs receive new data from controllers as soon as sensors send new sampling data. Based on Assumptions 5.3 and 5.4, it is easy to see that the proposed ZOHs are time-driven and are not updated immediately. Here, we define the time elapsed as $\rho^z(t)$, and we have [10]

$$0 \leq \rho_i^z(t) < \bar{z}_i. \tag{5.78}$$

Thus, it is easy to see that

$$\begin{aligned} \rho_i^{zs}(t) &= \rho_i^z(t) + t - s_i^k \\ &= \rho_i^z(t) + \rho_i^s(t), t \in [z_i^k, z_i^{k+1}) \end{aligned} \tag{5.79}$$

where $0 \leq \rho_i^s(t) < \bar{s}_i$, $0 \leq \rho_i^{zs}(t) < \bar{z}_i + \bar{s}_i$.

Now, consider the following decentralized sampled-data fuzzy controller:

Controller Rule $\mathscr{R}_i^s$: **IF** $z_{i1}(z_i^k)$ is F_{i1}^s and $z_{i2}(z_i^k)$ is F_{i2}^s and $\cdots$ and $z_{ig}(z_i^k)$ is F_{ig}^s, **THEN**

$$u_i(t) = K_{is}x_i(z_i^k), t \in [z_i^k, z_i^{k+1}) \tag{5.80}$$

where $K_{is} \in \Re^{n_{ui} \times n_{xi}}, s \in \mathscr{L}_i, i \in N$ are controller gains to be determined; $z_i(t_k^i) := [z_{i1}(z_i^k), z_{i2}(z_i^k), \ldots, z_{ig}(z_i^k)]$; $z_i(z_i^k)$ and $x_i(z_i^k)$ denote the updating signals in the fuzzy controller.

Similarly, the overall decentralized fuzzy controller is

$$u_i(t) = K_i(\hat{\mu}_i)x_i(z_i^k), t \in [z_i^k, z_i^{k+1}) \tag{5.81}$$

where

$$\begin{cases} K_i(\hat{\mu}_i) := \sum_{s=1}^{r_i} \hat{\mu}_{is}\left[z_i(z_i^k)\right] K_{is}, \sum_{s=1}^{r_i} \hat{\mu}_{is}\left[z_i(z_i^k)\right] = 1, \\ \hat{\mu}_{is}\left[z_i(z_i^k)\right] := \dfrac{\Pi_{\phi=1}^{g} \hat{\mu}_{is\phi}\left[z_{i\phi}(z_i^k)\right]}{\Sigma_{\varsigma=1}^{r_i} \Pi_{\phi=1}^{g} \hat{\mu}_{i\varsigma\phi}\left[z_{i\phi}(z_i^k)\right]} \geq 0. \end{cases} \tag{5.82}$$

In the following, we will denote $\hat{\mu}_{is} := \hat{\mu}_{is}\left[z_i(z_i^k)\right]$ for brevity.

Here, without loss of generality, we just consider the asynchronous sampled-data measurement. However, the obtained result can be easily extended to the synchronous case.

We further define

$$x_i(v) = x_i(z_i^k) - x_i(t). \tag{5.83}$$

Submitting (5.83) into the fuzzy controller (5.81) and combining it with the fuzzy system (5.18), the closed-loop fuzzy control system is

$$\begin{aligned}\dot{x}_i(t) &= (A_i(\mu_i) + B_i(\mu_i)K_i(\hat{\mu}_i))x_i(t) \\ &\quad + B_i(\mu_i)K_i(\hat{\mu}_i)x_i(v) + \sum_{\substack{j=1 \\ j\neq i}}^{N} A_{ij}(\mu_i)x_j(t).\end{aligned} \tag{5.84}$$

Note: The centralized control requires faster computers with larger memory to implement excessive information processing. Currently the decentralized control has attracted considerable interest in the field of large-scale systems. Instead of developing a single controller, a cluster of controllers, which are mutually independent, can be designed to execute the overall control task.

Note: The decentralized fuzzy controller (5.81) reduces to a primary domain controller (PDC) when $\mu_{il} = \hat{\mu}_{il}$. However, the premise variable $z(t)$ undergoes time-driven sensors and ZOHs, and is implemented by the proposed fuzzy controller in (5.81). In such circumstances, the asynchronous variables between μ_{il} and $\hat{\mu}_{il}$ are more realistic. As pointed out in [7], when the relationship between μ_{il} and $\hat{\mu}_{il}$ is unavailable, the condition $\mu_{il} \neq \hat{\mu}_{il}$ generally leads to a linear controller instead of a fuzzy one, which degrades the stabilization ability of the controller. When the relationship on μ_{il} and $\hat{\mu}_{il}$ is available, the design conservatism can be improved, and obtaining the corresponding fuzzy controller.

Note: z_i^k is relative to the clock on the i-th subsystem. In other words, the sampled-data clocks and the ZOH clocks can be different among all subsystems.

Now, we introduce the following LKF:

$$\begin{aligned}V(t) &= \sum_{i=1}^{N} x_i^T(t)P_i x_i(t) + \sum_{i=1}^{N} [\bar{z}_i + \bar{s}_i]^2 \int_{z_i^k}^{t} \dot{x}_i^T(\alpha)Q_i\dot{x}_i(\alpha)d\alpha \\ &\quad - \sum_{i=1}^{N} \frac{\pi^2}{4} \int_{z_i^k}^{t} \left[x_i(\alpha) - x_i\left(z_i^k\right)\right]^T Q_i \left[x_i(\alpha) - x_i\left(z_i^k\right)\right] d\alpha,\end{aligned} \tag{5.85}$$

where $\{P_i, Q_i\} \in \Re^{n_{xi}\times n_{xi}}$ are positive definite symmetric matrices.

By using Wirtinger's inequality in [4], it can be known that $V(t) > 0$. Based on the new model in (5.84) and the LKF in (5.85), a sufficient condition for existing a decentralized sampled-data controller can be given as below:

Lemma 5.3: Stability Analysis of Decentralized Sampled-Data Control

The closed-loop fuzzy system in (5.84) using a decentralized sampled-data fuzzy controller (5.81), is asymptotically stable, if there exist the symmetric positive definite matrices $\{P_i, W_i\} \in \Re^{n_{xi} \times n_{xi}}$, $K_i(\hat{\mu}_i) \in \Re^{n_{ui} \times n_{xi}}$, and the positive scalars $\{\bar{z}_i, \bar{s}_i\}$, such that the following matrix inequalities hold:

$$\Theta_i + \mathrm{Sym}\left(\mathbb{G}_i \mathbb{A}_i(\mu_i, \hat{\mu}_i)\right) + \sum_{\substack{j=1 \\ j \neq i}}^{N} \mathbb{G}_i A_{ij}(\mu_i) W_i A_{ij}^T(\mu_i) \mathbb{G}_i^T < 0, \tag{5.86}$$

where

$$\Theta_i = \begin{bmatrix} [\bar{z}_i + \bar{s}_i]^2 Q_i & P_i & 0 \\ \star & \sum\limits_{\substack{j=1 \\ j \neq i}}^{N} W_j^{-1} & 0 \\ \star & \star & -\frac{\pi^2}{4} Q_i \end{bmatrix}. \tag{5.87}$$

■

Proof. By taking the time derivative of $V(t)$ in (5.85), we have

$$\dot{V}(t) = \sum_{i=1}^{N} 2x_i^T(t) P_i \dot{x}_i(t) + \sum_{i=1}^{N} [\bar{z}_i + \bar{s}_i]^2 \dot{x}_i^T(t) Q_i \dot{x}_i(t) - \sum_{i=1}^{N} \frac{\pi^2}{4} x_i^T(v) Q_i x_i(v). \tag{5.88}$$

Define the matrix $\mathbb{G}_i \in \Re^{3n_{xi} \times n_{xi}}$ and $\chi_i(t) = \begin{bmatrix} \dot{x}_i^T(t) & x_i^T(t) & x_i^T(v) \end{bmatrix}^T$, and it follows from (5.84) that

$$0 = \sum_{i=1}^{N} 2\chi_i^T(t) \mathbb{G}_i \mathbb{A}_i(\mu_i, \hat{\mu}_i) \chi_i(t) + \sum_{i=1}^{N} 2\chi_i^T(t) \mathbb{G}_i \sum_{\substack{j=1 \\ j \neq i}}^{N} A_{ij}(\mu_i) x_j(t), \tag{5.89}$$

where $\mathbb{A}_i(\mu_i, \hat{\mu}_i) = \begin{bmatrix} -I & A_i(\mu_i) + B_i(\mu_i) K_i(\hat{\mu}_i) & B_i(\mu_i) K_i(\hat{\mu}_i) \end{bmatrix}$.

By introducing the symmetric positive definite matrices $W_i \in \Re^{n_{xi} \times n_{xi}}$, and using the relation of (5.28),

$$\begin{aligned} &\sum_{i=1}^{N} 2\chi_i^T(t) \mathbb{G}_i \sum_{\substack{j=1 \\ j \neq i}}^{N} A_{ij}(\mu_i) x_j(t) \\ &\leq \sum_{i=1}^{N} \sum_{\substack{j=1 \\ j \neq i}}^{N} \chi_i^T(t) \mathbb{G}_i A_{ij}(\mu_i) W_i A_{ij}^T(\mu_i) \mathbb{G}_i^T \chi_i(t) + \sum_{i=1}^{N} \sum_{\substack{j=1 \\ j \neq i}}^{N} x_j^T(t) W_i^{-1} x_j(t) \\ &= \sum_{i=1}^{N} \sum_{\substack{j=1 \\ j \neq i}}^{N} \chi_i^T(t) \mathbb{G}_i A_{ij}(\mu_i) W_i A_{ij}^T(\mu_i) \mathbb{G}_i^T \chi_i(t) + \sum_{i=1}^{N} \sum_{\substack{j=1 \\ j \neq i}}^{N} x_i^T(t) W_j^{-1} x_i(t). \end{aligned} \tag{5.90}$$

It follows from (5.88)-(5.90) that the result on (5.86) can be obtained directly.

It is noted that the condition on (5.86) is not an LMI-based result. When the asynchronized information of μ_{il} and $\hat{\mu}_{il}$ is unknown, the designed result generally leads to the linear controller instead of the fuzzy one [6]. From a practical perspective, obtaining a priori knowledge of μ_{il} and $\hat{\mu}_{il}$ is possible. Thus, we assume that the asynchronized condition is subject to

$$\underline{\rho}_{il} \leq \frac{\hat{\mu}_{il}}{\mu_{il}} \leq \bar{\rho}_{il}, \tag{5.91}$$

where $\underline{\rho}_{il}$ and $\bar{\rho}_{il}$ are positive scalars.

It follows from (5.86) and (5.91) that the design result on the decentralized sampled-data fuzzy controller can be summarized as below:

Theorem 5.6: Design of Decentralized Sampled-Data Fuzzy Control Using Asynchronized Method

The closed-loop fuzzy system in (5.84) using a decentralized sampled-data fuzzy controller (5.81), is asymptotically stable, if there exist the symmetric positive definite matrices $\{\bar{P}_i, W_i, W_0, \bar{Q}_i\} \in \Re^{n_{xi} \times n_{xi}}$, $W_0 \leq W_i$, and matrices $G_i \in \Re^{n_{xi} \times n_{xi}}$, $M_{ils} = M_{isl}^T \in \Re^{4n_{xi} \times 4n_{xi}}$, $\bar{K}_{is} \in \Re^{n_{ui} \times n_{xi}}$, and the positive scalars $\{\bar{z}_i, \bar{s}_i, \underline{\rho}_l, \bar{\rho}_l\}$, such that for all $(l,s) \in \mathscr{L}_i$, the following LMIs hold:

$$\bar{\rho}_{il}\Sigma_{ill} + M_{ill} < 0, \tag{5.92}$$

$$\underline{\rho}_{il}\Sigma_{ill} + M_{ill} < 0, \tag{5.93}$$

$$\bar{\rho}_{is}\Sigma_{ils} + \bar{\rho}_{il}\Sigma_{isl} + M_{ils} + M_{isl} < 0, \tag{5.94}$$

$$\underline{\rho}_{is}\Sigma_{ils} + \underline{\rho}_{il}\Sigma_{isl} + M_{ils} + M_{isl} < 0, \tag{5.95}$$

$$\underline{\rho}_{is}\Sigma_{ils} + \bar{\rho}_{il}\Sigma_{isl} + M_{ils} + M_{isl} < 0, \tag{5.96}$$

$$\bar{\rho}_{is}\Sigma_{ils} + \underline{\rho}_{il}\Sigma_{isl} + M_{ils} + M_{isl} < 0, \tag{5.97}$$

$$\begin{bmatrix} M_{i11} & \cdots & M_{i1r} \\ \vdots & \ddots & \vdots \\ M_{ir1} & \cdots & M_{irr} \end{bmatrix} > 0, \tag{5.98}$$

where

$$\Sigma_{ils} = \begin{bmatrix} \bar{\Theta}_i + \mathrm{Sym}\left(\bar{\mathbb{I}}_i \tilde{\mathbb{A}}_{ils}\right) + \sum_{\substack{j=1 \\ j\neq i}}^{N} \bar{\mathbb{I}}_i A_{ijl} W_i A_{ijl}^T \bar{\mathbb{I}}_i^T & \bar{G}_i \\ \star & -\frac{1}{(N-1)} W_0 \end{bmatrix},$$

$$\bar{\Theta}_i = \begin{bmatrix} [\bar{z}_i + \bar{s}_i]^2 \bar{Q}_i & \bar{P}_i & 0 \\ \star & 0 & 0 \\ \star & \star & -\frac{\pi^2}{4} \bar{Q}_i \end{bmatrix}, \bar{G}_i = \begin{bmatrix} 0 \\ G_i^T \\ 0 \end{bmatrix}, \bar{\mathbb{I}}_i = \begin{bmatrix} I \\ I \\ 0 \end{bmatrix},$$

$$\tilde{\mathbb{A}}_{ils} = \begin{bmatrix} -G_i & A_{il} G_i + B_{il} \bar{K}_{is} & B_{il} \bar{K}_{is} \end{bmatrix}. \tag{5.99}$$

In that case, the proposed fuzzy controller gains can be calculated by

$$K_{is} = \bar{K}_{is} G_i^{-1}, s \in \mathscr{L}_i. \tag{5.100}$$

■

Proof. The result of defining $W_0 \leq W_i, i \in N$ is

$$W_0^{-1} \geq W_i^{-1}, \tag{5.101}$$

which implies $(N-1) W_0^{-1} \geq \sum_{\substack{j=1 \\ j\neq i}}^{N} W_j^{-1}$.

It is easy to see that the following inequality holds, which implies (5.86),

$$\begin{bmatrix} \bar{\Theta}_i + \mathrm{Sym}\left(\mathbb{G}_i \mathbb{A}_i(\mu_i, \hat{\mu}_i)\right) + \sum_{\substack{j=1 \\ j\neq i}}^{N} \mathbb{G}_i A_{ij}(\mu_i) W_i A_{ij}^T(\mu_i) \mathbb{G}_i^T & \mathbb{I} \\ \star & -\frac{1}{(N-1)} W_0 \end{bmatrix} < 0, \tag{5.102}$$

where

$$\bar{\Theta}_i = \begin{bmatrix} [\bar{z}_i + \bar{s}_i]^2 Q_i & P_i & 0 \\ \star & 0 & 0 \\ \star & \star & -\frac{\pi^2}{4} Q_i \end{bmatrix}, \mathbb{I} = \begin{bmatrix} 0 \\ I \\ 0 \end{bmatrix}. \tag{5.103}$$

It follows from (5.92) and (5.99) that

$$[\bar{z}_i + \bar{s}_i]^2 Q_i - \mathrm{Sym}\{G_i\} < 0, i \in \mathscr{N} \tag{5.104}$$

which implies that $G_i, i \in \mathscr{N}$ are nonsingular matrices.

We further define

$$\begin{cases} \mathbb{G}_i = \begin{bmatrix} G_i^{-1} & G_i^{-1} & 0 \end{bmatrix}^T, \Gamma_1 := \mathrm{diag}\left\{ G_i \quad G_i \quad G_i \quad I \right\}, \\ \bar{P}_i = G_i^T P_i G_i, \bar{Q}_i = G_i^T Q_i G_i. \end{cases} \tag{5.105}$$

By substituting (5.105) into (5.102), and performing a congruence transformation by Γ_1, and extracting the fuzzy membership functions, we have

$$\sum_{l=1}^{r_i}\sum_{f=1}^{r_i}\sum_{s=1}^{r_i}\mu_{il}\mu_{if}\hat{\mu}_{is}\Sigma_{ilfs} < 0, \tag{5.106}$$

where

$$\Sigma_{ilfs} = \begin{bmatrix} \bar{\Theta}_i + \text{Sym}\left(\bar{\mathbb{I}}_i\bar{\mathbb{A}}_{ils}\right) + \sum_{j=1, j\neq i}^{N} \bar{\mathbb{I}}_i A_{ijl} W_i A_{ijf}^T \bar{\mathbb{I}}_i^T & \bar{G}_i \\ \star & -\frac{1}{(N-1)} W_0 \end{bmatrix},$$

$$\bar{\Theta}_i = \begin{bmatrix} [\bar{z}_i + \bar{s}_i]^2 \bar{Q}_i & \bar{P}_i & 0 \\ \star & 0 & 0 \\ \star & \star & -\frac{\pi^2}{4}\bar{Q}_i \end{bmatrix}, \bar{G}_i = \begin{bmatrix} 0 \\ G_i^T \\ 0 \end{bmatrix}, \bar{\mathbb{I}}_i = \begin{bmatrix} I \\ I \\ 0 \end{bmatrix},$$

$$\bar{\mathbb{A}}_{ils} = \begin{bmatrix} -G_i & A_{il}G_i + B_{il}K_{is}G_i & B_{il}K_{is}G_i \end{bmatrix}. \tag{5.107}$$

We then define $\bar{K}_{is} = K_{is}G_i$, and by using Lemma 5.1, the following inequality implies (5.106),

$$\sum_{l=1}^{r_i}\sum_{s=1}^{r_i}\mu_{il}\hat{\mu}_{is}\Sigma_{ils} < 0, \tag{5.108}$$

where Σ_{ils} is defined in (5.99).

By taking the relation in (5.91) and using the asynchronous method proposed in [7], the inequality in (5.108) holds if the inequalities (5.92)-(5.98) hold, thus completing this proof.

It is worth noting that the number of LMIs on Thoerem 5.4 is large. It is also noted that the existing relaxation technique $\sum_{l=1}^{r_i}[\mu_{il}]^2\Sigma_{ill} + \sum_{l=1}^{r_i}\sum_{l<s\le r_i}^{r_i}\mu_{il}\mu_{is}\Sigma_{ls} < 0$ is no longer applicable to fuzzy controller synthesis, since $\mu_{is} \neq \hat{\mu}_{is}$. Similarly to the relaxation technique in [8], it is assumed that

$$|\mu_{il} - \hat{\mu}_{il}| \le \delta_{il}, l \in \mathscr{L}_i, \tag{5.109}$$

where δ_{il} is a positive scalar. If $\Sigma_{ils} + M_{il} \geq 0$, where M_{il} is a symmetric matrix, we have

$$\begin{aligned} &\sum_{l=1}^{r_i}\sum_{s=1}^{r_i}\mu_{il}\hat{\mu}_{is}\Sigma_{ils} \\ &= \sum_{l=1}^{r_i}\sum_{s=1}^{r_i}\mu_{il}\mu_{is}\Sigma_{ils} + \sum_{l=1}^{r_i}\sum_{s=1}^{r_i}\mu_{il}\left(\hat{\mu}_{is} - \mu_{is}\right)\left(\Sigma_{ils} + M_{il}\right) \\ &\le \sum_{l=1}^{r_i}\sum_{s=1}^{r_i}\mu_{il}\mu_{is}\left[\Sigma_{ils} + \sum_{f=1}^{r_i}\delta_{if}\left(\Sigma_{ilf} + M_{il}\right)\right]. \end{aligned} \tag{5.110}$$

Therefore, upon defining $\Sigma_{ils} = \Phi_{ils} + \sum_{s=1}^{r_i}\delta_{is}\left(\Phi_{ils} + M_{il}\right)$, the existing relaxation technique from [9] can be applied to (5.108).

Based on the assumption in (5.109), and the result of (5.108), (5.110), the design result for a fuzzy controller is proposed as below:

Theorem 5.7: Design of Decentralized Sampled-Data Fuzzy Control Using Synchronized Method

The closed-loop fuzzy system in (5.84) using a decentralized sampled-data fuzzy controller (5.81), is asymptotically stable, if there exist the symmetric positive definite matrices $\{\bar{P}_i, W_i, W_0, \bar{Q}_i\} \in \Re^{n_{xi} \times n_{xi}}$, $W_0 \leq W_i$, and matrices $G_i \in \Re^{n_{xi} \times n_{xi}}$, $M_{il} = M_{il}^T \in \Re^{4n_{xi} \times 4n_{xi}}$, $\bar{K}_{is} \in \Re^{n_{ui} \times n_{xi}}$, and the positive scalars $\{\bar{z}_i, \bar{s}_i, \delta_{il}\}$, such that for all $(l,s) \in \mathscr{L}_i$, the following LMIs hold:

$$\Sigma_{ill} < 0, l \in \mathscr{L}_i \tag{5.111}$$

$$\Sigma_{ils} + \Sigma_{isl} < 0, 1 \leq l < s \leq r_i \tag{5.112}$$

where

$$\Sigma_{ils} = \Phi_{ils} + \sum\nolimits_{s=1}^{r_i} \delta_{is} \left(\Phi_{ils} + M_{il} \right),$$

$$\Phi_{ils} = \begin{bmatrix} \bar{\Theta}_i + \operatorname{Sym}\left(\bar{\mathbb{I}}_i \tilde{\mathbb{A}}_{ils}\right) + \sum\limits_{\substack{j=1 \\ j \neq i}}^{N} \bar{\mathbb{I}}_i A_{ijl} W_i A_{ijl}^T \bar{\mathbb{I}}_i^T & \bar{G}_i \\ \star & -\frac{1}{(N-1)} W_0 \end{bmatrix},$$

$$\bar{\Theta}_i = \begin{bmatrix} [\bar{z}_i + \bar{s}_i]^2 \bar{Q}_i & \bar{P}_i & 0 \\ \star & 0 & 0 \\ \star & \star & -\frac{\pi^2}{4} \bar{Q}_i \end{bmatrix}, \bar{G}_i = \begin{bmatrix} 0 \\ G_i^T \\ 0 \end{bmatrix}, \bar{\mathbb{I}}_i = \begin{bmatrix} I \\ I \\ 0 \end{bmatrix},$$

$$\tilde{\mathbb{A}}_{ils} = \begin{bmatrix} -G_i & A_{il} G_i + B_{il} \bar{K}_{is} & B_{il} \bar{K}_{is} \end{bmatrix}. \tag{5.113}$$

In that case, the controller gains can be calculated by

$$K_{il} = \bar{K}_{il} G_i^{-1}. \tag{5.114}$$

■

Note that when the asynchronized information of μ_{il} and $\hat{\mu}_{il}$ is unknown, the design result for a decentralized sampled-data linear controller can be directly derived as below:

Theorem 5.8: Decentralized Sampled-Data Linear Controller Design

The closed-loop fuzzy system in (5.84) using a decentralized sampled-data linear controller $u_i(t) = K_i x_i(t_k^i), t \in [t_k^i, t_{k+1}^i)$, is asymptotically stable, if there exist the symmetric positive definite matrices $\{\bar{P}_i, W_i, W_0, \bar{Q}_i\} \in \Re^{n_{xi} \times n_{xi}}$, $W_0 \leq W_i$, and matrices $G_i \in \Re^{n_{xi} \times n_{xi}}$, $\bar{K}_i \in \Re^{n_{ui} \times n_{xi}}$, and the positive scalars $\{\bar{z}_i, \bar{s}_i, \delta_{il}\}$, such that for all

$l \in \mathcal{L}_i$, the following LMIs hold:

$$\begin{bmatrix} \bar{\Theta}_i + \text{Sym}\left(\bar{\mathbb{I}}_i \tilde{\mathbb{A}}_{il}\right) + \sum_{\substack{j=1 \\ j \neq i}}^{N} \bar{\mathbb{I}}_i A_{ijl} W_i A_{ijl}^T \bar{\mathbb{I}}_i^T & \bar{G}_i \\ \star & -\frac{1}{(N-1)} W_0 \end{bmatrix} < 0, \tag{5.115}$$

where

$$\bar{\Theta}_i = \begin{bmatrix} [\bar{z}_i + \bar{s}_i]^2 \bar{Q}_i & \bar{P}_i & 0 \\ \star & 0 & 0 \\ \star & \star & -\frac{\pi^2}{4} \bar{Q}_i \end{bmatrix}, \bar{G}_i = \begin{bmatrix} 0 \\ G_i^T \\ 0 \end{bmatrix}, \bar{\mathbb{I}}_i = \begin{bmatrix} I \\ I \\ 0 \end{bmatrix},$$

$$\tilde{\mathbb{A}}_{il} = \begin{bmatrix} -G_i & A_{il} G_i + B_{il} \bar{K}_i & B_{il} \bar{K}_i \end{bmatrix}. \tag{5.116}$$

In that case, the proposed linear controller gains can be calculated by

$$K_{is} = \bar{K}_i G_i^{-1}. \tag{5.117}$$

■

5.4 SIMULATION STUDIES

Consider the multi-PV power system with DC load in (5.4) using the decentralized fuzzy controller (5.23). Here, the parameters of the multi-PV power are chosen as below: $L_{(1)} = 0.0526\text{H}, L_{(2)} = 0.0525\text{H}, C_{0(1)} = 0.0471\text{F}, C_{0(2)} = 0.0472\text{F}, C_{PV(1)} = 0.0101\text{F}, C_{PV(2)} = 0.0102\text{F}, R_{0(1)} = 1.1\Omega, R_{0(2)} = 1.2\Omega, R_{L(1)} = 1.7\Omega, R_{L(2)} = 1.8\Omega, R_{M(1)} = 0.84\Omega, R_{M(2)} = 0.86\Omega, R_{load} = 2.1\Omega, V_{D(1)} = 9.1\text{V}, V_{D(2)} = 9.1\text{V}$. Now, we choose $z_{i1}(t) = \frac{\phi_{PV(i)}}{v_{PV(i)}}, z_{i2}(t) = \phi_{L(i)}, z_{i3}(t) = v_{PV(i)}$, and $z_{i4}(t) = \frac{\phi_{0(i)}}{\phi_{L(i)}}$ as fuzzy premise variables, and linearize the first PV system around $\{0.037895, 1, 9.5, 0.008333\}$ and $\{0.037895, 0.5, 9.5, 0.008333\}$, and linearize the second PV power around $\{0.037895, 0.2, 9.5, 0.008333\}$ and $\{0.037895, 0.6, 9.5, 0.008333\}$. Then, the succeeding system matrices of T-S fuzzy model can be obtained as below:

$$A_{11} = \begin{bmatrix} 3.752 & 0 & 0 \\ 0 & -1783.3 & -20.913 \\ 0 & 9.6506 & -9.6506 \end{bmatrix}, A_{12} = \begin{bmatrix} 3.752 & 0 & 0 \\ 0 & -399.24 & -20.913 \\ 0 & 9.6506 & -9.6506 \end{bmatrix},$$

$$A_{21} = \begin{bmatrix} 3.715 & 0 & 0 \\ 0 & -923.81 & -19.048 \\ 0 & 9.6302 & -9.6302 \end{bmatrix}, A_{22} = \begin{bmatrix} 3.715 & 0 & 0 \\ 0 & -346.03 & -19.048 \\ 0 & 9.6302 & -9.6302 \end{bmatrix},$$

$$B_{11} = \begin{bmatrix} -49.505 \\ 345.63 \\ 0 \end{bmatrix}, B_{12} = \begin{bmatrix} -19.608 \\ 351.01 \\ 0 \end{bmatrix},$$

$$B_{21} = \begin{bmatrix} -49.505 \\ 345.63 \\ 0 \end{bmatrix}, B_{22} = \begin{bmatrix} -58.824 \\ 344.46 \\ 0 \end{bmatrix}.$$

Here, by applying Theorem 5.2, the fuzzy controller gains and observer gains are given by

$$\begin{aligned} K_{11} &= \begin{bmatrix} 9.8743 & -18.1840 & -1.7446 \end{bmatrix}, \\ K_{12} &= \begin{bmatrix} 10.0760 & -15.7490 & -1.749 \end{bmatrix}, \\ K_{21} &= \begin{bmatrix} 5.6711 & -9.8485 & -0.8076 \end{bmatrix}, \\ K_{22} &= \begin{bmatrix} 5.7024 & -8.5991 & -0.7938 \end{bmatrix}. \end{aligned}$$

It is noted that the open-loop T-S fuzzy system is unstable, as shown in Figures 5.1 and 5.2. Based on the above solutions, Figures 5.3 and 5.4 indicate that the state responses for the multi-PV power system converge to zero.

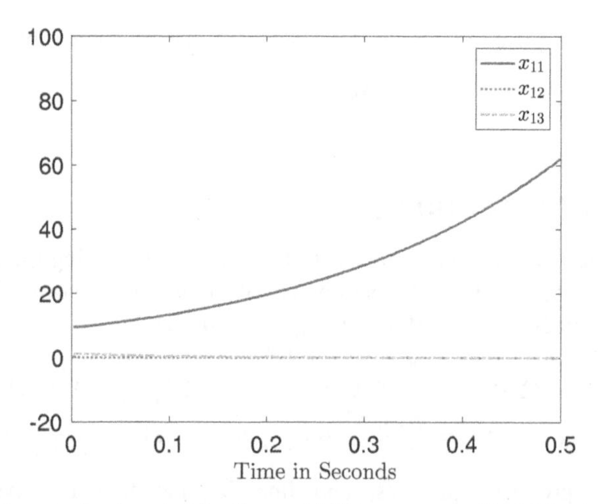

Figure 5.1 Decentralized control for first multi-PV generator with DC load. Note instability.

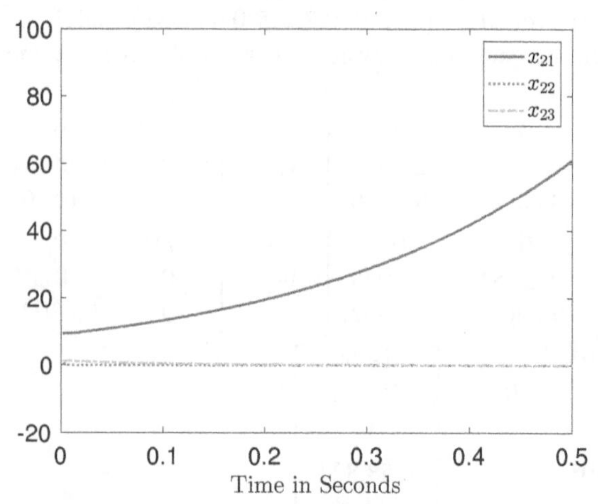

Figure 5.2 Decentralized control for second multi-PV generator with DC load. Note instability.

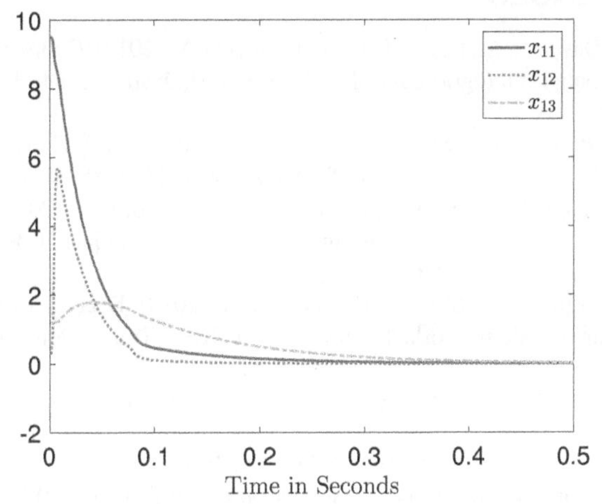

Figure 5.3 Decentralized control for first multi-PV generator with DC load. Note convergence to zero.

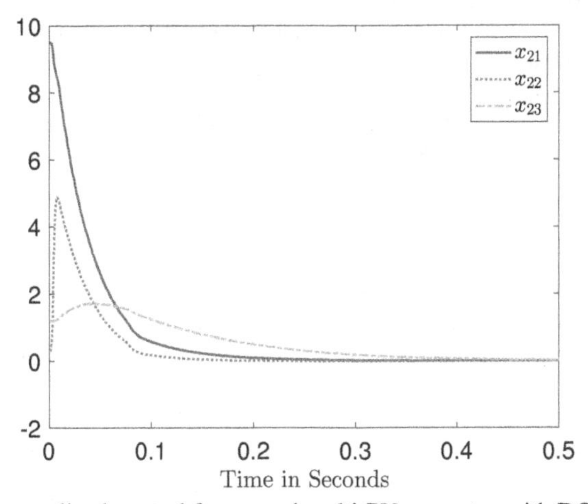

Figure 5.4 Decentralized control for second multi-PV generator with DC load. Note convergence to zero.

5.5 REFERENCES

1. Bidram, A., Davoudi, A., Lewis, F. L., and Nasirian, V. (2017). Cooperative synchronization in distributed microgrid control. Advances in Industrial Control. Berlin: Springer-Verlag.
2. Lin, W. and Bitar, E. (2018). Decentralized stochastic control of distributed energy resources. IEEE Transactions on Power Systems, 33(1): 888-900.
3. Tomislav Dragicevic, Lu, X., Vasquez, J. C., and Guerrero, J. M. (2016). DC microgrids. Part I: A review of control strategies and stabilization techniques. IEEE Transactions on Power Electronics, 31(7), 4876-4891.
4. Zhong, Z.X., Lin, C. M., Shao, Z. H., and Xu, M. (2018). Decentralized event-triggered control for large-scale networked fuzzy systems. IEEE Transactions on Fuzzy Systems, 26(1): 29-50.
5. Fridman, E. (2010). A refined input delay approach to sampled-data control. Automatica, 46(2), 421-427.
6. Lam, H. K. and Narimani, M. (2009). Stability analysis and performance design for fuzzy-model-based control system under imperfect premise matching. IEEE Transactions on Fuzzy Systems, 17(4), 949-961.
7. Arino, C. and Sala, A. (2008). Extensions to stability analysis of fuzzy control systems subject to uncertain grades of membership. IEEE Transactions on Systems, Man, and Cybernetics Part B. 38(2), 558-563.
8. Zhang, D., Han, Q. L., and Jia, X. (2017). Network-based output tracking control for a class of T-S fuzzy systems that can not be stabilized by nondelayed output feedback controllers. IEEE Transactions on Cybernetics, 45(8), 1511-1524.
9. Zhong, Z.X. and Lin C.M. (2017). Large-Scale Fuzzy Interconnected Control Systems Design and Analysis, Pennsylvania: IGI Global.
10. Moarref, M. and Rodrigues, L. (2014). Stability and stabilization of linear sampled-data systems with multi-rate samplers and time driven zero order holds. Automatica, 50(10), 2685-2691.
11. Liu, K. and Fridman, E. (2012). Wirtinger's inequality and Lyapunov-based sampled-data stabilization . Automatica, 48(1), 102-108.

6 Distributed Fuzzy Control

Nowadays, the use of renewable resources in the smart grid is distinct by its distributed nature in comparison with the grids currently used by large centralized power plants [1]. The centralized scheme will not be able to operate efficiently in the face of significantly increasing computational burdens. Therefore noncentralized techniques are better suited to provide the required functionality. Noncentralized techniques may be decentralized or distributed. Decentralized techniques ignore the interaction between units. Such techniques are not limited to the smart grid; rather they are already employed as a natural yet simplistic way to deal with large systems. As mentioned, decentralized methods assume that the interaction between subsystems is negligible. This assumption, however, is not always valid and can result in poor system-wide performance [2]. Unlike decentralized control techniques, distributed control techniques do consider the interactions among units. Assigning the control task to different units on the basis of operation in different time frames constitutes the idea of control hierarchy. Within the higher control levels the need for distributed approaches arises because of the desire and need for higher reliability, security, and situational awareness. The promise of the smart grid and developments made in the realms of communication and computation have brought such distributed control strategies closer to reality. The objective of this chapter is to provide a study of existing distributed control and management strategies [3].

In this chapter, the tracking problem of voltage synchronization of the microgrid is considered. We now consider a microgrid that includes several generators. Its nonlinear dynamics may be represented by a T-S fuzzy model:

$$\dot{x}_i(t) = A_{ii}(\mu_i)x_i(t) + B_i(\mu_i)u_i(t) + \sum_{\substack{j=1 \\ j\neq i}}^{N} A_{ij}(\mu_i)x_j(t) + \omega_i(t), \tag{6.1}$$

where $A_{ii}(\mu_i) := \sum_{l=1}^{r_i} \mu_{il}A_{iil}, B_i(\mu_i) := \sum_{l=1}^{r_i} \mu_{il}B_{il}, A_{ij}(\mu_i) := \sum_{l=1}^{r_i} \mu_{il}A_{ijl}$.

6.1 DISTRIBUTED CONTROL OF TRACKING SYNCHRONIZATION

6.1.1 DESIGN OF DISTRIBUTED FUZZY CONTROLLER

If we assume that the premise variables between the fuzzy system and controller are synchronous, the distributed fuzzy controller is given by

$$u_i(t) = K_{ii}(\mu_i)x_i(t) + \sum_{\substack{j=1 \\ j\neq i}}^{N} K_{ij}(\mu_i)x_j(t), \tag{6.2}$$

where $K_{ii}(\mu_i) := \sum_{l=1}^{r_i} \mu_{il} K_{iil}, K_{ij}(\mu_i) := \sum_{l=1}^{r_i} \mu_{il} K_{ijl}, K_{iil} \in \Re^{n_{ui} \times n_{xi}}$ and $K_{ijl} \in \Re^{n_{ui} \times n_{xj}}$ are controller gains to be determined.

Combining (6.1) and (6.2), the i-th closed-loop fuzzy control system is given by

$$\dot{x}_i(t) = \bar{A}_{ii}(\mu_i)x_i(t) + \sum_{\substack{j=1 \\ j \neq i}}^{N} \bar{A}_{ij}(\mu_i)x_j(t), \tag{6.3}$$

where $\bar{A}_{ii}(\mu_i) = A_{ii}(\mu_i) + B_i(\mu_i)K_{ii}(\mu_i), \bar{A}_{ij}(\mu_i) = A_{ij}(\mu_i) + B_i(\mu_i)K_{ij}(\mu_i)$.

Here, a sufficient condition for the stability of the closed-loop fuzzy control system in (6.3) is given by the following theorem.

Lemma 6.1: Stability Analysis of Distributed Fuzzy Control

Given the large-scale T-S fuzzy system in (6.1) and a distributed fuzzy controller in the form of (6.2), the closed-loop fuzzy control system in (6.3) is asymptotically stable, if there exist the symmetric positive definite matrices $\{P_i, W_i\} \in \Re^{n_{xi} \times n_{xi}}$, and the matrices $\bar{K}_{ii}(\mu_i) \in \Re^{n_{ui} \times n_{xi}}$, and the matrices $\bar{K}_{ij}(\mu_i) \in \Re^{n_{ui} \times n_{xj}}$, such that the following matrix inequalities hold:

$$\mathrm{Sym}\left(P_i\bar{A}_{ii}(\mu_i)\right) + \sum_{\substack{j=1 \\ j \neq i}}^{N} P_i\bar{A}_{ij}(\mu_i)W_i\bar{A}_{ij}^T(\mu_i)P_i + \sum_{\substack{j=1 \\ j \neq i}}^{N} W_j^{-1} < 0, \tag{6.4}$$

where $\bar{A}_{ii}(\mu_i) = A_{ii}(\mu_i) + B_i(\mu_i)K_i(\mu_i), \bar{A}_{ij}(\mu_i) = A_{ij}(\mu_i) + B_i(\mu_i)K_{ij}(\mu_i)$. ■

Proof. Consider the following Lyapunov functional,

$$\begin{aligned} V(t) &= \sum_{i=1}^{N} V_i(t) \\ &= \sum_{i=1}^{N} x_i^T(t)P_ix_i(t), \end{aligned} \tag{6.5}$$

where $P_i \in \Re^{n_{xi} \times n_{xi}}$ is the symmetric positive definite matrix.

By taking the time derivative of $V(t)$ along the trajectory of the system in (6.3),

$$\dot{V}(t) = 2\sum_{i=1}^{N} x_i^T(t)P_i\left[\bar{A}_{ii}(\mu_i)x_i(t) + \sum_{\substack{j=1\\ j\neq i}}^{N}\bar{A}_{ij}(\mu_i)x_j(t)\right]$$
$$= \sum_{i=1}^{N} x_i^T(t)\left\{\text{Sym}\left(P_i\bar{A}_{ii}(\mu_i)\right)\right\}x_i(t) + 2\sum_{i=1}^{N}\sum_{\substack{j=1\\ j\neq i}}^{N} x_i^T(t)P_i\bar{A}_{ij}(\mu_i)x_j(t). \tag{6.6}$$

Note that

$$2\bar{x}^T\bar{y} \leq \bar{x}^T M^{-1}\bar{x} + \bar{y}^T M\bar{y}, \tag{6.7}$$

where $\bar{x}, \bar{y} \in \Re^n$ and symmetric matrix $M > 0$.

Define $W_i = W_i^T > 0$, and by using (6.7), we have

$$2\sum_{i=1}^{N}\sum_{\substack{j=1\\ j\neq i}}^{N} x_i^T(t)P_i\bar{A}_{ij}(\mu_i)x_j(t)$$
$$\leq \sum_{i=1}^{N}\sum_{\substack{j=1\\ j\neq i}}^{N} x_i^T(t)P_i\bar{A}_{ij}(\mu_i)W_i\bar{A}_{ij}^T(\mu_i)P_ix_i(t) + \sum_{i=1}^{N}\sum_{\substack{j=1\\ j\neq i}}^{N} x_j^T(t)W_i^{-1}x_j(t). \tag{6.8}$$

It is easy to obtain the stability result on (6.4). Thus, the proof is completed.

Note that the result is not LMI-based. In the following, by using some linearization techniques of matrix inequality, the distributed sampled-data controller design result is derived in terms of LMIs.

Theorem 6.1: Design of Distributed Fuzzy Controller

Consider the large-scale T-S fuzzy system in (6.1) and a distributed fuzzy controller in the form of (6.2). The closed-loop fuzzy control system in (6.3) is asymptotically stable, if there exist the symmetric positive definite matrices $\{P_i, W_i, W_0\} \in \Re^{n_{xi}\times n_{xi}}$, $W_0 \leq W_i$, and the matrices $\bar{K}_{iil} \in \Re^{n_{ui}\times n_{xi}}$, and the matrices $\bar{K}_{ijl} \in \Re^{n_{ui}\times n_{xj}}$, such that the following LMIs hold:

$$\Sigma_{ill} < 0, l \in \mathscr{L}_i \tag{6.9}$$
$$\Sigma_{ils} + \Sigma_{isl} < 0, 1 \leq l < s \leq r_i \tag{6.10}$$

where

$$\Sigma_{ils} = \begin{bmatrix} \text{Sym}\left(A_{iil}X_i + B_{il}\bar{K}_{iis}\right) & X_i & \bar{\mathbb{A}}_{ijls} \\ \star & -\frac{1}{(N-1)}W_0 & 0 \\ \star & \star & \mathbb{W}_i - 2\mathbb{X}_i \end{bmatrix},$$

$$\mathbb{W}_i = \text{diag}\underbrace{\{W_1 \cdots W_{j,i\neq j} \cdots W_N\}}_{N-1}, \mathbb{X}_i = \text{diag}\underbrace{\{X_i \cdots X_i \cdots X_i\}}_{N-1},$$

$$\bar{\mathbb{A}}_{ijls} = \underbrace{\left[\bar{A}_{i1ls} \cdots \bar{A}_{ijls,i\neq j} \cdots \bar{A}_{iNls}\right]}_{N-1}, \bar{A}_{ijls} = B_{il}\bar{K}_{ijs} + A_{ijl}X_i. \tag{6.11}$$

In that case, the decentralized controller gains can be calculated by

$$K_{iis} = \bar{K}_{iis}X_i^{-1}, K_{ijs} = \bar{K}_{ijs}X_i^{-1}, s \in \mathscr{L}_i \tag{6.12}$$

■

Proof. By defining $W_0 \leq W_i, i \in \mathscr{N}$, we have

$$W_0^{-1} \geq W_i^{-1}, \tag{6.13}$$

which implies $(N-1)W_0^{-1} \geq \sum_{\substack{j=1 \\ j\neq i}}^{N} W_j^{-1}$.

By using Schur complement lemma, it is easy to see that the following inequality implies (6.4),

$$\begin{bmatrix} \text{Sym}\left(P_i\bar{A}_{ii}(\mu_i)\right) & I & \bar{\mathbb{A}}_{ij}(\mu_i) \\ \star & -\frac{1}{(N-1)}W_0 & 0 \\ \star & \star & -\mathbb{W}_i^{-1} \end{bmatrix} < 0, \tag{6.14}$$

where $\mathbb{W}_i = \text{diag}\underbrace{\{W_1 \cdots W_{j,i\neq j} \cdots W_N\}}_{N-1}, \bar{\mathbb{A}}_{ij}(\mu_i) = \underbrace{\left[P_i\bar{A}_{ij}(\mu_i) \cdots P_i\bar{A}_{ij}(\mu_i) \cdots P_i\bar{A}_{iN}(\mu_i)\right]}_{N-1}$.

Note that

$$W_i - X_i - X_i^T + X_i^T W_i^{-1} X_i = (W_i - X_i)^T W_i^{-1}(W_i - X_i) \geq 0, \tag{6.15}$$

where $X_i = X_i^T > 0$, which implies that

$$-X_i^T W_i^{-1} X_i \leq W_i - X_i - X_i^T. \tag{6.16}$$

Define $X_i = P_i^{-1}$, and by performing the congruence transformation to (6.14) by $\Gamma_i = \text{diag}\{X_i, I, \underbrace{X_i, X_i, X_i}_{N-1}\}$, and extracting the fuzzy premise variables, the results on (6.9) and (6.10) are easy to obtain, thus completing this proof.

6.1.2 DESIGN OF DISTRIBUTED SAMPLED-DATA CONTROLLER

Before moving on, the following assumptions must be considered [4].

Assumption 6.1. The sampler in each subsystem is clock-driven. If we let h_i denote the upper bound of sampling intervals, we have

$$t_{k+1}^i - t_k^i \leq h_i, k \in \mathbb{N} \tag{6.17}$$

where $h_i > 0$.

Assumption 6.2. The zero-order-hold (ZOH) is event-driven, and it uses the latest sampled-data signals and holds them until the next transmitted data are received.

Define $\rho_i^s(t) = t - t_k^i$, it has

$$0 \leq \rho_i^s(t) < h_i. \tag{6.18}$$

It is noted that in the context of digital control systems, both $z_i(t)$ and $x_i(t)$ are involved in the sampled-data measurement. Now, without loss of generality, we further assume that both $z_i(t)$ and $x_i(t)$ are packed, transmitted, and updated at the same time. Then, a distributed sampled-data fuzzy controller can be given by

Controller Rule $\mathscr{R}_i^s$: **IF** $z_{i1}(t_k^i)$ is $\mathscr{F}_{i1}^s$ and $z_{i2}(t_k^i)$ is $\mathscr{F}_{i2}^s$ and $\cdots$ and $z_{ig}(t_k^i)$ is $\mathscr{F}_{ig}^s$, **THEN**

$$u_i(t) = K_{iis}x_i(t_k^i) + \sum_{j=1, i\neq j}^{N} K_{ijs}x_j(t_k^j), t \in [t_k^i, t_{k+1}^i) \tag{6.19}$$

where $K_{iis} \in \Re^{n_{ui}\times n_{xi}}, K_{ijs} \in \Re^{n_{ui}\times n_{xj}}, s \in \mathscr{L}_i, i \in N$ are controller gains to be determined; $z_i(t_k^i) := [z_{i1}(t_k^i), z_{i2}(t_k^i), \ldots, z_{iG}(t_k^i)]$; $z_i(t_k^i)$ and $x_i(t_k^i)$ denote the updating signals in the fuzzy controller.

Similarly, the overall distributed fuzzy controller is

$$u_i(t) = K_{ii}(\hat{\mu}_i)x_i(t_k^i) + \sum_{j=1, i\neq j}^{N} K_{ij}(\hat{\mu}_i)x_j(t_k^j), t \in [t_k^i, t_{k+1}^i) \tag{6.20}$$

where

$$\begin{cases} K_{ii}(\hat{\mu}_i) := \sum_{s=1}^{r_i} \hat{\mu}_{is}\left[z_i(t_k^i)\right] K_{iis}, \sum_{s=1}^{r_i} \hat{\mu}_{is}\left[z_i(t_k^i)\right] = 1, \\ K_{ij}(\hat{\mu}_i) := \sum_{s=1}^{r_i} \hat{\mu}_{is}\left[z_i(t_k^i)\right] K_{ijs}, \\ \hat{\mu}_{is}\left[z_i(t_k^i)\right] := \frac{\Pi_{\phi=1}^{g}\hat{\mu}_{is\phi}\left[z_{i\phi}(t_k^i)\right]}{\Sigma_{\varsigma=1}^{r_i}\Pi_{\phi=1}^{g}\hat{\mu}_{i\varsigma\phi}\left[z_{i\phi}(t_k^i)\right]} \geq 0. \end{cases} \tag{6.21}$$

In the following, we will denote $\hat{\mu}_{is} := \hat{\mu}_{is}\left[z_i(t_k^i)\right]$ for brevity.

We further define

$$x_i(v) = x_i(t_k^i) - x_i(t), x_j(v) = x_j(t_k^j) - x_j(t). \tag{6.22}$$

Submitting (6.22) into the fuzzy controller (6.20) and combining it with the fuzzy system (6.1), the closed-loop fuzzy control system is

$$\dot{x}_i(t) = \bar{A}_{ii}(\mu_i,\hat{\mu}_i)x_i(t) + B_i(\mu_i)K_{ii}(\hat{\mu}_i)x_i(v) + \sum_{j=1,i\neq j}^{N} \bar{A}_{ij}(\mu_i,\hat{\mu}_i)x_j(t) + \sum_{\substack{j=1\\ j\neq i}}^{N} B_i(\mu_i)K_{ij}(\hat{\mu}_i)x_j(v), \tag{6.23}$$

where $\bar{A}_{ii}(\mu_i,\hat{\mu}_i) = A_{ii}(\mu_i) + B_i(\mu_i)K_{ii}(\hat{\mu}_i), \bar{A}_{ij}(\mu_i,\hat{\mu}_i) = A_{ij}(\mu_i) + B_i(\mu_i)K_{ij}(\hat{\mu}_i)$.

Note: When the premise variables on controller (6.20) are not considered in sampled-data measurement, the premise variables between the fuzzy system and the fuzzy controller are synchronous. Here, we just consider the asynchronous sampled-data measurement. However, the obtained result can be easily extended to the synchronous case.

Note: The premise variable $z_i(t)$ undergoes time-driven sensors and event-driven ZOHs, and is implemented by the proposed fuzzy controller in (6.20). Hence the premise variable spaces in asynchronous form between $z_i(t)$ and $z_i(t_k^i)$ are more practical.

Note: As pointed out in [5], when the knowledge between μ_{il} and $\hat{\mu}_{il}$ is unavailable, the condition $\mu_{il} \neq \hat{\mu}_{il}$ generally leads to a linear controller instead of a fuzzy one, which degrades the stabilization ability of the controller. When the knowledge on μ_{il} and $\hat{\mu}_{il}$ is available, the design conservatism can be improved and produce the corresponding fuzzy controller.

Note: The t_k^i value is relative to the clock on the i-th subsystem. In other words, the sampled-data clocks can be different among all subsystems.

Now, we introduce the following Lyapunov function:

$$\begin{aligned} V(t) &= \sum_{i=1}^{N} V_i(t) \\ &= \sum_{i=1}^{N} x_i^T(t) P_i x_i(t) + \sum_{i=1}^{N} h_i^2 \int_{t_k^i}^{t} \dot{x}_i^T(\alpha) Q_i \dot{x}_i(\alpha)\, d\alpha \\ &\quad - \sum_{i=1}^{N} \frac{\pi^2}{4} \int_{t_k^i}^{t} \left[x_i(\alpha) - x_i(t_k^i)\right]^T Q_i \left[x_i(\alpha) - x_i(t_k^i)\right] d\alpha, \end{aligned} \tag{6.24}$$

where $\{P_i, Q_i\} \in \Re^{n_{xi}\times n_{xi}}$ are positive definite symmetric matrices.

By using Wirtinger's inequality in [4], $V(t) > 0$. Based on the new model in (6.23) and the Lyapunov function in (6.24), a sufficient condition for existing a distributed sampled-data controller can be given as below:

Lemma 6.2: Stability Analysis of Distributed Sampled-Data Control

The closed-loop fuzzy system in (6.23) using a distributed sampled-data fuzzy controller (6.20), is asymptotically stable, if there exist the symmetric positive definite matrices $\{P_i, W_{i1}, W_{i2}\} \in \Re^{n_{xi} \times n_{xi}}$, $K_{ii}(\hat{\mu}_i) \in \Re^{n_{ui} \times n_{xi}}$, $K_{ij}(\hat{\mu}_i) \in \Re^{n_{ui} \times n_{xj}}$, and the positive scalars h_i, such that the following matrix inequalities hold:

$$\Theta_i + \mathrm{Sym}\left(\mathbb{G}_i \mathbb{A}_i(\mu_i, \hat{\mu}_i)\right) + \sum_{\substack{j=1 \\ j \neq i}}^{N} \mathbb{G}_i \bar{A}_{ij}(\mu_i, \hat{\mu}_i) W_{i1} \bar{A}_{ij}^T(\mu_i, \hat{\mu}_i) \mathbb{G}_i^T$$

$$+ \sum_{\substack{j=1 \\ j \neq i}}^{N} \mathbb{G}_i B_i(\mu_i) K_{ij}(\hat{\mu}_i) W_{i2} K_{ij}^T(\hat{\mu}_i) B_i^T(\mu_i) \mathbb{G}_i^T < 0, \tag{6.25}$$

where

$$\Theta_i = \begin{bmatrix} h_i^2 Q_i & P_i & 0 \\ \star & \sum_{\substack{j=1 \\ j \neq i}}^{N} W_{j1}^{-1} & 0 \\ \star & \star & -\frac{\pi^2}{4} Q_i + \sum_{\substack{j=1 \\ j \neq i}}^{N} W_{j2}^{-1} \end{bmatrix},$$

$$\mathbb{A}_i(\mu_i, \hat{\mu}_i) = \begin{bmatrix} -I & \bar{A}_{ii}(\mu_i, \hat{\mu}_i) & B_i(\mu_i) K_{ii}(\hat{\mu}_i) \end{bmatrix}. \tag{6.26}$$

■

Proof. By taking the time derivative of $V(t)$ in (6.24), we have

$$\dot{V}(t) = \sum_{i=1}^{N} \{2 x_i^T(t) P_i \dot{x}_i(t) + h_i^2 \dot{x}_i^T(t) Q_i \dot{x}_i(t) - \frac{\pi^2}{4} x_i^T(v) Q_i x_i(v)\}. \tag{6.27}$$

Define the matrix $\mathbb{G}_i \in \Re^{3n_{xi} \times n_{xi}}$ and $\chi_i(t) = \begin{bmatrix} \dot{x}_i^T(t) & x_i^T(t) & x_i^T(v) \end{bmatrix}^T$, and it follows from (6.23) that

$$0 = \sum_{i=1}^{N} 2\chi_i^T(t) \mathbb{G}_i \mathbb{A}_i(\mu_i, \hat{\mu}_i) \chi_i(t) + \sum_{i=1}^{N} 2\chi_i^T(t) \mathbb{G}_i \sum_{j=1, i \neq j}^{N} \bar{A}_{ij}(\mu_i, \hat{\mu}_i) x_j(t)$$

$$+ \sum_{i=1}^{N} 2\chi_i^T(t) \mathbb{G}_i \sum_{\substack{j=1 \\ j \neq i}}^{N} B_i(\mu_i) K_{ij}(\hat{\mu}_i) x_j(v), \tag{6.28}$$

where $\mathbb{A}_i(\mu_i, \hat{\mu}_i) = \begin{bmatrix} -I & \bar{A}_{ii}(\mu_i, \hat{\mu}_i) & B_i(\mu_i) K_{ii}(\hat{\mu}_i) \end{bmatrix}$.

By introducing matrix $0 < W_{i1} = W_{i1}^T \in \Re^{n_{xi} \times n_{xi}}$, and $0 < W_{i2} = W_{i2}^T \in \Re^{n_{xi} \times n_{xi}}$, and using the relation of (6.7), we have

$$\begin{aligned}
&\sum_{i=1}^{N} 2\chi_i^T(t)\mathbb{G}_i \sum_{\substack{j=1\\ j\neq i}}^{N} \bar{A}_{ij}(\mu_i,\hat{\mu}_i)x_j(t) \\
&\leq \sum_{i=1}^{N}\sum_{\substack{j=1\\ j\neq i}}^{N} \chi_i^T(t)\mathbb{G}_i\bar{A}_{ij}(\mu_i,\hat{\mu}_i)W_{i1}\bar{A}_{ij}^T(\mu_i,\hat{\mu}_i)\mathbb{G}_i^T\chi_i(t) + \sum_{i=1}^{N}\sum_{\substack{j=1\\ j\neq i}}^{N} x_j^T(t)W_{i1}^{-1}x_j(t) \\
&= \sum_{i=1}^{N}\sum_{\substack{j=1\\ j\neq i}}^{N} \chi_i^T(t)\mathbb{G}_i\bar{A}_{ij}(\mu_i,\hat{\mu}_i)W_{i1}\bar{A}_{ij}^T(\mu_i,\hat{\mu}_i)\mathbb{G}_i^T\chi_i(t) + \sum_{i=1}^{N}\sum_{\substack{j=1\\ j\neq i}}^{N} x_i^T(t)W_{j1}^{-1}x_i(t),
\end{aligned} \tag{6.29}$$

and

$$\begin{aligned}
&\sum_{i=1}^{N} 2\chi_i^T(t)\mathbb{G}_i \sum_{\substack{j=1\\ j\neq i}}^{N} B_i(\mu_i)K_{ij}(\hat{\mu}_i)x_j(v) \\
&\leq \sum_{i=1}^{N}\sum_{\substack{j=1\\ j\neq i}}^{N} \chi_i^T(t)\mathbb{G}_iB_i(\mu_i)K_{ij}(\hat{\mu}_i)W_{i2}K_{ij}^T(\hat{\mu}_i)B_i^T(\mu_i)\mathbb{G}_i^T\chi_i(t) + \sum_{i=1}^{N}\sum_{\substack{j=1\\ j\neq i}}^{N} x_j^T(v)W_{i2}^{-1}x_j(v) \\
&= \sum_{i=1}^{N}\sum_{\substack{j=1\\ j\neq i}}^{N} \chi_i^T(t)\mathbb{G}_iB_i(\mu_i)K_{ij}(\hat{\mu}_i)W_{i2}K_{ij}^T(\hat{\mu}_i)B_i^T(\mu_i)\mathbb{G}_i^T\chi_i(t) + \sum_{i=1}^{N}\sum_{\substack{j=1\\ j\neq i}}^{N} x_i^T(v)W_{j2}^{-1}x_i(v).
\end{aligned} \tag{6.30}$$

It follows from (6.27)-(6.30) that the result on (6.25) can be obtained directly.

The result from (6.25) is not LMI-based. When the asynchronized information of μ_{il} and $\hat{\mu}_{il}$ is unknown, the designed result generally leads to the linear controller instead of the fuzzy one [6]. From a practical perspective, obtaining a priori knowledge of μ_{il} and $\hat{\mu}_{il}$ is possible. Thus, we assume that the asynchronized condition is subject to

$$\underline{\rho}_{il} \leq \frac{\hat{\mu}_{il}}{\mu_{il}} \leq \bar{\rho}_{il}, \tag{6.31}$$

where $\underline{\rho}_{il}$ and $\bar{\rho}_{il}$ are positive scalars.

It follows (6.25) and (6.31) that the design result on the distributed sampled-data fuzzy controller can be summarized as below:

Theorem 6.2: Design of Distributed Sampled-Data Fuzzy Control Using Asynchronized Method

The closed-loop fuzzy system in (6.23) using a distributed sampled-data fuzzy controller (6.20), is asymptotically stable, if there exist the symmetric positive definite matrices $\{\bar{P}_i, W_{i1}, W_{i2}, W_{01}, W_{02}, \bar{Q}_i\} \in \Re^{n_{xi} \times n_{xi}}$, $W_{01} \leq W_{i1}$, $W_{02} \leq W_{i2}$, and matrices $G_i \in \Re^{n_{xi} \times n_{xi}}$, $M_{ils} = M_{isl}^T \in \Re^{4n_{xi} \times 4n_{xi}}$, $\bar{K}_{iis} \in \Re^{n_{ui} \times n_{xi}}$, $\bar{K}_{ijs} \in \Re^{n_{ui} \times n_{xj}}$, and the positive scalars $\{h_i, \underline{\rho}_{il}, \bar{\rho}_{il}\}$, such that for all $(l,s) \in \mathscr{L}_i$, the following LMIs hold:

$$\bar{\rho}_{il}\Sigma_{ill} + M_{ill} < 0, \tag{6.32}$$

$$\underline{\rho}_{il}\Sigma_{ill} + M_{ill} < 0, \tag{6.33}$$

$$\bar{\rho}_{is}\Sigma_{ils} + \bar{\rho}_{il}\Sigma_{isl} + M_{ils} + M_{isl} < 0, \tag{6.34}$$

$$\underline{\rho}_{is}\Sigma_{ils} + \underline{\rho}_{il}\Sigma_{isl} + M_{ils} + M_{isl} < 0, \tag{6.35}$$

$$\underline{\rho}_{is}\Sigma_{ils} + \bar{\rho}_{il}\Sigma_{isl} + M_{ils} + M_{isl} < 0, \tag{6.36}$$

$$\bar{\rho}_{is}\Sigma_{ils} + \underline{\rho}_{il}\Sigma_{isl} + M_{ils} + M_{isl} < 0, \tag{6.37}$$

$$\begin{bmatrix} M_{i11} & \cdots & M_{i1r} \\ \vdots & \ddots & \vdots \\ M_{ir1} & \cdots & M_{irr} \end{bmatrix} > 0, \tag{6.38}$$

where

$$\begin{aligned}
&\Sigma_{ils} = \begin{bmatrix} \tilde{\Theta}_{i1} + \mathrm{Sym}\left(\mathbb{I}_i \bar{\mathbb{A}}_{ils}\right) & \bar{G}_{i1} & \bar{G}_{i2} & \bar{\Theta}_{i2ls} & \bar{\Theta}_{i3ls} \\ \star & -\frac{1}{(N-1)} W_{01} & 0 & 0 & 0 \\ \star & \star & -\frac{1}{(N-1)} W_{02} & 0 & 0 \\ \star & \star & \star & \bar{\mathbb{W}}_1 & 0 \\ \star & \star & \star & \star & \bar{\mathbb{W}}_2 \end{bmatrix}, \\
&\tilde{\Theta}_{i1} = \begin{bmatrix} h_i^2 \bar{Q}_i & \bar{P}_i & 0 \\ \star & 0 & 0 \\ \star & \star & -\frac{\pi^2}{4} \bar{Q}_i \end{bmatrix}, \bar{G}_{i1} = \begin{bmatrix} 0 \\ G_i^T \\ 0 \end{bmatrix}, \bar{G}_{i2} = \begin{bmatrix} 0 \\ 0 \\ G_i^T \end{bmatrix}, \mathbb{I}_i = \begin{bmatrix} I \\ I \\ 0 \end{bmatrix}, \\
&\bar{\Theta}_{i2ls} := \underbrace{\begin{bmatrix} \bar{\Theta}_{i12ls} & \cdots & \bar{\Theta}_{ij2ls} & \cdots & \bar{\Theta}_{iN2ls} \end{bmatrix}}_{N-1}, \bar{\Theta}_{ij2ls} = \mathbb{I}_i \left(A_{ijl} G_i + B_{il} \bar{K}_{ijs}\right), \\
&\bar{\Theta}_{i3ls} := \underbrace{\begin{bmatrix} \bar{\Theta}_{i13ls} & \cdots & \bar{\Theta}_{ij3ls} & \cdots & \bar{\Theta}_{iN3ls} \end{bmatrix}}_{N-1}, \bar{\Theta}_{ij3ls} = \mathbb{I}_i B_{il} \bar{K}_{ijs}, \\
&\bar{\mathbb{A}}_{ils} = \begin{bmatrix} -G_i & A_{iil} G_i + B_{il} \bar{K}_{iis} & B_{il} \bar{K}_{iis} \end{bmatrix}, \bar{\mathbb{W}}_1 = \mathbb{W}_1 - \bar{\mathbb{G}}_i - \bar{\mathbb{G}}_i^T, \bar{\mathbb{W}}_2 = \mathbb{W}_2 - \bar{\mathbb{G}}_i - \bar{\mathbb{G}}_i^T, \\
&\mathbb{W}_1 := \mathrm{diag}\underbrace{\{ W_{i1} \ \cdots \ W_{i1} \ \cdots \ W_{i1} \}}_{N-1}, \\
&\mathbb{W}_2 := \mathrm{diag}\underbrace{\{ W_{i2} \ \cdots \ W_{i2} \ \cdots \ W_{i2} \}}_{N-1}, \bar{\mathbb{G}}_i := \mathrm{diag}\underbrace{\{ G_i \ \cdots \ G_i \ \cdots \ G_i \}}_{N-1}.
\end{aligned} \tag{6.39}$$

In that case, the proposed sampled-data fuzzy controller gains can be calculated by

$$K_{iis} = \bar{K}_{iis} G_i^{-1}, K_{ijs} = \bar{K}_{ijs} G_i^{-1}, s \in \mathscr{L}_i. \tag{6.40}$$

■

Proof. Define $W_{01} \leq W_{i1}$ and $W_{02} \leq W_{i2}, i \in N$, it has

$$W_{01}^{-1} \geq W_{i1}^{-1}, W_{02}^{-1} \geq W_{i2}^{-1}, \tag{6.41}$$

which implies $(N-1)W_{01}^{-1} \geq \sum_{j=1, j\neq i}^{N} W_{j1}^{-1}$ and $(N-1)W_{02}^{-1} \geq \sum_{j=1, j\neq i}^{N} W_{j2}^{-1}$, respectively.

By using the Schur complement lemma and the relation (6.41), it is easy to see that the following inequality implies (6.25),

$$\begin{bmatrix} \bar{\Theta}_{i1} + \mathrm{Sym}(\mathbb{G}_i \mathbb{A}_i(\mu_i, \hat{\mu}_i)) + \sum_{j=1, j\neq i}^{N} \Pi_{ij}(\mu_i, \hat{\mu}_i) & \bar{I}_1 & \bar{I}_2 \\ \star & -\frac{1}{(N-1)} W_{01} & 0 \\ \star & \star & -\frac{1}{(N-1)} W_{02} \end{bmatrix} < 0, \tag{6.42}$$

where

$$\begin{cases} \Pi_{ij}(\mu_i, \hat{\mu}_i) = \bar{\Theta}_{ij2}(\mu_i, \hat{\mu}_i) W_{i1} \bar{\Theta}_{ij2}^T(\mu_i, \hat{\mu}_i) + \bar{\Theta}_{ij3}(\mu_i, \hat{\mu}_i) W_{i2} \bar{\Theta}_{ij3}^T(\mu_i, \hat{\mu}_i), \\ \bar{\Theta}_{i1} = \begin{bmatrix} h_i^2 Q_i & P_i & 0 \\ \star & 0 & 0 \\ \star & \star & -\frac{\pi^2}{4} Q_i \end{bmatrix}, \bar{I}_1 = \begin{bmatrix} 0 \\ I \\ 0 \end{bmatrix}, \bar{I}_2 = \begin{bmatrix} 0 \\ 0 \\ I \end{bmatrix}, \\ \bar{\Theta}_{ij2}(\mu_i, \hat{\mu}_i) = \mathbb{G}_i \bar{A}_{ij}(\mu_i, \hat{\mu}_i), \bar{\Theta}_{ij3}(\mu_i, \hat{\mu}_i) = \mathbb{G}_i B_i(\mu_i) K_{ij}(\hat{\mu}_i). \end{cases} \tag{6.43}$$

It follows from (6.32) that

$$h_i^2 Q_i - \mathrm{Sym}\{G_i\} < 0, i \in N \tag{6.44}$$

which implies that $G_i, i \in N$ are nonsingular matrices.

We further define

$$\begin{cases} \mathbb{G}_i = \begin{bmatrix} G_i^{-1} & G_i^{-1} & 0 \end{bmatrix}^T, \Gamma_1 := \mathrm{diag}\{ G_i \ G_i \ G_i \ I \ I \ \bar{\mathbb{G}}_i \ \bar{\mathbb{G}}_i \}, \\ \mathbb{I} := \mathrm{diag}\underbrace{\{ I \ \cdots \ I \ \cdots \ I \}}_{N-1}, \bar{P}_i = G_i^T P_i G_i, \bar{Q}_i = G_i^T Q_i G_i, \bar{K}_{ijs} = K_{ijs} G_i, \\ \mathbb{W}_1 := \mathrm{diag}\underbrace{\{ W_{i1} \ \cdots \ W_{i1} \ \cdots \ W_{i1} \}}_{N-1}, \bar{K}_{iis} = K_{iis} G_i, \\ \mathbb{W}_2 := \mathrm{diag}\underbrace{\{ W_{i2} \ \cdots \ W_{i2} \ \cdots \ W_{i2} \}}_{N-1}, \\ \bar{\mathbb{G}}_i := \mathrm{diag}\underbrace{\{ G_i \ \cdots \ G_i \ \cdots \ G_i \}}_{N-1}, \\ \bar{\Theta}_{i2}(\mu_i, \hat{\mu}_i) := \underbrace{\begin{bmatrix} \bar{\Theta}_{i12}(\mu_i, \hat{\mu}_i) & \cdots & \bar{\Theta}_{ij2}(\mu_i, \hat{\mu}_i) & \cdots & \bar{\Theta}_{iN2}(\mu_i, \hat{\mu}_i) \end{bmatrix}}_{N-1}, \\ \bar{\Theta}_{i3}(\mu_i, \hat{\mu}_i) := \underbrace{\begin{bmatrix} \bar{\Theta}_{i13}(\mu_i, \hat{\mu}_i) & \cdots & \bar{\Theta}_{ij3}(\mu_i, \hat{\mu}_i) & \cdots & \bar{\Theta}_{iN3}(\mu_i, \hat{\mu}_i) \end{bmatrix}}_{N-1}, \\ \bar{\Theta}_{ij2}(\mu_i, \hat{\mu}_i) = \mathbb{G}_i \bar{A}_{ij}(\mu_i, \hat{\mu}_i), \bar{\Theta}_{ij3}(\mu_i, \hat{\mu}_i) = \mathbb{G}_i B_i(\mu_i) K_{ij}(\hat{\mu}_i). \end{cases} \tag{6.45}$$

By substituting (6.45) into (6.42), and using the Schur complement lemma, one has

$$\begin{bmatrix} \bar{\Theta}_{i1}+\mathrm{Sym}(\mathbb{G}_i\mathbb{A}_i(\mu_i,\hat{\mu}_i)) & \mathbb{I}_1 & \mathbb{I}_2 & \bar{\Theta}_{i2}(\mu_i,\hat{\mu}_i) & \bar{\Theta}_{i3}(\mu_i,\hat{\mu}_i) \\ \star & -\frac{W_{01}}{(N-1)} & 0 & 0 & 0 \\ \star & \star & -\frac{W_{02}}{(N-1)} & 0 & 0 \\ \star & \star & \star & -\mathbb{W}_1^{-1} & 0 \\ \star & \star & \star & \star & -\mathbb{W}_2^{-1} \end{bmatrix} < 0. \tag{6.46}$$

Note that

$$W_i - G_i - G_i^T + G_i^T W_i^{-1} G_i = (W_i - G_i)^T W_i^{-1} (W_i - G_i) \geq 0, \tag{6.47}$$

where $W_i = W_i^T > 0$, which implies that

$$-G_i^T W_i^{-1} G_i \leq W_i - G_i - G_i^T. \tag{6.48}$$

By performing a congruence transformation by Γ_1, and extracting the fuzzy membership functions, and using the relationship (6.47), we have

$$\sum_{l=1}^{r_i}\sum_{s=1}^{r_i} \mu_{il}\hat{\mu}_{is}\Sigma_{ils} < 0, \tag{6.49}$$

where Σ_{ils} is defined in (6.39).

By using the asynchronous method proposed in [5], the inequalities (6.32)-(6.38) are obtained. Thus, completing this proof.

Note that the number of LMIs on Thoerem 6.2 is large. It is also noted that the existing relaxation technique $\sum_{l=1}^{r_i}[\mu_{il}]^2\Sigma_{ill}+\sum_{l=1}^{r_i}\sum_{l<s\leq r_i}^{r_i}\mu_{il}\mu_{is}\Sigma_{ls}<0$ is no longer applicable to fuzzy controller synthesis, since $\mu_{is} \neq \hat{\mu}_{is}$. Similarly to the relaxation technique in [7], it is assumed that

$$|\mu_{il}-\hat{\mu}_{il}| \leq \delta_{il}, l \in \mathscr{L}_i, \tag{6.50}$$

where δ_{il} is a positive scalar. If $\Sigma_{ils}+M_{il} \geq 0$, where M_{il} is a symmetric matrix, we have

$$\begin{aligned} &\sum_{l=1}^{r_i}\sum_{s=1}^{r_i} \mu_{il}\hat{\mu}_{is}\Sigma_{ils} \\ &= \sum_{l=1}^{r_i}\sum_{s=1}^{r_i} \mu_{il}\mu_{is}\Sigma_{ils} + \sum_{l=1}^{r_i}\sum_{s=1}^{r_i} \mu_{il}(\hat{\mu}_{is}-\mu_{is})(\Sigma_{ils}+M_{il}) \\ &\leq \sum_{l=1}^{r_i}\sum_{s=1}^{r_i} \mu_{il}\mu_{is}\left[\Sigma_{ils}+\sum_{f=1}^{r_i}\delta_{if}(\Sigma_{ilf}+M_{il})\right]. \end{aligned} \tag{6.51}$$

Therefore, upon defining $\Sigma_{ils} = \Phi_{ils}+\sum_{s=1}^{r_i}\delta_{is}(\Phi_{ils}+M_{il})$, the existing relaxation technique from [7] can be applied to (6.51).

Based on the assumption in (6.50), and the result of (6.51), the design result on fuzzy control is proposed as below:

Theorem 6.3: Design of Distributed Sampled-Data Fuzzy Control Using Synchronized Method

The closed-loop fuzzy system in (6.23) using a distributed sampled-data fuzzy controller (6.20), is asymptotically stable, if there exist the symmetric positive definite matrices $\{\bar{P}_i, W_{i1}, W_{01} W_{i2}, W_{02}, \bar{Q}_i\} \in \Re^{n_{xi} \times n_{xi}}$, $W_{01} \leq W_{i1}$, $W_{02} \leq W_{i2}$, and matrices $G_i \in \Re^{n_{xi} \times n_{xi}}$, $M_{il} = M_{il}^T \in \Re^{4n_{xi} \times 4n_{xi}}$, $\bar{K}_{iis} \in \Re^{n_{ui} \times n_{xi}}$, $\bar{K}_{ijs} \in \Re^{n_{ui} \times n_{xj}}$, and the positive scalars $\{h_i, \delta_{il}\}$, such that for all $(l,s) \in \mathscr{L}_i$, the following LMIs hold:

$$\bar{\Sigma}_{ill} < 0, l \in \mathscr{L}_i \tag{6.52}$$

$$\bar{\Sigma}_{ils} + \bar{\Sigma}_{isl} < 0, 1 \leq l < s \leq r_i \tag{6.53}$$

where

$$\begin{aligned}
&\bar{\Sigma}_{ils} = \Sigma_{ils} + \sum\nolimits_{f=1}^{r_i} \delta_{if} \left(\Sigma_{ilf} + M_{il}\right), \\
&\Sigma_{ils} = \begin{bmatrix} \tilde{\Theta}_{i1} + \mathrm{Sym}\left(\mathbb{I}_i \bar{\mathbb{A}}_{ils}\right) & \bar{G}_{i1} & \bar{G}_{i2} & \bar{\Theta}_{i2ls} & \bar{\Theta}_{i3ls} \\ \star & -\frac{W_{01}}{(N-1)} & 0 & 0 & 0 \\ \star & \star & -\frac{W_{02}}{(N-1)} & 0 & 0 \\ \star & \star & \star & \bar{\mathbb{W}}_1 & 0 \\ \star & \star & \star & \star & \bar{\mathbb{W}}_2 \end{bmatrix}, \\
&\tilde{\Theta}_{i1} = \begin{bmatrix} h_i^2 \bar{Q}_i & \bar{P}_i & 0 \\ \star & 0 & 0 \\ \star & \star & -\frac{\pi^2}{4}\bar{Q}_i \end{bmatrix}, \bar{G}_{i1} = \begin{bmatrix} 0 \\ G_i^T \\ 0 \end{bmatrix}, \bar{G}_{i2} = \begin{bmatrix} 0 \\ 0 \\ G_i^T \end{bmatrix}, \mathbb{I}_i = \begin{bmatrix} I \\ I \\ 0 \end{bmatrix}, \\
&\bar{\Theta}_{i2ls} := \underbrace{\begin{bmatrix} \bar{\Theta}_{i12ls} & \cdots & \bar{\Theta}_{ij2ls} & \cdots & \bar{\Theta}_{iN2ls} \end{bmatrix}}_{N-1}, \bar{\Theta}_{ij2ls} = \mathbb{I}_i \left(A_{ijl} G_i + B_{il} \bar{K}_{ijs}\right), \\
&\bar{\Theta}_{i3ls} := \underbrace{\begin{bmatrix} \bar{\Theta}_{i13ls} & \cdots & \bar{\Theta}_{ij3ls} & \cdots & \bar{\Theta}_{iN3ls} \end{bmatrix}}_{N-1}, \bar{\Theta}_{ij3ls} = \mathbb{I}_i B_{il} \bar{K}_{ijs}, \\
&\bar{\mathbb{A}}_{ils} = \begin{bmatrix} -G_i & A_{iil} G_i + B_{il} K_{iis} G_i & B_{il} K_{is} G_i \end{bmatrix}, \\
&\bar{\mathbb{W}}_1 = \mathbb{W}_1 - \bar{\mathbb{G}}_i - \bar{\mathbb{G}}_i^T, \bar{\mathbb{W}}_2 = \mathbb{W}_2 - \bar{\mathbb{G}}_i - \bar{\mathbb{G}}_i^T, \\
&\mathbb{W}_1 := \mathrm{diag}\underbrace{\{ W_{i1} \quad \cdots \quad W_{i1} \quad \cdots \quad W_{i1} \}}_{N-1}, \\
&\mathbb{W}_2 := \mathrm{diag}\underbrace{\{ W_{i2} \quad \cdots \quad W_{i2} \quad \cdots \quad W_{i2} \}}_{N-1}, \bar{\mathbb{G}}_i := \mathrm{diag}\underbrace{\{ G_i \quad \cdots \quad G_i \quad \cdots \quad G_i \}}_{N-1}.
\end{aligned} \tag{6.54}$$

In that case, the controller gains can be calculated by

$$K_{iis} = \bar{K}_{iis} G_i^{-1}, K_{ijs} = \bar{K}_{ijs} G_i^{-1}, s \in \mathscr{L}_i. \tag{6.55}$$

■

Note that when the asynchronized information of μ_{il} and $\hat{\mu}_{il}$ is unknown, the design result on a distributed sampled-data linear controller can be directly derived as below:

Theorem 6.4: Distributed Sampled-Data Linear Controller Design

The closed-loop fuzzy system in (6.23) using a distributed sampled-data fuzzy controller (6.20), is asymptotically stable, if there exist the symmetric positive definite matrices $\{\bar{P}_i, W_{i1}, W_{01} W_{i2}, W_{02}, \bar{Q}_i\} \in \Re^{n_{xi}\times n_{xi}}$, $W_{01} \leq W_{i1}$, $W_{02} \leq W_{i2}$, and matrices $G_i \in \Re^{n_{xi}\times n_{xi}}$, $M_{il} = M_{il}^T \in \Re^{4n_{xi}\times 4n_{xi}}$, $\bar{K}_{ii} \in \Re^{n_{ui}\times n_{xi}}$, $\bar{K}_{ij} \in \Re^{n_{ui}\times n_{xj}}$, and the positive scalars $\{h_i, \delta_{il}\}$, such that for all $l \in \mathscr{L}_i$, the following LMIs hold:

$$\begin{bmatrix} \tilde{\Theta}_{i1} + \mathrm{Sym}\left(\mathbb{I}_i \bar{\mathbb{A}}_{il}\right) & \bar{G}_{i1} & \bar{G}_{i2} & \bar{\Theta}_{i2l} & \bar{\Theta}_{i3l} \\ \star & -\frac{W_{01}}{(N-1)} & 0 & 0 & 0 \\ \star & \star & -\frac{W_{02}}{(N-1)} & 0 & 0 \\ \star & \star & \star & \bar{\mathbb{W}}_1 & 0 \\ \star & \star & \star & \star & \bar{\mathbb{W}}_2 \end{bmatrix} < 0, \tag{6.56}$$

where

$$\tilde{\Theta}_{i1} = \begin{bmatrix} h_i^2 \bar{Q}_i & \bar{P}_i & 0 \\ \star & 0 & 0 \\ \star & \star & -\frac{\pi^2}{4}\bar{Q}_i \end{bmatrix}, \bar{G}_{i1} = \begin{bmatrix} 0 \\ G_i^T \\ 0 \end{bmatrix}, \bar{G}_{i2} = \begin{bmatrix} 0 \\ 0 \\ G_i^T \end{bmatrix}, \mathbb{I}_i = \begin{bmatrix} I \\ I \\ 0 \end{bmatrix},$$

$$\bar{\Theta}_{i2l} := \underbrace{\begin{bmatrix} \bar{\Theta}_{i12l} & \cdots & \bar{\Theta}_{ij2l} & \cdots & \bar{\Theta}_{iN2l} \end{bmatrix}}_{N-1}, \bar{\Theta}_{ij2l} = \mathbb{I}_i \left(A_{ijl} G_i + B_{il} \bar{K}_{ij}\right),$$

$$\bar{\Theta}_{i3l} := \underbrace{\begin{bmatrix} \bar{\Theta}_{i13l} & \cdots & \bar{\Theta}_{ij3l} & \cdots & \bar{\Theta}_{iN3l} \end{bmatrix}}_{N-1}, \bar{\Theta}_{ij3l} = \mathbb{I}_i B_{il} \bar{K}_{ij},$$

$$\bar{\mathbb{A}}_{il} = \begin{bmatrix} -G_i & A_{iil} G_i + B_{il} \bar{K}_{ii} & B_{il} \bar{K}_{ii} \end{bmatrix},$$

$$\bar{\mathbb{W}}_1 = \mathbb{W}_1 - \bar{\mathbb{G}}_i - \bar{\mathbb{G}}_i^T, \bar{\mathbb{W}}_2 = \mathbb{W}_2 - \bar{\mathbb{G}}_i - \bar{\mathbb{G}}_i^T,$$

$$\mathbb{W}_1 := \mathrm{diag}\underbrace{\{ W_{i1} \ \cdots \ W_{i1} \ \cdots \ W_{i1} \}}_{N-1},$$

$$\mathbb{W}_2 := \mathrm{diag}\underbrace{\{ W_{i2} \ \cdots \ W_{i2} \ \cdots \ W_{i2} \}}_{N-1}, \bar{\mathbb{G}}_i := \mathrm{diag}\underbrace{\{ G_i \ \cdots \ G_i \ \cdots \ G_i \}}_{N-1}. \tag{6.57}$$

In that case, the controller gains can be calculated by

$$K_{ii} = \bar{K}_{ii} G_i^{-1}, K_{ij} = \bar{K}_{ij} G_i^{-1}. \tag{6.58}$$

■

6.1.3 DISTRIBUTED SAMPLED-DATA CONTROL WITH TIME-DRIVEN ZOH

Before moving on, the assumptions are made as below:

Assumption 6.3. Sensors are time-driven, and satisfy the set,

$$\underline{s}_i \le s_i^{k+1} - s_i^k \le \bar{s}_i, k \in \mathbb{N}_i. \tag{6.59}$$

Assumption 6.4. ZOHs are time-driven, and satisfy the set,

$$\underline{z}_i \le z_i^{k+1} - z_i^k \le \bar{z}_i, k \in \mathbb{N}_i. \tag{6.60}$$

Note: The work in [7] considered even-based ZOHs in the sense that control signals are applied as soon as new data becomes available. However, that does not always happen in engineering applications when actuators have low updated rates. For example, according to the work in [8], the refresh rates in actuators such as electric cylinders, solenoids, shape memory alloys, and electroactive polymers are 10 Hz or less. Those actuators do not immediately refresh the control signals and can induce time-delays, which can affect stability and performance of the closed-loop system. Thus, this section considers the time-driven ZOHs in (6.60).

Generally, ZOHs receive new data from controllers as soon as sensors send new sampling data when considering event-driven ZOHs. Based on the Assumptions 6.3 and 6.4, it is easy to see that, the proposed ZOHs are time-driven and are not updated immediately. Motivated by [8], we define the time elapsed for the j-th sampler as $\rho_i^s(t) = t - \mathrm{s}_k^i$. Thus, it yields

$$0 \le \rho_i^s(t) < \bar{\mathrm{s}}_i, t \in [s_k^i, s_{k+1}^i), \tag{6.61}$$

according to Assumption 6.3. Then the time elapsed for the j-th ZOH is defined as $\rho_i^z(t) = t - \mathrm{z}_k^i$. Based on the Assumption 6.4, it yields

$$0 \le \rho_i^z(t) < \bar{\mathrm{z}}_i, t \in [z_k^i, z_{k+1}^i). \tag{6.62}$$

Based on the relations in (6.61) and (6.62) and defining as $\rho_{ji}^{sz}(t)$ as the total elapsed time from the sampling instant in the j-th subsystem to the updated one in the i-th subsystem, it yields

$$\rho_{ji}^{sz}(t) = \rho_j^s(t) + \rho_i^z(t), \tag{6.63}$$

which implies $0 \le \rho_{ji}^{sz}(t) < \rho_{ji}$, where $\rho_{ji} = \bar{\mathrm{s}}_j + \bar{\mathrm{z}}_i$.

Now, consider the following decentralized sampled-data fuzzy controller:

Controller Rule $\mathscr{R}_i^s$: **IF** $z_{i1}(z_i^k)$ is $\mathscr{F}_{i1}^s$ and $z_{i2}(z_i^k)$ is $\mathscr{F}_{i2}^s$ and $\cdots$ and $z_{ig}(z_i^k)$ is $\mathscr{F}_{ig}^s$, **THEN**

$$u_i(t) = K_{iis} x_i(z_i^k) + \sum_{\substack{j=1 \\ j \neq i}}^{N} K_{ijs} x_j(z_i^k), t \in [z_i^k, z_i^{k+1}) \tag{6.64}$$

where $K_{iis} \in \Re^{n_{ui} \times n_{xi}}, K_{ijs} \in \Re^{n_{ui} \times n_{xj}}, s \in \mathscr{L}_i, i \in \mathscr{N}$ are controller gains to be determined; $z_i(t_k^i) := [z_{i1}(z_i^k), z_{i2}(z_i^k), \ldots, z_{ig}(z_i^k)]$; $z_i(z_i^k)$ and $x_i(z_i^k)$ denote the updating signals in the fuzzy controller.

Similarly, the overall distributed fuzzy controller is

$$u_i(t) = K_{ii}(\hat{\mu}_i)x_i(z_i^k) + \sum_{\substack{j=1 \\ j\neq i}}^{N} K_{ij}(\hat{\mu}_i)x_j(z_i^k), t \in [z_i^k, z_i^{k+1}) \tag{6.65}$$

where

$$\begin{cases} K_{ii}(\hat{\mu}_i) := \sum_{s=1}^{r_i} \hat{\mu}_{is}\left[z_i(z_i^k)\right] K_{iis}, \sum_{s=1}^{r_i} \hat{\mu}_{is}\left[z_i(z_i^k)\right] = 1, \\ K_{ij}(\hat{\mu}_i) := \sum_{s=1}^{r_i} \hat{\mu}_{is}\left[z_i(z_i^k)\right] K_{ijs}, \\ \hat{\mu}_{is}\left[z_i(z_i^k)\right] := \frac{\Pi_{\phi=1}^{g} \hat{\mu}_{is\phi}\left[z_{i\phi}(z_i^k)\right]}{\Sigma_{\varsigma=1}^{r_i}\Pi_{\phi=1}^{g}\hat{\mu}_{i\varsigma\phi}\left[z_{i\phi}(z_i^k)\right]} \geq 0. \end{cases} \tag{6.66}$$

In the following, we will denote $\hat{\mu}_{is} := \hat{\mu}_{is}\left[z_i(z_i^k)\right]$ for brevity.

We further define

$$x_i(v) = x_i(z_i^k) - x_i(t), x_j(v) = x_j(z_i^k) - x_j(t). \tag{6.67}$$

Submitting (6.67) into the fuzzy controller (6.65) and combining it with the fuzzy system (6.1), the closed-loop fuzzy control system is

$$\begin{aligned} \dot{x}_i(t) &= \bar{A}_{ii}(\mu_i, \hat{\mu}_i)x_i(t) + B_i(\mu_i)K_{ii}(\hat{\mu}_i)x_i(v) \\ &+ \sum_{\substack{j=1 \\ j\neq i}}^{N} \bar{A}_{ij}(\mu_i, \hat{\mu}_i)x_j(t) + \sum_{\substack{j=1 \\ j\neq i}}^{N} B_i(\mu_i)K_{ij}(\hat{\mu}_i)x_j(v), \end{aligned} \tag{6.68}$$

where $\bar{A}_{ii}(\mu_i, \hat{\mu}_i) = A_{ii}(\mu_i) + B_i(\mu_i)K_{ii}(\hat{\mu}_i), \bar{A}_{ij}(\mu_i, \hat{\mu}_i) = A_{ij}(\mu_i) + B_i(\mu_i)K_{ij}(\hat{\mu}_i)$.

Note: The centralized control requires faster computers with larger memory to implement excessive information processing. Currently the decentralized control has attracted considerable interest in the field of large-scale systems. Instead of developing a single controller, a cluster of controllers, which are mutually independent, can be designed to execute the overall control task.

Note: The distributed fuzzy controller (6.65) reduces to a parallel distributed compensation (PDC) one when $\mu_{il} = \hat{\mu}_{il}$. However, the premise variable $z(t)$ undergoes time-driven sensors and ZOHs, and is implemented by the proposed fuzzy controller in (6.65). In such circumstances, the asynchronous variables between μ_{il} and $\hat{\mu}_{il}$ are more realistic. As pointed out in [5], when the information between μ_{il} and $\hat{\mu}_{il}$ is unavailable, the condition $\mu_{il} \neq \hat{\mu}_{il}$ generally leads to a linear controller instead of a fuzzy one, which degrades the stabilization ability of the controller. When the information on μ_{il} and $\hat{\mu}_{il}$ is available, the design conservatism can be improved, and the corresponding fuzzy controller can be obtained.

Note: z_i^k is relative to the clock on the i-th subsystem. In other words, the sampled-data clocks and the ZOH clocks can be different among all subsystems.

Now, we introduce the following LKF:

$$V(t) = \sum_{i=1}^{N} x_i^T(t) P_i x_i(t) + \sum_{i=1}^{N} [\bar{s}_j + \bar{z}_i]^2 \int_{z_i^k}^{t} \dot{x}_i^T(\alpha) Q_i \dot{x}_i(\alpha) d\alpha$$
$$- \sum_{i=1}^{N} \frac{\pi^2}{4} \int_{z_i^k}^{t} \left[x_i(\alpha) - x_i\left(z_i^k\right)\right]^T Q_i \left[x_i(\alpha) - x_i\left(z_i^k\right)\right] d\alpha, \tag{6.69}$$

where $\{P_i, Q_i\} \in \Re^{n_{xi} \times n_{xi}}$ are positive definite symmetric matrices.

By using Wirtinger's inequality in [9], it can be known that $V(t) > 0$. Based on the new model in (6.68) and the LKF in (6.69), a sufficient condition for a distributed sampled-data controller can be given as below:

Lemma 6.3: Stability Analysis of Distributed Sampled-Data Control

The closed-loop fuzzy system in (6.68) using a distributed sampled-data fuzzy controller (6.65), is asymptotically stable, if there exist the symmetric positive definite matrices $\{P_i, W_{i1}, W_{i2}\} \in \Re^{n_{xi} \times n_{xi}}$, $K_{ii}(\hat{\mu}_i) \in \Re^{n_{ui} \times n_{xi}}$, $K_{ij}(\hat{\mu}_i) \in \Re^{n_{ui} \times n_{xj}}$, and the positive scalars $\{\bar{z}_i, \bar{s}_i\}$, such that the following matrix inequalities hold:

$$\Theta_i + \mathrm{Sym}(\mathbb{G}_i \mathbb{A}_i(\mu_i, \hat{\mu}_i)) + \sum_{\substack{j=1 \\ j \neq i}}^{N} \mathbb{G}_i \bar{A}_{ij}(\mu_i, \hat{\mu}_i) W_{i1} \bar{A}_{ij}^T(\mu_i, \hat{\mu}_i) \mathbb{G}_i^T$$
$$+ \sum_{\substack{j=1 \\ j \neq i}}^{N} \mathbb{G}_i B_i(\mu_i) K_{ij}(\hat{\mu}_i) W_{i2} K_{ij}^T(\hat{\mu}_i) B_i^T(\mu_i) \mathbb{G}_i^T < 0, \tag{6.70}$$

where

$$\Theta_i = \begin{bmatrix} [\bar{z}_i + \bar{s}_j]^2 Q_i & P_i & 0 \\ \star & \sum_{\substack{j=1 \\ j \neq i}}^{N} W_j^{-1} & 0 \\ \star & \star & -\frac{\pi^2}{4} Q_i + \sum_{\substack{j=1 \\ j \neq i}}^{N} W_j^{-1} \end{bmatrix}. \tag{6.71}$$

■

Proof. By taking the time derivative of $V(t)$ in (6.69), we have

$$\dot{V}(t) = \sum_{i=1}^{N} 2x_i^T(t) P_i \dot{x}_i(t) + \sum_{i=1}^{N} [\bar{s}_j + \bar{z}_i]^2 \dot{x}_i^T(t) Q_i \dot{x}_i(t) - \sum_{i=1}^{N} \frac{\pi^2}{4} x_i^T(v) Q_i x_i(v). \tag{6.72}$$

Define the matrix $\mathbb{G}_i \in \Re^{3n_{xi} \times n_{xi}}$ and $\chi_i(t) = \begin{bmatrix} \dot{x}_i^T(t) & x_i^T(t) & x_i^T(v) \end{bmatrix}^T$, and it follows from (6.68) that

$$0 = \sum_{i=1}^{N} 2\chi_i^T(t)\mathbb{G}_i\mathbb{A}_i(\mu_i,\hat{\mu}_i)\chi_i(t) + \sum_{i=1}^{N}\sum_{\substack{j=1\\ j\neq i}}^{N} 2\chi_i^T(t)\mathbb{G}_i\bar{A}_{ij}(\mu_i,\hat{\mu}_i)x_j(t)$$
$$+ \sum_{i=1}^{N}\sum_{\substack{j=1\\ j\neq i}}^{N} 2\chi_i^T(t)\mathbb{G}_iB_i(\mu_i)K_{ij}(\hat{\mu}_i)x_j(v), \quad (6.73)$$

where $\mathbb{A}_i(\mu_i,\hat{\mu}_i) = \begin{bmatrix} -I & \bar{A}_{ii}(\mu_i,\hat{\mu}_i) & B_i(\mu_i)K_{ii}(\hat{\mu}_i) \end{bmatrix}$.

By introducing the symmetric positive definite matrices $\{W_{i1}, W_{i2}\} \in \Re^{n_{xi} \times n_{xi}}$, and using the relation of (6.7),

$$\sum_{i=1}^{N} 2\chi_i^T(t)\mathbb{G}_i \sum_{\substack{j=1\\ j\neq i}}^{N} \bar{A}_{ij}(\mu_i,\hat{\mu}_i)x_j(t)$$
$$\leq \sum_{i=1}^{N}\sum_{\substack{j=1\\ j\neq i}}^{N} \chi_i^T(t)\mathbb{G}_i\bar{A}_{ij}(\mu_i,\hat{\mu}_i)W_{i1}\bar{A}_{ij}^T(\mu_i,\hat{\mu}_i)\mathbb{G}_i^T\chi_i(t) + \sum_{i=1}^{N}\sum_{\substack{j=1\\ j\neq i}}^{N} x_j^T(t)W_{i1}^{-1}x_j(t)$$
$$= \sum_{i=1}^{N}\sum_{\substack{j=1\\ j\neq i}}^{N} \chi_i^T(t)\mathbb{G}_i\bar{A}_{ij}(\mu_i,\hat{\mu}_i)W_{i1}\bar{A}_{ij}^T(\mu_i,\hat{\mu}_i)\mathbb{G}_i^T\chi_i(t) + \sum_{i=1}^{N}\sum_{\substack{j=1\\ j\neq i}}^{N} x_i^T(t)W_{j1}^{-1}x_i(t), \quad (6.74)$$

and

$$\sum_{i=1}^{N} 2\chi_i^T(t)\mathbb{G}_i \sum_{\substack{j=1\\ j\neq i}}^{N} B_i(\mu_i)K_{ij}(\hat{\mu}_i)x_j(v)$$
$$\leq \sum_{i=1}^{N}\sum_{\substack{j=1\\ j\neq i}}^{N} \chi_i^T(t)\mathbb{G}_iB_i(\mu_i)K_{ij}(\hat{\mu}_i)W_{i2}K_{ij}^T(\hat{\mu}_i)B_i^T(\mu_i)\mathbb{G}_i^T\chi_i(t) + \sum_{i=1}^{N}\sum_{\substack{j=1\\ j\neq i}}^{N} x_j^T(v)W_{i2}^{-1}x_j(v)$$
$$= \sum_{i=1}^{N}\sum_{\substack{j=1\\ j\neq i}}^{N} \chi_i^T(t)\mathbb{G}_iB_i(\mu_i)K_{ij}(\hat{\mu}_i)W_{i2}K_{ij}^T(\hat{\mu}_i)B_i^T(\mu_i)\mathbb{G}_i^T\chi_i(t) + \sum_{i=1}^{N}\sum_{\substack{j=1\\ j\neq i}}^{N} x_i^T(v)W_{j2}^{-1}x_i(v). \quad (6.75)$$

It follows from (6.72)-(6.75) that the result on (6.70) can be obtained directly.

The result from (6.70) is not LMI-based. When the asynchronized information is unknown, the designed result generally leads to the linear controller instead of the fuzzy one [6]. From a practical perspective, obtaining a priori knowledge of μ_{il} and $\hat{\mu}_{il}$ is possible. Thus, we assume that the asynchronized condition is subject to

$$\underline{\rho}_{il} \leq \frac{\hat{\mu}_{il}}{\mu_{il}} \leq \bar{\rho}_{il}, \tag{6.76}$$

where $\underline{\rho}_{il}$ and $\bar{\rho}_{il}$ are positive scalars.

It follows from (6.70) and (6.76) that the design result on the decentralized sampled-data fuzzy controller can be summarized as below:

Theorem 6.5: Design of Distributed Sampled-Data Fuzzy Control Using Asynchronized Method

The closed-loop fuzzy system in (6.68) using a distributed sampled-data fuzzy controller (6.65), is asymptotically stable, if there exist the symmetric positive definite matrices $\{\bar{P}_i, W_{i1}, W_{i2}, W_{01}, W_{02}, \bar{Q}_i\} \in \Re^{n_{xi} \times n_{xi}}$, $W_{01} \leq W_{i1}$, $W_{02} \leq W_{i2}$, and matrices $G_i \in \Re^{n_{xi} \times n_{xi}}$, $M_{ils} = M_{isl}^T \in \Re^{4n_{xi} \times 4n_{xi}}$, $\bar{K}_{iis} \in \Re^{n_{ui} \times n_{xi}}$, $\bar{K}_{ijs} \in \Re^{n_{ui} \times n_{xj}}$, and the positive scalars $\{\bar{z}_i, \bar{s}_i, \underline{\rho}_{il}, \bar{\rho}_{il}\}$, such that for all $(l, s) \in \mathscr{L}_i$, the following LMIs hold:

$$\bar{\rho}_{il}\Sigma_{ill} + M_{ill} < 0, \tag{6.77}$$

$$\underline{\rho}_{il}\Sigma_{ill} + M_{ill} < 0, \tag{6.78}$$

$$\bar{\rho}_{is}\Sigma_{ils} + \bar{\rho}_{il}\Sigma_{isl} + M_{ils} + M_{isl} < 0, \tag{6.79}$$

$$\underline{\rho}_{is}\Sigma_{ils} + \underline{\rho}_{il}\Sigma_{isl} + M_{ils} + M_{isl} < 0, \tag{6.80}$$

$$\underline{\rho}_{is}\Sigma_{ils} + \bar{\rho}_{il}\Sigma_{isl} + M_{ils} + M_{isl} < 0, \tag{6.81}$$

$$\bar{\rho}_{is}\Sigma_{ils} + \underline{\rho}_{il}\Sigma_{isl} + M_{ils} + M_{isl} < 0, \tag{6.82}$$

$$\begin{bmatrix} M_{i11} & \cdots & M_{i1r} \\ \vdots & \ddots & \vdots \\ M_{ir1} & \cdots & M_{irr} \end{bmatrix} > 0, \tag{6.83}$$

where

$$\Sigma_{ils} = \begin{bmatrix} \tilde{\Theta}_{i1} + \mathrm{Sym}\left(\bar{\mathbb{I}}_i \bar{\mathbb{A}}_{ils}\right) & \bar{G}_{i1} & \bar{G}_{i2} & \bar{\Theta}_{i2ls} & \bar{\Theta}_{i3ls} \\ \star & -\frac{1}{(N-1)} W_{01} & 0 & 0 & 0 \\ \star & \star & -\frac{1}{(N-1)} W_{02} & 0 & 0 \\ \star & \star & \star & \bar{\mathbb{W}}_1 & 0 \\ \star & \star & \star & \star & \bar{\mathbb{W}}_2 \end{bmatrix},$$

$$\tilde{\Theta}_{i1} = \begin{bmatrix} [\bar{z}_i + \bar{s}_j]^2 \bar{Q}_i & \bar{P}_i & 0 \\ \star & 0 & 0 \\ \star & \star & -\frac{\pi^2}{4}\bar{Q}_i \end{bmatrix}, \bar{G}_{i1} = \begin{bmatrix} 0 \\ G_i^T \\ 0 \end{bmatrix}, \bar{G}_{i2} = \begin{bmatrix} 0 \\ 0 \\ G_i^T \end{bmatrix}, \bar{\mathbb{I}}_i = \begin{bmatrix} I \\ I \\ 0 \end{bmatrix},$$

$$\bar{\Theta}_{i2ls} := \underbrace{\begin{bmatrix} \bar{\Theta}_{i12ls} & \cdots & \bar{\Theta}_{ij2ls} & \cdots & \bar{\Theta}_{iN2ls} \end{bmatrix}}_{N-1}, \bar{\Theta}_{ij2ls} = \bar{\mathbb{I}}_i \left(A_{ijl} G_i + B_{il} \bar{K}_{ijs}\right),$$

$$\bar{\Theta}_{i3ls} := \underbrace{\begin{bmatrix} \bar{\Theta}_{i13ls} & \cdots & \bar{\Theta}_{ij3ls} & \cdots & \bar{\Theta}_{iN3ls} \end{bmatrix}}_{N-1}, \bar{\Theta}_{ij3ls} = \bar{\mathbb{I}}_i B_{il} \bar{K}_{ijs},$$

$$\bar{\mathbb{A}}_{ils} = \begin{bmatrix} -G_i & A_{iil} G_i + B_{il} \bar{K}_{iis} & B_{il} \bar{K}_{iis} \end{bmatrix},$$

$$\bar{\mathbb{W}}_1 = \mathbb{W}_1 - \bar{\mathbb{G}}_i - \bar{\mathbb{G}}_i^T, \bar{\mathbb{W}}_2 = \mathbb{W}_2 - \bar{\mathbb{G}}_i - \bar{\mathbb{G}}_i^T,$$

$$\mathbb{W}_1 := \mathrm{diag}\underbrace{\{ W_{i1} \ \cdots \ W_{i1} \ \cdots \ W_{i1} \}}_{N-1},$$

$$\mathbb{W}_2 := \mathrm{diag}\underbrace{\{ W_{i2} \ \cdots \ W_{i2} \ \cdots \ W_{i2} \}}_{N-1}, \bar{\mathbb{G}}_i := \mathrm{diag}\underbrace{\{ G_i \ \cdots \ G_i \ \cdots \ G_i \}}_{N-1}. \tag{6.84}$$

In that case, the proposed sampled-data fuzzy controller gains can be calculated by

$$K_{iis} = \bar{K}_{iis} G_i^{-1}, K_{ijs} = \bar{K}_{ijs} G_i^{-1}, s \in \mathcal{L}_i. \tag{6.85}$$

■

Proof. Define $W_{01} \leq W_{i1}$ and $W_{02} \leq W_{i2}, i \in \mathcal{N}$, it has

$$W_{01}^{-1} \geq W_{i1}^{-1}, W_{02}^{-1} \geq W_{i2}^{-1}, \tag{6.86}$$

which implies $(N-1) W_{01}^{-1} \geq \sum_{j=1, j\neq i}^{N} W_{j1}^{-1}$ and $(N-1) W_{02}^{-1} \geq \sum_{j=1, j\neq i}^{N} W_{j2}^{-1}$, respectively.

By using the Schur complement lemma and the relation in (6.86), it is easy to see that the following inequality implies (6.70),

$$\begin{bmatrix} \bar{\Theta}_{i1} + \mathrm{Sym}\left(\mathbb{G}_i \mathbb{A}_i(\mu_i, \hat{\mu}_i)\right) + \sum_{j=1, j\neq i}^{N} \Pi_{ij}(\mu_i, \hat{\mu}_i) & \bar{I}_1 & \bar{I}_2 \\ \star & -\frac{W_{01}}{(N-1)} & 0 \\ \star & \star & -\frac{W_{02}}{(N-1)} \end{bmatrix} < 0, \tag{6.87}$$

where

$$\bar{\Theta}_{i1} = \begin{bmatrix} [\bar{z}_i+\bar{s}_j]^2 Q_i & P_i & 0 \\ \star & 0 & 0 \\ \star & \star & -\frac{\pi^2}{4}Q_i \end{bmatrix}, \bar{I}_1 = \begin{bmatrix} 0 \\ I \\ 0 \end{bmatrix}, \bar{I}_2 = \begin{bmatrix} 0 \\ 0 \\ I \end{bmatrix},$$

$$\Pi_{ij}(\mu_i,\hat{\mu}_i) = \bar{\Theta}_{ij2}(\mu_i,\hat{\mu}_i)W_{i1}\bar{\Theta}_{ij2}^T(\mu_i,\hat{\mu}_i) + \bar{\Theta}_{ij3}(\mu_i,\hat{\mu}_i)W_{i2}\bar{\Theta}_{ij3}^T(\mu_i,\hat{\mu}_i),$$
$$\bar{\Theta}_{ij2}(\mu_i,\hat{\mu}_i) = \mathbb{G}_i\bar{A}_{ij}(\mu_i,\hat{\mu}_i), \bar{\Theta}_{ij3}(\mu_i,\hat{\mu}_i) = \mathbb{G}_iB_i(\mu_i)K_{ij}(\hat{\mu}_i). \tag{6.88}$$

It follows from (6.77) and (6.84) that

$$[\bar{z}_i+\bar{s}_j]^2 Q_i - \mathrm{Sym}\{G_i\} < 0, i \in \mathcal{N} \tag{6.89}$$

which implies that $G_i, i \in \mathcal{N}$ are nonsingular matrices.

We further define

$$\begin{cases} \mathbb{G}_i = \begin{bmatrix} G_i^{-1} & G_i^{-1} & 0 \end{bmatrix}^T, \Gamma_1 := \mathrm{diag}\{ G_i \quad G_i \quad G_i \quad I \quad I \quad \bar{\mathbb{G}}_i \quad \bar{\mathbb{G}}_i \}, \\ \mathbb{I} := \mathrm{diag}\underbrace{\{ I \quad \cdots \quad I \quad \cdots \quad I \}}_{N-1}, \bar{P}_i = G_i^T P_i G_i, \bar{Q}_i = G_i^T Q_i G_i, \bar{K}_{ijs} = K_{ijs}G_i, \\ \mathbb{W}_1 := \mathrm{diag}\underbrace{\{ W_{i1} \quad \cdots \quad W_{i1} \quad \cdots \quad W_{i1} \}}_{N-1}, \bar{K}_{iis} = K_{iis}G_i, \\ \mathbb{W}_2 := \mathrm{diag}\underbrace{\{ W_{i2} \quad \cdots \quad W_{i2} \quad \cdots \quad W_{i2} \}}_{N-1}, \\ \bar{\mathbb{G}}_i := \mathrm{diag}\underbrace{\{ G_i \quad \cdots \quad G_i \quad \cdots \quad G_i \}}_{N-1}, \\ \bar{\Theta}_{i2}(\mu_i,\hat{\mu}_i) := \underbrace{\begin{bmatrix} \bar{\Theta}_{i12}(\mu_i,\hat{\mu}_i) & \cdots & \bar{\Theta}_{ij2}(\mu_i,\hat{\mu}_i) & \cdots & \bar{\Theta}_{iN2}(\mu_i,\hat{\mu}_i) \end{bmatrix}}_{N-1}, \\ \bar{\Theta}_{i3}(\mu_i,\hat{\mu}_i) := \underbrace{\begin{bmatrix} \bar{\Theta}_{i13}(\mu_i,\hat{\mu}_i) & \cdots & \bar{\Theta}_{ij3}(\mu_i,\hat{\mu}_i) & \cdots & \bar{\Theta}_{iN3}(\mu_i,\hat{\mu}_i) \end{bmatrix}}_{N-1}, \\ \bar{\Theta}_{ij2}(\mu_i,\hat{\mu}_i) = \mathbb{G}_i\bar{A}_{ij}(\mu_i,\hat{\mu}_i), \bar{\Theta}_{ij3}(\mu_i,\hat{\mu}_i) = \mathbb{G}_iB_i(\mu_i)K_{ij}(\hat{\mu}_i). \end{cases} \tag{6.90}$$

By substituting (6.90) into (6.87), and using the Schur complement lemma, we have

$$\begin{bmatrix} \bar{\Theta}_{i1} + \mathrm{Sym}(\mathbb{G}_i\mathbb{A}_i(\mu_i,\hat{\mu}_i)) & \mathbb{I}_1 & \mathbb{I}_2 & \bar{\Theta}_{i2}(\mu_i,\hat{\mu}_i) & \bar{\Theta}_{i3}(\mu_i,\hat{\mu}_i) \\ \star & -\frac{W_{01}}{(N-1)} & 0 & 0 & 0 \\ \star & \star & -\frac{W_{02}}{(N-1)} & 0 & 0 \\ \star & \star & \star & -\mathbb{W}_1^{-1} & 0 \\ \star & \star & \star & \star & -\mathbb{W}_2^{-1} \end{bmatrix} < 0. \tag{6.91}$$

Note that

$$W_i - G_i - G_i^T + G_i^T W_i^{-1} G_i = (W_i - G_i)^T W_i^{-1} (W_i - G_i) \geq 0, \tag{6.92}$$

where $W_i = W_i^T > 0$, which implies that

$$-G_i^T W_i^{-1} G_i \leq W_i - G_i - G_i^T. \tag{6.93}$$

By performing a congruence transformation by Γ_1, and extracting the fuzzy membership functions, and using (6.93), one has

$$\sum_{l=1}^{r_i}\sum_{s=1}^{r_i}\mu_{il}\hat{\mu}_{is}\Sigma_{ils}<0, \tag{6.94}$$

where Σ_{ils} is defined in (6.84).

By using the asynchronous method proposed in [5], the inequalities (6.77)-(6.83) are obtained, thus, completing this proof.

It is worth nothing that the number of LMIs on Theorem 6.5 is large. The existing relaxation technique $\sum_{l=1}^{r_i}[\mu_{il}]^2\Sigma_{ill}+\sum_{l=1}^{r_i}\sum_{l<s\leq r_i}^{r_i}\mu_{il}\mu_{is}\Sigma_{ls}<0$ is no longer applicable to fuzzy controller synthesis, since $\mu_{is}\neq\hat{\mu}_{is}$. Similarly to the relaxation technique in [7], it is assumed that

$$|\mu_{il}-\hat{\mu}_{il}|\leq\delta_{il}, l\in\mathscr{L}_i, \tag{6.95}$$

where δ_{il} is a positive scalar. If $\Sigma_{ils}+M_{il}\geq 0$, where M_{il} is a symmetric matrix, we have

$$\begin{aligned}&\sum_{l=1}^{r_i}\sum_{s=1}^{r_i}\mu_{il}\hat{\mu}_{is}\Sigma_{ils}\\&=\sum_{l=1}^{r_i}\sum_{s=1}^{r_i}\mu_{il}\mu_{is}\Sigma_{ils}+\sum_{l=1}^{r_i}\sum_{s=1}^{r_i}\mu_{il}\left(\hat{\mu}_{is}-\mu_{is}\right)\left(\Sigma_{ils}+M_{il}\right)\\&\leq\sum_{l=1}^{r_i}\sum_{s=1}^{r_i}\mu_{il}\mu_{is}\left[\Sigma_{ils}+\sum_{f=1}^{r_i}\delta_{if}\left(\Sigma_{ilf}+M_{il}\right)\right].\end{aligned} \tag{6.96}$$

Therefore, upon defining $\Sigma_{ils}=\Phi_{ils}+\sum_{s=1}^{r_i}\delta_{is}\left(\Phi_{ils}+M_{il}\right)$, the existing relaxation technique from [7] can be applied to (6.96).

Based on the assumption in (6.95), and the result of (6.96), the design result on fuzzy control is proposed as below:

Theorem 6.6: Design of Distributed Sampled-Data Fuzzy Control Using Synchronized Method

The closed-loop fuzzy system in (6.68) using a distributed sampled-data fuzzy controller (6.65), is asymptotically stable, if there exist the symmetric positive definite matrices $\{\bar{P}_i, W_{i1}, W_{01}, W_{i2}, W_{02}, \bar{Q}_i\}\in\Re^{n_{xi}\times n_{xi}}$, $W_{01}\leq W_{i1}$, $W_{02}\leq W_{i2}$, and matrices $G_i\in\Re^{n_{xi}\times n_{xi}}$, $M_{il}=M_{il}^T\in\Re^{4n_{xi}\times 4n_{xi}}$, $\bar{K}_{iis}\in\Re^{n_{ui}\times n_{xi}}$, $\bar{K}_{ijs}\in\Re^{n_{ui}\times n_{xj}}$, and the positive scalars $\{\bar{z}_i,\bar{s}_j,\delta_{il}\}$, such that for all $(l,s)\in\mathscr{L}_i$, the following LMIs hold:

$$\bar{\Sigma}_{ill}<0, l\in\mathscr{L}_i \tag{6.97}$$

$$\bar{\Sigma}_{ils}+\bar{\Sigma}_{isl}<0, 1\leq l<s\leq r_i \tag{6.98}$$

where

$$
\begin{aligned}
&\bar{\Sigma}_{ils} = \Sigma_{ils} + \sum_{f=1}^{r_i} \delta_{if}\left(\Sigma_{ilf} + M_{il}\right),\\
&\Sigma_{ils} = \begin{bmatrix} \tilde{\Theta}_{i1} + \mathrm{Sym}\left(\mathbb{I}_i \bar{\mathbb{A}}_{ils}\right) & \bar{G}_{i1} & \bar{G}_{i2} & \bar{\Theta}_{i2ls} & \bar{\Theta}_{i3ls} \\ \star & -\frac{W_{01}}{(N-1)} & 0 & 0 & 0 \\ \star & \star & -\frac{W_{02}}{(N-1)} & 0 & 0 \\ \star & \star & \star & \bar{\mathbb{W}}_1 & 0 \\ \star & \star & \star & \star & \bar{\mathbb{W}}_2 \end{bmatrix},\\
&\tilde{\Theta}_{i1} = \begin{bmatrix} [\bar{z}_i + \bar{s}_j]^2 \bar{Q}_i & \bar{P}_i & 0 \\ \star & 0 & 0 \\ \star & \star & -\frac{\pi^2}{4}\bar{Q}_i \end{bmatrix}, \bar{G}_{i1} = \begin{bmatrix} 0 \\ G_i^T \\ 0 \end{bmatrix}, \bar{G}_{i2} = \begin{bmatrix} 0 \\ 0 \\ G_i^T \end{bmatrix}, \mathbb{I}_i = \begin{bmatrix} I \\ I \\ 0 \end{bmatrix},\\
&\bar{\Theta}_{i2ls} := \underbrace{\begin{bmatrix} \bar{\Theta}_{i12ls} & \cdots & \bar{\Theta}_{ij2ls} & \cdots & \bar{\Theta}_{iN2ls} \end{bmatrix}}_{N-1}, \bar{\Theta}_{ij2ls} = \mathbb{I}_i\left(A_{ijl}G_i + B_{il}\bar{K}_{ijs}\right),\\
&\bar{\Theta}_{i3ls} := \underbrace{\begin{bmatrix} \bar{\Theta}_{i13ls} & \cdots & \bar{\Theta}_{ij3ls} & \cdots & \bar{\Theta}_{iN3ls} \end{bmatrix}}_{N-1}, \bar{\Theta}_{ij3ls} = \mathbb{I}_i B_{il}\bar{K}_{ijs},\\
&\bar{\mathbb{A}}_{ils} = \begin{bmatrix} -G_i & A_{iil}G_i + B_{il}\bar{K}_{iis} & B_{il}\bar{K}_{iis} \end{bmatrix},\\
&\bar{\mathbb{W}}_1 = \mathbb{W}_1 - \mathbb{G}_i - \mathbb{G}_i^T, \bar{\mathbb{W}}_2 = \mathbb{W}_2 - \mathbb{G}_i - \mathbb{G}_i^T,\\
&\mathbb{W}_1 := \mathrm{diag}\underbrace{\left\{ W_{i1} \ \cdots \ W_{i1} \ \cdots \ W_{i1} \right\}}_{N-1},\\
&\mathbb{W}_2 := \mathrm{diag}\underbrace{\left\{ W_{i2} \ \cdots \ W_{i2} \ \cdots \ W_{i2} \right\}}_{N-1}, \mathbb{G}_i := \mathrm{diag}\underbrace{\left\{ G_i \ \cdots \ G_i \ \cdots \ G_i \right\}}_{N-1}.
\end{aligned}
\tag{6.99}
$$

In that case, the controller gains can be calculated by

$$K_{iis} = \bar{K}_{iis}G_i^{-1}, K_{ijs} = \bar{K}_{ijs}G_i^{-1}, s \in \mathscr{L}_i. \tag{6.100}$$

■

Note that when the asynchronized information of μ_{il} and $\hat{\mu}_{il}$ is unknown, the design result on a distributed sampled-data linear controller can be directly derived as below:

Theorem 6.7: Distributed Sampled-Data Linear Controller Design

The closed-loop fuzzy system in (6.68) using a distributed sampled-data linear controller $u_i(t) = K_{ii}x_i(t_k^i) + \sum_{j=1, j\neq i}^{N} K_{ij}x_j(t_k^i), t \in [t_k^i, t_{k+1}^i)$, is asymptotically stable, if there exist the symmetric positive definite matrices $\{\bar{P}_i, W_{i1}, W_{i2}, W_{01}, W_{02}, \bar{Q}_i\} \in \Re^{n_{xi} \times n_{xi}}$, $W_{01} \leq W_{i1}$, $W_{02} \leq W_{i2}$, and matrices $G_i \in \Re^{n_{xi} \times n_{xi}}$, $\bar{K}_{ii} \in \Re^{n_{ui} \times n_{xi}}$, $\bar{K}_{ij} \in \Re^{n_{ui} \times n_{xj}}$, and the positive scalars $\bar{z}_i, \bar{s}_j$, such that for all $l \in \mathscr{L}_i$, the following LMIs hold:

$$\begin{bmatrix} \tilde{\Theta}_{i1} + \mathrm{Sym}\left(\bar{\mathbb{I}}_i \bar{\mathbb{A}}_{il}\right) & \bar{G}_{i1} & \bar{G}_{i2} & \bar{\Theta}_{i2l} & \bar{\Theta}_{i3l} \\ \star & -\frac{W_{01}}{(N-1)} & 0 & 0 & 0 \\ \star & \star & -\frac{W_{02}}{(N-1)} & 0 & 0 \\ \star & \star & \star & \bar{\mathbb{W}}_1 & 0 \\ \star & \star & \star & \star & \bar{\mathbb{W}}_2 \end{bmatrix} < 0, l \in \mathscr{L}_i \tag{6.101}$$

where

$$\begin{aligned} &\tilde{\Theta}_{i1} = \begin{bmatrix} [\bar{z}_i + \bar{s}_j]^2 \bar{Q}_i & \bar{P}_i & 0 \\ \star & 0 & 0 \\ \star & \star & -\frac{\pi^2}{4}\bar{Q}_i \end{bmatrix}, \bar{G}_{i1} = \begin{bmatrix} 0 \\ G_i^T \\ 0 \end{bmatrix}, \bar{G}_{i2} = \begin{bmatrix} 0 \\ 0 \\ G_i^T \end{bmatrix}, \bar{\mathbb{I}}_i = \begin{bmatrix} I \\ I \\ 0 \end{bmatrix}, \\ &\bar{\Theta}_{i2l} := \underbrace{\begin{bmatrix} \bar{\Theta}_{i12l} & \cdots & \bar{\Theta}_{ij2l} & \cdots & \bar{\Theta}_{iN2l} \end{bmatrix}}_{N-1}, \bar{\Theta}_{ij2l} = \bar{\mathbb{I}}_i \left(A_{ijl} G_i + B_{il} \bar{K}_{ij}\right), \\ &\bar{\Theta}_{i3l} := \underbrace{\begin{bmatrix} \bar{\Theta}_{i13l} & \cdots & \bar{\Theta}_{ij3l} & \cdots & \bar{\Theta}_{iN3l} \end{bmatrix}}_{N-1}, \bar{\Theta}_{ij3l} = \bar{\mathbb{I}}_i B_{il} \bar{K}_{ij}, \\ &\bar{\mathbb{A}}_{il} = \begin{bmatrix} -G_i & A_{iil} G_i + B_{il} \bar{K}_{ii} & B_{il} \bar{K}_{ii} \end{bmatrix}, \\ &\bar{\mathbb{W}}_1 = \mathbb{W}_1 - \mathbb{G}_i - \mathbb{G}_i^T, \bar{\mathbb{W}}_2 = \mathbb{W}_2 - \mathbb{G}_i - \mathbb{G}_i^T, \\ &\mathbb{W}_1 := \mathrm{diag}\underbrace{\{ W_{i1} \ \cdots \ W_{i1} \ \cdots \ W_{i1} \}}_{N-1}, \\ &\mathbb{W}_2 := \mathrm{diag}\underbrace{\{ W_{i2} \ \cdots \ W_{i2} \ \cdots \ W_{i2} \}}_{N-1}, \mathbb{G}_i := \mathrm{diag}\underbrace{\{ G_i \ \cdots \ G_i \ \cdots \ G_i \}}_{N-1}. \end{aligned} \tag{6.102}$$

In that case, the proposed linear controller gains can be calculated by

$$K_{ii} = \bar{K}_{ii} G_i^{-1}, K_{ij} = \bar{K}_{ij} G_j^{-1}. \tag{6.103}$$

■

6.2 SIMULATION STUDIES

Consider the multi-PV power with DC load in (6.1) using the distributed fuzzy controller (6.2). Here, the parameters of the multi-PV power are chosen as below: $L_{(1)} = 0.0526\text{H}, L_{(2)} = 0.0525\text{H}, C_{O(1)} = 0.0471\text{F}, C_{O(2)} = 0.0472\text{F}, C_{PV(1)} = 0.0101\text{F}, C_{PV(2)} = 0.0102\text{F}, R_{O(1)} = 1.1\Omega, R_{O(2)} = 1.2\Omega, R_{L(1)} = 1.7\Omega, R_{L(2)} = 1.8\Omega, R_{M(1)} = 0.84\Omega, R_{M(2)} = 0.86\Omega, R_{load} = 2.1\Omega, V_{D(1)} = 9.1\text{V}, V_{D(2)} = 9.1\text{V}$. Now, we choose $z_{i1}(t) = \frac{\phi_{PV(i)}}{v_{PV(i)}}, z_{i2}(t) = \phi_{L(i)}, z_{i3}(t) = v_{PV(i)}$, and $z_{i4}(t) =$

$\frac{\dot{\phi}_{0(i)}}{\varphi_{L(i)}}$ as fuzzy premise variables, and linearize the first PV system around $\{0.037895, 1, 9.5, 0.008333\}$ and $\{0.037895, 0.5, 9.5, 0.008333\}$, and linearize the second PV system around $\{0.037895, 0.2, 9.5, 0.008333\}$ and $\{0.037895, 0.6, 9.5, 0.008333\}$. Then, the succeeding system matrices of T-S fuzzy model can be obtained as below:

$$A_{11} = \begin{bmatrix} 3.752 & 0 & 0 \\ 0 & -1783.3 & -20.913 \\ 0 & 9.6506 & -9.6506 \end{bmatrix}, A_{12} = \begin{bmatrix} 3.752 & 0 & 0 \\ 0 & -399.24 & -20.913 \\ 0 & 9.6506 & -9.6506 \end{bmatrix},$$

$$A_{21} = \begin{bmatrix} 3.715 & 0 & 0 \\ 0 & -923.81 & -19.048 \\ 0 & 9.6302 & -9.6302 \end{bmatrix}, A_{22} = \begin{bmatrix} 3.715 & 0 & 0 \\ 0 & -346.03 & -19.048 \\ 0 & 9.6302 & -9.6302 \end{bmatrix},$$

$$B_{11} = \begin{bmatrix} -49.505 \\ 345.63 \\ 0 \end{bmatrix}, B_{12} = \begin{bmatrix} -19.608 \\ 351.01 \\ 0 \end{bmatrix},$$

$$B_{21} = \begin{bmatrix} -49.505 \\ 345.63 \\ 0 \end{bmatrix}, B_{22} = \begin{bmatrix} -58.824 \\ 344.46 \\ 0 \end{bmatrix}.$$

Here, by applying Theorem 6.1, the fuzzy controller gains for the first subsystem are given by

$$\begin{aligned} K_{111} &= \begin{bmatrix} 0.033085 & 4.9991 & 0.028456 \end{bmatrix}, \\ K_{112} &= \begin{bmatrix} 0.038745 & 1.0949 & 0.031981 \end{bmatrix}, \\ K_{121} &= \begin{bmatrix} 7.4278e-06 & -4.6241e-06 & 0.1075 \end{bmatrix}, \\ K_{122} &= \begin{bmatrix} -8.7141e-06 & -2.1076e-06 & 0.11995 \end{bmatrix}. \end{aligned}$$

The fuzzy controller gains for the second subsystem are given by

$$\begin{aligned} K_{221} &= \begin{bmatrix} 0.036917 & 2.5938 & 0.013441 \end{bmatrix}, \\ K_{222} &= \begin{bmatrix} 0.044221 & 0.96749 & 0.014912 \end{bmatrix}, \\ K_{211} &= \begin{bmatrix} -1.554e-08 & 2.6607e-06 & 0.11815 \end{bmatrix}, \\ K_{212} &= \begin{bmatrix} -1.8147e-06 & -8.2614e-06 & 0.11815 \end{bmatrix}. \end{aligned}$$

It is noted that the open-loop T-S fuzzy system is unstable, as shown in Figures 6.1 and 6.2. Based on the above solutions, Figures 6.3 and 6.4 indicate that the state responses for the multi-PV power system converge to zero.

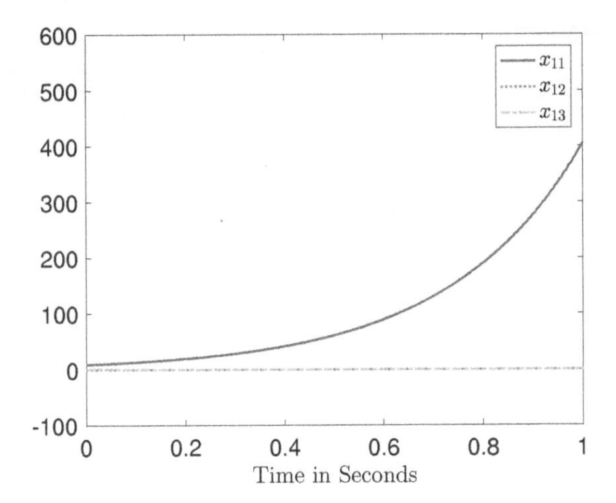

Figure 6.1 Instability of distributed control for first multi-PV power system with DC load.

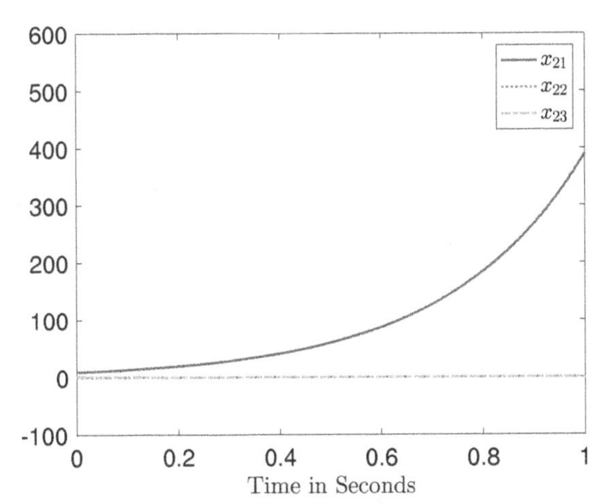

Figure 6.2 Instability of distributed control for second multi-PV power system with DC load.

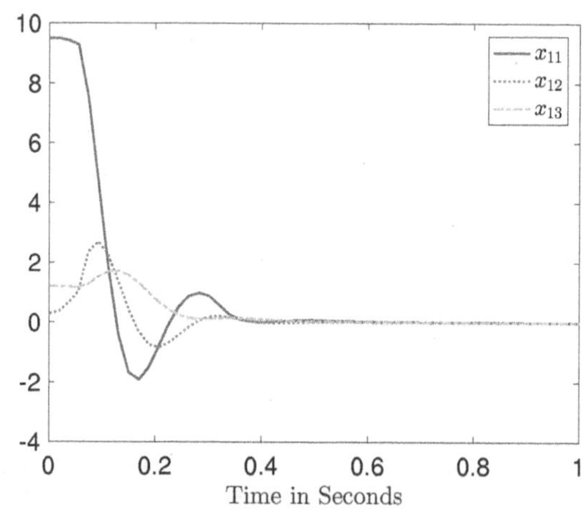

Figure 6.3 Distributed control for first multi-PV power system with DC load showing convergence to zero.

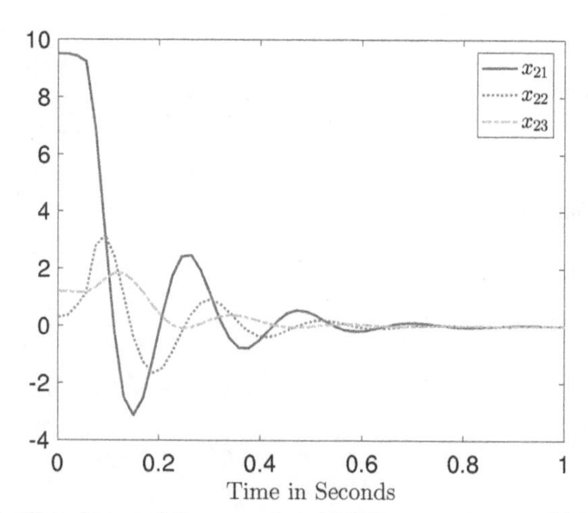

Figure 6.4 Distributed control for second multi-PV power system with DC load showing convergence to zero.

6.3 REFERENCES

1. Tomislav Dragicevic, Lu, X., Vasquez, J. C., and Guerrero, J. M. (2016). DC microgrids. Part I: A review of control strategies and stabilization techniques. IEEE Transactions on Power Electronics, 31(7), 4876-4891.
2. Lin, W. and Bitar, E. (2016). Decentralized stochastic control of distributed energy resources. IEEE Transactions on Power Systems, 33(1), 888-900.
3. Bidram, A., Davoudi, A., Lewis, F. L., and Nasirian, V. (2017). Cooperative synchronization in distributed microgrid control. Advances in Industrial Control. Berlin: Springer-Verlag.
4. Fridman, E. (2010). A refined input delay approach to sampled-data control. Automatica, 46(2), 421-427.
5. Arino, C. and Sala, A. (2008). Extensions to stability analysis of fuzzy control systems subject to uncertain grades of membership. IEEE Transactions on Systems, Man, and Cybernetics, 39(2), 558-563.
6. Lam, H. K. and Narimani, M. (2009). Stability analysis and performance design for fuzzy-model-based control system under imperfect premise matching. IEEE Transactions on Fuzzy Systems, 17(4), 949-961.
7. Zhong, Z.X. and Lin C.M. (2017). Large-Scale Fuzzy Interconnected Control Systems Design and Analysis, Pennsylvania: IGI Global.
8. Moarref, M. and Rodrigues, L. (2014). Stability and stabilization of linear sampled-data systems with multi-rate samplers and time driven zero order holds. Automatica, 50(10), 2685-2691.
9. Liu, K. and Fridman, E. (2012). Wirtinger's inequality and Lyapunov-based sampled-data stabilization. Automatica, 48(1), 102-108.

Part III

Energy Management for Microgrids

Preview

In microgrids, distributed energy resources consist of distributed generation units and distributed energy storage systems, which could be installed at electric utility facilities and/or electricity consumers' premises. The energy storage systems support the connection between distributed energy resources and loads in the microgrid by connecting and disconnecting line flows, so that the power balance can be achieved. The coexistence of distributed energy resources and storage systems with differing dynamic properties has raised concerns about stability, control and efficiency. Several functions such as the switching operations, load changes, power fluctuation, energy storage reserve, and maximization of the use of renewable energy options, are challenging because of intermittent and fluctuating power and low or absent inertial, dispatchable and non-dispatchable, and other changing characteristics of distributed energy resources.

For several decades, utilities tried partially adjust loads to market constraints (economic) and network (technical) through a demand side management. This type of management includes scenarios that allow consumers to reduce energy consumption during peak or other critical energy use periods. In microgrids, loads are commonly categorized into two types: Fixed and flexible. Fixed loads cannot be altered and must be satisfied under normal operating conditions while flexible loads are responsive to controlling signals. Flexible loads could be curtailed or deferred in response to economic incentives or islanding requirements.

In microgrids, PV and wind generators usually have a unique maximum power points (MPPs). Some intrinsic characteristics and disturbances on PV or wind generators, such as aging, irradiance intensity and temperature conditions, and disturbance uncertainties, generally lead to inefficient implementations on maximum power point tracking (MPPT) control. Traditional MPPT control aims at locate the MPP for online operation in steady state conditions. Unfortunately, those methods often result in a slower convergence. Recently, dynamic MPPT control methods have been proposed to improve the transient performance. However, those methods are not generally suited for strict convergence analysis. Optimal power system operation requires intensive numerical analysis aimed at studying and improving both steady and transient states.

This part is organized as follows. In Chapter 7, the detailed model and fuzzy logical formulations regarding for the microgrid with the PV system and wind generator and the energy storage systems are given. To maintain the power balance, three switching operations are proposed. After that, the model formulations and switching operations are investigated for microgrids handling PV and wind power and the energy storage systems. Chapter 8 focuses on the power management strategy, where the multi-model operators are investigated in the framework of the switching systems for both steady and transient states.

7 Operation of Microgrids

During the past two decades, microgrids have been regarded as major enablers of smart grids for the integration of the distributed energy resource units into the electricity grids. Microgrids offer promising cost-effective solutions for the integration of renewable energy with reduced losses, lower transmission and distribution costs, higher energy efficiency, and a number of environmental and economic benefits. However, the coexistence of multiple energy resources with differing dynamic properties, such as no or low inertial, dispatchable and non-dispatchable, and nonfirm characteristics, has raised concerns over the stability, control, and efficiency of microgrids. Maintaining the power balance is a challenging problem for microgrids. Microgrids provide flexibility by allowing different generator units to be operated to adjust to demand changes.

This chapter is organized as follows. First, the detailed model and fuzzy logical formulations regarding the microgrid with the PV system and wind generator and the energy storage systems are given. To maintain the power balance, three switching operations are proposed. After that, the model formulations and switching operations are investigated for the microgrid with PV and wind and the energy storage systems.

7.1 PHOTOVOLTAIC SYSTEM FOR DC LOAD

The section considers a small-scale DC power system, which consists of the solar PV power subsystem, the lead-acid battery and the DC load. The solar PV power system uses an DC-DC buck converter to a DC-bus, and a bi-directional buck-boost converter will be used to interface the battery and the DC-bus. This system, as illustrated in Figure 7.1, utilizes a bi-directional converter operating as a boost converter during discharging mode of the battery and as a buck converter during charging.

7.1.1 OPERATION MODES

In order to ensure the reliable and stable operations of the small-scale DC power system under different conditions, the power balance of supply and demand must be maintained under all operating conditions. Now, we define P_{PV}, $P_{LD}^{\min}$, and $P_{LD}^{\max}$ as the PV power supply, the minimum load demand, and the maximum load demand, respectively. There exist these three possible conditions of the system as below [1].

1) Deficit power mode$\left(P_{PV} \leq P_{LD}^{\min}\right)$

In this mode, maximum available power is less than power demand. Specifically, the solar PV power is not sufficient to satisfy the DC loads. The solar PV converter works with the MPPT algorithm, and the lead-acid battery converters regulate the output voltage by battery discharging operation. Thus, they work as voltage sources. To obtain both accurate current sharing and desirable voltage regulation using the

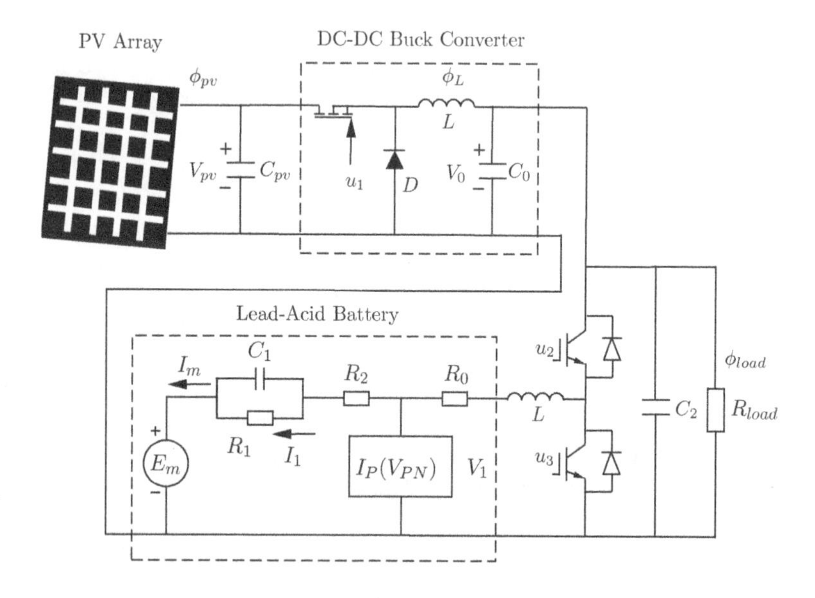

Figure 7.1 DC microgrid with solar PV.

fuzzy control method, the DC-bus voltage reaches its minimum acceptable value at the maximum discharge current of the batteries.

2) Floating power mode $\left(P_{LD}^{\min} \leq P_{PV} \leq P_{LD}^{\max}\right)$

The DC system operates in an islanded mode. The solar PV power is basically sufficient to satisfy the DC loads, and the batteries do not work in this case. The PV converter works with the MPPT algorithm, and regulates the DC bus voltage by using fuzzy tracking control.

3) Excess power mode$(P_{PV} \geq P_{LD}^{\max})$

The PV converter works with the MPPT algorithm. Since the PV power is greater than the load power, the DC voltage increases. The excess power is used to charge the lead-acid batteries. Therefore, the batteries regulate the DC bus voltage by battery charging operations.

7.1.2 DYNAMIC MODELING

Recall the system models of the PV with buck converter in (1.1), the boost converter in (1.6), and the lead-acid battery in (3.1) as below:

$$\begin{cases} \dot{v}_{PV} = \frac{1}{C_{PV}}\left(\phi_{PV} - \phi_{L,PV}u_1\right), \\ \dot{\phi}_{L,PV} = \frac{1}{L_{PV}}R_{0,PV}\left(\left(\phi_{0,PV} - \phi_{L,PV}\right) - R_{L,PV}\phi_{L,PV} - v_{0,PV}\right) \\ \qquad + \frac{1}{L_{PV}}\left(V_{D,PV} + v_{PV} - R_{M,PV}\phi_{L,PV}\right)u_1 - \frac{V_{D,PV}}{L_{PV}}, \\ \dot{v}_{0,PV} = \frac{1}{C_{0,PV}}\left(\phi_{L,PV} - \phi_{0,PV}\right), \end{cases} \tag{7.1}$$

and

$$\begin{cases} \dot{\phi}_{1,LA} = \frac{1}{R_{1,LA}C_{1,LA}}\left(\phi_{m,LA} - \phi_{1,LA}\right), \\ \dot{Q}_{e,LA} = -\phi_{m,LA}, \\ \dot{\theta}_{LA} = \frac{1}{C_{\theta,LA}}\left[P_{s,LA} - \frac{\left(\theta_{LA} - \theta_{a,LA}\right)}{R_{\theta,LA}}\right], \\ \phi_{p,LA} = G_{p,LA}v_{PN,LA}, \\ G_{p,LA} = G_{p0,LA}\exp\left(v_{PN,LA}/v_{p0,LA} + A_{p,LA}\left(1 - \theta_{LA}/\theta_{f,LA}\right)\right), \end{cases} \tag{7.2}$$

and

$$\begin{cases} \dot{\phi}_{LA} = -\frac{1}{L_{LA}}(1-u)\,v_{dc} + \frac{1}{L_{LA}}v_{LA}, \\ \dot{v}_{dc} = \frac{1}{C_{0,LA}}(1-u)\,\phi_{LA} - \frac{1}{C_{0,LA}}\phi_{0,LA}, \end{cases} \tag{7.3}$$

where the subscripts PV and LA stand for the PV power and the lead-acid battery, respectively, and all notations are defined in (1.1), (1.6), and (3.1), respectively.

1) Deficit power mode

First, consider the dynamic model of the lead-acid battery with the boost converter as below:

$$\begin{cases} \dot{\phi}_{1,LA} = \frac{1}{R_{1,LA}C_{1,LA}}\left(\phi_{m,LA} - \phi_{1,LA}\right), \\ \dot{\phi}_{m,LA} = -\frac{1}{R_{0,LA}L_{LA}}(1-u_3)\,v_{0,LA} + \frac{1}{R_{0,LA}L_{LA}}v_{LA} - \dot{G}_{p,LA}v_{PN,LA} - G_{p,LA}\dot{v}_{PN,LA}, \\ \dot{v}_{0,LA} = \frac{1}{C_{0,LA}}(1-u_3)\left(\phi_{m,LA} + \phi_{p,LA}\right) - \frac{1}{R_{0,LA}C_{0,LA}}\phi_{0,LA}, \end{cases} \tag{7.4}$$

and the PV with buck converter is

$$\begin{cases} \dot{v}_{PV} = \frac{1}{C_{PV}}\left(\phi_{PV} - \phi_{L,PV}u_1\right), \\ \dot{\phi}_{L,PV} = \frac{1}{L_{PV}}R_{0,PV}\left(\phi_{0,PV} - \phi_{L,PV} - R_{L,PV}\phi_{L,PV} - v_{0,PV}\right) \\ \qquad + \frac{1}{L_{PV}}\left(V_{D,PV} + v_{PV} - R_{M,PV}\phi_{L,PV}\right)u_1 - \frac{V_{D,PV}}{L_{PV}}, \\ \dot{v}_{0,PV} = \frac{1}{C_{0,PV}}\left(\phi_{L,PV} - \phi_{0,PV}\right). \end{cases} \tag{7.5}$$

Under the deficit power mode, both the PV power and the lead-acid battery can be supplied to the DC load. Based on the Thevenin's theorem,

$$v_{0,LA} = \phi_{0,LA}R_{line,LA} + \left(\phi_{0,LA} + \phi_{0,PV}\right)R_{load}, \tag{7.6}$$

$$v_{0,PV} = \phi_{0,PV}R_{line,PV} + \left(\phi_{0,LA} + \phi_{0,PV}\right)R_{load}, \tag{7.7}$$

where $R_{line,LA}$ and $R_{line,PV}$ are the resistances of power line on the lead-acid battery and on the PV system, respectively, and R_{load} is the load resistance.

Here, the objective is to design a supervisory controller, such the MPPT can be obtained and the output voltage tracks v_{ref}. The voltage reference v^* of the MPPT can be calculated by [2]

$$0 = \phi_{pv} - n_p\gamma I_{rs}v^*e^{\gamma v^*}. \tag{7.8}$$

Define $e_{0,LA} = v_{0,LA} - v_{ref}, e_{0,PV} = v_{0,PV} - v_{ref}, e_{PV} = v_{PV} - v^*$. Based on the relationships (7.4)-(7.8), the interconnected system is given by

$$\begin{cases} \dot{\phi}_{1,LA} = -\frac{1}{R_{1,LA}C_{1,LA}}\phi_{1,LA} + \frac{1}{R_{1,LA}C_{1,LA}}\phi_{m,LA}, \\ \dot{\phi}_{m,LA} = -\frac{e_{0,LA}}{R_{0,LA}L_{LA}} + \frac{(e_{0,LA}+v_{ref})}{R_{0,LA}L_{LA}}u_3 + \frac{1}{R_{0,LA}L_{LA}}v_{LA} \\ \quad - \dot{G}_{p,LA}v_{PN,LA} - G_{p,LA}\dot{v}_{PN,LA} - \frac{v_{ref}}{R_{0,LA}L_{LA}}, \\ \dot{e}_{0,LA} = \frac{1}{C_{0,LA}}\phi_{m,LA} - \frac{1}{C_{0,LA}}(\phi_{m,LA} + \phi_{p,LA})u_3 - \frac{e_{0,PV}}{R_{0,LA}C_{0,LA}R_{load}} \\ \quad + \frac{R_{line,PV}+R_{load}}{R_{0,LA}C_{0,LA}R_{load}}\phi_{0,PV} - \frac{v_{ref}}{R_{0,LA}C_{0,LA}R_{load}} + \frac{1}{C_{0,LA}}\phi_{p,LA}, \\ \dot{e}_{PV} = \frac{1}{C_{PV}}\phi_{PV} - \frac{1}{C_{PV}}\phi_{L,PV}u_1 - \dot{v}^*, \\ \dot{\phi}_{L,PV} = \frac{R_{0,PV}}{L_{PV}}(\phi_{0,PV} - \phi_{L,PV} - R_{L,PV}\phi_{L,PV} - e_{0,PV} - v_{ref}) \\ \quad + \frac{1}{L_{PV}}(V_{D,PV} + v_{PV} - R_{M,PV}\phi_{L,PV})u_1 - \frac{V_{D,PV}}{L_{PV}}, \\ \dot{e}_{0,PV} = \frac{1}{C_{0,PV}}\phi_{L,PV} - \frac{e_{0,LA}}{C_{0,PV}R_{load}} - \frac{v_{ref}}{C_{0,PV}R_{load}} + \frac{1}{C_{0,PV}}\left(\frac{R_{line,LA}}{R_{load}} + 1\right)\phi_{0,LA}. \end{cases} \tag{7.9}$$

Define $x(t) = \begin{bmatrix} \phi_{1,LA} & \phi_{m,LA} & e_{0,LA} & e_{pv} & \phi_{L,PV} & e_{0,PV} \end{bmatrix}^T$, and choose $\{R_{1,LA}, R_{0,LA}, e_{0,LA}, \phi_{m,LA}, \phi_{p,LA}, \frac{\phi_{0,PV}}{\phi_{L,PV}}, \frac{\phi_{PV}}{\phi_{L,PV}}, \frac{\phi_{0,LA}}{\phi_{m,LA}}, v_{PV}, \phi_{L,PV}\}$ as the fuzzy premise variables, the augmented fuzzy system is given by

Plant Rule $\mathscr{R}^l$: **IF** z_1 is $\mathscr{F}_1^l$ and z_2 is $\mathscr{F}_2^l$ and $\cdots$ and z_{10} is $\mathscr{F}_{10}^l$, **THEN**

$$\dot{x}(t) = A_l x(t) + B_l u(t) + \omega(t), l \in \mathscr{L} := \{1, 2, \ldots, r\} \tag{7.10}$$

where $\mathscr{R}^l$ denotes the l-th fuzzy inference rule; $\mathscr{L} := \{1, 2, \ldots, r\}$, r is the number of inference rules; $\mathscr{F}_\theta^l(\theta = 1, 2, \cdots, 10)$ is the fuzzy set; $x(t) \in \Re^{n_x}$ and $u(t) \in \Re^{n_u}$ denote the system state and control input, respectively; $z(t) \triangleq [z_1, z_2, \ldots, z_{10}]$ are the measurable variables; The l-th local model $\{A_l, B_l\}$ and the disturbance $\omega(t)$ are defined as below:

$$A_l = \begin{bmatrix} -\frac{1}{\mathscr{F}_1^l C_{1,LA}} & \frac{1}{\mathscr{F}_1^l C_{1,LA}} & 0 & 0 & 0 & 0 \\ 0 & 0 & -\frac{1}{\mathscr{F}_2^l L_{LA}} & 0 & 0 & 0 \\ 0 & \frac{1}{C_{0,LA}} & 0 & 0 & \frac{\mathscr{F}_6^l k_4}{\mathscr{F}_2^l k_1} & -\frac{1}{\mathscr{F}_2^l k_1} \\ 0 & 0 & 0 & 0 & \frac{\mathscr{F}_7^l}{C_{PV}} & 0 \\ 0 & 0 & 0 & 0 & k_5 & -\frac{R_{0,PV}}{L_{PV}} \\ 0 & \frac{\mathscr{F}_8^l k_3}{k_2} & -\frac{1}{k_2} & 0 & \frac{1}{C_{0,PV}} & 0 \end{bmatrix},$$

$$B_l = \begin{bmatrix} 0 & 0 \\ \frac{\mathscr{F}_3^l + v_{ref}}{\mathscr{F}_2^l L_{LA}} & 0 \\ -\frac{\mathscr{F}_4^l + \mathscr{F}_5^l}{C_{0,LA}} & 0 \\ 0 & -\frac{\mathscr{F}_{10}^l}{C_{PV}} \\ 0 & k_6 \\ 0 & 0 \end{bmatrix}, \omega(t) = \begin{bmatrix} 0 \\ k_7(t) \\ k_8(t) \\ -\dot{v}^* \\ -\frac{V_{D,PV}}{L_{PV}} \\ -\frac{v_{ref}}{C_{0,PV}R_{load}} \end{bmatrix},$$

$$
\begin{aligned}
&k_1 = C_{0,LA}R_{load}, k_2 = C_{0,PV}R_{load}, k_3 = R_{line,LA} + R_{load}, k_4 = R_{line,PV} + R_{load},\\
&k_5 = \frac{R_{0,PV}\left(\mathscr{F}_6^l - 1 - R_{L,PV}\right)}{L_{PV}}, k_6 = \frac{V_{D,PV} + \mathscr{F}_9^l - R_{M,PV}\mathscr{F}_{10}^l}{L_{PV}},\\
&k_7(t) = \frac{v_{LA} - v_{ref}}{R_{0,LA}L_{LA}} - \dot{G}_{p,LA}v_{PN,LA} - G_{p,LA}\dot{v}_{PN,LA},\\
&k_8(t) = -\frac{v_{ref}}{R_{0,LA}C_{0,LA}R_{load}} + \frac{\phi_{p,LA}}{C_{0,LA}}.
\end{aligned} \tag{7.11}
$$

2) Floating power mode

The lead–acid battery does not work. In other words, the power balance can be maintained between the PV system and the loading. Hence, the system model is

$$
\begin{cases}
\dot{v}_{PV} = \frac{1}{C_{PV}}\left(\phi_{PV} - \phi_{L,PV}u_1\right),\\
\dot{\phi}_{L,PV} = \frac{1}{L_{PV}}R_{0,PV}\left(\phi_{0,PV} - \phi_{L,PV} - R_{L,PV}\phi_{L,PV} - v_{0,PV}\right)\\
\qquad + \frac{1}{L_{PV}}\left(V_{D,PV} + v_{PV} - R_{M,PV}\phi_{L,PV}\right)u_1 - \frac{V_{D,PV}}{L_{PV}},\\
\dot{v}_{0,PV} = \frac{1}{C_{0,PV}}\left(\phi_{L,PV} - \phi_{0,PV}\right).
\end{cases} \tag{7.12}
$$

Define $e_{0,PV} = v_{0,PV} - v_{ref}$ and $e_{PV} = v_{PV} - v^*$, where v^* can be determined by (7.8), and v_{ref} is output voltage reference. Define $x(t) = \begin{bmatrix} e_{PV} & \phi_{L,PV} & e_{0,PV} \end{bmatrix}^T$, and choose $\left\{\frac{\phi_{PV}}{v_{PV}}, \phi_{L,PV}, \frac{\phi_{0,PV}}{\phi_{L,PV}}, v_{PV}\right\}$ as the fuzzy premise variables, then the augmented fuzzy system is given by

Plant Rule $\mathscr{R}^l$: **IF** z_1 is $\mathscr{F}_1^l$ and z_2 is $\mathscr{F}_2^l$ and $\cdots$ and z_4 is $\mathscr{F}_4^l$, **THEN**

$$
\dot{x}(t) = A_l x(t) + B_l u(t) + \omega(t), l \in \mathscr{L} \tag{7.13}
$$

where $\mathscr{R}^l$ denotes the l-th fuzzy inference rule; $\mathscr{L} := \{1, 2, \ldots, r\}$, r is the number of inference rules; $\mathscr{F}_\theta^l(\theta = 1, 2, 3, 4)$ is the fuzzy set; $x(t) \in \Re^{n_x}$ and $u(t) \in \Re^{n_u}$ denote the system state and control input, respectively; $z(t) \triangleq [z_1, z_2, \ldots, z_4]$ are the measurable variables; The l-th local model $\{A_l, B_l\}$ and the disturbance $\omega(t)$ are given as below:

$$
\begin{aligned}
A_l &= \begin{bmatrix} \frac{\mathscr{F}_1^l}{C_{PV}} & 0 & 0 \\ 0 & \frac{R_{0,PV}\left(\mathscr{F}_3^l - 1 - R_{L,PV}\right)}{L_{PV}} & -\frac{R_{0,PV}}{L_{PV}} \\ 0 & \frac{1-\mathscr{F}_3^l}{C_{0,PV}} & 0 \end{bmatrix},\\
B_l &= \begin{bmatrix} -\frac{\mathscr{F}_2^l}{C_{PV}} \\ \frac{V_{D,PV} + \mathscr{F}_4^l - R_{M,PV}\mathscr{F}_2^l}{L_{PV}} \\ 0 \end{bmatrix}, \omega(t) = \begin{bmatrix} -\dot{v}^* + \frac{\phi_{PV}}{C_{PV}v_{PV}}v^* \\ -\frac{R_{0,PV}v_{ref}}{L_{PV}} - \frac{V_{D,PV}}{L_{PV}} \\ 0 \end{bmatrix}.
\end{aligned} \tag{7.14}
$$

3) Excess power mode

The lead–acid battery works as the loading, hence the system model can be given by

$$\begin{cases} \dot{\phi}_{1,LA} = \frac{1}{R_{1,LA}C_{1,LA}}\left(\phi_{m,LA} - \phi_{1,LA}\right), \\ \dot{\phi}_{m,LA} = \frac{1}{R_{0,LA}C_{0,LA}}\left(\phi_{L,LA} - \phi_{m,LA}\right) - \frac{\dot{v}_{PN,LA}}{R_{0,LA}}, \\ \dot{\phi}_{L,LA} = \frac{1}{L_{LA}}R_{0,LA}\left(\phi_{m,LA} - \phi_{L,LA} - R_{L,LA}\phi_{L,LA} - \left(\phi_{m,LA}R_{0,LA} + v_{PN,LA}\right)\right) \\ \qquad + \frac{1}{L_{LA}}\left(V_{D,LA} + v_{0,PV} - R_{M,LA}\phi_{L,LA}\right)u_2 - \frac{V_{D,LA}}{L_{LA}}, \end{cases} \tag{7.15}$$

and

$$\begin{cases} \dot{v}_{PV} = \frac{1}{C_{PV}}\left(\phi_{PV} - \phi_{L,PV}u_1\right), \\ \dot{\phi}_{L,PV} = \frac{1}{L_{PV}}R_{0,PV}\left(\phi_{0,PV} - \phi_{L,PV} - R_{L,PV}\phi_{L,PV} - v_{0,PV}\right) \\ \qquad + \frac{1}{L_{PV}}\left(V_{D,PV} + v_{PV} - R_{M,PV}\phi_{L,PV}\right)u_1 - \frac{V_{D,PV}}{L_{PV}}, \\ \dot{v}_{0,PV} = \frac{1}{C_{0,PV}}\left(\phi_{L,PV} - \phi_{0,PV}\right). \end{cases} \tag{7.16}$$

Under the excess power mode, the PV power can be supplied to the DC load and lead-acid battery. Based on the Kirchhoff's current law,

$$\phi_{0,PV} = \phi_{0,LA} + \frac{v_{0,PV} - \phi_{0,PV}R_{line,PV}}{R_{load}}. \tag{7.17}$$

Here, the objective is to design a supervisory controller, such that the MPPT can be obtained and the output voltage tracks the reference v_{ref}. The voltage reference v^* of the MPPT can be calculated by (7.8). Define $e_{0,PV} = v_{0,PV} - v_{ref}, e_{PV} = v_{PV} - v^*$. Based on the relationships (7.15)-(7.17), the interconnected system is given by

$$\begin{cases} \dot{\phi}_{1,LA} = \frac{1}{R_{1,LA}C_{1,LA}}\left(\phi_{m,LA} - \phi_{1,LA}\right), \\ \dot{\phi}_{m,LA} = \frac{1}{R_{0,LA}C_{0,LA}}\left(\phi_{L,LA} - \phi_{m,LA}\right) - \frac{\dot{v}_{PN,LA}}{R_{0,LA}}, \\ \dot{\phi}_{L,LA} = \frac{R_{0,LA}}{L_{LA}}\left(\phi_{m,LA} - \phi_{L,LA} - R_{L,LA}\phi_{L,LA} - \left(\phi_{m,LA}R_{0,LA} + v_{PN,LA}\right)\right) \\ \qquad + \frac{1}{L_{LA}}\left(V_{D,LA} + e_{0,PV} + v_{ref} - R_{M,LA}\phi_{L,LA}\right)u_2 - \frac{V_{D,LA}}{L_{LA}}, \\ \dot{e}_{PV} = \frac{\phi_{PV}}{C_{PV}v_{PV}}e_{PV} - \frac{1}{C_{PV}}\phi_{L,PV}u_1 - \dot{v}^* + \frac{1}{C_{PV}}\frac{\phi_{PV}}{v_{PV}}v^*, \\ \dot{\phi}_{L,PV} = \frac{R_{0,PV}}{L_{PV}}\left(\frac{\phi_{0,PV}}{\phi_{L,PV}}\phi_{L,PV} - \phi_{L,PV} - R_{L,PV}\phi_{L,PV} - e_{0,PV} - v_{ref}\right) \\ \qquad + \frac{1}{L_{PV}}\left(V_{D,PV} + v_{PV} - R_{M,PV}\phi_{L,PV}\right)u_1 - \frac{V_{D,PV}}{L_{PV}}, \\ \dot{e}_{0,PV} = \frac{1}{C_{0,PV}}\left(\phi_{L,PV} - \frac{\phi_{0,LA}}{\phi_{m,LA}}\phi_{m,LA} - \frac{e_{0,PV}+v_{ref}}{R_{load}} + \frac{R_{line,PV}}{R_{load}}\phi_{0,PV}\right). \end{cases} \tag{7.18}$$

Define $x(t) = \begin{bmatrix} \phi_{1,LA} & \phi_{m,LA} & \phi_{L,LA} & e_{PV} & \phi_{L,PV} & e_{0,PV} \end{bmatrix}^T$, and choose $\left\{R_{1,LA}, R_{0,LA}, \frac{\phi_{PV}}{v_{PV}}, \frac{\phi_{0,PV}}{\phi_{L,PV}}, \frac{\phi_{0,LA}}{\phi_{m,LA}}, e_{0,PV}, \phi_{L,LA}, \phi_{L,PV}, R_{0,LA}^2\right\}$ as the fuzzy premise variables, then the augmented fuzzy system is given by

Plant Rule $\mathscr{R}^l$: **IF** z_1 is $\mathscr{F}_1^l$ and z_2 is $\mathscr{F}_2^l$ and $\cdots$ and z_9 is $\mathscr{F}_9^l$, **THEN**

$$\dot{x}(t) = A_l x(t) + B_l u(t) + \omega(t), l \in \mathscr{L} \tag{7.19}$$

where $\mathscr{R}^l$ denotes the l-th fuzzy inference rule; $\mathscr{L} := \{1, 2, \ldots, r\}$, and r is the number of inference rules; $\mathscr{F}_\theta^l(\theta = 1, 2, 3, \cdots, 9)$ is the fuzzy set; $x(t) \in \Re^{n_x}$ and $u(t) \in$

$\Re^{n_u}$ denote the system state and control input, respectively; $z(t) \triangleq [z_1, z_2, \ldots, z_9]$ are the measurable variables; The l-th local model $\{A_l, B_l\}$ and the disturbance $\omega(t)$ are given as below:

$$A_l = \begin{bmatrix} \frac{-1}{\mathscr{F}_1^l C_{1,LA}} & \frac{1}{\mathscr{F}_1^l C_{1,LA}} & 0 & 0 & 0 & 0 \\ 0 & \frac{-1}{\mathscr{F}_2^l C_{0,LA}} & \frac{1}{\mathscr{F}_2^l C_{0,LA}} & 0 & 0 & 0 \\ 0 & \frac{\mathscr{F}_2^l - \mathscr{F}_9^l}{L_{LA}} & \frac{-\mathscr{F}_2^l (1+R_{L,LA})}{L_{LA}} & 0 & 0 & 0 \\ 0 & 0 & 0 & \frac{\mathscr{F}_3^l}{C_{PV}} & 0 & 0 \\ 0 & 0 & 0 & 0 & k_1 & \frac{-R_{0,PV}}{L_{PV}} \\ 0 & \frac{-\mathscr{F}_5^l}{C_{0,PV}} & 0 & 0 & k_2 & \frac{-1}{C_{0,PV} R_{load}} \end{bmatrix},$$

$$B_l = \begin{bmatrix} 0 & 0 \\ 0 & 0 \\ k_3 & 0 \\ 0 & \frac{-\mathscr{F}_8^l}{C_{PV}} \\ 0 & 0 \\ 0 & 0 \end{bmatrix}, \omega(t) = \begin{bmatrix} 0 \\ \frac{-\dot{v}_{PN,LA}}{R_{0,LA}} \\ \frac{-R_{0,LA} v_{PN,LA}}{L_{LA}} \\ \frac{\phi_{PV} v^*}{C_{PV} v_{PV}} - \dot{v}^* \\ \frac{-R_{0,PV} v_{ref}}{L_{PV}} \\ \frac{-v_{ref}}{C_{0,PV} R_{load}} \end{bmatrix},$$

$$k_1 = \frac{R_{0,PV}\left(\mathscr{F}_4^l - 1 - R_{L,PV}\right)}{L_{PV}}, k_2 = \frac{R_{load} + R_{line,PV}\mathscr{F}_4^l}{C_{0,PV} R_{load}},$$

$$k_3 = \frac{V_{D,LA} + \mathscr{F}_6^l + v_{ref} - R_{M,LA}\mathscr{F}_7^l}{L_{LA}}. \tag{7.20}$$

7.2 PHOTOVOLTAIC SYSTEM FOR AC LOAD

This section considers a small-scale AC power system, which consists of the solar PV power subsystem, the lead-acid battery and the AC load. The solar PV power system uses a DC-DC buck converter to the DC-bus, and a bi-directional buck-boost converter is used to interface the battery and the DC-bus, and the DC-bus is connected to the AC-bus by using an DC-AC converter. In this system, as illustrated in Figure 7.2, the bi-directional converter operates as a boost converter during the discharging mode of the battery and as a buck converter during the charging mode of the battery.

7.2.1 OPERATION MODES

In order to ensure the reliable and stable operations of the small-scale AC power system under different conditions, the power balance among the supply and the demand must be maintained under all operating conditions. Now, we define P_{PV}, $P_{LD}^{\min}$, and $P_{LD}^{\max}$ as the PV power supply, the minimum load demand, and the maximum load demand, respectively. There exist these three possible conditions of the system as shown in Figure 7.2 [1].

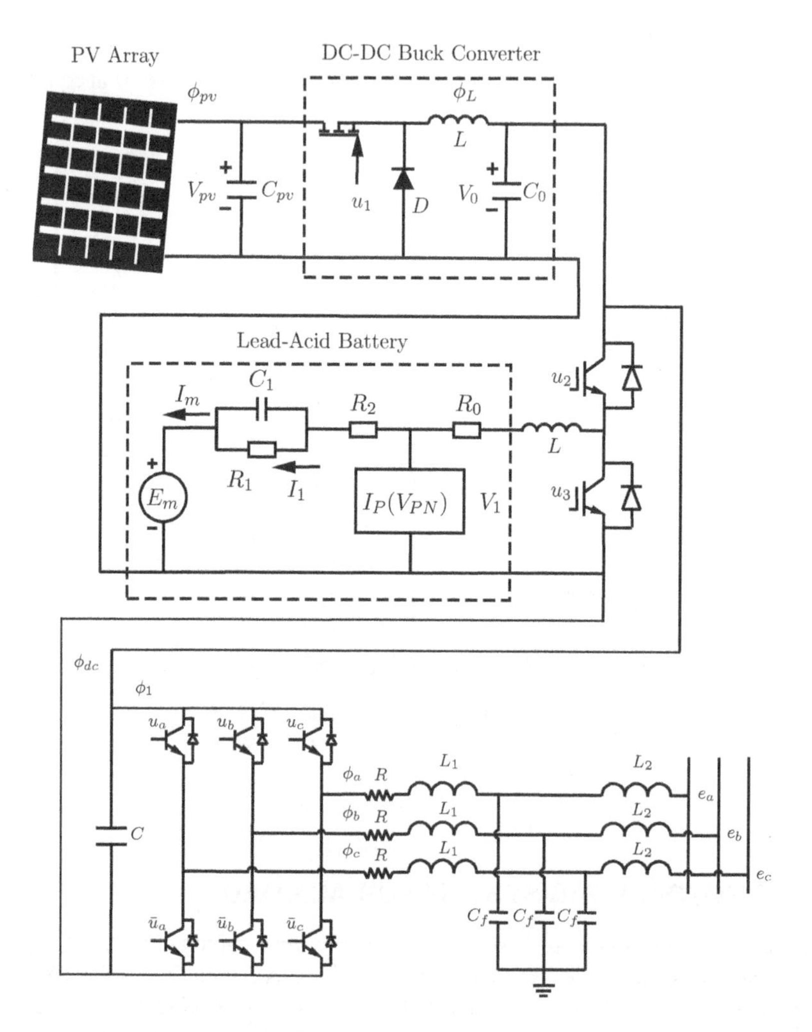

Figure 7.2 AC microgrid with solar PV.

1) Deficit power mode$\left(P_{PV} \leq P_{LD}^{\min}\right)$

In this mode, the maximum available power is less than the power demand. Specifically, the solar PV power is not sufficient to satisfy the AC load. The solar PV converter works with the MPPT algorithm, and the lead-acid battery converters regulate the output voltage by the battery discharging operation. Thus, they work as the voltage sources. To obtain both the accurate current sharing and the desirable voltage regulation using the fuzzy control method, the voltage of AC bus reaches its minimum acceptable value at the maximum discharge current of the batteries.

2) Floating power mode$\left(P_{LD}^{\min} \leq P_{PV} \leq P_{LD}^{\max}\right)$

In this mode, the DC bus operates in an islanded mode. The solar PV power is basically sufficient to satisfy AC load, and the batteries do not work in this case. The PV converter works with the MPPT algorithm, and regulates the voltage of AC bus by using the fuzzy tracking control.

3) Excess power mode$(P_{PV} \geq P_{LD}^{\max})$

In this mode, the PV converter works with the MPPT algorithm. Since the PV power is greater than the load power, the AC voltage increases. The excess power is utilized to charge the lead-acid batteries. Therefore, the batteries regulate the voltage of AC bus by the battery charging operations.

7.2.2 DYNAMIC MODELING

In the synchronous dq-reference frame, the instantaneous active power P and reactive power Q can be calculated as [3]

$$\begin{cases} P = \frac{3}{2}\left(u_d\phi_d + u_q\phi_q\right), \\ Q = \frac{3}{2}\left(u_q\phi_d - u_d\phi_q\right). \end{cases} \tag{7.21}$$

Assume that the phase-locked loop (PLL) is aligned with the load voltage vector to the d-axis of the dq-reference frame (i.e., $u_q = 0$), the transfer functions from the d- and q-axis output currents to the active and reactive power can be calculated as

$$\begin{cases} \phi_d^* = \frac{2P^*}{3u_d}, \\ \phi_q^* = -\frac{2Q^*}{3u_d}, \end{cases} \tag{7.22}$$

where P^* and Q^* are the instantaneous active power and the reactive power, respectively.

The three-phase DC-AC converter is controlled as a current source in order to track a certain current reference in the synchronous dq-reference frame. The d- and q-axis current references (ϕ_d^* and ϕ_q^*) can be injected directly or calculated from the desired active and the reactive power as shown in (7.21).

Here, recall the DC-AC converter in (1.13) as below [4]:

$$\begin{cases} \dot{v}_{dc} = \frac{1}{C_{pv}}\left(\phi_{dc} - \frac{1.5u_d}{v_{dc}}\phi_d\right), \\ \dot{\phi}_d = -\frac{R_1}{L_1}\phi_d - \omega\phi_q + \frac{1}{L_1}e_d, \\ \dot{\phi}_q = \omega\phi_d - \frac{R_1}{L_1}\phi_q + \frac{1}{L_1}e_q. \end{cases} \tag{7.23}$$

Neglecting the conversion loss of the converters, the active power pg transferred between the DC subgrid and the AC grid can be expressed by

$$\begin{aligned} P &= v_{0,LA}\phi_{0,LA} + v_{0,PV}\phi_{0,PV} \\ &= 1.5u_d\phi_d. \end{aligned} \tag{7.24}$$

1) Deficit power mode

First, consider the system model of the lead-acid battery with the boost converter as below:

$$\begin{cases} \dot{\phi}_{1,LA} = \frac{1}{R_{1,LA}C_{1,LA}}(\phi_{m,LA} - \phi_{1,LA}), \\ \dot{\phi}_{m,LA} = -\frac{1}{R_{0,LA}L_{LA}}(1-u_3)v_{0,LA} + \frac{1}{R_{0,LA}L_{LA}}v_{LA} - \dot{G}_{p,LA}v_{PN,LA} - G_{p,LA}\dot{v}_{PN,LA}, \\ \dot{v}_{0,LA} = \frac{1}{C_{0,LA}}(1-u_3)(\phi_{m,LA}+\phi_{p,LA}) - \frac{1}{R_{0,LA}C_{0,LA}}\phi_{0,LA}, \end{cases} \tag{7.25}$$

and the PV with the buck converter is

$$\begin{cases} \dot{v}_{PV} = \frac{1}{C_{PV}}(\phi_{PV} - \phi_{L,PV}u_1), \\ \dot{\phi}_{L,PV} = \frac{1}{L_{PV}}R_{0,PV}(\phi_{0,PV} - \phi_{L,PV} - R_{L,PV}\phi_{L,PV} - v_{0,PV}) \\ \qquad + \frac{1}{L_{PV}}(V_{D,PV} + v_{PV} - R_{M,PV}\phi_{L,PV})u_1 - \frac{V_{D,PV}}{L_{PV}}, \\ \dot{v}_{0,PV} = \frac{1}{C_{0,PV}}(\phi_{L,PV} - \phi_{0,PV}), \end{cases} \tag{7.26}$$

and the DC-AC converter is shown in (7.23).

Under this operation mode, the objective is to design a supervisory controller, so that the MPPT can be obtained and the output voltage conforms to the reference. The voltage reference v^* of the MPPT can be calculated by (7.8). Then, define the voltage reference v_{ref}, $e_{0,LA} = v_{0,LA} - v^*_{dc}$, $e_{0,PV} = v_{0,PV} - v^*_{dc}$, $\varepsilon_d = \phi_d - \phi^*_d$, $\varepsilon_q = \phi_q - \phi^*_q$, and it follows (7.21)-(7.26) that

$$\begin{cases} \dot{\phi}_{1,LA} = \frac{1}{R_{1,LA}C_{1,LA}}(\phi_{m,LA} - \phi_{1,LA}), \\ \dot{\phi}_{m,LA} = -\frac{1}{R_{0,LA}L_{LA}}e_{0,LA} + \frac{e_{0,LA}+v^*_{dc}}{R_{0,LA}L_{LA}}u_3 + \frac{1}{R_{0,LA}L_{LA}}v_{LA} - \dot{G}_{p,LA}v_{PN,LA} \\ \qquad - G_{p,LA}\dot{v}_{PN,LA} - \frac{1}{R_{0,LA}L_{LA}}v^*_{dc}, \\ \dot{e}_{0,LA} = \frac{1}{C_{0,LA}}\phi_{m,LA} + \frac{1}{C_{0,LA}}\phi_{p,LA} - \frac{\phi_{m,LA}+\phi_{p,LA}}{C_{0,LA}}u_3 - \frac{1}{R_{0,LA}C_{0,LA}}\phi_{0,LA}, \\ \dot{v}_{PV} = \frac{1}{C_{PV}}(\phi_{PV} - \phi_{L,PV}u_1), \\ \dot{\phi}_{L,PV} = \frac{R_{0,PV}}{L_{PV}}\phi_{0,PV} - \frac{R_{0,PV}+R_{0,PV}R_{L,PV}}{L_{PV}}\phi_{L,PV} - \frac{R_{0,PV}}{L_{PV}}e_{0,PV} \\ \qquad + \frac{1}{L_{PV}}(V_{D,PV} + v_{PV} - R_{M,PV}\phi_{L,PV})u_1 - \frac{V_{D,PV}}{L_{PV}} - \frac{R_{0,PV}}{L_{PV}}v^*_{dc}, \\ \dot{e}_{0,PV} = \frac{1}{C_{0,PV}}(\phi_{L,PV} - \phi_{0,PV}), \\ \dot{v}_{dc} = \frac{1}{C_{pv}}\phi_{0,PV} + \frac{1}{C_{pv}}\phi_{0,LA} - \frac{\phi_{0,LA}}{v_{dc}}e_{0,LA} - \frac{\phi_{0,PV}}{v_{dc}}e_{0,PV} - \frac{\phi_{0,PV}}{v_{dc}}v^*_{dc} - \frac{\phi_{0,LA}}{v_{dc}}v^*_{dc}, \\ \dot{\varepsilon}_d = -\frac{R_1}{L_1}\varepsilon_d - \omega\varepsilon_q + \frac{e_d}{L_1} - \dot{\phi}^*_d - \omega\phi^*_q - \frac{R_1}{L_1}\phi^*_d, \\ \dot{\varepsilon}_q = \omega\varepsilon_d - \frac{R_1}{L_1}\varepsilon_q + \frac{e_q}{L_1} - \dot{\phi}^*_q + \omega\phi^*_d - \frac{R_1}{L_1}\phi^*_q. \end{cases} \tag{7.27}$$

Define $x(t) = \begin{bmatrix} \phi_{1,LA} & \phi_{m,LA} & e_{0,LA} & v_{PV} & \phi_{L,PV} & e_{0,PV} & v_{dc} & \varepsilon_d & \varepsilon_q \end{bmatrix}^T$, and choose $\{\frac{v_{LA}}{\phi_{m,LA}}, \frac{\phi_{p,LA}}{\phi_{m,LA}}, \frac{\phi_{0,PV}}{\phi_{L,PV}}, \frac{\phi_{PV}}{v_{PV}}, \frac{\phi_{0,LA}}{\phi_{m,LA}}, e_{0,LA}, \phi_{m,LA}, \phi_{p,LA}, \phi_{L,PV}, v_{PV}, e^{\gamma v_{pv}}, \omega, \frac{\phi_{0,LA}}{v_{dc}}, \frac{\phi_{0,PV}}{v_{dc}}\}$ as the fuzzy premise variables, then the augmented fuzzy system is given by

Plant Rule $\mathscr{R}^l$: **IF** z_1 is $\mathscr{F}_1^l$ and z_2 is $\mathscr{F}_2^l$ and $\cdots$ and z_{14} is $\mathscr{F}_{14}^l$, **THEN**

$$\dot{x}(t) = A_l x(t) + B_l u(t) + \omega(t), l \in \mathscr{L} \tag{7.28}$$

where $\mathscr{R}^l$ denotes the l-th fuzzy inference rule; $\mathscr{L} := \{1, 2, \ldots, r\}$, and r is the number of inference rules; $\mathscr{F}_\theta^l(\theta = 1, 2, \cdots, 14)$ is the fuzzy set; $x(t) \in \Re^{n_x}$ and $u(t) \in \Re^{n_u}$ denote the system state and the control input, respectively; $z(t) \triangleq [z_1, z_2, \ldots, z_{14}]$ are the measurable variables; The l-th local model $\{A_l, B_l\}$ and the disturbance $\omega(t)$ are given as below:

$$A_l = \begin{bmatrix} \frac{-1}{k_1} & \frac{1}{k_1} & 0 & 0 & 0 & 0 & 0 & 0 & 0 \\ 0 & \frac{\mathscr{F}_1^l}{k_2} & -\frac{1}{k_2} & 0 & 0 & 0 & 0 & 0 & 0 \\ 0 & k_3 & 0 & 0 & 0 & 0 & 0 & 0 & 0 \\ 0 & 0 & 0 & \frac{\mathscr{F}_4^l}{C_{PV}} & 0 & 0 & 0 & 0 & 0 \\ 0 & 0 & 0 & 0 & k_4 & -\frac{R_{0,PV}}{L_{PV}} & 0 & 0 & 0 \\ 0 & 0 & 0 & 0 & \frac{1-\mathscr{F}_3^l}{C_{0,PV}} & 0 & 0 & 0 & 0 \\ 0 & \frac{\mathscr{F}_5^l}{C_{pv}} & -\mathscr{F}_{13}^l & 0 & \frac{\mathscr{F}_3^l}{C_{pv}} & -\mathscr{F}_{14}^l & 0 & 0 & 0 \\ 0 & 0 & 0 & 0 & 0 & 0 & 0 & -\frac{R_1}{L_1} & -\mathscr{F}_{12}^l \\ 0 & 0 & 0 & 0 & 0 & 0 & 0 & \mathscr{F}_{12}^l & -\frac{R_1}{L_1} \end{bmatrix},$$

$$B_l = \begin{bmatrix} 0 & 0 & 0 & 0 \\ \frac{\mathscr{F}_6^l + v_{dc}^*}{R_{0,LA}L_{LA}} & 0 & 0 & 0 \\ -\frac{\mathscr{F}_7^l + \mathscr{F}_8^l}{C_{0,LA}} & 0 & 0 & 0 \\ 0 & -\frac{\mathscr{F}_9^l}{C_{PV}} & 0 & 0 \\ 0 & k_5 & 0 & 0 \\ 0 & 0 & 0 & 0 \\ 0 & 0 & 0 & 0 \\ 0 & 0 & \frac{1}{L_1} & 0 \\ 0 & 0 & 0 & \frac{1}{L_1} \end{bmatrix}, \omega(t) = \begin{bmatrix} 0 \\ k_6(t) \\ 0 \\ 0 \\ -\frac{V_{D,PV}}{L_{PV}} - \frac{R_{0,PV}}{L_{PV}} v_{dc}^* \\ 0 \\ -\frac{\phi_{0,PV} + \phi_{0,LA}}{v_{dc}} v_{dc}^* \\ -\dot{\phi}_d^* - \omega\phi_q^* - \frac{R_1}{L_1}\phi_d^* \\ -\dot{\phi}_q^* + \omega\phi_d^* - \frac{R_1}{L_1}\phi_q^* \end{bmatrix},$$

$$k_1 = R_{1,LA}C_{1,LA}, k_2 = R_{0,LA}L_{LA}, k_3 = \frac{1 + \mathscr{F}_2^l}{C_{0,LA}} - \frac{\mathscr{F}_5^l}{R_{0,LA}C_{0,LA}},$$

$$k_4 = \frac{R_{0,PV}\mathscr{F}_3^l - R_{0,PV} - R_{0,PV}R_{L,PV}}{L_{PV}}, k_5 = \frac{V_{D,PV} + \mathscr{F}_{10}^l - R_{M,PV}\mathscr{F}_9^l}{L_{PV}},$$

$$k_6(t) = -\dot{G}_{p,LA}v_{PN,LA} - G_{p,LA}\dot{v}_{PN,LA} - \frac{1}{R_{0,LA}L_{LA}} v_{dc}^*. \tag{7.29}$$

2) Floating power mode

In this mode, the lead–acid battery does not work. In other words, the power can be maintained between the PV system and the loading. Hence, the system model is

$$\begin{cases} \dot{v}_{PV} = \frac{1}{C_{PV}}\left(\phi_{PV} - \phi_{L,PV}u_1\right), \\ \dot{\phi}_{L,PV} = \frac{1}{L_{PV}}R_{0,PV}\left(\left(\phi_{0,PV} - \phi_{L,PV}\right) - R_{L,PV}\phi_{L,PV} - v_{0,PV}\right) \\ \qquad + \frac{1}{L_{PV}}\left(V_{D,PV} + v_{PV} - R_{M,PV}\phi_{L,PV}\right)u_1 - \frac{V_{D,PV}}{L_{PV}}, \\ \dot{v}_{0,PV} = \frac{1}{C_{0,PV}+C_{pv}}\left(\phi_{L,PV} - \phi_1\right), \\ \dot{\phi}_d = -\frac{R_1}{L_1}\phi_d - \omega\phi_q + \frac{1}{L_1}e_d, \\ \dot{\phi}_q = \omega\phi_d - \frac{R_1}{L_1}\phi_q + \frac{1}{L_1}e_q. \end{cases} \tag{7.30}$$

Then, define $e_{0,PV} = v_{0,PV} - v_{dc}^*, \varepsilon_d = \phi_d - \phi_d^*, \varepsilon_q = \phi_q - \phi_q^*$, and $x(t) = \left[\begin{array}{ccccc} v_{PV} & \phi_{L,PV} & e_{0,PV} & \varepsilon_d & \varepsilon_q \end{array}\right]^T$, and choose $\left\{\frac{\phi_{PV}}{v_{PV}}, \phi_{L,PV}, \frac{\phi_{0,PV}}{\phi_{L,PV}}, \frac{\phi_1}{\phi_{L,PV}}, \omega, v_{PV}, e^{\gamma v_{pv}}\right\}$ as the fuzzy premise variables, then the augmented fuzzy system is given by

Plant Rule $\mathscr{R}^l$: **IF** z_1 is $\mathscr{F}_1^l$ and z_2 is $\mathscr{F}_2^l$ and $\cdots$ and z_7 is $\mathscr{F}_7^l$, **THEN**

$$\dot{x}(t) = A_l x(t) + B_l u(t) + \omega(t), l \in \mathscr{L} \tag{7.31}$$

where $\mathscr{R}^l$ denotes the l-th fuzzy inference rule; $\mathscr{L} := \{1, 2, \ldots, r\}$, and r is the number of inference rules; $\mathscr{F}_\theta^l(\theta = 1, 2, \cdots, 7)$ is the fuzzy set; $x(t) \in \mathfrak{R}^{n_x}$ and $u(t) \in \mathfrak{R}^{n_u}$ denote the system state and control input, respectively; $z(t) \triangleq [z_1, z_2, \ldots, z_7]$ are the measurable variables; The l-th local model $\{A_l, B_l\}$ and the disturbance $\omega(t)$ are given as below:

$$A_l = \begin{bmatrix} \frac{\mathscr{F}_1^l}{C_{PV}} & 0 & 0 & 0 & 0 \\ 0 & k_1 & -\frac{R_{0,PV}}{L_{PV}} & 0 & 0 \\ 0 & \frac{1-\mathscr{F}_4^l}{C_{0,PV}+C_{pv}} & 0 & 0 & 0 \\ 0 & 0 & 0 & -\frac{R_1}{L_1} & -\mathscr{F}_5^l \\ 0 & 0 & 0 & \omega & -\frac{R_1}{L_1} \end{bmatrix},$$

$$B_l = \begin{bmatrix} -\frac{\mathscr{F}_2^l}{C_{PV}} & 0 & 0 \\ k_2 & 0 & 0 \\ 0 & 0 & 0 \\ 0 & \frac{1}{L_1} & 0 \\ 0 & 0 & \frac{1}{L_1} \end{bmatrix}, \omega(t) = \begin{bmatrix} 0 \\ -\frac{V_{D,PV}}{L_{PV}} - \frac{R_{0,PV}}{L_{PV}}v_{dc}^* \\ 0 \\ -\frac{R_1}{L_1}\phi_d^* - \omega\phi_q^* - \dot{\phi}_d^* \\ \omega\phi_d^* - \frac{R_1}{L_1}\phi_q^* - \dot{\phi}_q^* \end{bmatrix},$$

$$k_1 = \frac{R_{0,PV}\mathscr{F}_3^l}{L_{PV}} - \frac{R_{0,PV}}{L_{PV}} - \frac{R_{0,PV}R_{L,PV}}{L_{PV}}, k_2 = \frac{V_{D,PV} + \mathscr{F}_6^l - R_{M,PV}\mathscr{F}_2^l}{L_{PV}}. \tag{7.32}$$

3) Excess power mode

In this model, the lead–acid battery works as loading, hence the system model can be given by

$$\begin{cases} \dot{v}_{PV} = \frac{1}{C_{PV}}\left(\phi_{PV} - \phi_{L,PV}u\right), \\ \dot{\phi}_{L,PV} = \frac{1}{L_{PV}}R_{0,PV}\left(\left(\phi_{LA} + \phi_{0,PV} - \phi_{L,PV}\right) - R_{L,PV}\phi_{L,PV} - v_{0,PV}\right) \\ \quad + \frac{1}{L_{PV}}\left(V_{D,PV} + v_{PV} - R_{M,PV}\phi_{L,PV}\right)u_1 - \frac{V_{D,PV}}{L_{PV}}, \\ \dot{v}_{0,PV} = \frac{1}{C_{0,PV}+C_{LA}+C_{pv}}\phi_{L,PV} - \frac{1}{C_{0,PV}+C_{LA}+C_{pv}}\phi_1 - \frac{\phi_{L,LA}}{C_{0,PV}+C_{LA}+C_{pv}}u_2, \\ \dot{\phi}_{1,LA} = \frac{1}{R_{1,LA}C_{1,LA}}\left(\phi_{m,LA} - \phi_{1,LA}\right), \\ \dot{\phi}_{m,LA} = \frac{1}{R_{0,LA}C_{0,LA}}\left(\phi_{L,LA} - \phi_{m,LA}\right) - \frac{\dot{v}_{PN,LA}}{R_{0,LA}}, \\ \dot{\phi}_{L,LA} = \frac{R_{0,LA}}{L_{LA}}\left(\phi_{m,LA} - \phi_{L,LA} - R_{L,LA}\phi_{L,LA} - \phi_{m,LA}R_{0,LA} - v_{PN,LA}\right) \\ \quad + \frac{1}{L_{LA}}\left(V_{D,LA} + v_{0,PV} - R_{M,LA}\phi_{L,LA}\right)u_2 - \frac{V_{D,LA}}{L_{LA}}, \\ \dot{\phi}_d = -\frac{R_1}{L_1}\phi_d - \omega\phi_q + \frac{1}{L_1}e_d, \\ \dot{\phi}_q = \omega\phi_d - \frac{R_1}{L_1}\phi_q + \frac{1}{L_1}e_q. \end{cases} \tag{7.33}$$

Under this operation mode, the objective is to design a supervisory controller, so that the MPPT can be obtained and the output voltage conforms to the reference. The voltage reference v^* of the MPPT can be calculated by (7.8). Define the voltage reference v_{ref}, and $x(t) = [\phi_{1,LA}\ \phi_{m,LA}\ e_{0,LA}\ v_{PV}\ \phi_{L,PV}\ e_{0,PV}]^T$, then the system in (7.33) can be rewritten as below:

$$\begin{cases} \dot{v}_{PV} = \frac{1}{C_{PV}}\left(\phi_{PV} - \phi_{L,PV}u\right), \\ \dot{\phi}_{L,PV} = \frac{R_{0,PV}}{L_{PV}}\left(\phi_{LA} + \phi_{0,PV} - \phi_{L,PV} - R_{L,PV}\phi_{L,PV} - e_{0,PV} - v_{dc}^*\right) \\ \quad + \frac{1}{L_{PV}}\left(V_{D,PV} + v_{PV} - R_{M,PV}\phi_{L,PV}\right)u_1 - \frac{V_{D,PV}}{L_{PV}}, \\ \dot{e}_{0,PV} = \frac{1}{C_{0,PV}+C_{LA}+C_{pv}}\phi_{L,PV} - \frac{1}{C_{0,PV}+C_{LA}+C_{pv}}\phi_1 - \frac{\phi_{L,LA}}{C_{0,PV}+C_{LA}+C_{pv}}u_2, \\ \dot{\phi}_{1,LA} = \frac{1}{R_{1,LA}C_{1,LA}}\left(\phi_{m,LA} - \phi_{1,LA}\right), \\ \dot{\phi}_{m,LA} = \frac{1}{R_{0,LA}C_{0,LA}}\left(\phi_{L,LA} - \phi_{m,LA}\right) - \frac{\dot{v}_{PN,LA}}{R_{0,LA}}, \\ \dot{\phi}_{L,LA} = \frac{R_{0,LA}}{L_{LA}}\left(\phi_{m,LA} - \phi_{L,LA} - R_{L,LA}\phi_{L,LA} - \phi_{m,LA}R_{0,LA} - v_{PN,LA}\right) \\ \quad + \frac{1}{L_{LA}}\left(V_{D,LA} + e_{0,PV} + v_{dc}^* - R_{M,LA}\phi_{L,LA}\right)u_2 - \frac{V_{D,LA}}{L_{LA}}, \\ \dot{\phi}_d = -\frac{R_1}{L_1}\phi_d - \omega\phi_q + \frac{1}{L_1}e_d, \\ \dot{\phi}_q = \omega\phi_d - \frac{R_1}{L_1}\phi_q + \frac{1}{L_1}e_q. \end{cases} \tag{7.34}$$

Then, choose $\{\frac{v_{LA}}{\phi_{m,LA}}, \frac{\phi_{p,LA}}{\phi_{m,LA}}, \frac{\phi_{0,PV}}{\phi_{L,PV}}, \frac{\phi_{PV}}{v_{PV}}, \frac{\phi_{0,LA}}{\phi_{m,LA}}, e_{0,LA}, \phi_{m,LA}, \phi_{p,LA}, \phi_{L,PV}, v_{PV}, e^{\gamma v_{pv}}\}$ as the fuzzy premise variables, then the augmented fuzzy system is given by

Plant Rule $\mathscr{R}^l$: **IF** z_1 is $\mathscr{F}_1^l$ and z_2 is $\mathscr{F}_2^l$ and $\cdots$ and z_{11} is $\mathscr{F}_{11}^l$, **THEN**

$$\dot{x}(t) = A_l x(t) + B_l u(t) + \omega(t), l \in \mathscr{L} \tag{7.35}$$

where $\mathscr{R}^l$ denotes the l-th fuzzy inference rule; $\mathscr{L} := \{1, 2, \ldots, r\}$, and r is the number of inference rules; $\mathscr{F}_\theta^l(\theta = 1, 2, \cdots, 11)$ is the fuzzy set; $x(t) \in \Re^{n_x}$ and $u(t) \in \Re^{n_u}$ denote the system state and control input, respectively; $z(t) \triangleq [z_1, z_2, \ldots, z_{11}]$ are the measurable variables; The l-th local model $\{A_l, B_l\}$ and the disturbance $\omega(t)$ are given as below:

$$A_l = \begin{bmatrix} \frac{\mathscr{F}_1^l}{C_{PV}} & 0 & 0 & 0 & 0 & 0 \\ \frac{R_{0,PV}\mathscr{F}_2^l}{L_{PV}} & -\frac{k_4}{L_{PV}} & \frac{k_5}{L_{PV}R_{load}} & 0 & 0 & 0 \\ 0 & \frac{1}{k_3} & -\frac{1}{R_{load}k_3} & 0 & 0 & 0 \\ 0 & 0 & 0 & -\frac{1}{k_1} & \frac{1}{k_1} & 0 \\ 0 & 0 & 0 & 0 & -\frac{1}{k_2} & \frac{1}{k_2} \\ 0 & 0 & 0 & 0 & k_7 & -\frac{k_6}{L_{LA}} \end{bmatrix},$$

$$B_l = \begin{bmatrix} -\frac{\mathscr{F}_3^l}{C_{PV}} & 0 \\ k_8 & 0 \\ 0 & -\frac{\mathscr{F}_5^l}{C_{0,PV}+C_{LA}} \\ 0 & 0 \\ 0 & 0 \\ 0 & k_9 \end{bmatrix}, \omega(t) = \begin{bmatrix} 0 \\ k_{10}(t) \\ -\frac{1}{R_{load}\left(C_{0,PV}+C_{LA}\right)}v_{ref} \\ 0 \\ -\frac{\dot{v}_{PN,LA}}{R_{0,LA}} \\ -\frac{R_{0,LA}}{L_{LA}}v_{PN,LA} - \frac{V_{D,LA}}{L_{LA}} \end{bmatrix},$$

$$\begin{aligned} & k_1 = R_{1,LA}C_{1,LA}, k_2 = R_{0,LA}C_{0,LA}, k_3 = C_{0,PV} + C_{LA}, \\ & k_4 = R_{0,PV} + R_{0,PV}R_{L,PV}, k_5 = R_{0,PV} - R_{0,PV}R_{load}, \\ & k_6 = R_{0,LA} + R_{0,LA}R_{L,LA}, k_7 = \frac{R_{0,LA} - R_{0,LA}^2}{L_{LA}}, \\ & k_8 = \frac{V_{D,PV} + \mathscr{F}_4^l - R_{M,PV}\mathscr{F}_3^l}{L_{PV}}, k_9 = \frac{V_{D,LA} + \mathscr{F}_6^l + v_{ref} - R_{M,LA}\mathscr{F}_5^l}{L_{LA}}, \\ & k_{10}(t) = \frac{R_{0,PV}}{L_{PV}R_{load}}v_{ref} - \frac{R_{0,PV}}{L_{PV}}v_{ref} - \frac{V_{D,PV}}{L_{PV}}. \end{aligned} \tag{7.36}$$

7.3 PMSG SYSTEM FOR DC LOAD

This section considers a small-scale DC power system, which consists of the PMSG subsystem, the lead-acid battery and the DC load. The PMSG subsystem uses an AC-DC buck converter to the DC-bus, and a bi-directional buck-boost converter will be used to interface the battery and the DC-bus. In this system, as illustrated in Figure 7.3, the bi-directional converter operates as a boost converter during the discharging mode of the battery and as a buck converter during the charging mode of the battery.

7.3.1 OPERATION MODES

In order to ensure the reliable and stable operations of the small-scale DC power system under the different conditions, the power balance among the supply and the demand must be maintained under all operating conditions. Now, we define P_{MG}, $P_{LD}^{\min}$, and $P_{LD}^{\max}$ as the PMSG supply, the minimum load demand, and the maximum

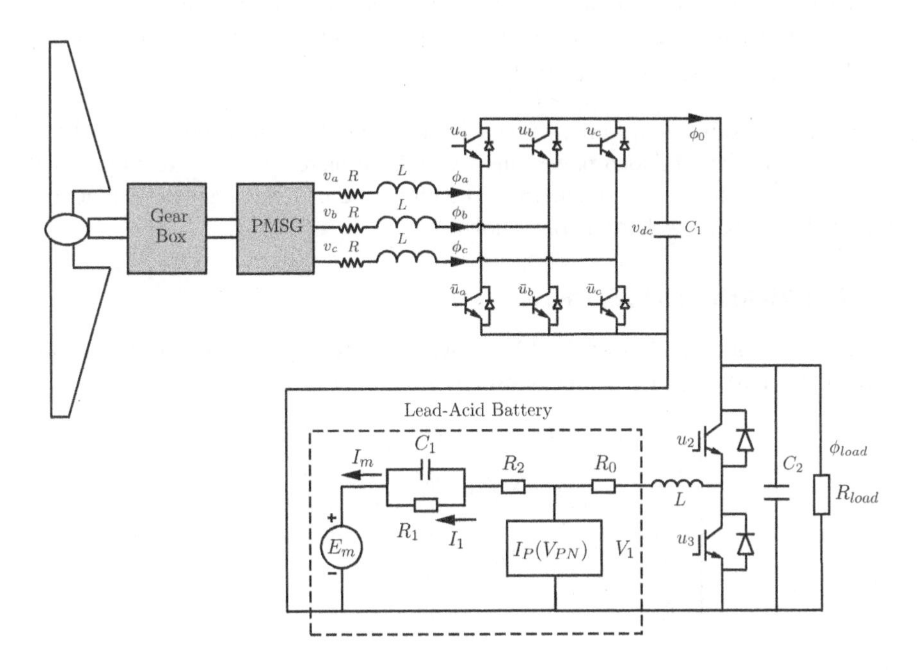

Figure 7.3 DC microgrid with PMSG.

load demand, respectively. There exist these three possible operation conditions of the system [1].

1) Deficit power mode $\left(P_{MG} \leq P_{LD}^{\min}\right)$

In this mode, the maximum available power is less than the power demand. Specifically, the PMSG is not sufficient to satisfy the DC loads. The PMSG subsystem works with the MPPT algorithm, and the lead-acid battery converters regulate the output voltage by the battery discharging operation. Thus, they work as the voltage sources. To obtain both the accurate current sharing and the desirable voltage regulation using the fuzzy control method, the voltage of DC-bus reaches its minimum acceptable value at the maximum discharge current of the batteries.

2) Floating power mode $\left(P_{LD}^{\min} \leq P_{MG} \leq P_{LD}^{\max}\right)$

The DC bus operates in an islanded mode. The PMSG subsystem is basically sufficient to satisfy the DC loads, and the batteries do not work in this case. The PMSG subsystem works with the MPPT algorithm, and regulates the voltage of DC-bus by using the fuzzy tracking control.

3) Excess power mode $\left(P_{MG} \geq P_{LD}^{\max}\right)$

The PMSG subsystem works with the MPPT algorithm. Since the PMSG subsystem is greater than the load power, the DC voltage increases. The excess power is used to charge the lead-acid batteries. Therefore, the batteries regulate the DC bus voltage by the battery charging operations.

7.3.2 DYNAMIC MODELING

Recall the system models of the PMSG with the AC-DC converter in (2.20), the lead-acid battery in (3.1), and the boost converter in (1.6) as below:

$$\begin{cases} \dot{P} = \omega v_{\alpha\beta}^T J^T \phi_{\alpha\beta} + \frac{1}{L} v_{\alpha\beta}^T v_{\alpha\beta} - v_{\alpha\beta}^T \frac{v_{dc}}{2L} u_{\alpha\beta}, \\ \dot{Q} = \omega v_{\alpha\beta}^T J^T J \phi_{\alpha\beta} + \frac{v_{\alpha\beta}^T J}{L} v_{\alpha\beta} - \frac{v_{\alpha\beta}^T J v_{dc}}{2L} u_{\alpha\beta}, \\ \dot{v}_{\alpha\beta} = \omega J v_{\alpha\beta}, \\ \dot{\phi}_{\alpha\beta} = \frac{1}{L} v_{\alpha\beta} - \frac{v_{dc}}{2L} u_{\alpha\beta}, \\ \dot{v}_{dc} = -\frac{1}{CR_L} v_{dc} + \frac{\phi_{\alpha\beta}^T}{2C} u_{\alpha\beta}, \end{cases} \tag{7.37}$$

and

$$\begin{cases} \dot{\phi}_{1,LA} = \frac{1}{R_{1,LA} C_{1,LA}} \left(\phi_{m,LA} - \phi_{1,LA}\right), \\ \dot{Q}_{e,LA} = -\phi_{m,LA}, \\ \dot{\theta}_{LA} = \frac{1}{C_{\theta,LA}} \left[P_{s,LA} - \frac{\left(\theta_{LA} - \theta_{a,LA}\right)}{R_{\theta,LA}} \right], \\ \phi_{p,LA} = G_{p,LA} v_{PN,LA}, \\ G_{p,LA} = G_{p0,LA} \exp\left(v_{PN,LA}/v_{p0,LA} + A_{p,LA}\left(1 - \theta_{LA}/\theta_{f,LA}\right)\right), \end{cases} \tag{7.38}$$

and

$$\begin{cases} \dot{\phi}_{LA} = -\frac{1}{L_{LA}} (1-u) v_{dc} + \frac{1}{L_{LA}} v_{LA}, \\ \dot{v}_{dc} = \frac{1}{C_{0,LA}} (1-u) \phi_{LA} - \frac{1}{C_{0,LA}} \phi_{0,LA}, \end{cases} \tag{7.39}$$

where the subscripts MG and LA stand for the PMSG and the lead-acid battery, respectively, and all notations are defined in (2.20), (3.1), and (1.6), respectively.

1) Deficit power mode

First, consider the system model of the lead-acid battery with the boost converter as below:

$$\begin{cases} \dot{\phi}_{1,LA} = \frac{1}{R_{1,LA}C_{1,LA}}(\phi_{m,LA} - \phi_{1,LA}), \\ \dot{\phi}_{m,LA} = -\frac{1}{R_{0,LA}L_{LA}}(1-u_3)v_{0,LA} + \frac{1}{R_{0,LA}L_{LA}}v_{LA} - \dot{G}_{p,LA}v_{PN,LA} - G_{p,LA}\dot{v}_{PN,LA}, \\ \dot{v}_{0,LA} = \frac{1}{C_{0,LA}}(1-u_3)(\phi_{m,LA} + \phi_{p,LA}) - \frac{1}{R_{0,LA}C_{0,LA}}\phi_{0,LA}, \end{cases} \tag{7.40}$$

and the PMSG with the AC-DC converter and the boost converter is

$$\begin{cases} \dot{P} = \omega v_{\alpha\beta}^T J^T \phi_{\alpha\beta} + \frac{1}{L} v_{\alpha\beta}^T v_{\alpha\beta} - v_{\alpha\beta}^T \frac{v_{0,MG}}{2L} u_{\alpha\beta}, \\ \dot{Q} = \omega v_{\alpha\beta}^T J^T J \phi_{\alpha\beta} + \frac{v_{\alpha\beta}^T J}{L} v_{\alpha\beta} - \frac{v_{\alpha\beta}^T J v_{0,MG}}{2L} u_{\alpha\beta}, \\ \dot{v}_{\alpha\beta} = \omega J v_{\alpha\beta}, \\ \dot{\phi}_{\alpha\beta} = \frac{1}{L} v_{\alpha\beta} - \frac{v_{0,MG}}{2L} u_{\alpha\beta}, \\ \dot{v}_{0,MG} = -\frac{1}{CR_L} v_{0,MG} + \frac{\phi_{\alpha\beta}^T}{2C} u_{\alpha\beta}. \end{cases} \tag{7.41}$$

Under the deficit power mode, both the PMSG and the lead-acid battery can be supplied to the DC load. Based on the Thevenin's theorem,

$$v_{0,LA} = \phi_{0,LA} R_{line,LA} + (\phi_{0,LA} + \phi_{0,MG}) R_{load}, \tag{7.42}$$

$$v_{0,MG} = \phi_{0,MG} R_{line,MG} + (\phi_{0,LA} + \phi_{0,MG}) R_{load}, \tag{7.43}$$

where the $R_{line,LA}$ and $R_{line,MG}$ are the resistances of power line on the lead-acid battery and the PMSG system, respectively, R_{load} is the load resistance.

The objective is to design a supervisory controller, such the MPPT can be obtained and the output voltage tracks the reference v_{ref}. The voltage reference v^* of the MPPT can be calculated by (7.8), and $P \to P^*, Q \to Q^*, v_{0,MG} \to v_{ref}, v_{0,LA} \to v_{ref}$. Define $P_e = P - P^*, Q_e = Q - Q^*, v_{e,MG} = v_{0,MG} - v_{ref}, v_{e,LA} = v_{0,LA} - v_{ref}$, and we get

$$\begin{cases} \dot{\phi}_{1,LA} = \frac{1}{R_{1,LA}C_{1,LA}}(\phi_{m,LA} - \phi_{1,LA}), \\ \dot{\phi}_{m,LA} = -\frac{1}{R_{0,LA}L_{LA}} v_{e,LA} + \frac{v_{0,LA}}{R_{0,LA}L_{LA}} u_3 + \frac{1}{R_{0,LA}L_{LA}} v_{LA} \\ \qquad - \dot{G}_{p,LA} v_{PN,LA} - G_{p,LA}\dot{v}_{PN,LA} - \frac{1}{R_{0,LA}L_{LA}} v_{ref}, \\ \dot{v}_{e,LA} = \frac{\phi_{m,LA}}{C_{0,LA}} + \frac{\phi_{p,LA}}{C_{0,LA}} - \frac{\phi_{m,LA}+\phi_{p,LA}}{C_{0,LA}} u_3 - \frac{1}{R_{0,LA}R_{load}C_{0,LA}} v_{e,LA} \\ \qquad + \frac{1}{R_{0,LA}C_{0,LA}} \phi_{0,MG} + \frac{R_{line,LA}}{R_{0,LA}C_{0,LA}R_{load}} \phi_{0,LA} - \frac{1}{R_{0,LA}R_{load}C_{0,LA}} v_{ref}, \\ \dot{P}_e = \omega v_{\alpha\beta}^T J^T \phi_{\alpha\beta} + \frac{1}{L} v_{\alpha\beta}^T v_{\alpha\beta} - v_{\alpha\beta}^T \frac{v_{0,MG}}{2L} u_{\alpha\beta}, \\ \dot{Q}_e = \omega v_{\alpha\beta}^T J^T J \phi_{\alpha\beta} + \frac{v_{\alpha\beta}^T J}{L} v_{\alpha\beta} - \frac{v_{\alpha\beta}^T J v_{0,MG}}{2L} u_{\alpha\beta}, \\ \dot{v}_{\alpha\beta} = \omega J v_{\alpha\beta}, \\ \dot{\phi}_{\alpha\beta} = \frac{1}{L} v_{\alpha\beta} - \frac{v_{0,MG}}{2L} u_{\alpha\beta}, \\ \dot{v}_{e,MG} = -\frac{1}{CR_L} v_{e,MG} + \frac{\phi_{\alpha\beta}^T}{2C} u_{\alpha\beta} - \frac{1}{CR_L} v_{ref}. \end{cases} \tag{7.44}$$

Define $x(t) = \begin{bmatrix} \phi_{1,LA} & \phi_{m,LA} & v_{e,LA} & P_e & Q_e & v_{\alpha\beta} & \phi_{\alpha\beta} & v_{e,MG} \end{bmatrix}^T$, and choose $\{\frac{v_{LA}}{\phi_{m,LA}}, v_{0,LA}, \frac{\phi_{p,LA}}{\phi_{m,LA}}, \frac{\phi_{0,MG}}{\phi_{m,LA}}, \frac{\phi_{0,LA}}{\phi_{m,LA}}, \phi_{m,LA}, \phi_{p,LA}, v_{\alpha\beta}^T, \omega v_{\alpha\beta}^T, v_{0,MG}, \phi_{\alpha\beta}^T, \omega\}$ as the fuzzy premise variables, then the augmented fuzzy system is given by

Plant Rule $\mathscr{R}^l$: **IF** z_1 is $\mathscr{F}_1^l$ and z_2 is $\mathscr{F}_2^l$ and $\cdots$ and z_{11} is $\mathscr{F}_{11}^l$, **THEN**

$$\dot{x}(t) = A_l x(t) + B_l u(t) + \omega(t), l \in \mathscr{L} := \{1, 2, \ldots, r\} \tag{7.45}$$

where $\mathscr{R}^l$ denotes the l-th fuzzy inference rule; $\mathscr{L} := \{1, 2, \ldots, r\}$, and r is the number of inference rules; $\mathscr{F}_\theta^l(\theta = 1, 2, \cdots, 11)$ is the fuzzy set; $x(t) \in \mathfrak{R}^{n_x}$ and $u(t) \in \mathfrak{R}^{n_u}$ denote the system state and control input, respectively; $z(t) \triangleq [z_1, z_2, \ldots, z_{11}]$ are the measurable variables; The l-th local model $\{A_l, B_l\}$ and the disturbance $\omega(t)$ are given as below:

$$A_l = \begin{bmatrix} -k_1 & k_1 & 0 & 0 & 0 & 0 & 0 & 0 \\ 0 & k_2\mathscr{F}_1^l & -k_2 & 0 & 0 & 0 & 0 & 0 \\ 0 & k_3 & -k_4 & 0 & 0 & 0 & 0 & 0 \\ 0 & 0 & 0 & 0 & 0 & \frac{\mathscr{F}_8^l}{L} & \mathscr{F}_9^l J^T & 0 \\ 0 & 0 & 0 & 0 & 0 & \frac{\mathscr{F}_8^l J}{L} & \mathscr{F}_9^l J^T J & 0 \\ 0 & 0 & 0 & 0 & 0 & \mathscr{F}_{12}^l J & 0 & 0 \\ 0 & 0 & 0 & 0 & 0 & \frac{1}{L} & 0 & 0 \\ 0 & 0 & 0 & 0 & 0 & 0 & 0 & -\frac{1}{CR_L} \end{bmatrix},$$

$$B_l = \begin{bmatrix} 0 & 0 \\ \frac{\mathscr{F}_2^l}{R_{0,LA}L_{LA}} & 0 \\ -\frac{\mathscr{F}_6^l + \mathscr{F}_7^l}{C_{0,LA}} & 0 \\ 0 & -\frac{\mathscr{F}_8^l \mathscr{F}_{10}^l}{2L} \\ 0 & -\frac{\mathscr{F}_8^l J \mathscr{F}_{10}^l}{2L} \\ 0 & 0 \\ 0 & -\frac{\mathscr{F}_{10}^l}{2L} \\ 0 & \frac{\mathscr{F}_{11}^l}{2C} \end{bmatrix}, \omega(t) = \begin{bmatrix} 0 \\ k_5(t) \\ k_6(t) \\ 0 \\ 0 \\ 0 \\ 0 \\ -\frac{v_{ref}}{CR_L} \end{bmatrix},$$

$$k_1 = \frac{1}{R_{1,LA}C_{1,LA}}, k_2 = \frac{1}{R_{0,LA}L_{LA}}, k_3 = \frac{1 + \mathscr{F}_3^l}{C_{0,LA}} + \frac{\mathscr{F}_4^l}{R_{0,LA}C_{0,LA}} + \frac{R_{line,LA}}{R_{0,LA}C_{0,LA}R_{load}}\mathscr{F}_5^l,$$
$$k_4 = \frac{1}{R_{0,LA}R_{load}C_{0,LA}}, k_5(t) = -\dot{G}_{p,LA}v_{PN,LA} - G_{p,LA}\dot{v}_{PN,LA} - \frac{1}{R_{0,LA}L_{LA}}v_{ref},$$
$$k_6(t) = -\frac{1}{R_{0,LA}R_{load}C_{0,LA}}v_{ref}. \tag{7.46}$$

2) Floating power mode

In this mode, the lead-acid battery does not work. In other words, the power can be maintained between the PMSG and the loading. Hence, the system model is

$$\left\{\begin{array}{l}\dot{P}=\omega v_{\alpha\beta}^{T}J^{T}\phi_{\alpha\beta}+\frac{1}{L}v_{\alpha\beta}^{T}v_{\alpha\beta}-v_{\alpha\beta}^{T}\frac{v_{0,MG}}{2L}u_{\alpha\beta},\\ \dot{Q}=\omega v_{\alpha\beta}^{T}J^{T}J\phi_{\alpha\beta}+\frac{v_{\alpha\beta}^{T}J}{L}v_{\alpha\beta}-\frac{v_{\alpha\beta}^{T}Jv_{0,MG}}{2L}u_{\alpha\beta},\\ \dot{v}_{\alpha\beta}=\omega Jv_{\alpha\beta},\\ \dot{\phi}_{\alpha\beta}=\frac{1}{L}v_{\alpha\beta}-\frac{v_{0,MG}}{2L}u_{\alpha\beta},\\ \dot{\phi}_{0,MG}=-\frac{1}{L_{MG}}(1-u)\left(v_{0,MG}-\phi_{0,MG}R_{line,MG}\right)+\frac{1}{L_{MG}}v_{0,MG},\\ \dot{v}_{0,MG}=-\frac{1}{CR_{L}}v_{0,MG}+\frac{\phi_{\alpha\beta}^{T}}{2C}u_{\alpha\beta}.\end{array}\right. \tag{7.47}$$

Under this operation mode, the objective is to design a supervisory controller, so that the MPPT can be obtained and the output voltage tracks the reference v_{ref}. The voltage reference v^* of the MPPT can be calculated by (7.8), and $P\to P^*, Q\to Q^*, v_{0,MG}\to v_{ref}, v_{0,LA}\to v_{ref}$. Define $P_e=P-P^*, Q_e=Q-Q^*, v_{e,MG}=v_{0,MG}-v_{ref}$,

$$\left\{\begin{array}{l}\dot{P}_e=\omega v_{\alpha\beta}^{T}J^{T}\phi_{\alpha\beta}+\frac{1}{L}v_{\alpha\beta}^{T}v_{\alpha\beta}-v_{\alpha\beta}^{T}\frac{v_{0,MG}}{2L}u_{\alpha\beta},\\ \dot{Q}_e=\omega v_{\alpha\beta}^{T}J^{T}J\phi_{\alpha\beta}+\frac{v_{\alpha\beta}^{T}J}{L}v_{\alpha\beta}-\frac{v_{\alpha\beta}^{T}Jv_{0,MG}}{2L}u_{\alpha\beta},\\ \dot{v}_{\alpha\beta}=\omega Jv_{\alpha\beta},\\ \dot{\phi}_{\alpha\beta}=\frac{1}{L}v_{\alpha\beta}-\frac{v_{0,MG}}{2L}u_{\alpha\beta},\\ \dot{v}_{e,MG}=-\frac{1}{CR_{L}}v_{e,MG}-\frac{1}{CR_{L}}v_{ref}+\frac{\phi_{\alpha\beta}^{T}}{2C}u_{\alpha\beta}.\end{array}\right. \tag{7.48}$$

Define $x(t)=\left[\begin{array}{ccccc}P_e & Q_e & v_{\alpha\beta} & \phi_{\alpha\beta} & v_{e,MG}\end{array}\right]^T$, and choose $\{v_{\alpha\beta}^T, \omega v_{\alpha\beta}^T, v_{0,MG}, \phi_{\alpha\beta}^T, \omega\}$ as the fuzzy premise variables, then the augmented fuzzy system is given by

Plant Rule $\mathscr{R}^l$: **IF** z_1 is $\mathscr{F}_1^l$ and z_2 is $\mathscr{F}_2^l$ and $\cdots$ and z_5 is $\mathscr{F}_5^l$, **THEN**

$$\dot{x}(t)=A_l x(t)+B_l u(t)+\omega(t), l\in\mathscr{L}:=\{1,2,\ldots,r\} \tag{7.49}$$

where $\mathscr{R}^l$ denotes the l-th fuzzy inference rule; $\mathscr{L}:=\{1,2,\ldots,r\}$, and r is the number of inference rules; $\mathscr{F}_\theta^l(\theta=1,2,\cdots,5)$ is the fuzzy set; $x(t)\in\Re^{n_x}$ and $u(t)\in\Re^{n_u}$ denote the system state and control input, respectively; $z(t)\triangleq[z_1,z_2,\ldots,z_5]$ are the measurable variables; The l-th local model $\{A_l,B_l\}$ and the disturbance $\omega(t)$ are given as below:

$$A_l=\left[\begin{array}{ccccc}0&0&\frac{\mathscr{F}_1^l}{L}&\mathscr{F}_2^lJ^T&0\\ 0&0&\frac{\mathscr{F}_1^lJ}{L}&\mathscr{F}_2^lJ^TJ&0\\ 0&0&\mathscr{F}_5^lJ&0&0\\ 0&0&\frac{1}{L}&0&0\\ 0&0&0&0&-\frac{1}{CR_L}\end{array}\right], B_l=\left[\begin{array}{c}-\frac{\mathscr{F}_1^l\mathscr{F}_3^l}{2L}\\ -\frac{\mathscr{F}_1^lJ\mathscr{F}_3^l}{2L}\\ 0\\ -\frac{\mathscr{F}_3^l}{2L}\\ \frac{\mathscr{F}_4^l}{2C}\end{array}\right], \omega(t)=\left[\begin{array}{c}0\\0\\0\\0\\-\frac{v_{ref}}{CR_L}\end{array}\right]. \tag{7.50}$$

3) Excess power mode

In this model, the lead–acid battery works as the loading, hence the system model can be given by

$$\begin{cases} \dot{\phi}_{1,LA} = \frac{1}{R_{1,LA}C_{1,LA}}\left(\phi_{m,LA} - \phi_{1,LA}\right), \\ \dot{\phi}_{m,LA} = \frac{1}{R_{0,LA}C_{0,LA}}\left(\phi_{L,LA} - \phi_{m,LA}\right) - \frac{\dot{v}_{PN,LA}}{R_{0,LA}}, \\ \dot{\phi}_{L,LA} = \frac{1}{L_{LA}}R_{0,LA}\left(\phi_{m,LA} - \phi_{L,LA} - R_{L,LA}\phi_{L,LA} - \left(\phi_{m,LA}R_{0,LA} + v_{PN,LA}\right)\right) \\ \quad + \frac{1}{L_{LA}}\left(V_{D,LA} + v_{0,MG} - R_{M,LA}\phi_{L,LA}\right)u_2 - \frac{V_{D,LA}}{L_{LA}}, \end{cases} \tag{7.51}$$

and the PMSG with the AC-DC converter and the boost converter is

$$\begin{cases} \dot{P} = \omega v_{\alpha\beta}^T J^T \phi_{\alpha\beta} + \frac{1}{L} v_{\alpha\beta}^T v_{\alpha\beta} - v_{\alpha\beta}^T \frac{v_{0,MG}}{2L} u_{\alpha\beta}, \\ \dot{Q} = \omega v_{\alpha\beta}^T J^T J \phi_{\alpha\beta} + \frac{v_{\alpha\beta}^T J}{L} v_{\alpha\beta} - \frac{v_{\alpha\beta}^T J v_{0,MG}}{2L} u_{\alpha\beta}, \\ \dot{v}_{\alpha\beta} = \omega J v_{\alpha\beta}, \\ \dot{\phi}_{\alpha\beta} = \frac{1}{L} v_{\alpha\beta} - \frac{v_{0,MG}}{2L} u_{\alpha\beta}, \\ \dot{v}_{0,MG} = -\frac{1}{CR_L} v_{0,MG} + \frac{\phi_{\alpha\beta}^T}{2C} u_{\alpha\beta}. \end{cases} \tag{7.52}$$

Under the deficit power mode, both the PMSG and the lead-acid battery can be supplied to the DC load. Based on the Kirchhoff's current law,

$$\phi_{0,MG} = \phi_{0,LA} + \frac{v_{0,MG} - \phi_{0,MG}R_{line,MG}}{R_{load}}. \tag{7.53}$$

The objective is to design a supervisory controller, such that the MPPT can be obtained and the output voltage tracks the reference v_{ref}. The voltage reference v^* of the MPPT can be calculated by (7.8), and $P \to P^*, Q \to Q^*, v_{0,MG} \to v_{ref}$. Define $P_e = P - P^*, Q_e = Q - Q^*, v_{e,MG} = v_{0,MG} - v_{ref}$,

$$\begin{cases} \dot{\phi}_{1,LA} = \frac{1}{R_{1,LA}C_{1,LA}}\phi_{m,LA} - \frac{1}{R_{1,LA}C_{1,LA}}\phi_{1,LA}, \\ \dot{\phi}_{m,LA} = \frac{1}{R_{0,LA}C_{0,LA}}\left(\phi_{L,LA} - \phi_{m,LA}\right) - \frac{\dot{v}_{PN,LA}}{R_{0,LA}}, \\ \dot{\phi}_{L,LA} = \frac{1}{L_{LA}}R_{0,LA}\left(\phi_{m,LA} - \phi_{L,LA} - R_{L,LA}\phi_{L,LA} - \left(\phi_{m,LA}R_{0,LA} + v_{PN,LA}\right)\right) \\ \quad + \frac{1}{L_{LA}}\left(V_{D,LA} + v_{0,MG} - R_{M,LA}\phi_{L,LA}\right)u_2 - \frac{V_{D,LA}}{L_{LA}}, \\ \dot{P}_e = \omega v_{\alpha\beta}^T J^T \phi_{\alpha\beta} + \frac{1}{L} v_{\alpha\beta}^T v_{\alpha\beta} - v_{\alpha\beta}^T \frac{v_{0,MG}}{2L} u_{\alpha\beta}, \\ \dot{Q}_e = \omega v_{\alpha\beta}^T J^T J \phi_{\alpha\beta} + \frac{v_{\alpha\beta}^T J}{L} v_{\alpha\beta} - \frac{v_{\alpha\beta}^T J v_{0,MG}}{2L} u_{\alpha\beta}, \\ \dot{v}_{\alpha\beta} = \omega J v_{\alpha\beta}, \\ \dot{\phi}_{\alpha\beta} = \frac{1}{L} v_{\alpha\beta} - \frac{v_{0,MG}}{2L} u_{\alpha\beta}, \\ \dot{v}_{e,MG} = -\frac{1}{CR_L}\left(\phi_{0,MG}R_{load} - \phi_{0,LA}R_{load} + \phi_{0,MG}R_{line,MG}\right) + \frac{\phi_{\alpha\beta}^T}{2C} u_{\alpha\beta}. \end{cases} \tag{7.54}$$

Define $x(t) = \begin{bmatrix} \phi_{1,LA} & \phi_{m,LA} & \phi_{L,LA} & P_e & Q_e & v_{\alpha\beta} & \phi_{\alpha\beta} & v_{e,MG} \end{bmatrix}^T$, and choose $\{v_{0,MG}, \phi_{L,LA}, v_{\alpha\beta}^T, \omega v_{\alpha\beta}^T, \frac{\phi_{0,MG}}{\phi_{\alpha\beta}}, \frac{\phi_{0,LA}}{\phi_{\alpha\beta}}, \phi_{\alpha\beta}^T\}$ as the fuzzy premise variables, then the augmented fuzzy system is given by

Plant Rule $\mathscr{R}^l$: **IF** z_1 is $\mathscr{F}_1^l$ and z_2 is $\mathscr{F}_2^l$ and $\cdots$ and z_7 is $\mathscr{F}_7^l$, **THEN**

$$\dot{x}(t) = A_l x(t) + B_l u(t) + \omega(t), l \in \mathscr{L} := \{1, 2, \ldots, r\} \tag{7.55}$$

where $\mathscr{R}^l$ denotes the l-th fuzzy inference rule; $\mathscr{L} := \{1,2,\ldots,r\}$, and r is the number of inference rules; $\mathscr{F}_\theta^l(\theta = 1,2,\cdots,7)$ is the fuzzy set; $x(t) \in \Re^{n_x}$ and $u(t) \in \Re^{n_u}$ denote the system state and the control input, respectively; $z(t) \triangleq [z_1, z_2, \cdots, z_7]$ are the measurable variables; The l-th local model $\{A_l, B_l\}$ and the disturbance $\omega(t)$ are given as below:

$$A_l = \begin{bmatrix} -k_1 & k_1 & 0 & 0 & 0 & 0 & 0 & 0 \\ 0 & -k_2 & k_2 & 0 & 0 & 0 & 0 & 0 \\ 0 & k_3 & -k_4 & 0 & 0 & 0 & 0 & 0 \\ 0 & 0 & 0 & 0 & 0 & \frac{\mathscr{F}_3^l}{L} & \mathscr{F}_4^l J^T & 0 \\ 0 & 0 & 0 & 0 & 0 & \frac{\mathscr{F}_3^l J}{L} & \mathscr{F}_4^l J^T J & 0 \\ 0 & 0 & 0 & 0 & 0 & \omega J & 0 & 0 \\ 0 & 0 & 0 & 0 & 0 & \frac{1}{L} & 0 & 0 \\ 0 & 0 & 0 & 0 & 0 & 0 & -k_5 \mathscr{F}_5^l & \frac{R_{load} \mathscr{F}_6^l}{CR_L} \end{bmatrix},$$

$$B_l = \begin{bmatrix} 0 & 0 \\ 0 & 0 \\ k_6 & 0 \\ 0 & -\frac{\mathscr{F}_3^l \mathscr{F}_1^l}{2L} \\ 0 & -\frac{\mathscr{F}_3^l J \mathscr{F}_1^l}{2L} \\ 0 & -\frac{\mathscr{F}_1^l}{2L} \\ 0 & 0 \\ 0 & \frac{\mathscr{F}_7^l}{2C} \end{bmatrix}, \omega(t) = \begin{bmatrix} 0 \\ -\frac{\dot{v}_{PN,LA}}{R_{0,LA}} \\ k_7(t) \\ 0 \\ 0 \\ 0 \\ 0 \\ 0 \end{bmatrix},$$

$$k_1 = \frac{1}{R_{1,LA}C_{1,LA}}, k_2 = \frac{1}{R_{0,LA}C_{0,LA}}, k_3 = \frac{R_{0,LA} - R_{0,LA}^2}{L_{LA}},$$
$$k_4 = \frac{R_{0,LA} + R_{0,LA}R_{L,LA}}{L_{LA}}, k_5 = \frac{R_{load} + R_{line,MG}}{CR_L},$$
$$k_6 = \frac{V_{D,LA} + \mathscr{F}_1^l - R_{M,LA}\mathscr{F}_2^l}{L_{LA}}, k_7(t) = -\frac{R_{0,LA}v_{PN,LA} + V_{D,LA}}{L_{LA}}. \tag{7.56}$$

7.4 PMSG SYSTEM FOR AC LOAD

The section considers a small-scale DC power system, which consists of the PMSG subsystem, the lead-acid battery and the DC load. The PMSG subsystem uses an AC-DC buck converter to the DC-bus, and a bi-directional buck-boost converter will be used to interface the battery and the DC-bus, and the DC-bus is connected to the AC-bus by using an DC-AC converter. In this system, as illustrated in Figure 7.4, the bi-directional converter operates as a boost converter during the discharging mode of the battery and as a buck converter during the charging mode of the battery.

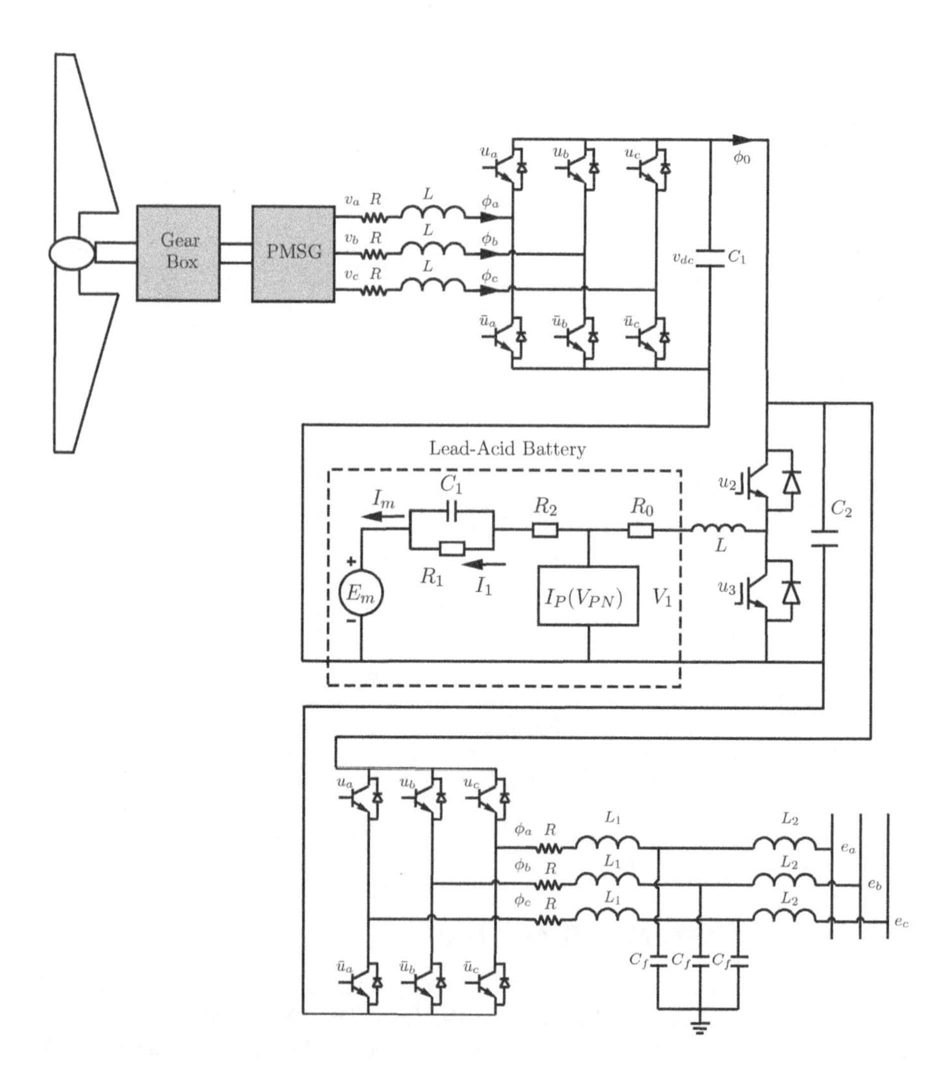

Figure 7.4 AC microgrid with PMSG.

7.4.1 OPERATION MODES

In order to ensure the reliable and stable operations of the small-scale AC power system under the different conditions, the power balance among the supply and the demand must be maintained under all operating conditions. Now, we define P_{MG}, $P_{LD}^{\min}$, and $P_{LD}^{\max}$ as the PMSG supply, the minimum load demand, and the maximum load demand, respectively. There exist these three possible conditions of the system [1].

1) Deficit power mode$\left(P_{MG} \leq P_{LD}^{\min}\right)$

In this mode, the maximum available power is less than the power demand. Specifically, the PMSG is not sufficient to satisfy the DC load. The PMSG subsystem works with the MPPT algorithm, and the lead-acid battery converters regulate the output voltage by the battery discharging operation. Thus, they work as the voltage sources. To obtain both the accurate current sharing and the desirable voltage regulation using the fuzzy control method, the DC-bus voltage reaches its minimum acceptable value at the maximum discharge current of the batteries.

2) Floating power mode$\left(P_{LD}^{\min} \leq P_{MG} \leq P_{LD}^{\max}\right)$

In this mode, the DC operates in an islanded mode. The PMSG subsystem is basically sufficient to satisfy the DC load, and the batteries do not work in this case. The PMSG subsystem works with the MPPT algorithm, and regulates the DC bus voltage by using the fuzzy tracking control.

3) Excess power mode$(P_{MG} \geq P_{LD}^{\max})$

In this mode, the PMSG subsystem works with the MPPT algorithm. Since the PMSG subsystem is greater than the load power, the DC voltage increases. The excess power is used to charge the lead-acid batteries. Therefore, the batteries regulate the DC bus voltage by the battery charging operations.

7.4.2 DYNAMIC MODELING

Let us recall the system models of the PMSG with the AC-DC converter in (2.20), the lead-acid battery in (3.1), and the boost converter in (1.6), and the DC-AC converter in (1.13) as below:

$$\begin{cases} \dot{P} = \omega v_{\alpha\beta}^T J^T \phi_{\alpha\beta} + \frac{1}{L} v_{\alpha\beta}^T v_{\alpha\beta} - v_{\alpha\beta}^T \frac{v_{dc}}{2L} u_{\alpha\beta}, \\ \dot{Q} = \omega v_{\alpha\beta}^T J^T J \phi_{\alpha\beta} + \frac{v_{\alpha\beta}^T J}{L} v_{\alpha\beta} - \frac{v_{\alpha\beta}^T J v_{dc}}{2L} u_{\alpha\beta}, \\ \dot{v}_{\alpha\beta} = \omega J v_{\alpha\beta}, \\ \dot{\phi}_{\alpha\beta} = \frac{1}{L} v_{\alpha\beta} - \frac{v_{dc}}{2L} u_{\alpha\beta}, \\ \dot{v}_{dc} = -\frac{1}{CR_L} v_{dc} + \frac{\phi_{\alpha\beta}^T}{2C} u_{\alpha\beta}, \end{cases} \tag{7.57}$$

and

$$\begin{cases} \dot{\phi}_{1,LA} = \frac{1}{R_{1,LA}C_{1,LA}} \left(\phi_{m,LA} - \phi_{1,LA}\right), \\ \dot{Q}_{e,LA} = -\phi_{m,LA}, \\ \dot{\theta}_{LA} = \frac{1}{C_{\theta,LA}} \left[P_{s,LA} - \frac{\left(\theta_{LA} - \theta_{a,LA}\right)}{R_{\theta,LA}}\right], \\ \phi_{p,LA} = G_{p,LA} v_{PN,LA}, \\ G_{p,LA} = G_{p0,LA} \exp\left(v_{PN,LA}/v_{p0,LA} + A_{p,LA}\left(1 - \theta_{LA}/\theta_{f,LA}\right)\right), \end{cases} \tag{7.58}$$

and

$$\begin{cases} \dot{\phi}_{LA} = -\frac{1}{L_{LA}}(1-u) v_{dc} + \frac{1}{L_{LA}} v_{LA}, \\ \dot{v}_{dc} = \frac{1}{C_{0,LA}}(1-u)\phi_{LA} - \frac{1}{C_{0,LA}}\phi_{0,LA}, \end{cases} \tag{7.59}$$

and

$$\begin{cases} \dot{v}_{pv} = \frac{1}{C_{pv}}(\phi_{pv} - \frac{1.5u_d}{v_{pv}}\phi_d), \\ \dot{\phi}_d = -\frac{R_1}{L_1}\phi_d - \bar{\omega}\phi_q + \frac{1}{L_1}e_d, \\ \dot{\phi}_q = \bar{\omega}\phi_d - \frac{R_1}{L_1}\phi_q + \frac{1}{L_1}e_q, \end{cases} \tag{7.60}$$

where the subscripts MG and LA stand for the PMSG and the lead-acid battery, respectively, and all notations are defined in (1.1), (3.1), (1.6), and (1.13), respectively.

1) Deficit power mode

First, consider the system model of the lead-acid battery with the boost converter as below:

$$\begin{cases} \dot{\phi}_{1,LA} = \frac{1}{R_{1,LA}C_{1,LA}}\left(\phi_{m,LA} - \phi_{1,LA}\right), \\ \dot{\phi}_{m,LA} = -\frac{1}{R_{0,LA}L_{LA}}(1-u_3)v_{0,LA} + \frac{1}{R_{0,LA}L_{LA}}v_{LA} - \dot{G}_{p,LA}v_{PN,LA} - G_{p,LA}\dot{v}_{PN,LA}, \\ \dot{v}_{0,LA} = \frac{1}{C_{0,LA}}(1-u_3)\left(\phi_{m,LA} + \phi_{p,LA}\right) - \frac{1}{R_{0,LA}C_{0,LA}}\phi_{0,LA}, \end{cases} \tag{7.61}$$

and the PMSG with the AC-DC converter and the boost converter is

$$\begin{cases} \dot{P} = \omega v_{\alpha\beta}^T J^T \phi_{\alpha\beta} + \frac{1}{L} v_{\alpha\beta}^T v_{\alpha\beta} - v_{\alpha\beta}^T \frac{v_{0,MG}}{2L} u_{\alpha\beta}, \\ \dot{Q} = \omega v_{\alpha\beta}^T J^T J \phi_{\alpha\beta} + \frac{v_{\alpha\beta}^T J}{L} v_{\alpha\beta} - \frac{v_{\alpha\beta}^T J v_{0,MG}}{2L} u_{\alpha\beta}, \\ \dot{v}_{\alpha\beta} = \omega J v_{\alpha\beta}, \\ \dot{\phi}_{\alpha\beta} = \frac{1}{L} v_{\alpha\beta} - \frac{v_{0,MG}}{2L} u_{\alpha\beta}, \\ \dot{v}_{0,MG} = -\frac{1}{CR_L} v_{0,MG} + \frac{\phi_{\alpha\beta}^T}{2C} u_{\alpha\beta}, \end{cases} \tag{7.62}$$

and

$$\begin{cases} \dot{v}_0 = \frac{1}{C_{pv}}(\phi_{0,LA} + \phi_{0,MG} - \frac{1.5u_d}{v_0}\phi_d), \\ \dot{\phi}_d = -\frac{R_1}{L_1}\phi_d - \bar{\omega}\phi_q + \frac{1}{L_1}e_d, \\ \dot{\phi}_q = \bar{\omega}\phi_d - \frac{R_1}{L_1}\phi_q + \frac{1}{L_1}e_q. \end{cases} \tag{7.63}$$

Under the deficit power mode, both the PMSG and the lead-acid battery can be supplied to the DC load. Based on the Thevenin's theorem,

$$v_{0,LA} = \phi_{0,LA} R_{line,LA} + \left(\phi_{0,LA} + \phi_{0,MG}\right) R_{load}, \tag{7.64}$$

$$v_{0,MG} = \phi_{0,MG} R_{line,MG} + \left(\phi_{0,LA} + \phi_{0,MG}\right) R_{load}, \tag{7.65}$$

where the $R_{line,LA}$ and $R_{line,MG}$ are the resistances of power lines on the lead-acid battery and the PMSG system, respectively, R_{load} is the load resistance.

The objective is to design a supervisory controller, so that the MPPT can be obtained and the output voltage tracks the reference v_{ref}. The voltage reference v^* of the MPPT can be calculated by (7.8), and $P \to P^*, Q \to Q^*, v_{0,MG} \to v_{ref}, v_{0,LA} \to v_{ref}$. Define $P_e = P - P^*, Q_e = Q - Q^*, v_{e,MG} = v_{0,MG} - v_{ref},, v_{e,LA} = v_{0,LA} - v_{ref},$

$$\begin{cases} \dot{\phi}_{1,LA} = \frac{1}{R_{1,LA}C_{1,LA}}\left(\phi_{m,LA} - \phi_{1,LA}\right), \\ \dot{\phi}_{m,LA} = -\frac{1}{R_{0,LA}L_{LA}}v_{e,LA} + \frac{v_{0,LA}}{R_{0,LA}L_{LA}}u_3 + \frac{1}{R_{0,LA}L_{LA}}v_{LA} \\ \quad - \dot{G}_{p,LA}v_{PN,LA} - G_{p,LA}\dot{v}_{PN,LA} - \frac{1}{R_{0,LA}L_{LA}}v_{ref}, \\ \dot{v}_{e,LA} = \frac{\phi_{m,LA}}{C_{0,LA}} + \frac{\phi_{p,LA}}{C_{0,LA}} - \frac{\phi_{m,LA}+\phi_{p,LA}}{C_{0,LA}}u_3 - \frac{1}{R_{0,LA}R_{load}C_{0,LA}}v_{e,LA} \\ \quad + \frac{1}{R_{0,LA}C_{0,LA}}\phi_{0,MG} + \frac{R_{line,LA}}{R_{0,LA}C_{0,LA}R_{load}}\phi_{0,LA} - \frac{1}{R_{0,LA}R_{load}C_{0,LA}}v_{ref}, \\ \dot{P}_e = \omega v_{\alpha\beta}^T J^T \phi_{\alpha\beta} + \frac{1}{L}v_{\alpha\beta}^T v_{\alpha\beta} - v_{\alpha\beta}^T \frac{v_{0,MG}}{2L}u_{\alpha\beta}, \\ \dot{Q}_e = \omega v_{\alpha\beta}^T J^T J\phi_{\alpha\beta} + \frac{v_{\alpha\beta}^T J}{L}v_{\alpha\beta} - \frac{v_{\alpha\beta}^T J v_{0,MG}}{2L}u_{\alpha\beta}, \\ \dot{v}_{\alpha\beta} = \omega J v_{\alpha\beta}, \\ \dot{\phi}_{\alpha\beta} = \frac{1}{L}v_{\alpha\beta} - \frac{v_{0,MG}}{2L}u_{\alpha\beta}, \\ \dot{v}_{e,MG} = -\frac{1}{CR_L}v_{e,MG} + \frac{\phi_{\alpha\beta}^T}{2C}u_{\alpha\beta} - \frac{1}{CR_L}v_{ref}, \\ \dot{v}_0 = \frac{1}{C_{pv}}\left(\phi_{0,LA} + \phi_{0,MG} - \frac{1.5u_d}{v_0}\phi_d\right), \\ \dot{\phi}_d = -\frac{R_1}{L_1}\phi_d - \bar{\omega}\phi_q + \frac{1}{L_1}e_d, \\ \dot{\phi}_q = \bar{\omega}\phi_d - \frac{R_1}{L_1}\phi_q + \frac{1}{L_1}e_q. \end{cases} \tag{7.66}$$

Define $x(t) = \begin{bmatrix} \phi_{1,LA} & \phi_{m,LA} & v_{e,LA} & P_e & Q_e & v_{\alpha\beta} & \phi_{\alpha\beta} & v_{e,MG} & v_0 & \phi_d & \phi_q \end{bmatrix}^T$, and choose $\{\frac{v_{LA}}{\phi_{m,LA}}, \frac{\phi_{0,LA}}{\phi_{m,LA}}, \omega v_{\alpha\beta}^T, v_{\alpha\beta}^T, v_{0,LA}, \phi_{m,LA}, \phi_{p,LA}, v_{0,MG}, \phi_{\alpha\beta}^T, \omega, \frac{u_d}{v_0}, \bar{\omega}\}$ as the fuzzy premise variables, then the augmented fuzzy system is given by

Plant Rule $\mathscr{R}^l$: **IF** z_1 is $\mathscr{F}_1^l$ and z_2 is $\mathscr{F}_2^l$ and $\cdots$ and z_{12} is $\mathscr{F}_{12}^l$, **THEN**

$$\dot{x}(t) = A_l x(t) + B_l u(t) + \omega(t), l \in \mathscr{L} := \{1, 2, \ldots, r\} \tag{7.67}$$

where $\mathscr{R}^l$ denotes the l-th fuzzy inference rule; $\mathscr{L} := \{1, 2, \ldots, r\}$, and r is the number of inference rules; $\mathscr{F}_\theta^l(\theta = 1, 2, \cdots, 12)$ is the fuzzy set; $x(t) \in \Re^{n_x}$ and $u(t) \in \Re^{n_u}$ denote the system state and control input, respectively; $z(t) \triangleq [z_1, z_2, \cdots, z_{12}]$ are the measurable variables; The l-th local model $\{A_l, B_l\}$ and the disturbance $\omega(t)$ are

given as below:

$$A_l = \begin{bmatrix} -k_1 & k_1 & 0 & 0 & 0 & 0 & 0 & 0 & 0 & 0 & 0 \\ 0 & k_2\mathscr{F}_1^l & -k_2 & 0 & 0 & 0 & 0 & 0 & 0 & 0 & 0 \\ 0 & k_3 & -k_4 & 0 & 0 & 0 & 0 & 0 & 0 & 0 & 0 \\ 0 & 0 & 0 & 0 & 0 & \frac{\mathscr{F}_4^l}{L} & \mathscr{F}_3^l J^T & 0 & 0 & 0 & 0 \\ 0 & 0 & 0 & 0 & 0 & \frac{\mathscr{F}_4^l J}{L} & \mathscr{F}_3^l J^T J & 0 & 0 & 0 & 0 \\ 0 & 0 & 0 & 0 & 0 & \mathscr{F}_{10}^l J & 0 & 0 & 0 & 0 & 0 \\ 0 & 0 & 0 & 0 & 0 & \frac{1}{L} & 0 & 0 & 0 & 0 & 0 \\ 0 & 0 & 0 & 0 & 0 & 0 & 0 & -\frac{1}{CR_L} & 0 & 0 & 0 \\ 0 & \frac{\mathscr{F}_2^l}{C_{pv}} & 0 & 0 & 0 & 0 & 0 & 0 & 0 & -\frac{1.5\mathscr{F}_{11}^l}{C_{pv}} & 0 \\ 0 & 0 & 0 & 0 & 0 & 0 & 0 & 0 & 0 & -\frac{R_1}{L_1} & -\mathscr{F}_{12}^l \\ 0 & 0 & 0 & 0 & 0 & 0 & 0 & 0 & 0 & \mathscr{F}_{12}^l & -\frac{R_1}{L_1} \end{bmatrix},$$

$$B_l = \begin{bmatrix} 0 & 0 & 0 \\ k_6 & 0 & 0 \\ -k_7 & 0 & 0 \\ 0 & -k_8 & 0 \\ 0 & -k_9 & 0 \\ 0 & 0 & 0 \\ 0 & -\frac{\mathscr{F}_8^l}{2L} & 0 \\ 0 & 0 & \frac{1}{L_1} \\ 0 & \frac{\mathscr{F}_9^l}{2C} & \frac{1}{L_1} \end{bmatrix}, \omega(t) = \begin{bmatrix} 0 \\ k_{10}(t) \\ -\frac{1}{R_{0,LA}R_{load}C_{0,LA}}v_{ref} \\ 0 \\ 0 \\ 0 \\ 0 \\ \frac{R_{line,MG}}{R_{load}L_{MG}}v_{ref} \\ -\frac{1}{CR_L}v_{ref} \\ 0 \\ 0 \end{bmatrix},$$

$$k_1 = \frac{1}{R_{1,LA}C_{1,LA}}, k_2 = \frac{1}{R_{0,LA}L_{LA}}, k_3 = \frac{1}{C_{0,LA}} + \frac{R_{line,LA}\mathscr{F}_2^l}{R_{0,LA}C_{0,LA}R_{load}},$$
$$k_4 = \frac{1}{R_{0,LA}R_{load}C_{0,LA}}, k_5 = \frac{R_{line,MG}}{R_{load}L_{MG}}, k_6 = \frac{\mathscr{F}_5^l}{R_{0,LA}L_{LA}},$$
$$k_7 = \frac{\mathscr{F}_6^l + \mathscr{F}_7^l}{C_{0,LA}}, k_8 = \frac{\mathscr{F}_4^l\mathscr{F}_8^l}{2L}, k_9 = \frac{\mathscr{F}_4^l J\mathscr{F}_8^l}{2L},$$
$$k_{10}(t) = -\dot{G}_{p,LA}v_{PN,LA} - G_{p,LA}\dot{v}_{PN,LA} - \frac{1}{R_{0,LA}L_{LA}}v_{ref}. \tag{7.68}$$

2) Floating power mode

In this mode, the lead–acid battery does not work. In other words, the power can be maintained between the PMSG system and the loading. Hence, the system model is

$$\begin{cases} \dot{P} = \omega v_{\alpha\beta}^T J^T \phi_{\alpha\beta} + \frac{1}{L} v_{\alpha\beta}^T v_{\alpha\beta} - v_{\alpha\beta}^T \frac{v_{0,MG}}{2L} u_{\alpha\beta}, \\ \dot{Q} = \omega v_{\alpha\beta}^T J^T J \phi_{\alpha\beta} + \frac{v_{\alpha\beta}^T J}{L} v_{\alpha\beta} - \frac{v_{\alpha\beta}^T J v_{0,MG}}{2L} u_{\alpha\beta}, \\ \dot{v}_{\alpha\beta} = \omega J v_{\alpha\beta}, \\ \dot{\phi}_{\alpha\beta} = \frac{1}{L} v_{\alpha\beta} - \frac{v_{0,MG}}{2L} u_{\alpha\beta}, \\ \dot{v}_{0,MG} = -\frac{1}{CR_L} v_{0,MG} + \frac{\phi_{\alpha\beta}^T}{2C} u_{\alpha\beta}, \\ \dot{v}_0 = \frac{1}{C_{pv}} (\phi_{0,MG} - \frac{1.5u_d}{v_0} \phi_d), \\ \dot{\phi}_d = -\frac{R_1}{L_1} \phi_d - \bar{\omega} \phi_q + \frac{1}{L_1} e_d, \\ \dot{\phi}_q = \bar{\omega} \phi_d - \frac{R_1}{L_1} \phi_q + \frac{1}{L_1} e_q. \end{cases} \tag{7.69}$$

Under this operation mode, the objective is to design a supervisory controller, so that the MPPT can be obtained and the output voltage tracks the reference v_{ref}. The voltage reference v^* of the MPPT can be calculated by (7.8), and $P \to P^*, Q \to Q^*, v_{0,MG} \to v_{ref}, v_{0,LA} \to v_{ref}$. Define $P_e = P - P^*, Q_e = Q - Q^*, v_{e,MG} = v_{0,MG} - v_{ref}$,

$$\begin{cases} \dot{P}_e = \omega v_{\alpha\beta}^T J^T \phi_{\alpha\beta} + \frac{1}{L} v_{\alpha\beta}^T v_{\alpha\beta} - v_{\alpha\beta}^T \frac{v_{0,MG}}{2L} u_{\alpha\beta}, \\ \dot{Q}_e = \omega v_{\alpha\beta}^T J^T J \phi_{\alpha\beta} + \frac{v_{\alpha\beta}^T J}{L} v_{\alpha\beta} - \frac{v_{\alpha\beta}^T J v_{0,MG}}{2L} u_{\alpha\beta}, \\ \dot{v}_{\alpha\beta} = \omega J v_{\alpha\beta}, \\ \dot{\phi}_{\alpha\beta} = \frac{1}{L} v_{\alpha\beta} - \frac{v_{0,MG}}{2L} u_{\alpha\beta}, \\ \dot{v}_{e,MG} = -\frac{1}{CR_L} v_{e,MG} - \frac{1}{CR_L} v_{ref} + \frac{\phi_{\alpha\beta}^T}{2C} u_{\alpha\beta}, \\ \dot{v}_0 = \frac{1}{C_{pv}} (\phi_{0,MG} - \frac{1.5u_d}{v_0} \phi_d), \\ \dot{\phi}_d = -\frac{R_1}{L_1} \phi_d - \bar{\omega} \phi_q + \frac{1}{L_1} e_d, \\ \dot{\phi}_q = \bar{\omega} \phi_d - \frac{R_1}{L_1} \phi_q + \frac{1}{L_1} e_q. \end{cases} \tag{7.70}$$

Define $x(t) = \begin{bmatrix} P_e & Q_e & v_{\alpha\beta} & \phi_{\alpha\beta} & v_{e,MG} & v_0 & \phi_d & \phi_q \end{bmatrix}^T$, and choose $\{\omega v_{\alpha\beta}^T, v_{\alpha\beta}^T, v_{0,MG}, \omega, \bar{\omega}\}$ as the fuzzy premise variables. The augmented fuzzy system is then given by

Plant Rule $\mathscr{R}^l$: **IF** z_1 is $\mathscr{F}_1^l$ and z_2 is $\mathscr{F}_2^l$ and $\cdots$ and z_5 is $\mathscr{F}_5^l$, **THEN**

$$\dot{x}(t) = A_l x(t) + B_l u(t) + \omega(t), l \in \mathscr{L} := \{1, 2, \ldots, r\} \tag{7.71}$$

where $\mathscr{R}^l$ denotes the l-th fuzzy inference rule; $\mathscr{L} := \{1, 2, \ldots, r\}$, and r is the number of inference rules; $\mathscr{F}_\theta^l (\theta = 1, 2, \cdots, 5)$ is the fuzzy set; $x(t) \in \Re^{n_x}$ and $u(t) \in \Re^{n_u}$ denote the system state and the control input, respectively; $z(t) \triangleq [z_1, z_2, \cdots, z_5]$ are the measurable variables; The l-th local model $\{A_l, B_l\}$ and the disturbance $\omega(t)$ are

given as below:

$$A_l = \begin{bmatrix} 0 & 0 & \frac{1}{L}\mathscr{F}_2^l & \mathscr{F}_1^l J^T & 0 & 0 & 0 & 0 \\ 0 & 0 & 0 & \mathscr{F}_1^l J^T J & 0 & 0 & 0 & 0 \\ 0 & 0 & \mathscr{F}_4^l J & 0 & 0 & 0 & 0 & 0 \\ 0 & 0 & \frac{1}{L} & 0 & 0 & 0 & 0 & 0 \\ 0 & 0 & 0 & 0 & -\frac{1}{CR_L} & 0 & 0 & 0 \\ 0 & 0 & 0 & 0 & 0 & 0 & -\frac{1.5u_d}{C_{pv}v_0} & 0 \\ 0 & 0 & 0 & 0 & 0 & 0 & -\frac{R_1}{L_1} & -\mathscr{F}_5^l \\ 0 & 0 & 0 & 0 & 0 & 0 & \mathscr{F}_5^l & -\frac{R_1}{L_1} \end{bmatrix},$$

$$B_l = \begin{bmatrix} -\frac{\mathscr{F}_2^l \mathscr{F}_3^l}{2L} & 0 & 0 \\ -\frac{\mathscr{F}_2^l J \mathscr{F}_3^l}{2L} & 0 & 0 \\ 0 & 0 & 0 \\ -\frac{\mathscr{F}_3^l}{2L} & 0 & 0 \\ 0 & 0 & 0 \\ \frac{\mathscr{F}_2^l}{2C} & 0 & 0 \\ 0 & \frac{1}{L_1} & 0 \\ 0 & 0 & \frac{1}{L_1} \end{bmatrix}, \omega(t) = \begin{bmatrix} 0 \\ 0 \\ 0 \\ 0 \\ 0 \\ -\frac{v_{ref}}{CR_L} \\ 0 \\ 0 \\ 0 \end{bmatrix}. \tag{7.72}$$

3) Excess power mode

First, consider the system model of lead-acid battery with the boost converter as below:

$$\begin{cases} \dot{\phi}_{1,LA} = \frac{1}{R_{1,LA}C_{1,LA}}\left(\phi_{m,LA} - \phi_{1,LA}\right), \\ \dot{\phi}_{m,LA} = \frac{1}{R_{0,LA}C_{0,LA}}\left(\phi_{L,LA} - \phi_{m,LA}\right) - \frac{\dot{v}_{PN,LA}}{R_{0,LA}}, \\ \dot{\phi}_{L,LA} = \frac{1}{L_{LA}}R_{0,LA}\left(\phi_{m,LA} - \phi_{L,LA} - R_{L,LA}\phi_{L,LA} - (\phi_{m,LA}R_{0,LA} + v_{PN,LA})\right) \\ \qquad + \frac{1}{L_{LA}}\left(V_{D,LA} + v_{0,MG} - R_{M,LA}\phi_{L,LA}\right)u_2 - \frac{V_{D,LA}}{L_{LA}}, \end{cases} \tag{7.73}$$

and the PMSG with the AC-DC converter and the boost converter is

$$\begin{cases} \dot{P} = \omega v_{\alpha\beta}^T J^T \phi_{\alpha\beta} + \frac{1}{L} v_{\alpha\beta}^T v_{\alpha\beta} - v_{\alpha\beta}^T \frac{v_{0,MG}}{2L} u_{\alpha\beta}, \\ \dot{Q} = \omega v_{\alpha\beta}^T J^T J \phi_{\alpha\beta} + \frac{v_{\alpha\beta}^T J}{L} v_{\alpha\beta} - \frac{v_{\alpha\beta}^T J v_{0,MG}}{2L} u_{\alpha\beta}, \\ \dot{v}_{\alpha\beta} = \omega J v_{\alpha\beta}, \\ \dot{\phi}_{\alpha\beta} = \frac{1}{L} v_{\alpha\beta} - \frac{v_{0,MG}}{2L} u_{\alpha\beta}, \\ \dot{v}_{0,MG} = -\frac{1}{CR_L} v_{0,MG} + \frac{\phi_{\alpha\beta}^T}{2C} u_{\alpha\beta}, \end{cases} \tag{7.74}$$

and

$$\begin{cases} \dot{v}_0 = \frac{1}{C_{pv}}\left(\phi_{0,MG} - \phi_{0,LA} - \frac{1.5u_d}{v_0}\phi_d\right), \\ \dot{\phi}_d = -\frac{R_1}{L_1}\phi_d - \bar{\omega}\phi_q + \frac{1}{L_1}e_d, \\ \dot{\phi}_q = \bar{\omega}\phi_d - \frac{R_1}{L_1}\phi_q + \frac{1}{L_1}e_q. \end{cases} \tag{7.75}$$

Under the deficit power mode, the PMSG can be supplied to the DC load and the lead-acid battery. Based on the Kirchhoff's current law, it has

$$\phi_{0,MG} = \phi_{0,LA} + \frac{v_{0,MG} - \phi_{0,MG} R_{line,MG}}{R_{load}}. \tag{7.76}$$

The objective is to design a supervisory controller, so that the MPPT can be obtained and the output voltage tracks the reference v^* of the MPPT calculated by (7.8), and $P \to P^*, Q \to Q^*, v_{0,MG} \to v_{ref}, v_{0,LA} \to v_{ref}$. Define $P_e = P - P^*, Q_e = Q - Q^*, v_{e,MG} = v_{0,MG} - v_{ref},, v_{e,LA} = v_{0,LA} - v_{ref},$

$$\begin{cases} \dot{\phi}_{1,LA} = \frac{1}{R_{1,LA}C_{1,LA}} \phi_{m,LA} - \frac{1}{R_{1,LA}C_{1,LA}} \phi_{1,LA}, \\ \dot{\phi}_{m,LA} = \frac{1}{R_{0,LA}C_{0,LA}} (\phi_{L,LA} - \phi_{m,LA}) - \frac{\dot{v}_{PN,LA}}{R_{0,LA}}, \\ \dot{\phi}_{L,LA} = \frac{1}{L_{LA}} R_{0,LA} (\phi_{m,LA} - \phi_{L,LA} - R_{L,LA}\phi_{L,LA} - (\phi_{m,LA}R_{0,LA} + v_{PN,LA})) \\ \quad + \frac{1}{L_{LA}} (V_{D,LA} + v_{0,MG} - R_{M,LA}\phi_{L,LA}) u_2 - \frac{V_{D,LA}}{L_{LA}}, \\ \dot{P}_e = \omega v_{\alpha\beta}^T J^T \phi_{\alpha\beta} + \frac{1}{L} v_{\alpha\beta}^T v_{\alpha\beta} - v_{\alpha\beta}^T \frac{v_{0,MG}}{2L} u_{\alpha\beta}, \\ \dot{Q}_e = \omega v_{\alpha\beta}^T J^T J \phi_{\alpha\beta} + \frac{v_{\alpha\beta}^T J}{L} v_{\alpha\beta} - \frac{v_{\alpha\beta}^T J v_{0,MG}}{2L} u_{\alpha\beta}, \\ \dot{v}_{\alpha\beta} = \omega J v_{\alpha\beta}, \\ \dot{\phi}_{\alpha\beta} = \frac{1}{L} v_{\alpha\beta} - \frac{v_{0,MG}}{2L} u_{\alpha\beta}, \\ \dot{v}_{e,MG} = -\frac{1}{CR_L} v_{e,MG} + \frac{\phi_{\alpha\beta}^T}{2C} u_{\alpha\beta} - \frac{1}{CR_L} v_{ref}, \\ \dot{v}_0 = \frac{1}{C_{pv}} (\phi_{0,MG} - \phi_{0,LA} - \frac{1.5u_d}{v_0} \phi_d), \\ \dot{\phi}_d = -\frac{R_1}{L_1} \phi_d - \bar{\omega}\phi_q + \frac{1}{L_1} e_d, \\ \dot{\phi}_q = \bar{\omega}\phi_d - \frac{R_1}{L_1} \phi_q + \frac{1}{L_1} e_q. \end{cases} \tag{7.77}$$

Define $x(t) = \begin{bmatrix} \phi_{1,LA} & \phi_{m,LA} & v_{e,LA} & P_e & Q_e & v_{\alpha\beta} & \phi_{\alpha\beta} & v_{e,MG} & v_0 & \phi_d & \phi_q \end{bmatrix}^T$, and choose $\{v_{0,MG}, \phi_{L,LA}, v_{\alpha\beta}^T, \omega v_{\alpha\beta}^T, \omega, \phi_{\alpha\beta}^T, \frac{\phi_{0,MG}}{\phi_{\alpha\beta}}, \frac{\phi_{0,LA}}{\phi_{\alpha\beta}}, \frac{u_d}{v_0}, \bar{\omega}\}$ as the fuzzy premise variables, then the augmented fuzzy system is given by

Plant Rule $\mathscr{R}^l$: **IF** z_1 is $\mathscr{F}_1^l$ and z_2 is $\mathscr{F}_2^l$ and $\cdots$ and z_{10} is $\mathscr{F}_{10}^l$, **THEN**

$$\dot{x}(t) = A_l x(t) + B_l u(t) + \omega(t), l \in \mathscr{L} := \{1, 2, \ldots, r\} \tag{7.78}$$

where $\mathscr{R}^l$ denotes the l-th fuzzy inference rule; $\mathscr{L} := \{1, 2, \ldots, r\}$, and r is the number of inference rules; $\mathscr{F}_\theta^l(\theta = 1, 2, \cdots, 10)$ is the fuzzy set; $x(t) \in \Re^{n_x}$ and $u(t) \in \Re^{n_u}$ denote the system state and control input, respectively; $z(t) \triangleq [z_1, z_2, \cdots, z_{10}]$ are the measurable variables; The l-th local model $\{A_l, B_l\}$ and the disturbance $\omega(t)$ are

given as below:

$$A_l = \begin{bmatrix} -k_1 & k_1 & 0 & 0 & 0 & 0 & 0 & 0 & 0 & 0 & 0 \\ 0 & -k_2 & k_2 & 0 & 0 & 0 & 0 & 0 & 0 & 0 & 0 \\ 0 & k_3 & -k_4 & 0 & 0 & 0 & 0 & 0 & 0 & 0 & 0 \\ 0 & 0 & 0 & 0 & 0 & \frac{\mathscr{F}_3^l}{L} & \mathscr{F}_4^l J^T & 0 & 0 & 0 & 0 \\ 0 & 0 & 0 & 0 & 0 & \frac{\mathscr{F}_3^l J}{L} & \mathscr{F}_4^l J^T J & 0 & 0 & 0 & 0 \\ 0 & 0 & 0 & 0 & 0 & \mathscr{F}_5^l J & 0 & 0 & 0 & 0 & 0 \\ 0 & 0 & 0 & 0 & 0 & \frac{1}{L} & 0 & 0 & 0 & 0 & 0 \\ 0 & 0 & 0 & 0 & 0 & 0 & 0 & -\frac{1}{CR_L} & 0 & 0 & 0 \\ 0 & 0 & 0 & 0 & 0 & 0 & 0 & \frac{\mathscr{F}_7^l - \mathscr{F}_8^l}{C_{pv}} & 0 & -\frac{1.5\mathscr{F}_9^l}{C_{pv}} & 0 \\ 0 & 0 & 0 & 0 & 0 & 0 & 0 & 0 & 0 & -\frac{R_1}{L_1} & -\mathscr{F}_{10}^l \\ 0 & 0 & 0 & 0 & 0 & 0 & 0 & 0 & 0 & \mathscr{F}_{10}^l & -\frac{R_1}{L_1} \end{bmatrix},$$

$$B_l = \begin{bmatrix} 0 & 0 & 0 & 0 \\ 0 & 0 & 0 & 0 \\ k_5 & 0 & 0 & 0 \\ 0 & -\frac{\mathscr{F}_3^l \mathscr{F}_1^l}{2L} & 0 & 0 \\ 0 & -\frac{\mathscr{F}_3^l J \mathscr{F}_1^l}{2L} & 0 & 0 \\ 0 & -\frac{\mathscr{F}_1^l}{2L} & 0 & 0 \\ 0 & 0 & 0 & 0 \\ 0 & \frac{\mathscr{F}_6^l}{2C} & 0 & 0 \\ 0 & 0 & 0 & 0 \\ 0 & 0 & \frac{1}{L_1} & 0 \\ 0 & 0 & 0 & \frac{1}{L_1} \end{bmatrix}, \omega(t) = \begin{bmatrix} 0 \\ -\frac{\dot{v}_{PN,LA}}{R_{0,LA}} \\ -\frac{R_{0,LA} v_{PN,LA} + V_{D,LA}}{L_{LA}} \\ 0 \\ 0 \\ 0 \\ 0 \\ -\frac{v_{ref}}{CR_L} \\ 0 \\ 0 \\ 0 \end{bmatrix},$$

$$k_1 = \frac{1}{R_{1,LA} C_{1,LA}}, k_2 = \frac{1}{R_{0,LA} C_{0,LA}}, k_3 = \frac{R_{0,LA} + R_{0,LA}^2}{L_{LA}},$$
$$k_4 = \frac{R_{0,LA} + R_{0,LA} R_{L,LA}}{L_{LA}}, k_5 = \frac{V_{D,LA} + \mathscr{F}_1^l - R_{M,LA} \mathscr{F}_2^l}{L_{LA}}. \tag{7.79}$$

7.5 PV SYSTEM AND PMSG SYSTEM FOR DC LOAD

The section considers a small-scale DC power system, which consists of the PMSG subsystem, the lead-acid battery and the DC load. The PMSG subsystem uses an AC-DC buck converter to the DC-bus, and a bi-directional buck-boost converter will be used to interface the battery and the DC-bus. In this system, as illustrated in Figure 7.5, the bi-directional converter operates as a boost converter during the discharging mode of the battery and as a buck converter during the charging mode of the battery.

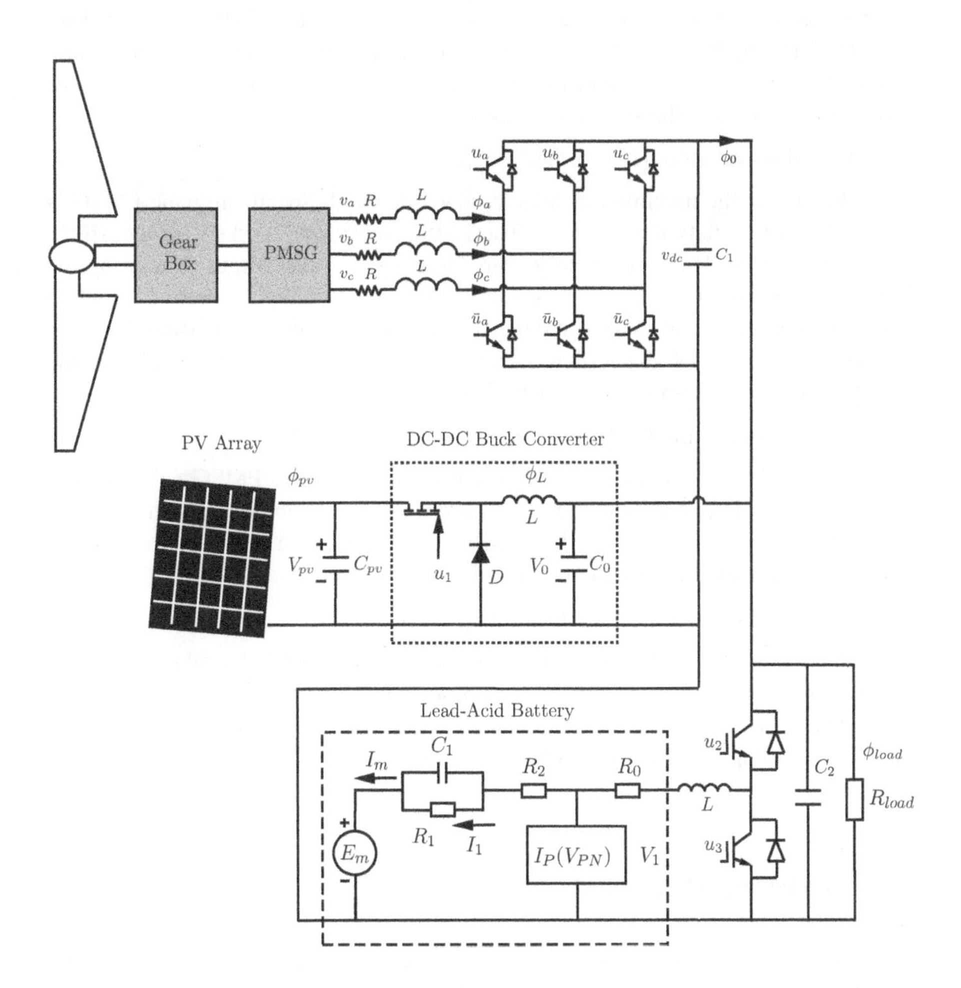

Figure 7.5 DC microgrid with PV and PMSG.

7.5.1 OPERATION MODES

In order to ensure the reliable and stable operations of the small-scale DC power system under the different conditions, the power balance among the supply and the demand must be maintained under all operating conditions. Now, we define P_{PV}, P_{MG}, $P_{LD}^{\min}$, and $P_{LD}^{\max}$ as the PV power supply, the PMSG power supply, the minimum load demand, and the maximum load demand, respectively. There exist these three possible conditions of the system as shown in Figure 7.3 [1].

1) Deficit power mode$\left(P_{MG}+P_{PV} \leq P_{LD}^{\min}\right)$

In this mode, the maximum available power demand. Specifically, the PMSG is not sufficient to satisfy the DC load. The PMSG subsystem works with the MPPT algorithm, and the lead-acid battery converters regulate the output voltage by the battery discharging operation. Thus, they work as the voltage sources. To obtain both the accurate current sharing and the desirable voltage regulation using the fuzzy control method, the DC-bus voltage reaches its minimum acceptable value at the maximum discharge current of the batteries.

2) Floating power mode$\left(P_{LD}^{\min} \leq P_{MG}+P_{PV} \leq P_{LD}^{\max}\right)$

In this mode, the DC operates in an islanded mode. The PMSG subsystem is basically sufficient to satisfy the DC load, and the batteries do not work in this case. The PMSG subsystem works with the MPPT algorithm, and regulates the DC bus voltage by using the fuzzy tracking control.

3) Excess power mode$(P_{MG}+P_{PV} \geq P_{LD}^{\max})$

In this mode, the PMSG subsystem works with the MPPT algorithm. Since the PMSG subsystem is greater than the load power, the DC voltage increases. The excess power is used to charge the lead-acid batteries. Therefore, the batteries regulate the DC bus voltage by the battery charging operations.

7.5.2 DYNAMIC MODELING

Recall the system models of the PV with the buck converter (1.3), the PMSG with the AC-DC converter in (2.20), the lead-acid battery in (3.1), and the boost converter in (1.6) as below:

$$\left\{\begin{array}{l} \dot{v}_{PV}=\frac{1}{C_{PV}}\left(\phi_{PV}-\phi_{L,PV}u_1\right), \\ \dot{\phi}_{L,PV}=\frac{1}{L_{PV}}R_{0,PV}\left(\left(\phi_{0,PV}-\phi_{L,PV}\right)-R_{L,PV}\phi_{L,PV}-v_{0,PV}\right) \\ \qquad +\frac{1}{L_{PV}}\left(V_{D,PV}+v_{PV}-R_{M,PV}\phi_{L,PV}\right)u_1-\frac{V_{D,PV}}{L_{PV}}, \\ \dot{v}_{0,PV}=\frac{1}{C_{0,PV}}\left(\phi_{L,PV}-\phi_{0,PV}\right), \end{array}\right. \tag{7.80}$$

and

$$\begin{cases} \dot{P} = \omega v_{\alpha\beta}^T J^T \phi_{\alpha\beta} + \frac{1}{L} v_{\alpha\beta}^T v_{\alpha\beta} - v_{\alpha\beta}^T \frac{v_{dc}}{2L} u_{\alpha\beta}, \\ \dot{Q} = \omega v_{\alpha\beta}^T J^T J \phi_{\alpha\beta} + \frac{v_{\alpha\beta}^T J}{L} v_{\alpha\beta} - \frac{v_{\alpha\beta}^T J v_{dc}}{2L} u_{\alpha\beta}, \\ \dot{v}_{\alpha\beta} = \omega J v_{\alpha\beta}, \\ \dot{\phi}_{\alpha\beta} = \frac{1}{L} v_{\alpha\beta} - \frac{v_{dc}}{2L} u_{\alpha\beta}, \\ \dot{v}_{dc} = -\frac{1}{CR_L} v_{dc} + \frac{\phi_{\alpha\beta}^T}{2C} u_{\alpha\beta}, \end{cases} \tag{7.81}$$

and

$$\begin{cases} \dot{\phi}_{1,LA} = \frac{1}{R_{1,LA} C_{1,LA}} (\phi_{m,LA} - \phi_{1,LA}), \\ \dot{Q}_{e,LA} = -\phi_{m,LA}, \\ \dot{\theta}_{LA} = \frac{1}{C_{\theta,LA}} \left[P_{s,LA} - \frac{(\theta_{LA} - \theta_{a,LA})}{R_{\theta,LA}} \right], \\ \phi_{p,LA} = G_{p,LA} v_{PN,LA}, \\ G_{p,LA} = G_{p0,LA} \exp\left(v_{PN,LA} / v_{p0,LA} + A_{p,LA} \left(1 - \theta_{LA}/\theta_{f,LA}\right)\right), \end{cases} \tag{7.82}$$

and

$$\begin{cases} \dot{\phi}_{LA} = -\frac{1}{L_{LA}} (1-u) v_{dc} + \frac{1}{L_{LA}} v_{LA}, \\ \dot{v}_{dc} = \frac{1}{C_{0,LA}} (1-u) \phi_{LA} - \frac{1}{C_{0,LA}} \phi_{0,LA}, \end{cases} \tag{7.83}$$

where the superscripts MG and LA stand for the PMSG and the lead-acid battery, respectively, and all notations are defined in (1.1), (3.1), (1.6), and (1.6), respectively.

1) Deficit power mode

First, consider the system model of the lead-acid battery with the boost converter as below:

$$\begin{cases} \dot{\phi}_{1,LA} = \frac{1}{R_{1,LA} C_{1,LA}} (\phi_{m,LA} - \phi_{1,LA}), \\ \dot{\phi}_{m,LA} = -\frac{1}{R_{0,LA} L_{LA}} (1-u_3) v_{0,LA} + \frac{1}{R_{0,LA} L_{LA}} v_{LA} - \dot{G}_{p,LA} v_{PN,LA} - G_{p,LA} \dot{v}_{PN,LA}, \\ \dot{v}_{0,LA} = \frac{1}{C_{0,LA}} (1-u_3)(\phi_{m,LA} + \phi_{p,LA}) - \frac{1}{R_{0,LA} C_{0,LA}} \phi_{0,LA}, \end{cases} \tag{7.84}$$

and the PMSG with the AC-DC converter and the boost converter is

$$\begin{cases} \dot{P} = \omega v_{\alpha\beta}^T J^T \phi_{\alpha\beta} + \frac{1}{L} v_{\alpha\beta}^T v_{\alpha\beta} - v_{\alpha\beta}^T \frac{v_{0,MG}}{2L} u_{\alpha\beta}, \\ \dot{Q} = \omega v_{\alpha\beta}^T J^T J \phi_{\alpha\beta} + \frac{v_{\alpha\beta}^T J}{L} v_{\alpha\beta} - \frac{v_{\alpha\beta}^T J v_{0,MG}}{2L} u_{\alpha\beta}, \\ \dot{v}_{\alpha\beta} = \omega J v_{\alpha\beta}, \\ \dot{\phi}_{\alpha\beta} = \frac{1}{L} v_{\alpha\beta} - \frac{v_{0,MG}}{2L} u_{\alpha\beta}, \\ \dot{v}_{0,MG} = -\frac{1}{CR_L} v_{0,MG} + \frac{\phi_{\alpha\beta}^T}{2C} u_{\alpha\beta}, \end{cases} \tag{7.85}$$

and

$$\begin{cases} \dot{v}_{PV} = \frac{1}{C_{PV}} (\phi_{PV} - \phi_{L,PV} u_1), \\ \dot{\phi}_{L,PV} = \frac{1}{L_{PV}} R_{0,PV} ((\phi_{0,PV} - \phi_{L,PV}) - R_{L,PV} \phi_{L,PV} - v_{0,PV}) \\ \qquad + \frac{1}{L_{PV}} (V_{D,PV} + v_{PV} - R_{M,PV} \phi_{L,PV}) u_1 - \frac{V_{D,PV}}{L_{PV}}, \\ \dot{v}_{0,PV} = \frac{1}{C_{0,PV}} (\phi_{L,PV} - \phi_{0,PV}). \end{cases} \tag{7.86}$$

Under the deficit power mode, both the PMSG and the lead-acid battery can be supplied to the DC load. Based on the Thevenin's theorem,

$$v_{0,LA} = \phi_{0,LA}R_{line,LA} + (\phi_{0,LA} + \phi_{0,MG} + \phi_{0,PV})R_{load}, \tag{7.87}$$

$$v_{0,MG} = \phi_{0,MG}R_{line,MG} + (\phi_{0,LA} + \phi_{0,MG} + \phi_{0,PV})R_{load}, \tag{7.88}$$

$$v_{0,PV} = \phi_{0,PV}R_{line,PV} + (\phi_{0,LA} + \phi_{0,MG} + \phi_{0,PV})R_{load}, \tag{7.89}$$

where the $R_{line,LA}$, $R_{line,MG}$, $R_{line,PV}$ are the resistances of power line on the lead-acid battery, and the one on the PMSG system, and the one on the PV system, respectively, R_{load} is the load resistance. Recall the MPPT of the PV in (7.8) as below:

$$0 = \phi_{pv} - n_p \gamma I_{rs} v^* e^{\gamma v^*}. \tag{7.90}$$

Under this operation mode, the objective is to design a supervisory controller, so that MPPT of the PV and the PMSG can be obtained and the output voltage tracks the reference v_{ref}. The PV voltage reference v^* of MPPT can be calculated by (7.8), and $v_{PV} \to v^*, P \to P^*, Q \to Q^*, v_{0,MG} \to v_{ref}, v_{0,LA} \to v_{ref}, v_{0,PV} \to v_{ref}$. Define $e_{PV} = v_{PV} - v^*, P_e = P - P^*, Q_e = Q - Q^*, v_{e,MG} = v_{0,MG} - v_{ref}, v_{e,LA} = v_{0,LA} - v_{ref}, v_{e,PV} = v_{0,PV} - v_{ref}$.

Based on the relationships (7.84)-(7.90), the interconnected system is given by

$$\left\{\begin{array}{l}
\dot{\phi}_{1,LA} = \frac{1}{R_{1,LA}C_{1,LA}}(\phi_{m,LA} - \phi_{1,LA}), \\
\dot{\phi}_{m,LA} = -\frac{1}{R_{0,LA}L_{LA}}(\phi_{0,LA}R_{line,LA} + (\phi_{0,LA} + \phi_{0,MG} + \phi_{0,PV})R_{load}) \\
\quad + \frac{v_{0,LA}}{R_{0,LA}L_{LA}}u_3 + \frac{1}{R_{0,LA}L_{LA}}v_{LA} - \dot{G}_{p,LA}v_{PN,LA} - G_{p,LA}\dot{v}_{PN,LA}, \\
\dot{v}_{e,LA} = \frac{1}{C_{0,LA}}(1 - u_3)(\phi_{m,LA} + \phi_{p,LA}) - \frac{v_{e,LA} + v_{ref} - (\phi_{0,MG} + \phi_{0,PV})R_{load}}{R_{0,LA}C_{0,LA}(R_{line,LA} + R_{load})} \\
\dot{P}_e = \omega v_{\alpha\beta}^T J^T \phi_{\alpha\beta} + \frac{1}{L}v_{\alpha\beta}^T v_{\alpha\beta} - v_{\alpha\beta}^T \frac{v_{0,MG}}{2L}u_{\alpha\beta}, \\
\dot{Q}_e = \omega v_{\alpha\beta}^T J^T J \phi_{\alpha\beta} + \frac{v_{\alpha\beta}^T J}{L}v_{\alpha\beta} - \frac{v_{\alpha\beta}^T J v_{0,MG}}{2L}u_{\alpha\beta}, \\
\dot{v}_{\alpha\beta} = \omega J v_{\alpha\beta}, \\
\dot{\phi}_{\alpha\beta} = \frac{1}{L}v_{\alpha\beta} - \frac{v_{0,MG}}{2L}u_{\alpha\beta}, \\
\dot{v}_{e,MG} = -\frac{1}{CR_L}v_{e,MG} + \frac{\phi_{\alpha\beta}^T}{2C}u_{\alpha\beta} - \frac{1}{CR_L}v_{ref}, \\
\dot{e}_{PV} = \frac{1}{C_{PV}}(\phi_{PV} - \phi_{L,PV}u_1), \\
\dot{\phi}_{L,PV} = \frac{R_{0,PV}}{L_{PV}}\phi_{0,PV} - \frac{R_{0,PV}}{L_{PV}}(1 + R_{L,PV})\phi_{L,PV} - \frac{R_{0,PV}}{L_{PV}}\phi_{0,PV}R_{line,PV} \\
\quad - \frac{R_{0,PV}}{L_{PV}}(\phi_{0,LA} + \phi_{0,MG} + \phi_{0,PV})R_{load} \\
\quad + \frac{1}{L_{PV}}(V_{D,PV} + v_{PV} - R_{M,PV}\phi_{L,PV})u_1 - \frac{V_{D,PV}}{L_{PV}}, \\
\dot{v}_{e,PV} = \frac{1}{C_{0,PV}}(\phi_{L,PV} - \phi_{0,PV}).
\end{array}\right. \tag{7.91}$$

If we define $x(t) = [\phi_{1,LA}\ \phi_{m,LA}\ v_{e,LA}\ P_e\ Q_e\ v_{\alpha\beta}\ \phi_{\alpha\beta}\ v_{e,MG}\ e_{PV}\ \phi_{L,PV}\ v_{e,PV}]^T$, and choose $\{\frac{\phi_{0,LA}}{\phi_{m,LA}}, v_{0,LA}, \frac{\phi_{p,LA}}{\phi_{m,LA}}, \phi_{m,LA}, \phi_{p,LA}, \frac{\phi_{0,PV}}{\phi_{L,PV}}, \omega v_{\alpha\beta}^T, v_{\alpha\beta}^T, v_{0,MG}, \phi_{\alpha\beta}^T, \frac{\phi_{PV}}{e_{PV}}, \frac{\phi_{0,PV}}{\phi_{L,PV}}, v_{PV}, \phi_{L,PV}, \omega\}$ as the fuzzy premise variables, the augmented fuzzy system is given by

Plant Rule $\mathscr{R}^l$: **IF** z_1 is $\mathscr{F}_1^l$ and z_2 is $\mathscr{F}_2^l$ and $\cdots$ and z_{15} is $\mathscr{F}_{15}^l$, **THEN**

$$\dot{x}(t) = A_l x(t) + B_l u(t) + \omega(t), l \in \mathscr{L} := \{1, 2, \ldots, r\} \tag{7.92}$$

where $\mathscr{R}^l$ denotes the l-th fuzzy inference rule; $\mathscr{L} := \{1, 2, \ldots, r\}$, and r is the number of inference rules; $\mathscr{F}_\theta^l(\theta = 1, 2, \cdots, 15)$ is the fuzzy set; $x(t) \in \Re^{n_x}$ and $u(t) \in \Re^{n_u}$ denote the system state and the control input, respectively; $z(t) \triangleq [z_1, z_2, \cdots, z_{15}]$ are the measurable variables; The l-th local model $\{A_l, B_l\}$ and the disturbance $\omega(t)$ are given as below:

$$A_l = \begin{bmatrix} -k_1 & k_1 & 0 & 0 & 0 & 0 & 0 & 0 & 0 & 0 & 0 \\ 0 & -k_2\mathscr{F}_1^l & -k_2 & 0 & 0 & 0 & 0 & 0 & 0 & 0 & 0 \\ 0 & \frac{1+\mathscr{F}_3^l}{C_{0,LA}} & -\frac{k_4}{R_{load}} & 0 & 0 & 0 & 0 & 0 & 0 & k_4\mathscr{F}_6^l & 0 \\ 0 & 0 & 0 & 0 & 0 & \frac{\mathscr{F}_8^l}{L} & \mathscr{F}_7^l J^T & 0 & 0 & 0 & 0 \\ 0 & 0 & 0 & 0 & 0 & \frac{\mathscr{F}_8^l J}{L} & \mathscr{F}_7^l J^T J & 0 & 0 & 0 & 0 \\ 0 & 0 & 0 & 0 & 0 & \mathscr{F}_{15} J & 0 & 0 & 0 & 0 & 0 \\ 0 & 0 & 0 & 0 & 0 & \frac{1}{L} & 0 & 0 & 0 & 0 & 0 \\ 0 & 0 & 0 & 0 & 0 & 0 & 0 & \frac{-1}{CR_L} & 0 & 0 & 0 \\ 0 & 0 & 0 & 0 & 0 & 0 & 0 & 0 & \frac{\mathscr{F}_{11}^l}{C_{PV}} & 0 & 0 \\ 0 & -k_6\mathscr{F}_1^l & 0 & 0 & 0 & 0 & 0 & 0 & 0 & k_7\mathscr{F}_{12}^l & 0 \\ 0 & 0 & 0 & 0 & 0 & 0 & 0 & 0 & 0 & \frac{1+\mathscr{F}_6^l}{C_{0,PV}} & 0 \end{bmatrix},$$

$$B_l = \begin{bmatrix} 0 & 0 & 0 \\ \frac{\mathscr{F}_2^l}{R_{0,LA} L_{LA}} & 0 & 0 \\ -\frac{\mathscr{F}_4^l + \mathscr{F}_5^l}{C_{0,LA}} & 0 & 0 \\ 0 & -\frac{\mathscr{F}_8^l \mathscr{F}_9^l}{2L} & 0 \\ 0 & -\frac{\mathscr{F}_8^l J \mathscr{F}_9^l}{2L} & 0 \\ 0 & 0 & 0 \\ 0 & -\frac{\mathscr{F}_9^l}{2L} & 0 \\ 0 & \frac{\mathscr{F}_{10}^l}{2C} & 0 \\ 0 & 0 & -\frac{\mathscr{F}_{14}^l}{C_{PV}} \\ 0 & 0 & k_8 \\ 0 & 0 & 0 \end{bmatrix}, \omega(t) = \begin{bmatrix} 0 \\ k_9(t) \\ -k_{10}(t) \\ 0 \\ 0 \\ 0 \\ 0 \\ -\frac{v_{ref}}{L_{MG}} \\ -\frac{v_{ref}}{CR_L} \\ 0 \\ 0 \end{bmatrix}, k_1 = \frac{1}{R_{1,LA} C_{1,LA}},$$

$$k_2 = \frac{R_{line,LA} + R_{load}}{R_{0,LA} L_{LA}}, k_3 = \frac{R_{load}}{R_{0,LA} L_{LA}}, k_4 = \frac{R_{load}}{R_{0,LA} C_{0,LA} \left(R_{line,LA} + R_{load}\right)},$$
$$k_5 = \frac{2R_{line,MG} + R_{load}}{L_{MG}}, k_6 = \frac{R_{0,PV} R_{load}}{L_{PV}}, k_7 = \frac{\left(R_{0,PV} - R_{line,PV} R_{0,PV} - R_{0,PV} R_{load}\right)}{L_{PV}},$$
$$k_8 = \frac{1}{L_{PV}} \left(\mathscr{F}^l_{13} - R_{M,PV} \mathscr{F}^l_{14}\right), k_9(t) = \frac{1}{R_{0,LA} L_{LA}} v_{LA} - \dot{G}_{p,LA} v_{PN,LA} - G_{p,LA} \dot{v}_{PN,LA},$$
$$k_{10}(t) = \frac{v_{ref}}{R_{0,LA} C_{0,LA} \left(R_{line,LA} + R_{load}\right)}. \tag{7.93}$$

2) Floating power mode

In this mode, the lead-acid battery does not work. In other words, the power can be maintained between the PMSG system and the loading. Hence, the system model is

$$\begin{cases} \dot{P} = \omega v_{\alpha\beta}^T J^T \phi_{\alpha\beta} + \frac{1}{L} v_{\alpha\beta}^T v_{\alpha\beta} - v_{\alpha\beta}^T \frac{v_{0,MG}}{2L} u_{\alpha\beta}, \\ \dot{Q} = \omega v_{\alpha\beta}^T J^T J \phi_{\alpha\beta} + \frac{v_{\alpha\beta}^T J}{L} v_{\alpha\beta} - \frac{v_{\alpha\beta}^T J v_{0,MG}}{2L} u_{\alpha\beta}, \\ \dot{v}_{\alpha\beta} = \omega J v_{\alpha\beta}, \\ \dot{\phi}_{\alpha\beta} = \frac{1}{L} v_{\alpha\beta} - \frac{v_{0,MG}}{2L} u_{\alpha\beta}, \\ \dot{v}_{0,MG} = -\frac{1}{CR_L} v_{0,MG} + \frac{\phi_{\alpha\beta}^T}{2C} u_{\alpha\beta}, \\ \dot{v}_{PV} = \frac{1}{C_{PV}} \left(\phi_{PV} - \phi_{L,PV} u_1\right), \\ \dot{\phi}_{L,PV} = \frac{1}{L_{PV}} R_{0,PV} \left(\left(\phi_{0,PV} - \phi_{L,PV}\right) - R_{L,PV} \phi_{L,PV} - v_{0,PV}\right) \\ \qquad + \frac{1}{L_{PV}} \left(V_{D,PV} + v_{PV} - R_{M,PV} \phi_{L,PV}\right) u_1 - \frac{V_{D,PV}}{L_{PV}}, \\ \dot{v}_{0,PV} = \frac{1}{C_{0,PV}} \left(\phi_{L,PV} - \phi_{0,PV}\right). \end{cases} \tag{7.94}$$

Under the deficit power mode, both the PMSG and the lead-acid battery can be supplied to the DC load. Based on the Thevenin's theorem,

$$v_{0,MG} = \phi_{0,MG} R_{line,MG} + \left(\phi_{0,MG} + \phi_{0,PV}\right) R_{load}, \tag{7.95}$$
$$v_{0,PV} = \phi_{0,PV} R_{line,PV} + \left(\phi_{0,MG} + \phi_{0,PV}\right) R_{load}. \tag{7.96}$$

Here, the objective is to design a supervisory controller, so that MPPT can be obtained and the output voltage tracks the reference v_{ref}. The PV voltage reference v^* of MPPT can be calculated by (7.8), and $P \to P^*, Q \to Q^*, v_{PV} \to v^*, v_{0,MG} \to v_{ref}, v_{0,PV} \to v_{ref}$. Define $P_e = P - P^*, Q_e = Q - Q^*, e_{PV} = v_{PV} - v^*, v_{e,MG} = v_{0,MG} -$

v_{ref}, $v_{e,PV} = v_{0,PV} - v_{ref}$,

$$\begin{cases} \dot{P}_e = \omega v_{\alpha\beta}^T J^T \phi_{\alpha\beta} + \frac{1}{L} v_{\alpha\beta}^T v_{\alpha\beta} - v_{\alpha\beta}^T \frac{v_{0,MG}}{2L} u_{\alpha\beta}, \\ \dot{Q}_e = \omega v_{\alpha\beta}^T J^T J \phi_{\alpha\beta} + \frac{v_{\alpha\beta}^T J}{L} v_{\alpha\beta} - \frac{v_{\alpha\beta}^T J v_{0,MG}}{2L} u_{\alpha\beta}, \\ \dot{v}_{\alpha\beta} = \omega J v_{\alpha\beta}, \\ \dot{\phi}_{\alpha\beta} = \frac{1}{L} v_{\alpha\beta} - \frac{v_{0,MG}}{2L} u_{\alpha\beta}, \\ \dot{v}_{e,MG} = -\frac{R_{line,MG}\phi_{0,MG} + R_{load}\phi_{0,MG} + R_{load}\phi_{0,PV}}{CR_L} + \frac{\phi_{\alpha\beta}^T}{2C} u_{\alpha\beta}, \\ \dot{e}_{PV} = \frac{1}{C_{PV}} \left(\phi_{PV} - \phi_{L,PV} u_1\right), \\ \dot{\phi}_{L,PV} = \frac{R_{0,PV}}{L_{PV}} \left(\left(1 - R_{line,PV} - R_{load}\right) \phi_{0,PV} - \left(1 + R_{L,PV}\right) \phi_{L,PV} - R_{load} \phi_{0,MG}\right) \\ \quad + \frac{1}{L_{PV}} \left(V_{D,PV} + v_{PV} - R_{M,PV} \phi_{L,PV}\right) u_1 - \frac{V_{D,PV}}{L_{PV}}, \\ \dot{v}_{e,PV} = \frac{1}{C_{0,PV}} \left(\phi_{L,PV} - \phi_{0,PV}\right), \end{cases} \tag{7.97}$$

where $R_{line,LA}$, $R_{line,MG}$, $R_{line,PV}$ are the resistances of power lines on the lead-acid battery, and the one on the PMSG system, and the one on the PV system, respectively, R_{load} is the load resistance.

We define $x(t) = [P_e \ Q_e \ v_{\alpha\beta} \ \phi_{\alpha\beta} \ v_{e,MG} \ e_{PV} \ \phi_{L,PV} \ v_{e,PV}]^T$, and choose $\{\frac{\phi_{0,PV}}{\phi_{L,PV}}, \omega v_{\alpha\beta}^T, v_{\alpha\beta}^T, v_{0,MG}, \phi_{\alpha\beta}^T, \frac{\phi_{PV}}{e_{PV}}, \frac{\phi_{0,PV}}{\phi_{L,PV}}, v_{PV}, \phi_{L,PV}, \omega\}$ as the fuzzy premise variables, then the augmented fuzzy system is given by

Plant Rule $\mathscr{R}^l$: **IF** z_1 is $\mathscr{F}_1^l$ and z_2 is $\mathscr{F}_2^l$ and $\cdots$ and z_{10} is $\mathscr{F}_{10}^l$, **THEN**

$$\dot{x}(t) = A_l x(t) + B_l u(t) + \omega(t), l \in \mathscr{L} := \{1, 2, \ldots, r\} \tag{7.98}$$

where $\mathscr{R}^l$ denotes the l-th fuzzy inference rule; $\mathscr{L} := \{1, 2, \ldots, r\}$, and r is the number of inference rules; $\mathscr{F}_\theta^l(\theta = 1, 2, \cdots, 10)$ is the fuzzy set; $x(t) \in \Re^{n_x}$ and $u(t) \in \Re^{n_u}$ denote the system state and the control input, respectively; $z(t) \triangleq [z_1, z_2, \cdots, z_{10}]$ are the measurable variables; The l-th local model $\{A_l, B_l\}$ and the disturbance $\omega(t)$ are given as below:

$$A_l = \begin{bmatrix} 0 & 0 & \frac{\mathscr{F}_3^l}{L} & \mathscr{F}_2^l J^T & 0 & 0 & 0 & 0 \\ 0 & 0 & \frac{\mathscr{F}_3^l J}{L} & \mathscr{F}_2^l J^T J & 0 & 0 & 0 & 0 \\ 0 & 0 & \mathscr{F}_{10} J & 0 & 0 & 0 & 0 & 0 \\ 0 & 0 & \frac{1}{L} & 0 & 0 & 0 & 0 & 0 \\ 0 & 0 & 0 & 0 & \frac{-1}{CR_L} & 0 & 0 & 0 \\ 0 & 0 & 0 & 0 & 0 & \frac{\mathscr{F}_6^l}{C_{PV}} & 0 & 0 \\ 0 & 0 & 0 & 0 & 0 & 0 & k\mathscr{F}_7^l & 0 \\ 0 & 0 & 0 & 0 & 0 & 0 & \frac{1+\mathscr{F}_1^l}{C_{0,PV}} & 0 \end{bmatrix},$$

$$B_l = \begin{bmatrix} -\frac{\mathscr{F}_3^l \mathscr{F}_4^l}{2L} & 0 \\ -\frac{\mathscr{F}_3^l J \mathscr{F}_4^l}{2L} & 0 \\ 0 & 0 \\ -\frac{\mathscr{F}_4^l}{2L} & 0 \\ \frac{\mathscr{F}_5^l}{2C} & 0 \\ 0 & -\frac{\mathscr{F}_9^l}{C_{PV}} \\ 0 & \frac{\mathscr{F}_8^l - R_{M,PV}\mathscr{F}_9^l}{L_{PV}} \\ 0 & 0 \end{bmatrix}, \omega(t) = \begin{bmatrix} 0 \\ 0 \\ 0 \\ 0 \\ -\frac{v_{ref}}{L_{MG}} \\ -\frac{v_{ref}}{CR_L} \\ 0 \\ 0 \end{bmatrix},$$

$$k = \frac{\left(R_{0,PV} - R_{line,PV}R_{0,PV} - R_{0,PV}R_{load}\right)}{L_{PV}}. \tag{7.99}$$

3) Excess power mode

First, consider the system model of lead-acid battery with the boost converter as below:

$$\begin{cases} \dot{\phi}_{1,LA} = \frac{1}{R_{1,LA}C_{1,LA}}\left(\phi_{m,LA} - \phi_{1,LA}\right), \\ \dot{\phi}_{m,LA} = \frac{1}{R_{0,LA}C_{0,LA}}\left(\phi_{L,LA} - \phi_{m,LA}\right) - \frac{\dot{v}_{PN,LA}}{R_{0,LA}}, \\ \dot{\phi}_{L,LA} = \frac{1}{L_{LA}}R_{0,LA}\left(\phi_{m,LA} - \phi_{L,LA} - R_{L,LA}\phi_{L,LA} - \left(\phi_{m,LA}R_{0,LA} + v_{PN,LA}\right)\right) \\ \quad + \frac{1}{L_{LA}}\left(V_{D,LA} + v_{0,MG} + -R_{M,LA}\phi_{L,LA}\right)u_2 - \frac{V_{D,LA}}{L_{LA}}, \end{cases} \tag{7.100}$$

and the PMSG with the AC-DC converter and the boost converter is

$$\begin{cases} \dot{P} = \omega v_{\alpha\beta}^T J^T \phi_{\alpha\beta} + \frac{1}{L} v_{\alpha\beta}^T v_{\alpha\beta} - v_{\alpha\beta}^T \frac{v_{0,MG}}{2L} u_{\alpha\beta}, \\ \dot{Q} = \omega v_{\alpha\beta}^T J^T J \phi_{\alpha\beta} + \frac{v_{\alpha\beta}^T J}{L} v_{\alpha\beta} - \frac{v_{\alpha\beta}^T J v_{0,MG}}{2L} u_{\alpha\beta}, \\ \dot{v}_{\alpha\beta} = \omega J v_{\alpha\beta}, \\ \dot{\phi}_{\alpha\beta} = \frac{1}{L} v_{\alpha\beta} - \frac{v_{0,MG}}{2L} u_{\alpha\beta}, \\ \dot{v}_{0,MG} = -\frac{1}{CR_L} v_{0,MG} + \frac{\phi_{\alpha\beta}^T}{2C} u_{\alpha\beta}, \end{cases} \tag{7.101}$$

and

$$\begin{cases} \dot{v}_{PV} = \frac{1}{C_{PV}}\left(\phi_{PV} - \phi_{L,PV}u_1\right), \\ \dot{\phi}_{L,PV} = \frac{1}{L_{PV}}R_{0,PV}\left(\left(\phi_{0,PV} - \phi_{L,PV}\right) - R_{L,PV}\phi_{L,PV} - v_{0,PV}\right) \\ \quad + \frac{1}{L_{PV}}\left(V_{D,PV} + v_{PV} - R_{M,PV}\phi_{L,PV}\right)u_1 - \frac{V_{D,PV}}{L_{PV}}, \\ \dot{v}_{0,PV} = \frac{1}{C_{0,PV}}\left(\phi_{L,PV} - \phi_{0,PV}\right). \end{cases} \tag{7.102}$$

Under the deficit power mode, the PMSG can be supplied to the DC load and the lead-acid battery. Based on the Kirchhoff's current law,

$$\phi_{0,MG} = \phi_{0,LA} + \frac{v_{0,MG} - \phi_{0,MG}R_{line,MG}}{R_{load}}. \tag{7.103}$$

Here, the objective is to design a supervisory controller, so that MPPT can be obtained and the output voltage tracks the reference v_{ref}. The PV voltage reference v^* of MPPT can be calculated by (7.8), and $v_{0,PV} \rightarrow v_{ref}$, $P \rightarrow P^*$, $Q \rightarrow Q^*$, $v_{0,MG} \rightarrow v_{ref}$,$v_{0,LA} \rightarrow v_{ref}$. Define $P_e = P - P^*, Q_e = Q - Q^*, v_{e,MG} = v_{0,MG} - v_{ref}, v_{e,LA} = v_{0,LA} - v_{ref}$, $v_{e,PV} = v_{0,PV} - v_{ref}$,

$$\begin{cases} \dot{\phi}_{1,LA} = \frac{1}{R_{1,LA}C_{1,LA}}\phi_{m,LA} - \frac{1}{R_{1,LA}C_{1,LA}}\phi_{1,LA}, \\ \dot{\phi}_{m,LA} = \frac{1}{R_{0,LA}C_{0,LA}}(\phi_{L,LA} - \phi_{m,LA}) - \frac{\dot{v}_{PN,LA}}{R_{0,LA}}, \\ \dot{\phi}_{L,LA} = \frac{1}{L_{LA}}R_{0,LA}(\phi_{m,LA} - \phi_{L,LA} - R_{L,LA}\phi_{L,LA} - (\phi_{m,LA}R_{0,LA} + v_{PN,LA})) \\ \qquad + \frac{1}{L_{LA}}(V_{D,LA} + v_{0,MG} + -R_{M,LA}\phi_{L,LA})u_2 - \frac{V_{D,LA}}{L_{LA}}, \\ \dot{P}_e = \omega v_{\alpha\beta}^T J^T \phi_{\alpha\beta} + \frac{1}{L}v_{\alpha\beta}^T v_{\alpha\beta} - v_{\alpha\beta}^T \frac{v_{0,MG}}{2L}u_{\alpha\beta}, \\ \dot{Q}_e = \omega v_{\alpha\beta}^T J^T J\phi_{\alpha\beta} + \frac{v_{\alpha\beta}^T J}{L}v_{\alpha\beta} - \frac{v_{\alpha\beta}^T J v_{0,MG}}{2L}u_{\alpha\beta}, \\ \dot{v}_{\alpha\beta} = \omega J v_{\alpha\beta}, \\ \dot{\phi}_{\alpha\beta} = \frac{1}{L}v_{\alpha\beta} - \frac{v_{0,MG}}{2L}u_{\alpha\beta}, \\ \dot{v}_{e,MG} = -\frac{1}{CR_L}v_{e,MG} + \frac{\phi_{\alpha\beta}^T}{2C}u_{\alpha\beta} - \frac{1}{CR_L}v_{ref}, \\ \dot{v}_{PV} = \frac{1}{C_{PV}}(\phi_{PV} - \phi_{L,PV}u_1), \\ \dot{\phi}_{L,PV} = \frac{1}{L_{PV}}R_{0,PV}((\phi_{0,PV} - \phi_{L,PV}) - R_{L,PV}\phi_{L,PV} - v_{0,PV}) \\ \qquad + \frac{1}{L_{PV}}(V_{D,PV} + v_{PV} - R_{M,PV}\phi_{L,PV})u_1 - \frac{V_{D,PV}}{L_{PV}}, \\ \dot{v}_{e,PV} = \frac{1}{C_{0,PV}}(\phi_{L,PV} - \phi_{0,PV}). \end{cases} \tag{7.104}$$

Define $x(t) = \begin{bmatrix} \phi_{1,LA} & \phi_{m,LA} & \phi_{L,LA} & P_e & Q_e & v_{\alpha\beta} & \phi_{\alpha\beta} & v_{e,MG} & v_{PV} & \phi_{L,PV} & v_{e,PV} \end{bmatrix}^T$, and choose $\{v_{\alpha\beta}^T, \omega v_{\alpha\beta}^T, \omega, \frac{\phi_{PV}}{\phi_{L,PV}}, \frac{\phi_{0,PV}}{\phi_{L,PV}}, v_{0,MG}, \phi_{L,LA}, \phi_{\alpha\beta}^T, \phi_{L,PV}, v_{PV}\}$ as the fuzzy premise variables, the augmented fuzzy system is given by

Plant Rule $\mathscr{R}^l$: **IF** z_1 is $\mathscr{F}_1^l$ and z_2 is $\mathscr{F}_2^l$ and $\cdots$ and z_{10} is $\mathscr{F}_{10}^l$, **THEN**

$$\dot{x}(t) = A_l x(t) + B_l u(t) + \omega(t), l \in \mathscr{L} := \{1, 2, \ldots, r\} \tag{7.105}$$

where $\mathscr{R}^l$ denotes the l-th fuzzy inference rule; $\mathscr{L} := \{1, 2, \ldots, r\}$, and r is the number of inference rules; $\mathscr{F}_\theta^l(\theta = 1, 2, \cdots, 10)$ is the fuzzy set; $x(t) \in \Re^{n_x}$ and $u(t) \in \Re^{n_u}$ denote the system state and control input, respectively; $z(t) \triangleq [z_1, z_2, \cdots, z_{10}]$ are the measurable variables; The l-th local model $\{A_l, B_l\}$ and the disturbance $\omega(t)$ are

given as below:

$$A_l = \begin{bmatrix} k_1 & k_2 & 0 & 0 & 0 & 0 & 0 & 0 & 0 & 0 & 0 \\ 0 & -k_3 & k_3 & 0 & 0 & 0 & 0 & 0 & 0 & 0 & 0 \\ 0 & k_4 & k_5 & 0 & 0 & 0 & 0 & 0 & 0 & 0 & 0 \\ 0 & 0 & 0 & 0 & 0 & \frac{\mathscr{F}_1^l}{L} & \mathscr{F}_2^l J^T & 0 & 0 & 0 & 0 \\ 0 & 0 & 0 & 0 & 0 & \frac{\mathscr{F}_1^l J}{L} & \mathscr{F}_2^l J^T J & 0 & 0 & 0 & 0 \\ 0 & 0 & 0 & 0 & 0 & \mathscr{F}_3^l J & 0 & 0 & 0 & 0 & 0 \\ 0 & 0 & 0 & 0 & 0 & \frac{1}{L} & 0 & 0 & 0 & 0 & 0 \\ 0 & 0 & 0 & 0 & 0 & 0 & 0 & \frac{-1}{CR_L} & 0 & 0 & 0 \\ 0 & 0 & 0 & 0 & 0 & 0 & 0 & 0 & 0 & \frac{\mathscr{F}_4^l}{C_{PV}} & 0 \\ 0 & 0 & 0 & 0 & 0 & 0 & 0 & 0 & 0 & k_6 & \frac{-R_{0,PV}}{L_{PV}} \\ 0 & 0 & 0 & 0 & 0 & 0 & 0 & 0 & 0 & \frac{1-\mathscr{F}_5^l}{C_{0,PV}} & 0 \end{bmatrix},$$

$$B_l = \begin{bmatrix} 0 & 0 & 0 \\ 0 & 0 & 0 \\ k_7 & 0 & 0 \\ 0 & -\frac{\mathscr{F}_1^l \mathscr{F}_6^l}{2L} & 0 \\ 0 & -\frac{\mathscr{F}_1^l J \mathscr{F}_6^l}{2L} & 0 \\ 0 & -\frac{\mathscr{F}_6^l}{2L} & 0 \\ 0 & \frac{\mathscr{F}_8^l}{2C} & 0 \\ 0 & 0 & 0 \\ 0 & 0 & \frac{-\mathscr{F}_9^l}{C_{PV}} \\ 0 & 0 & k_8 \\ 0 & 0 & 0 \end{bmatrix}, \omega(t) = \begin{bmatrix} 0 \\ -\frac{\dot{v}_{PN,LA}}{R_{0,LA}} \\ -\frac{R_{0,LA} v_{PN,LA} - V_{D,LA}}{L_{LA}} \\ 0 \\ 0 \\ 0 \\ 0 \\ -\frac{v_{ref}}{CR_L} \\ 0 \\ -\frac{R_{0,PV} v_{ref} + V_{D,PV}}{L_{PV}} \\ 0 \end{bmatrix},$$

$$\begin{aligned} k_1 &= \frac{-1}{R_{1,LA}C_{1,LA}}, k_2 = \frac{1}{R_{1,LA}C_{1,LA}}, k_3 = \frac{1}{R_{0,LA}C_{0,LA}}, \\ k_4 &= \frac{R_{0,LA} - R_{0,LA}R_{0,LA}}{L_{LA}}, k_5 = -\frac{R_{0,LA} + R_{0,LA}R_{L,LA}}{L_{LA}}, \\ k_6 &= \frac{R_{0,PV}\mathscr{F}_5^l - R_{0,PV} - R_{0,PV}R_{L,PV}}{L_{PV}}, \\ k_7 &= \frac{V_{D,LA} + \mathscr{F}_6^l + -R_{M,LA}\mathscr{F}_7^l}{L_{LA}}, k_8 = \frac{V_{D,PV} + \mathscr{F}_{10}^l - R_{M,PV}\mathscr{F}_9^l}{L_{PV}}. \end{aligned} \tag{7.106}$$

7.6 PMSG SYSTEM AND PV SYSTEM FOR AC LOAD

The section considers a small-scale DC power system, which consists of the PMSG subsystem, the lead-acid battery and the DC load. The PMSG subsystem uses an AC-DC buck converter to the DC-bus, and a bi-directional buck-boost converter will be used to interface the battery and the DC-bus. This system, as illustrated in Figure 7.3, utilizes a bi-directional converter operating as a boost converter during the discharging mode of the battery and as a buck converter during the charging mode.

7.6.1 OPERATION MODES

In order to ensure the reliable and stable operations of the small-scale DC power system under different conditions, the power balance among the supply and the demand must be maintained under all operating conditions. Now, we define P_{PV}, $P_{LD}^{\min}$, and $P_{LD}^{\max}$ as the PMSG supply, the minimum load demand, and the maximum load demand, respectively. There exist these three possible conditions of the system as shown in Figure 7.6 [1].

1) Deficit power mode$\left(P_{MG}+P_{PV} \leq P_{LD}^{\min}\right)$

The maximum available power is less than the power demand in this operating mode. Specifically, the PMSG is not sufficient to satisfy the DC loads. The PMSG subsystem works with the MPPT algorithm, and the lead-acid battery converters regulate the output voltage by the battery discharging operation. Thus, they work as the voltage sources. To obtain both the accurate current sharing and the desirable voltage regulation using the fuzzy control method, the DC-bus voltage reaches its minimum acceptable value at the maximum discharge current of the batteries.

2) Floating power mode$\left(P_{LD}^{\min} \leq P_{MG}+P_{PV} \leq P_{LD}^{\max}\right)$

In this mode, the DC system operates in an islanded mode. The PMSG subsystem is basically sufficient to satisfy the DC loads, and the batteries do not work in this case. The PMSG subsystem works with the MPPT algorithm, and regulates the DC bus voltage by using the fuzzy tracking control.

3) Excess power mode$(P_{MG}+P_{PV} \geq P_{LD}^{\max})$

In this mode, the PMSG subsystem works with the MPPT algorithm. Since the PMSG subsystem is greater than the load power, the DC voltage increases. The excess power is used to charge the lead-acid batteries. Therefore, the batteries regulate the DC bus voltage by the battery charging operations.

7.6.2 DYNAMIC MODELING

Recall the system models of the PV with the buck converter in (1.3), the PMSG with the AC-DC converter in (2.20), the lead-acid battery in (3.1), and the boost converter in (1.6), and the DC-AC converter in (1.13) as below:

$$\left\{\begin{array}{l}
\dot{v}_{PV}=\frac{1}{C_{PV}}\left(\phi_{PV}-\phi_{L,PV}u_1\right), \\
\dot{\phi}_{L,PV}=\frac{1}{L_{PV}}R_{0,PV}\left(\left(\phi_{0,PV}-\phi_{L,PV}\right)-R_{L,PV}\phi_{L,PV}-v_{0,PV}\right) \\
\qquad +\frac{1}{L_{PV}}\left(V_{D,PV}+v_{PV}-R_{M,PV}\phi_{L,PV}\right)u_1-\frac{V_{D,PV}}{L_{PV}}, \\
\dot{v}_{0,PV}=\frac{1}{C_{0,PV}}\left(\phi_{L,PV}-\phi_{0,PV}\right),
\end{array}\right. \tag{7.107}$$

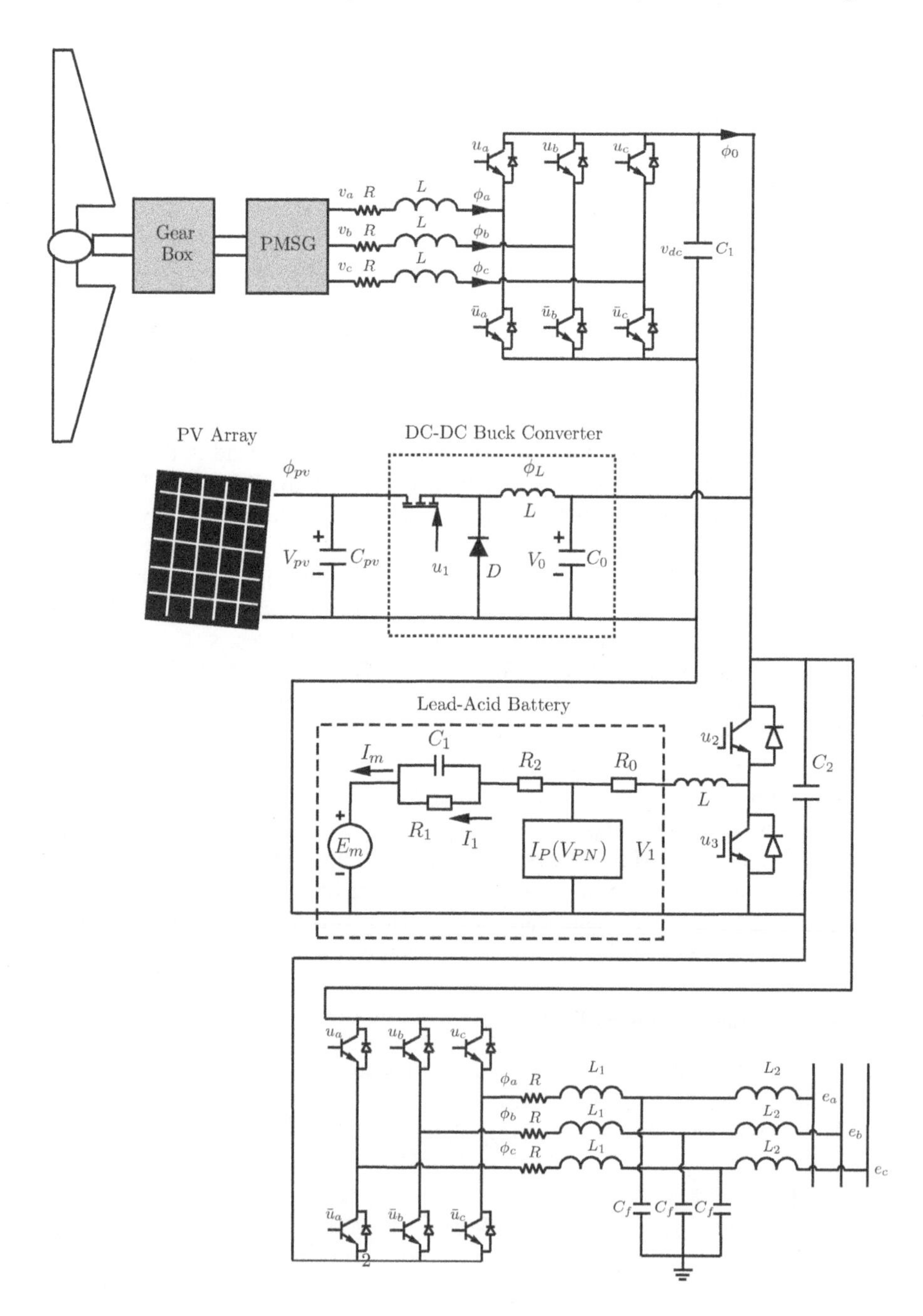

Figure 7.6 AC microgrid with PV and PMSG.

and

$$\begin{cases} \dot{P} = \omega v_{\alpha\beta}^T J^T \phi_{\alpha\beta} + \frac{1}{L} v_{\alpha\beta}^T v_{\alpha\beta} - v_{\alpha\beta}^T \frac{v_{dc}}{2L} u_{\alpha\beta}, \\ \dot{Q} = \omega v_{\alpha\beta}^T J^T J \phi_{\alpha\beta} + \frac{v_{\alpha\beta}^T J}{L} v_{\alpha\beta} - \frac{v_{\alpha\beta}^T J v_{dc}}{2L} u_{\alpha\beta}, \\ \dot{v}_{\alpha\beta} = \omega J v_{\alpha\beta}, \\ \dot{\phi}_{\alpha\beta} = \frac{1}{L} v_{\alpha\beta} - \frac{v_{dc}}{2L} u_{\alpha\beta}, \\ \dot{v}_{dc} = -\frac{1}{CR_L} v_{dc} + \frac{\phi_{\alpha\beta}^T}{2C} u_{\alpha\beta}, \end{cases} \tag{7.108}$$

and

$$\begin{cases} \dot{\phi}_{1,LA} = \frac{1}{R_{1,LA} C_{1,LA}} \left(\phi_{m,LA} - \phi_{1,LA}\right), \\ \dot{Q}_{e,LA} = -\phi_{m,LA}, \\ \dot{\theta}_{LA} = \frac{1}{C_{\theta,LA}} \left[P_{s,LA} - \frac{(\theta_{LA} - \theta_{a,LA})}{R_{\theta,LA}} \right], \\ \phi_{p,LA} = G_{p,LA} v_{PN,LA}, \\ G_{p,LA} = G_{p0,LA} \exp\left(v_{PN,LA}/v_{p0,LA} + A_{p,LA}\left(1 - \theta_{LA}/\theta_{f,LA}\right)\right), \end{cases} \tag{7.109}$$

and

$$\begin{cases} \dot{\phi}_{LA} = -\frac{1}{L_{LA}} (1-u) v_{dc} + \frac{1}{L_{LA}} v_{LA}, \\ \dot{v}_{dc} = \frac{1}{C_{0,LA}} (1-u) \phi_{LA} - \frac{1}{C_{0,LA}} \phi_{0,LA}, \end{cases} \tag{7.110}$$

and

$$\begin{cases} \dot{v}_{dc} = \frac{1}{C_{pv}} \left(\phi_{dc} - \frac{1.5 u_d}{v_{dc}} \phi_d \right), \\ \dot{\phi}_d = -\frac{R_1}{L_1} \phi_d - \omega \phi_q + \frac{1}{L_1} e_d, \\ \dot{\phi}_q = \omega \phi_d - \frac{R_1}{L_1} \phi_q + \frac{1}{L_1} e_q, \end{cases} \tag{7.111}$$

where the subscripts MG and LA stand for the PMSG system and the lead-acid battery, respectively, and all notations are defined in (1.3), (2.20), (1.6), and (1.13), respectively.

1) Deficit power mode

First, consider the system model of the lead-acid battery with the boost converter as below:

$$\begin{cases} \dot{\phi}_{1,LA} = \frac{1}{R_{1,LA} C_{1,LA}} \left(\phi_{m,LA} - \phi_{1,LA}\right), \\ \dot{\phi}_{m,LA} = -\frac{1}{R_{0,LA} L_{LA}} (1-u_3) v_{0,LA} + \frac{1}{R_{0,LA} L_{LA}} v_{LA} - \dot{G}_{p,LA} v_{PN,LA} - G_{p,LA} \dot{v}_{PN,LA}, \\ \dot{v}_{0,LA} = \frac{1}{C_{0,LA}} (1-u_3) \left(\phi_{m,LA} + \phi_{p,LA}\right) - \frac{1}{R_{0,LA} C_{0,LA}} \phi_{0,LA}, \end{cases} \tag{7.112}$$

and the PMSG with the AC-DC converter and the boost converter is

$$\begin{cases} \dot{P} = \omega v_{\alpha\beta}^T J^T \phi_{\alpha\beta} + \frac{1}{L} v_{\alpha\beta}^T v_{\alpha\beta} - v_{\alpha\beta}^T \frac{v_{0,MG}}{2L} u_{\alpha\beta}, \\ \dot{Q} = \omega v_{\alpha\beta}^T J^T J \phi_{\alpha\beta} + \frac{v_{\alpha\beta}^T J}{L} v_{\alpha\beta} - \frac{v_{\alpha\beta}^T J v_{0,MG}}{2L} u_{\alpha\beta}, \\ \dot{v}_{\alpha\beta} = \omega J v_{\alpha\beta}, \\ \dot{\phi}_{\alpha\beta} = \frac{1}{L} v_{\alpha\beta} - \frac{v_{0,MG}}{2L} u_{\alpha\beta}, \\ \dot{v}_{0,MG} = -\frac{1}{CR_L} v_{0,MG} + \frac{\phi_{\alpha\beta}^T}{2C} u_{\alpha\beta}, \end{cases} \tag{7.113}$$

and

$$\begin{cases} \dot{v}_{PV} = \frac{1}{C_{PV}}\left(\phi_{PV} - \phi_{L,PV}u_1\right), \\ \dot{\phi}_{L,PV} = \frac{1}{L_{PV}}R_{0,PV}\left(\left(\phi_{0,PV} - \phi_{L,PV}\right) - R_{L,PV}\phi_{L,PV} - v_{0,PV}\right) \\ \quad + \frac{1}{L_{PV}}\left(V_{D,PV} + v_{PV} - R_{M,PV}\phi_{L,PV}\right)u_1 - \frac{V_{D,PV}}{L_{PV}}, \\ \dot{v}_{0,PV} = \frac{1}{C_{0,PV}}\left(\phi_{L,PV} - \phi_{0,PV}\right), \end{cases} \tag{7.114}$$

and

$$\begin{cases} \dot{v}_{dc} = \frac{1}{C_{pv}}\left(\phi_{dc} - \frac{1.5u_d}{v_{dc}}\phi_d\right), \\ \dot{\phi}_d = -\frac{R_1}{L_1}\phi_d - \omega\phi_q + \frac{1}{L_1}e_d, \\ \dot{\phi}_q = \omega\phi_d - \frac{R_1}{L_1}\phi_q + \frac{1}{L_1}e_q. \end{cases} \tag{7.115}$$

Under the deficit power mode, both the PMSG and the lead-acid battery can be supplied to the DC load. Based on the Thevenin's theorem,

$$v_{0,LA} = \phi_{0,LA}R_{line,LA} + \left(\phi_{0,LA} + \phi_{0,MG} + \phi_{0,PV}\right)R_{load}, \tag{7.116}$$

$$v_{0,MG} = \phi_{0,MG}R_{line,MG} + \left(\phi_{0,LA} + \phi_{0,MG} + \phi_{0,PV}\right)R_{load}, \tag{7.117}$$

$$v_{0,PV} = \phi_{0,PV}R_{line,PV} + \left(\phi_{0,LA} + \phi_{0,MG} + \phi_{0,PV}\right)R_{load}, \tag{7.118}$$

where the $R_{line,LA}$, $R_{line,MG}$, $R_{line,PV}$ are the resistances of power lines on lead–acid battery, and the one on the PMSG system, and the one on the PV system, respectively, R_{load} is the load resistance.

Here, the objective is to design a supervisory controller, so that MPPT of the PV and the PMSG can be obtained and the output voltage tracks the reference v_{ref}. The PV voltage reference v^* of MPPT can be calculated by (7.8), and $v_{PV} \to v^*, P \to P^*, Q \to Q^*, v_{0,MG} \to v_{ref}, v_{0,LA} \to v_{ref}, v_{0,PV} \to v_{ref}$. Define $e_{PV} = v_{PV} - v^*$, $P_e = P - P^*, Q_e = Q - Q^*, v_{e,MG} = v_{0,MG} - v_{ref}, v_{e,LA} = v_{0,LA} - v_{ref}, v_{e,PV} = v_{0,PV} - v_{ref}$.

Based on the relationships (7.112)-(7.118), the interconnected system is given by

$$\begin{cases} \dot{\phi}_{1,LA} = \frac{1}{R_{1,LA}C_{1,LA}}(\phi_{m,LA}-\phi_{1,LA}), \\ \dot{\phi}_{m,LA} = -\frac{1}{R_{0,LA}L_{LA}}(\phi_{0,LA}R_{line,LA}+(\phi_{0,LA}+\phi_{0,MG}+\phi_{0,PV})R_{load}) \\ \quad +\frac{v_{0,LA}}{R_{0,LA}L_{LA}}u_3+\frac{1}{R_{0,LA}L_{LA}}v_{LA}-\dot{G}_{p,LA}v_{PN,LA}-G_{p,LA}\dot{v}_{PN,LA}, \\ \dot{v}_{e,LA} = \frac{1}{C_{0,LA}}(1-u_3)(\phi_{m,LA}+\phi_{p,LA})-\frac{v_{e,LA}+v_{ref}-(\phi_{0,MG}+\phi_{0,PV})R_{load}}{R_{0,LA}C_{0,LA}(R_{line,LA}+R_{load})} \\ \dot{P}_e = \omega v_{\alpha\beta}^T J^T\phi_{\alpha\beta}+\frac{1}{L}v_{\alpha\beta}^T v_{\alpha\beta}-v_{\alpha\beta}^T\frac{v_{0,MG}}{2L}u_{\alpha\beta}, \\ \dot{Q}_e = \omega v_{\alpha\beta}^T J^T J\phi_{\alpha\beta}+\frac{v_{\alpha\beta}^T J}{L}v_{\alpha\beta}-\frac{v_{\alpha\beta}^T J v_{0,MG}}{2L}u_{\alpha\beta}, \\ \dot{v}_{\alpha\beta} = \omega J v_{\alpha\beta}, \\ \dot{\phi}_{\alpha\beta} = \frac{1}{L}v_{\alpha\beta}-\frac{v_{0,MG}}{2L}u_{\alpha\beta}, \\ \dot{v}_{e,MG} = -\frac{1}{CR_L}v_{e,MG}+\frac{\phi_{\alpha\beta}^T}{2C}u_{\alpha\beta}-\frac{1}{CR_L}v_{ref}, \\ \dot{e}_{PV} = \frac{1}{C_{PV}}(\phi_{PV}-\phi_{L,PV}u_1), \\ \dot{\phi}_{L,PV} = \frac{R_{0,PV}}{L_{PV}}\phi_{0,PV}-\frac{R_{0,PV}}{L_{PV}}(1+R_{L,PV})\phi_{L,PV}-\frac{R_{0,PV}}{L_{PV}}\phi_{0,PV}R_{line,PV} \\ \quad -\frac{R_{0,PV}}{L_{PV}}(\phi_{0,LA}+\phi_{0,MG}+\phi_{0,PV})R_{load} \\ \quad +\frac{1}{L_{PV}}(V_{D,PV}+v_{PV}-R_{M,PV}\phi_{L,PV})u_1-\frac{V_{D,PV}}{L_{PV}}, \\ \dot{v}_{e,PV} = \frac{1}{C_{0,PV}}(\phi_{L,PV}-\phi_{0,PV}), \\ \dot{v}_{dc} = \frac{1}{C_{pv}}\left(\phi_{0,LA}+\phi_{0,MG}+\phi_{0,PV}-\frac{1.5u_d}{v_{dc}}\phi_d\right), \\ \dot{\phi}_d = -\frac{R_1}{L_1}\phi_d-\bar{\omega}\phi_q+\frac{1}{L_1}e_d, \\ \dot{\phi}_q = \bar{\omega}\phi_d-\frac{R_1}{L_1}\phi_q+\frac{1}{L_1}e_q. \end{cases} \tag{7.119}$$

If we define $x(t) = [\phi_{1,LA}, \phi_{m,LA}, v_{e,LA}, P_e, Q_e\ v_{\alpha\beta}\ \phi_{\alpha\beta}\ v_{e,MG}\ e_{PV}\ \phi_{L,PV}\ v_{e,PV}, v_{dc}, \phi_d, \phi_q]^T$, and choose $\{\frac{\phi_{0,PV}}{\phi_{L,PV}}, \omega v_{\alpha\beta}^T, v_{\alpha\beta}^T, v_{0,MG}, \phi_{0,MG}, \phi_{\alpha\beta}^T, \frac{\phi_{PV}}{e_{PV}}, \frac{\phi_{0,MG}}{\phi_{\alpha\beta}}, v_{PV}, \phi_{L,PV}, \frac{u_d}{v_{dc}}, \omega, \bar{\omega}\}$ as the fuzzy premise variables, then the augmented fuzzy system is given by

Plant Rule $\mathscr{R}^l$: IF z_1 is $\mathscr{F}_1^l$ and z_2 is $\mathscr{F}_2^l$ and $\cdots$ and z_{13} is $\mathscr{F}_{13}^l$, **THEN**

$$\dot{x}(t) = A_l x(t) + B_l u(t) + \omega(t), l \in \mathscr{L} := \{1,2,\ldots,r\} \tag{7.120}$$

where $\mathscr{R}^l$ denotes the l-th fuzzy inference rule; $\mathscr{L} := \{1,2,\ldots,r\}$, and r is the number of inference rules; $\mathscr{F}_\theta^l(\theta = 1,2,\cdots,13)$ is the fuzzy set; $x(t) \in \Re^{n_x}$ and $u(t) \subset \Re^{n_u}$ denote the system state and the control input, respectively; $z(t) \triangleq [z_1, z_2, \cdots, z_{13}]$ are the measurable variables; The l-th local model $\{A_l, B_l\}$ and the disturbance $\omega(t)$

are given as below:

$$A_l = \begin{bmatrix} A_{l(11)} & A_{l(12)} & 0 \\ A_{l(21)} & A_{l(22)} & A_{l(23)} \\ 0 & 0 & A_{l(33)} \end{bmatrix}, B_l = \begin{bmatrix} B_{l(1)} & 0 & 0 \\ 0 & B_{l(2)} & 0 \\ 0 & 0 & B_{l(3)} \end{bmatrix}, \omega(t) = \begin{bmatrix} \omega_{(1)} \\ \omega_{(2)} \\ \omega_{(3)} \end{bmatrix},$$

$$A_{l(11)} = \begin{bmatrix} -k_1 & k_1 & 0 & 0 & 0 & 0 & 0 & 0 & 0 \\ 0 & -k_2\mathscr{F}_1^l & -k_2 & 0 & 0 & 0 & 0 & 0 & 0 \\ 0 & \frac{1+\mathscr{F}_3^l}{C_{0,LA}} & \frac{-k_4}{R_{load}} & 0 & 0 & 0 & 0 & 0 & 0 \\ 0 & 0 & 0 & 0 & 0 & \frac{\mathscr{F}_8^l}{L} & \mathscr{F}_7^l J^T & 0 & 0 \\ 0 & 0 & 0 & 0 & 0 & \frac{\mathscr{F}_8^l J}{L} & \mathscr{F}_7^l J^T J & 0 & 0 \\ 0 & 0 & 0 & 0 & 0 & \mathscr{F}_{12}^l J & 0 & 0 & 0 \\ 0 & 0 & 0 & 0 & 0 & \frac{1}{L} & 0 & 0 & 0 \\ 0 & 0 & 0 & 0 & 0 & 0 & 0 & \frac{-1}{CR_L} & 0 \\ 0 & 0 & 0 & 0 & 0 & 0 & 0 & 0 & \frac{\mathscr{F}_{12}^l}{C_{PV}} \end{bmatrix},$$

$$A_{l(12)} = \begin{bmatrix} 0 & 0 & 0 \\ 0 & 0 & 0 \\ k_5 & 0 & 0 \\ 0 & 0 & 0 \\ 0 & 0 & 0 \\ 0 & 0 & 0 \\ 0 & 0 & 0 \\ 0 & 0 & 0 \\ 0 & 0 & 0 \end{bmatrix}, A_{l(21)} = \begin{bmatrix} 0 & -k_6\mathscr{F}_1^l & 0 & 0 & 0 & 0 & 0 & 0 & 0 \\ 0 & 0 & 0 & 0 & 0 & 0 & 0 & 0 & 0 \\ 0 & \frac{\mathscr{F}_1^l}{C_{pv}} & 0 & 0 & 0 & 0 & \frac{\mathscr{F}_{13}^l}{C_{pv}} & 0 & 0 \end{bmatrix},$$

$$A_{l(22)} = \begin{bmatrix} k_7 & 0 & 0 \\ k_9 & 0 & 0 \\ \frac{\mathscr{F}_6^l}{C_{pv}} & 0 & 0 \end{bmatrix}, A_{l(23)} = \begin{bmatrix} 0 & 0 \\ 0 & 0 \\ k_{10} & 0 \end{bmatrix}, A_{l(33)} = \begin{bmatrix} \frac{-R_1}{L_1} & -\mathscr{F}_{13}^l \\ \mathscr{F}_{13}^l & \frac{-R_1}{L_1} \end{bmatrix},$$

$$B_{l(1)} = \begin{bmatrix} 0 & 0 \\ \frac{\mathscr{F}_2^l}{R_{0,LA}L_{LA}} & 0 \\ \frac{-\mathscr{F}_4^l-\mathscr{F}_5^l}{C_{0,LA}} & 0 \\ 0 & \frac{-\mathscr{F}_8^l\mathscr{F}_9^l}{2L} \\ 0 & \frac{-\mathscr{F}_8^l J\mathscr{F}_9^l}{2L} \\ 0 & 0 \\ 0 & \frac{-\mathscr{F}_9^l}{2L} \\ 0 & 0 \\ 0 & \frac{\mathscr{F}_{11}^l}{2C} \end{bmatrix}, \omega_{(1)} = \begin{bmatrix} 0 \\ k_{11}(t) \\ -k_{12}(t) \\ 0 \\ 0 \\ 0 \\ 0 \\ 0 \\ 0 \end{bmatrix}, B_{l(2)} = \begin{bmatrix} \frac{-\mathscr{F}_{15}^l}{C_{PV}} \\ k_8 \\ 0 \\ 0 \end{bmatrix},$$

$$B_{l(3)} = \begin{bmatrix} \frac{1}{L_1} & 0 \\ 0 & \frac{1}{L_1} \end{bmatrix}, \omega_{(2)} = \begin{bmatrix} 0 \\ \frac{-v_{ref}}{L_{MG}} \\ \frac{-v_{ref}}{CR_L} \\ 0 \end{bmatrix}, \omega_{(3)} = \begin{bmatrix} 0 \\ 0 \end{bmatrix},$$

$$k_1 = \frac{1}{R_{1,LA}C_{1,LA}}, k_2 = \frac{R_{line,LA}+R_{load}}{R_{0,LA}L_{LA}}, k_3 = \frac{R_{load}}{R_{0,LA}L_{LA}},$$
$$k_4 = \frac{R_{load}}{R_{0,LA}C_{0,LA}\left(R_{line,LA}+R_{load}\right)}, k_5 = \frac{R_{load}\mathscr{F}_6^l}{R_{0,LA}C_{0,LA}\left(R_{line,LA}+R_{load}\right)},$$
$$k_6 = \frac{R_{0,PV}R_{load}}{L_{PV}}, k_7 = \frac{\left(R_{0,PV}-R_{line,PV}R_{0,PV}-R_{0,PV}R_{load}\right)}{L_{PV}}\mathscr{F}_{13}^l,$$
$$k_8 = \frac{1}{L_{PV}}\left(\mathscr{F}_{14}^l - R_{M,PV}\mathscr{F}_{15}^l\right), k_9 = \frac{1+\mathscr{F}_6^l}{C_{0,PV}}, k_{10} = \frac{-1.5\mathscr{F}_{16}^l}{C_{pv}},$$
$$k_{11}(t) = \frac{v_{LA}}{R_{0,LA}L_{LA}} - \dot{G}_{p,LA}v_{PN,LA} - G_{p,LA}\dot{v}_{PN,LA},$$
$$k_{12}(t) = \frac{v_{ref}}{R_{0,LA}C_{0,LA}\left(R_{line,LA}+R_{load}\right)}. \tag{7.121}$$

2) Floating power mode

In this mode, the lead-acid battery does not work. In other words, the power can be maintained between the PMSG system and the loading. Hence, the system model is

$$\begin{cases} \dot{P} = \omega v_{\alpha\beta}^T J^T \phi_{\alpha\beta} + \frac{1}{L}v_{\alpha\beta}^T v_{\alpha\beta} - v_{\alpha\beta}^T \frac{v_{0,MG}}{2L}u_{\alpha\beta}, \\ \dot{Q} = \omega v_{\alpha\beta}^T J^T J \phi_{\alpha\beta} + \frac{v_{\alpha\beta}^T J}{L}v_{\alpha\beta} - \frac{v_{\alpha\beta}^T J v_{0,MG}}{2L}u_{\alpha\beta}, \\ \dot{v}_{\alpha\beta} = \omega J v_{\alpha\beta}, \\ \dot{\phi}_{\alpha\beta} = \frac{1}{L}v_{\alpha\beta} - \frac{v_{0,MG}}{2L}u_{\alpha\beta}, \\ \dot{v}_{0,MG} = -\frac{1}{CR_L}v_{0,MG} + \frac{\phi_{\alpha\beta}^T}{2C}u_{\alpha\beta}, \\ \dot{v}_{PV} = \frac{1}{C_{PV}}\left(\phi_{PV} - \phi_{L,PV}u_1\right), \\ \dot{\phi}_{L,PV} = \frac{1}{L_{PV}}R_{0,PV}\left(\left(\phi_{0,PV} - \phi_{L,PV}\right) - R_{L,PV}\phi_{L,PV} - v_{0,PV}\right) \\ \qquad + \frac{1}{L_{PV}}\left(V_{D,PV} + v_{PV} - R_{M,PV}\phi_{L,PV}\right)u_1 - \frac{V_{D,PV}}{L_{PV}}, \\ \dot{v}_{0,PV} = \frac{1}{C_{0,PV}}\left(\phi_{L,PV} - \phi_{0,PV}\right), \\ \dot{v}_{dc} = \frac{1}{C_{pv}}\left(\phi_{0,PV} + \phi_{0,MG} - \frac{1.5u_d}{v_{dc}}\phi_d\right), \\ \dot{\phi}_d = -\frac{R_1}{L_1}\phi_d - \omega\phi_q + \frac{1}{L_1}e_d, \\ \dot{\phi}_q = \omega\phi_d - \frac{R_1}{L_1}\phi_q + \frac{1}{L_1}e_q. \end{cases} \tag{7.122}$$

Under the floating power mode, both the PMSG and the PV systems can be supplied to the AC load. Based on the Thevenin's theorem,

$$v_{0,MG} = \phi_{0,MG}R_{line,MG} + \left(\phi_{0,MG} + \phi_{0,PV}\right)R_{load}, \tag{7.123}$$
$$v_{0,PV} = \phi_{0,PV}R_{line,PV} + \left(\phi_{0,MG} + \phi_{0,PV}\right)R_{load}. \tag{7.124}$$

Here, the objective is to design a supervisory controller, so that the MPPT can be obtained and the output voltage tracks the reference v_{ref}. The PV voltage reference v^* of MPPT can be calculated by (7.8), and $P \to P^*, Q \to Q^*, v_{PV} \to v^*, v_{0,MG} \to v_{ref}, v_{0,PV} \to v_{ref}$. Define $P_e = P - P^*, Q_e = Q - Q^*, e_{PV} = v_{PV} - v^*, v_{e,MG} = v_{0,MG} - v_{ref}$, $v_{e,PV} = v_{0,PV} - v_{ref}$,

$$
\left\{
\begin{aligned}
&\dot{P}_e = \omega v_{\alpha\beta}^T J^T \phi_{\alpha\beta} + \tfrac{1}{L} v_{\alpha\beta}^T v_{\alpha\beta} - v_{\alpha\beta}^T \tfrac{v_{0,MG}}{2L} u_{\alpha\beta},\\
&\dot{Q}_e = \omega v_{\alpha\beta}^T J^T J \phi_{\alpha\beta} + \frac{v_{\alpha\beta}^T J}{L} v_{\alpha\beta} - \frac{v_{\alpha\beta}^T J v_{0,MG}}{2L} u_{\alpha\beta},\\
&\dot{v}_{\alpha\beta} = \omega J v_{\alpha\beta},\\
&\dot{\phi}_{\alpha\beta} = \tfrac{1}{L} v_{\alpha\beta} - \tfrac{v_{0,MG}}{2L} u_{\alpha\beta},\\
&\dot{v}_{e,MG} = -\frac{R_{line,MG}\phi_{0,MG} + R_{load}\phi_{0,MG} + R_{load}\phi_{0,PV}}{CR_L} + \frac{\phi_{\alpha\beta}^T}{2C} u_{\alpha\beta},\\
&\dot{e}_{PV} = \tfrac{1}{C_{PV}} \left(\phi_{PV} - \phi_{L,PV} u_1\right),\\
&\dot{\phi}_{L,PV} = \tfrac{R_{0,PV}}{L_{PV}} \left(\left(1 - R_{line,PV} - R_{load}\right) \phi_{0,PV} - \left(1 + R_{L,PV}\right) \phi_{L,PV} - R_{load} \phi_{0,MG} \right)\\
&\qquad + \tfrac{1}{L_{PV}} \left(V_{D,PV} + v_{PV} - R_{M,PV} \phi_{L,PV}\right) u_1 - \tfrac{V_{D,PV}}{L_{PV}},\\
&\dot{v}_{e,PV} = \tfrac{1}{C_{0,PV}} \left(\phi_{L,PV} - \phi_{0,PV}\right),\\
&\dot{v}_{dc} = \tfrac{1}{C_{pv}} \left(\phi_{0,PV} + \phi_{0,MG} - \tfrac{1.5u_d}{v_{dc}} \phi_d \right),\\
&\dot{\phi}_d = -\tfrac{R_1}{L_1} \phi_d - \omega \phi_q + \tfrac{1}{L_1} e_d,\\
&\dot{\phi}_q = \omega \phi_d - \tfrac{R_1}{L_1} \phi_q + \tfrac{1}{L_1} e_q,
\end{aligned}
\right.
\tag{7.125}
$$

where the $R_{line,LA}$, $R_{line,MG}$, $R_{line,PV}$ are the resistances of power lines on lead-acid battery, and the one on the PMSG, and the one on the PV systems, respectively, R_{load} is the load resistance.

Define $x(t) = \left[P_e,\ Q_e\ v_{\alpha\beta}\ \phi_{\alpha\beta}\ v_{e,MG}\ e_{PV}\ \phi_{L,PV}\ v_{e,PV}, v_{dc}, \phi_d, \phi_q\right]^T$, and choose $\{\frac{\phi_{0,PV}}{\phi_{L,PV}}, \omega v_{\alpha\beta}^T, v_{\alpha\beta}^T, v_{0,MG}, \phi_{0,MG},\ \phi_{\alpha\beta}^T, \frac{\phi_{PV}}{e_{PV}}, \frac{\phi_{0,MG}}{\phi_{\alpha\beta}},\ v_{PV}, \phi_{L,PV}, \frac{u_d}{v_{dc}}, \omega, \bar{\omega}\}$ as the fuzzy premise variables, the augmented fuzzy system is given by

Plant Rule $\mathscr{R}^l$: **IF** z_1 is $\mathscr{F}_1^l$ and z_2 is $\mathscr{F}_2^l$ and $\cdots$ and z_{13} is $\mathscr{F}_{13}^l$, **THEN**

$$
\dot{x}(t) = A_l x(t) + B_l u(t) + \omega(t), l \in \mathscr{L} := \{1, 2, \ldots, r\} \tag{7.126}
$$

where $\mathscr{R}^l$ denotes the l-th fuzzy inference rule; $\mathscr{L} := \{1, 2, \ldots, r\}$, and r is the number of inference rules; $\mathscr{F}_\theta^l(\theta = 1, 2, \cdots, 13)$ is the fuzzy set; $x(t) \in \Re^{n_x}$ and $u(t) \in \Re^{n_u}$ denote the system state and the control input, respectively; $z(t) \triangleq [z_1, z_2, \cdots, z_{13}]$ are the measurable variables; The l-th local model $\{A_l, B_l\}$ and the disturbance $\omega(t)$

are given as below:

$$A_l = \begin{bmatrix} 0 & 0 & \frac{\mathscr{F}_3^l}{L} & \mathscr{F}_2^l J^T & 0 & 0 & 0 & 0 & 0 & 0 & 0 \\ 0 & 0 & \frac{\mathscr{F}_3^l J}{L} & \mathscr{F}_2^l J^T J & 0 & 0 & 0 & 0 & 0 & 0 & 0 \\ 0 & 0 & \mathscr{F}_{12}^l J & 0 & 0 & 0 & 0 & 0 & 0 & 0 & 0 \\ 0 & 0 & \frac{1}{L} & 0 & 0 & 0 & 0 & 0 & 0 & 0 & 0 \\ 0 & 0 & 0 & 0 & \frac{-1}{CR_L} & 0 & 0 & 0 & 0 & 0 & 0 \\ 0 & 0 & 0 & 0 & 0 & \frac{\mathscr{F}_7^l}{C_{PV}} & 0 & 0 & 0 & 0 & 0 \\ 0 & 0 & 0 & 0 & 0 & 0 & k_1 & 0 & 0 & 0 & 0 \\ 0 & 0 & 0 & 0 & 0 & 0 & k_3 & 0 & 0 & 0 & 0 \\ 0 & 0 & 0 & \frac{\mathscr{F}_8^l}{C_{pv}} & 0 & 0 & \frac{\mathscr{F}_1^l}{C_{pv}} & 0 & 0 & k_4 & 0 \\ 0 & 0 & 0 & 0 & 0 & 0 & 0 & 0 & 0 & \frac{-R_1}{L_1} & -\mathscr{F}_{13}^l \\ 0 & 0 & 0 & 0 & 0 & 0 & 0 & 0 & 0 & \mathscr{F}_{13}^l & \frac{-R_1}{L_1} \end{bmatrix},$$

$$B_l = \begin{bmatrix} \frac{-\mathscr{F}_3^l \mathscr{F}_4^l}{2L} & 0 & 0 & 0 \\ \frac{-\mathscr{F}_3^l J \mathscr{F}_4^l}{2L} & 0 & 0 & 0 \\ 0 & 0 & 0 & 0 \\ \frac{-\mathscr{F}_4^l}{2L} & 0 & 0 & 0 \\ \frac{\mathscr{F}_6^l}{2C} & 0 & 0 & 0 \\ 0 & \frac{-\mathscr{F}_9^l}{C_{PV}} & 0 & 0 \\ 0 & k_2 & 0 & 0 \\ 0 & 0 & 0 & 0 \\ 0 & 0 & 0 & 0 \\ 0 & 0 & \frac{1}{L_1} & 0 \\ 0 & 0 & 0 & \frac{1}{L_1} \end{bmatrix}, \omega(t) = \begin{bmatrix} 0 \\ 0 \\ 0 \\ 0 \\ 0 \\ \frac{-v_{ref}}{L_{MG}} \\ \frac{-v_{ref}}{CR_L} \\ 0 \\ 0 \\ 0 \\ 0 \end{bmatrix},$$

$$k_1 = \frac{\left(R_{0,PV} - R_{line,PV} R_{0,PV} - R_{0,PV} R_{load}\right)}{L_{PV}} \mathscr{F}_8^l,$$

$$k_2 = \frac{1}{L_{PV}} \left(\mathscr{F}_9^l - R_{M,PV} \mathscr{F}_{10}^l\right), k_3 = \frac{1 + \mathscr{F}_1^l}{C_{0,PV}}, k_4 = \frac{-1.5 \mathscr{F}_{11}^l}{C_{pv}}. \tag{7.127}$$

3) Excess power mode

First, consider the system model of the lead-acid battery with the boost converter as below:

$$\begin{cases} \dot{\phi}_{1,LA} = \frac{1}{R_{1,LA} C_{1,LA}} \left(\phi_{m,LA} - \phi_{1,LA}\right), \\ \dot{\phi}_{m,LA} = \frac{1}{R_{0,LA} C_{0,LA}} \left(\phi_{L,LA} - \phi_{m,LA}\right) - \frac{\dot{v}_{PN,LA}}{R_{0,LA}}, \\ \dot{\phi}_{L,LA} = \frac{1}{L_{LA}} R_{0,LA} \left(\phi_{m,LA} - \phi_{L,LA} - R_{L,LA} \phi_{L,LA} - \left(\phi_{m,LA} R_{0,LA} + v_{PN,LA}\right)\right) \\ \qquad + \frac{1}{L_{LA}} \left(V_{D,LA} + v_{0,MG} + -R_{M,LA} \phi_{L,LA}\right) u_2 - \frac{V_{D,LA}}{L_{LA}}, \end{cases} \tag{7.128}$$

and the PMSG with the AC-DC converter and the boost converter is

$$\begin{cases} \dot{P} = \omega v_{\alpha\beta}^T J^T \phi_{\alpha\beta} + \frac{1}{L} v_{\alpha\beta}^T v_{\alpha\beta} - v_{\alpha\beta}^T \frac{v_{0,MG}}{2L} u_{\alpha\beta}, \\ \dot{Q} = \omega v_{\alpha\beta}^T J^T J \phi_{\alpha\beta} + \frac{v_{\alpha\beta}^T J}{L} v_{\alpha\beta} - \frac{v_{\alpha\beta}^T J v_{0,MG}}{2L} u_{\alpha\beta}, \\ \dot{v}_{\alpha\beta} = \omega J v_{\alpha\beta}, \\ \dot{\phi}_{\alpha\beta} = \frac{1}{L} v_{\alpha\beta} - \frac{v_{0,MG}}{2L} u_{\alpha\beta}, \\ \dot{v}_{0,MG} = -\frac{1}{CR_L} v_{0,MG} + \frac{\phi_{\alpha\beta}^T}{2C} u_{\alpha\beta}, \end{cases} \tag{7.129}$$

and

$$\begin{cases} \dot{v}_{PV} = \frac{1}{C_{PV}} \left(\phi_{PV} - \phi_{L,PV} u_1 \right), \\ \dot{\phi}_{L,PV} = \frac{1}{L_{PV}} R_{0,PV} \left(\left(\phi_{0,PV} - \phi_{L,PV} \right) - R_{L,PV} \phi_{L,PV} - v_{0,PV} \right) \\ \quad + \frac{1}{L_{PV}} \left(V_{D,PV} + v_{PV} - R_{M,PV} \phi_{L,PV} \right) u_1 - \frac{V_{D,PV}}{L_{PV}}, \\ \dot{v}_{0,PV} = \frac{1}{C_{0,PV}} \left(\phi_{L,PV} - \phi_{0,PV} \right), \end{cases} \tag{7.130}$$

and

$$\begin{cases} \dot{v}_{dc} = \frac{1}{C_{pv}} \left(\phi_{dc} - \frac{1.5 u_d}{v_{dc}} \phi_d \right), \\ \dot{\phi}_d = -\frac{R_1}{L_1} \phi_d - \omega \phi_q + \frac{1}{L_1} e_d, \\ \dot{\phi}_q = \omega \phi_d - \frac{R_1}{L_1} \phi_q + \frac{1}{L_1} e_q. \end{cases} \tag{7.131}$$

Under the deficit power mode, the PMSG can be supplied to the DC load and the lead-acid battery. Based on the Kirchhoff's current law,

$$\phi_{0,MG} = \phi_{0,LA} + \frac{v_{0,MG} - \phi_{0,MG} R_{line,MG}}{R_{load}}. \tag{7.132}$$

The objective is to design a supervisory controller, so that the MPPT can be obtained and the output voltage tracks the reference v^* of MPPT can be calculated by (7.8), and $P \to P^*, Q \to Q^*, v_{0,MG} \to v_{ref}, v_{0,LA} \to v_{ref}$. Define $P_e = P - P^*, Q_e =$

$Q - Q^*, v_{e,MG} = v_{0,MG} - v_{ref}, v_{e,LA} = v_{0,LA} - v_{ref}$, it has

$$\begin{cases} \dot{\phi}_{1,LA} = \frac{1}{R_{1,LA}C_{1,LA}}\phi_{m,LA} - \frac{1}{R_{1,LA}C_{1,LA}}\phi_{1,LA}, \\ \dot{\phi}_{m,LA} = \frac{1}{R_{0,LA}C_{0,LA}}(\phi_{L,LA} - \phi_{m,LA}) - \frac{\dot{v}_{PN,LA}}{R_{0,LA}}, \\ \dot{\phi}_{L,LA} = \frac{1}{L_{LA}}R_{0,LA}(\phi_{m,LA} - \phi_{L,LA} - R_{L,LA}\phi_{L,LA} - (\phi_{m,LA}R_{0,LA} + v_{PN,LA})) \\ \quad + \frac{1}{L_{LA}}(V_{D,LA} + v_{0,MG} + -R_{M,LA}\phi_{L,LA})u_2 - \frac{V_{D,LA}}{L_{LA}}, \\ \dot{P}_e = \omega v_{\alpha\beta}^T J^T \phi_{\alpha\beta} + \frac{1}{L}v_{\alpha\beta}^T v_{\alpha\beta} - v_{\alpha\beta}^T \frac{v_{0,MG}}{2L}u_{\alpha\beta}, \\ \dot{Q}_e = \omega v_{\alpha\beta}^T J^T J\phi_{\alpha\beta} + \frac{v_{\alpha\beta}^T J}{L}v_{\alpha\beta} - \frac{v_{\alpha\beta}^T J v_{0,MG}}{2L}u_{\alpha\beta}, \\ \dot{v}_{\alpha\beta} = \omega J v_{\alpha\beta}, \\ \dot{\phi}_{\alpha\beta} = \frac{1}{L}v_{\alpha\beta} - \frac{v_{0,MG}}{2L}u_{\alpha\beta}, \\ \dot{v}_{e,MG} = -\frac{1}{CR_L}v_{e,MG} + \frac{\phi_{\alpha\beta}^T}{2C}u_{\alpha\beta} - \frac{1}{CR_L}v_{ref}, \\ \dot{v}_{PV} = \frac{1}{C_{PV}}(\phi_{PV} - \phi_{L,PV}u_1), \\ \dot{\phi}_{L,PV} = \frac{1}{L_{PV}}R_{0,PV}((\phi_{0,PV} - \phi_{L,PV}) - R_{L,PV}\phi_{L,PV} - v_{0,PV}) \\ \quad + \frac{1}{L_{PV}}(V_{D,PV} + v_{PV} - R_{M,PV}\phi_{L,PV})u_1 - \frac{V_{D,PV}}{L_{PV}}, \\ \dot{v}_{0,PV} = \frac{1}{C_{0,PV}}(\phi_{L,PV} - \phi_{0,PV}), \\ \dot{v}_{dc} = \frac{1}{C_{pv}}\left(\phi_{dc} - \frac{1.5u_d}{v_{dc}}\phi_d\right), \\ \dot{\phi}_d = -\frac{R_1}{L_1}\phi_d - \omega\phi_q + \frac{1}{L_1}e_d, \\ \dot{\phi}_q = \omega\phi_d - \frac{R_1}{L_1}\phi_q + \frac{1}{L_1}e_q. \end{cases} \quad (7.133)$$

If we define $x(t) = [\phi_{1,LA}\ \phi_{m,LA}\ \phi_{L,LA}\ P_e\ Q_e\ v_{\alpha\beta}\ \phi_{\alpha\beta}\ v_{e,MG}\ v_{PV}\ \phi_{L,PV}\ v_{e,PV}\ v_{dc}\ \phi_d\ \phi_q]^T$, and choose $\{v_{\alpha\beta}^T, \omega v_{\alpha\beta}^T, \omega, \frac{\phi_{PV}}{\phi_{L,PV}}, \frac{\phi_{0,PV}}{\phi_{L,PV}}, v_{0,MG}, \phi_{L,LA}, \phi_{\alpha\beta}^T, \phi_{L,PV}, v_{PV}, \frac{\phi_{dc}}{\phi_d}, \frac{u_d}{v_{dc}}, \bar{\omega}\}$ as the fuzzy premise variables, then the augmented fuzzy system is given by

Plant Rule $\mathscr{R}^l$: **IF** z_1 is $\mathscr{F}_1^l$ and z_2 is $\mathscr{F}_2^l$ and $\cdots$ and z_{13} is $\mathscr{F}_{13}^l$, **THEN**

$$\dot{x}(t) = A_l x(t) + B_l u(t) + \omega(t), l \in \mathscr{L} := \{1, 2, \ldots, r\} \quad (7.134)$$

where $\mathscr{R}^l$ denotes the l-th fuzzy inference rule; $\mathscr{L} := \{1, 2, \ldots, r\}$, and r is the number of inference rules; $\mathscr{F}_\theta^l(\theta = 1, 2, \cdots, 13)$ is the fuzzy set; $x(t) \in \Re^{n_x}$ and $u(t) \in \Re^{n_u}$ denote the system state and the control input, respectively; $z(t) \triangleq [z_1, z_2, \cdots, z_{13}]$ are the measurable variables; The l-th local model $\{A_l, B_l\}$ and the disturbance $\omega(t)$

are given as below:

$$A_l = \begin{bmatrix} A_{l(11)} & 0 & 0 \\ 0 & A_{l(22)} & 0 \\ 0 & 0 & A_{l(33)} \end{bmatrix}, A_{l(11)} = \begin{bmatrix} k_1 & k_2 & 0 & 0 & 0 & 0 & 0 & 0 \\ 0 & -k_3 & k_3 & 0 & 0 & 0 & 0 & 0 \\ 0 & k_4 & k_5 & 0 & 0 & 0 & 0 & 0 \\ 0 & 0 & 0 & 0 & 0 & \frac{\mathscr{F}_1^l}{L} & \mathscr{F}_2^l J^T & 0 \\ 0 & 0 & 0 & 0 & 0 & \frac{\mathscr{F}_1^l J}{L} & \mathscr{F}_2^l J^T J & 0 \\ 0 & 0 & 0 & 0 & 0 & \mathscr{F}_3^l J & 0 & 0 \\ 0 & 0 & 0 & 0 & 0 & \frac{1}{L} & 0 & 0 \\ 0 & 0 & 0 & 0 & 0 & 0 & 0 & \frac{-1}{CR_L} \end{bmatrix},$$

$$A_{l(22)} = \begin{bmatrix} 0 & \frac{\mathscr{F}_4^l}{C_{PV}} & 0 \\ 0 & k_6 & \frac{-R_{0,PV}}{L_{PV}} \\ 0 & \frac{1-\mathscr{F}_5^l}{C_{0,PV}} & 0 \end{bmatrix}, A_{l(33)} = \begin{bmatrix} 0 & \frac{\mathscr{F}_{11}^l}{C_{pv}} - \frac{1.5\mathscr{F}_{12}^l}{C_{pv}} & 0 \\ 0 & -\frac{R_1}{L_1} & -\mathscr{F}_{13}^l \\ 0 & \mathscr{F}_{13}^l & -\frac{R_1}{L_1} \end{bmatrix},$$

$$B_l = \begin{bmatrix} B_{l(1)} & 0 & 0 \\ 0 & B_{l(2)} & 0 \\ 0 & 0 & B_{l(3)} \end{bmatrix}, B_{l(1)} = \begin{bmatrix} 0 & 0 \\ 0 & 0 \\ k_7 & 0 \\ 0 & -\frac{\mathscr{F}_1^l \mathscr{F}_6^l}{2L} \\ 0 & -\frac{\mathscr{F}_1^l J \mathscr{F}_6^l}{2L} \\ 0 & -\frac{\mathscr{F}_6^l}{2L} \\ 0 & \frac{\mathscr{F}_8^l}{2C} \\ 0 & 0 \end{bmatrix}, B_{l(2)} = \begin{bmatrix} \frac{-\mathscr{F}_9^l}{C_{PV}} \\ k_8 \\ 0 \end{bmatrix},$$

$$B_{l(3)} = \begin{bmatrix} 0 & 0 \\ \frac{1}{L_1} & 0 \\ 0 & \frac{1}{L_1} \end{bmatrix}, \omega(t) = \begin{bmatrix} \omega_{(1)} \\ \omega_{(2)} \\ 0 \end{bmatrix}, \omega_{(1)} = \begin{bmatrix} 0 \\ -\frac{\dot{v}_{PN,LA}}{R_{0,LA}} \\ -\frac{R_{0,LA} v_{PN,LA} - V_{D,LA}}{L_{LA}} \\ 0 \\ 0 \\ 0 \\ 0 \\ -\frac{v_{ref}}{CR_L} \end{bmatrix},$$

$$\omega_{(2)} = \begin{bmatrix} 0 \\ -\frac{R_{0,PV} v_{ref} + V_{D,PV}}{L_{PV}} \\ 0 \end{bmatrix}, k_1 = \frac{-1}{R_{1,LA} C_{1,LA}}, k_2 = \frac{1}{R_{1,LA} C_{1,LA}}, k_3 = \frac{1}{R_{0,LA} C_{0,LA}},$$

$$k_4 = \frac{R_{0,LA} - R_{0,LA} R_{0,LA}}{L_{LA}}, k_5 = -\frac{R_{0,LA} + R_{0,LA} R_{L,LA}}{L_{LA}},$$

$$k_6 = \frac{R_{0,PV} \mathscr{F}_5^l - R_{0,PV} - R_{0,PV} R_{L,PV}}{L_{PV}},$$

$$k_7 = \frac{V_{D,LA} + \mathscr{F}_6^l + -R_{M,LA} \mathscr{F}_7^l}{L_{LA}}, k_8 = \frac{V_{D,PV} + \mathscr{F}_{10}^l - R_{M,PV} \mathscr{F}_9^l}{L_{PV}}. \quad (7.135)$$

7.7 REFERENCES

1. Kotra, S. and Mishra, M. (2017). A supervisory power management system for a hybrid microgrid with HESS, 64(5), 3640-3649.
2. Zolfaghari, M., Hosseinian, S., Fathi, S., Abedi, M., and Gharehpetian, G. (2018). A new power management scheme for parallel-connected PV systems in microgrids. IEEE Transactions on Sustainable Energy, doi: 10.1109/TSTE.2018.2799972.
3. Sangwongwanich, A., Abdelhakim, A., Yang, Y., and Zhou, K. (2018). Control of single-phase and three-phase DC/AC converters. In: Blaabjerg, F. (ed.) Control of Power Electronic Converters and Systems. London: Academic Press.
4. Wang, C., Li, X., Guo, L., and Li, Y. W. (2014). A nonlinear-disturbance-observer-based DC-bus voltage control for a hybrid AC/DC microgrid. IEEE Transactions on Power Electronics, 29(11), 6162-6177.

8 Optimization of Microgrids

Optimal operation of power systems requires intensive numerical analysis aimed at studying and improving system stability and control performance. To this aim, the switching impact among different subsystems, and system uncertainties associated with load and generation variations, especially from solar and wind power sources, have been extensively investigated [1, 2].

For a specific operating condition, PV systems usually have a unique maximum power point (MPP). It is noted that some intrinsic characteristics and disturbances on PV systems, such as aging, irradiance intensity and temperature conditions, generally lead to inefficient implementations on maximum power point tracking (MPPT) control [3]. Traditional MPPT control of PV systems aims to locate the MPP for online operation in steady state conditions, such as the perturbation and observe methods [4] and the hill climbing method [5], etc. Unfortunately, the above mentioned methods often result in a slower convergence. Recently, dynamic MPPT control methods have been proposed to improve the transient performance, such as cuckoo search control [6], neural network-based control [7], online system identification [8], etc. However, those methods generally lack strict convergence analysis.

Recently, nonlinear control techniques have been developed for wind turbines, such as variable structure control. This method is robust against parametric uncertainties, external disturbances, and unmodeled dynamics [9]. A reachable set of dynamic systems that can handle external disturbances is required to ensure power stability. Reachable estimation is also vital for ensuring robust control and also maintaining safe operations via controllers. More recently, engineers have recognized that finite time stable (FTS) systems provide two advantages: i) Faster convergence speed around the equilibrium point; ii) Better disturbance rejection performance [10]. Given a bound on the initial condition, system states do not exceed a certain threshold within a specified time interval. In that case, the system is said to be finite-time stable. In practice, compared with Lyapunov asymptotic stability (LAS), FTS is more suitable for analyzing the transient dynamics of the controlled system within a specified finite interval. Therefore, FTS is used to avoid excitations or saturations of nonlinear dynamics during the transients [11].

In this chapter, the renewable energy sources are considered as distributed connections to a common bus in a microgrid. To maintain power balances, the power management system decides the operating mode of energy systems based on the measured currents. To achieve system stability and maintain transient and steady performances in the microgrid with three switched modes, optimization of microgrid using reachable set estimation and finite-time control will be investigated in the framework of the switching system.

8.1 POWER MANAGEMENT STRATEGY

Renewable energy sources serve as distributed connections to a common bus in a microgrid. To maintain power balances, the power management system decides the operating mode of energy systems based on the measured currents. The power management system is formulated with following operational objectives.

1) Identify the operating mode of the system and make decisions based on the measured currents.

2) Achieve system stability in every operating mode.

3) Maintain transient and steady performances in every operating mode.

In the following, optimization of microgrid on transient and steady performances will be proposed, respectively.

8.2 TRANSIENT PERFORMANCE ANALYSIS

8.2.1 MPPT OPTIMAL ALGORITHM FOR SINGLE GENERATOR

Recall the MPPT modeling of the PV system in (1.27) as below:

$$\begin{cases} \dot{\phi}_L = \frac{1}{L}R_0\left((\phi_0-\phi_L)-R_L\phi_L-v_0\right)+\frac{1}{L}\left(V_D+v_{PV}-R_M\phi_L\right)u-\frac{V_D}{L}, \\ \dot{v}_0 = \frac{1}{C_0}\left(\phi_L-\phi_0\right), \\ \dot{e}_{pv} = \frac{1}{C_{PV}}\left(\phi_{PV}-\phi_L u\right)-\dot{v}^*_{pv}, \\ 0\dot{v}^*_{pv} = \phi_{PV}-n_p\gamma I_{rs}v^*_{pv}e^{\gamma v^*_{pv}}. \end{cases} \tag{8.1}$$

Here, choose $z_1=\frac{\phi_0}{\phi_L}, z_2=\phi_L, z_3=v_{pv}, z_4=\frac{\phi_{pv}}{e_{pv}}, z_5=\frac{\dot{v}^*_{pv}}{v^*_{pv}}$ and $z_6=e^{\gamma v^*_{pv}}$ as the fuzzy premise variables then the MPPT of the PV power fuzzy system in (1.27) is given by the following T-S model,

$$E\dot{x}(t)=A(\mu)x(t)+B(\mu)u(t)+D\omega(t), \tag{8.2}$$

where $A(\mu):=\sum_{l=1}^{r}\mu_l A_l, B(\mu):=\sum_{l=1}^{r}\mu_l B_l$,

$$E=\begin{bmatrix} 1&0&0&0\\ 0&1&0&0\\ 0&0&1&0\\ 0&0&0&0 \end{bmatrix}, A_l=\begin{bmatrix} \frac{R_0(\mathscr{F}_1^l-1-R_L)}{L} & -\frac{R_0}{L} & 0 & 0\\ \frac{1-\mathscr{F}_1^l}{C_0} & 0 & 0 & 0\\ 0 & 0 & \frac{\mathscr{F}_4^l}{C_{PV}} & -\mathscr{F}_5^l\\ 0 & 0 & \mathscr{F}_4^l & -n_p\gamma I_{rs}\mathscr{F}_6^l \end{bmatrix},$$

$$B_l=\begin{bmatrix} \frac{V_D+\mathscr{F}_3^l-R_M\mathscr{F}_2^l}{L}\\ 0\\ -\frac{\mathscr{F}_2^l}{C_{PV}}\\ 0 \end{bmatrix}, D=\begin{bmatrix} -1\\ 0\\ 0\\ 0 \end{bmatrix}, \omega(t)=\frac{V_D}{L}. \tag{8.3}$$

Consider a solar PV power system using the DC-DC boost converter as shown in (1.9) whose control problem is reformulated into the following descriptor system:

$$\begin{cases} \dot{\phi}_{pv} = -\frac{1}{L}(1-u)v_{dc} + \frac{1}{L}v_{pv}, \\ \dot{v}_{dc} = \frac{1}{C_0}(1-u)\phi_{pv} - \frac{1}{C_0}\phi_0, \\ 0 \cdot \dot{\varepsilon}_{pv} = \phi_{pv} - n_p\gamma I_{rs}e^{\gamma v^*_{pv}}\varepsilon_{pv} - n_p\gamma I_{rs}v_{pv}e^{\gamma v^*_{pv}}. \end{cases} \tag{8.4}$$

Define $x(t) = \begin{bmatrix} \phi_{pv} & v_{dc} & \varepsilon_{pv} \end{bmatrix}^T$, and choose $z_1 = \frac{v_{pv}}{\phi_{pv}}$, $z_2 = \frac{\phi_0}{v_{dc}}$, $z_3 = e^{\gamma v^*_{pv}}$, $z_4 = \frac{v_{pv}}{\phi_{pv}}e^{\gamma v^*_{pv}}$, $z_5 = v_{dc}$, and $z_6 = \phi_{pv}$ as fuzzy premise variables. The PV nonlinear system in (8.4) is represented by the following descriptor T-S model:

$$E\dot{x}(t) = A(\mu)x(t) + B(\mu)u(t), \tag{8.5}$$

where $A(\mu) := \sum_{l=1}^{r} \mu_l A_l, B(\mu) := \sum_{l=1}^{r} \mu_l B_l,$

$$A_l = \begin{bmatrix} \frac{\mathscr{F}_1^l}{L} & -\frac{1}{L} & 0 \\ \frac{1}{C_0} & -\frac{\mathscr{F}_2^l}{C_0} & 0 \\ 1 - n_p\gamma I_{rs}\mathscr{F}_4^l & 0 & -n_p\gamma I_{rs}\mathscr{F}_3^l \end{bmatrix},$$

$$E = \begin{bmatrix} 1 & 0 & 0 \\ 0 & 1 & 0 \\ 0 & 0 & 0 \end{bmatrix}, B_l = \begin{bmatrix} \frac{\mathscr{F}_5^l}{L} \\ -\frac{\mathscr{F}_6^l}{C_0} \\ 0 \end{bmatrix}. \tag{8.6}$$

Then, we consider, without loss of generality, only the class of norm-bounded square integrable disturbance that acts on the output voltage v_{dc} over the time interval $[t_1, t_2]$, which is defined as below:

$$\mathbb{W}_{[t_1,t_2],\delta} \triangleq \left\{ \omega \in \mathscr{L}_2[t_1,t_2] : \int_{t_1}^{t_2} \omega^2(s)ds \leq \delta \right\}, \tag{8.7}$$

where δ is a positive scalar.

Consider a fuzzy controller, which shares the same premise variables in (8.2) as follows:

Controller Rule $\mathscr{R}^l$: **IF** z_1 is $\mathscr{F}_1^l$ and z_2 is $\mathscr{F}_2^l$ and $\cdots$ and z_7 is $\mathscr{F}_7^l$, **THEN**

$$u(t) = K_l x(t), l \in \mathscr{L} \tag{8.8}$$

where $K_l \in \Re^{n_u \times n_x}$ are controller gains to be designed.

Likewise, the global fuzzy controller is given by

$$u(t) = K(\mu)x(t), \tag{8.9}$$

where $K(\mu) := \sum_{l=1}^{r} \mu_l K_l$.

By submitting the controller into the fuzzy system (8.2),

$$E\dot{x}(t) = \bar{A}(\mu)x(t) + D\omega(t), \tag{8.10}$$

where $\bar{A}(\mu) = A(\mu) + B(\mu)K(\mu)$.

Before moving on, we extend the definition of the finite time boundedness (FTB) in [12] to the descriptor fuzzy system (8.2) as follows:

Definition 8.1. For a given time interval $[t_1, t_2]$, two scalars c_1, c_2 subject to $0 < c_1 < c_2$, and a symmetrical matrix $R > 0$, the descriptor fuzzy system (8.2) with $u(t) = 0$ is said to be the FTB with respect to $(c_1, c_2, [t_1, t_2], R, \mathrm{W}_{[t_1,t_2],\delta})$, if it satisfies

$$x^T(t_1)E^T REx(t_1) \le c_1 \Longrightarrow x^T(t_2)E^T REx(t_2) < c_2, \forall t \in [t_1, t_2],$$

for all $\omega(t) \in \mathrm{W}_{[t_1,t_2],\delta}$.

In the following theorem, we derive a sufficient condition for the FTB of closed-loop system (8.10) in finite time interval $[0, T^*]$.

Theorem 8.1: Finite-Time Controller Design

Consider the fuzzy law (8.9). The resulting closed-loop control system in (8.10) is the FTB with respect to $(c_1, c^*, [0, T^*], R, \mathrm{W}_{[0,T^*],\delta})$, if there exist matrices $X = \begin{bmatrix} X_{(1)} & 0 \\ X_{(2)} & X_{(3)} \end{bmatrix}$, $0 < X_{(1)} = X_{(1)}^T \in \Re^{(n_x-1)\times(n_x-1)}, X_{(2)} \in \Re^{1\times(n_x-1)}$, $X_{(3)}$ is a scalar, $\bar{K}_l \in \Re^{n_u \times n_x}$, and the positive scalars η, such that the following LMIs hold:

$$\Phi_{ll} < 0, 1 \le l \le r \tag{8.11}$$

$$\Phi_{lp} + \Phi_{pl} < 0, 1 \le l < p \le r \tag{8.12}$$

where

$$\Phi_{lp} = \begin{bmatrix} \mathrm{Sym}(A_l X + B_l \bar{K}_p) - \eta EX & D \\ \star & -\eta I \end{bmatrix}. \tag{8.13}$$

Furthermore, the controller gains can be calculated by

$$\bar{K}_p = \bar{K}_p X^{-1}. \tag{8.14}$$

■

Proof. Consider the following Lyapunov functional

$$V(t) = x^T(t)E^T Px(t), \forall t \in [0, T^*]. \tag{8.15}$$

It is easy to see from (8.19) that $E^T P = P^T E \ge 0$.

Along the trajectory of system (8.10), we have

$$\begin{aligned}\dot{V}(t) &= 2\left[E\dot{x}(t)\right]^T Px(t)\\ &= 2\left[\bar{A}(\mu)x(t)+D\omega(t)\right]^T Px(t).\end{aligned} \tag{8.16}$$

An auxiliary function is introduced as below:

$$J(t) = \dot{V}(t) - \eta V(t) - \eta\omega^2(t), \tag{8.17}$$

where η is a positive scalar.

It follows from (8.16) and (8.17) that

$$\begin{aligned}J(t) &= 2\left[\bar{A}(\mu)x(t)+D\omega(t)\right]^T Px(t) - \eta x^T(t)E^T Px(t) - \eta\omega^2(t)\\ &= \chi^T(t)\Phi(\mu)\chi(t),\end{aligned} \tag{8.18}$$

where $\chi(t) = \left[\begin{array}{cc} x^T(t) & \omega^T(t)\end{array}\right]^T$, and $\Phi(\mu) = \left[\begin{array}{cc}\mathrm{Sym}\left(P^T\bar{A}(\mu)\right) - \eta P^T E & P^T D\\ \star & -\eta I\end{array}\right]$.

It is easy to see that $\Phi(\mu) < 0$ implies $J(t) < 0$. Now, define

$$P = \left[\begin{array}{cc} P_{(1)} & 0\\ P_{(2)} & P_{(3)}\end{array}\right], \tag{8.19}$$

where $0 < P_{(1)} = P_{(1)}^T \in \Re^{(n_x-1)\times(n_x-1)}, P_{(2)} \in \Re^{1\times(n_x-1)}$, $P_{(3)}$ is a scalar. It is easy to see that the inequality $E^T P = P^T E \geq 0$ holds.

In order to derive an LMI-based result, we define $X = P^{-1}$, that is

$$X = \left[\begin{array}{cc} X_{(1)} & 0\\ X_{(2)} & X_{(3)}\end{array}\right], \tag{8.20}$$

where $0 < X_{(1)} = X_{(1)}^T \in \Re^{(n_x-1)\times(n_x-1)}, X_{(2)} \in \Re^{1\times(n_x-1)}$, $X_{(3)}$ is a scalar.

By performing a congruence transformation to $\Phi(\mu) < 0$ by $\Gamma = \mathrm{diag}\left\{X^T, I\right\}, X = P^{-1}$, we have

$$\left[\begin{array}{cc}\mathrm{Sym}\left(\bar{A}(\mu)X\right) - \eta EX & D\\ \star & -\eta I\end{array}\right] < 0. \tag{8.21}$$

By defining $\bar{K}(\mu) = K(\mu)X$, and extracting the fuzzy premise variables, the inequality in (8.11) and (8.12) can be obtained.

Due to $J(t) < 0$, it has

$$\dot{V}(t) < \eta V(t) + \eta\omega^2(t). \tag{8.22}$$

By multiplying both sides of (8.22) by $e^{-\eta t}$ and integrating the resulting inequality from 0 to t with $t \in [0, T^*]$,

$$\begin{aligned}e^{-\eta t}V(t) &< V(0) + \eta\int_0^t e^{-\eta s}\omega^2(s)ds\\ &\leq x^T(0)E^T Px(0) + \eta\delta.\end{aligned} \tag{8.23}$$

On the other hand, it follows from (8.15) that

$$e^{-\eta t}V_2(t) \geq e^{-\eta T^*}x^T(t)E^T Px(t), \tag{8.24}$$

which implies that

$$\begin{aligned} e^{-\eta T^*}x^T(t)E^T Px(t) < x^T(0)E^T Px(0) + \eta T^*\left(1+\varepsilon^2\right)\bar{\rho}^2 \\ + \eta\left(1+\varepsilon^2\right)\delta + \eta\delta. \end{aligned} \tag{8.25}$$

Now, we partition $x(t)$ as

$$x(t) = \begin{bmatrix} \bar{x}(t) \\ x_3(t) \end{bmatrix}, \tag{8.26}$$

where $\bar{x}(t) = \begin{bmatrix} x_1(t) \\ x_2(t) \end{bmatrix}$.

It follows from (8.22)-(8.26) that

$$e^{-\eta T^*}\bar{x}^T(t)P_1\bar{x}(t) < \bar{x}^T(0)P_1\bar{x}(0) + \eta\delta, \tag{8.27}$$

where P_1 is defined in (8.19).

Now, we introduce the matrix $0 < R_1^T = R_1 \in \Re^{(n_x-1)\times(n_x-1)}$, and further define

$$\begin{aligned} &c_1 = \bar{x}^T(0)R_1\bar{x}(0), \\ &\bar{\sigma}_{P1} = \lambda_{\max}\left(R_1^{-\frac{1}{2}}P_1R_1^{-\frac{1}{2}}\right), \underline{\sigma}_{P1} = \lambda_{\min}\left(R_1^{-\frac{1}{2}}P_1R_1^{-\frac{1}{2}}\right). \end{aligned} \tag{8.28}$$

Based on the relationships (8.27) and (8.28), we have

$$\bar{x}^T(t)R_1\bar{x}(t) < \frac{\bar{\sigma}_{P1}c_1 + \eta\delta}{e^{-\eta T^*}\underline{\sigma}_{P1}}. \tag{8.29}$$

From the Definition 8.1, the descriptor fuzzy system in (8.10) is the FTB. This completes the proof.

It follows from (8.28) and (8.29) that

$$\|\bar{x}(t)\| < \frac{\bar{\sigma}_{P1}c_1 + \eta\delta}{e^{-\eta T^*}\underline{\sigma}_{P1}\lambda_{\min}(R_1)}, t \in [0, T^*]. \tag{8.30}$$

8.2.2 MPPT OPTIMAL ALGORITHM FOR MULTI-MACHINE GENERATORS

Consider an interconnected multi-PV generator with the DC load. Based on the Thevenin's theorem,

$$v_{0(i)} = \phi_{0(1)}R_{load} + \phi_{0(2)}R_{load} + \cdots + \phi_{0(N)}R_{load}, \tag{8.31}$$

where the subscript i denotes the i-th subsystem, $i \in \mathscr{N} := \{1,2,\ldots,N\}$, N denotes the number of subsystems, R_{load} is the load resistance, $\phi_{0(i)}$ is the line current of the i-th subsystem.

It follows from (8.1) and (8.31) that

$$\begin{cases} \dot{\phi}_{L(i)} = \frac{R_{0(i)}}{L_{(i)}}\left(\left(\phi_{0(i)} - \phi_{L(i)}\right) - R_{L(i)}\phi_{L(i)} - v_{0(i)}\right) + \frac{V_{D(i)}+v_{PV(i)}-R_{M(i)}\phi_{L(i)}}{L_{(i)}}u_{(i)} - \frac{V_{D(i)}}{L_{(i)}}, \\ \dot{v}_{0(i)} = \frac{1}{C_{0(i)}}\left(\phi_{L(i)} - \frac{v_{0(i)}}{R_{load}} + \sum_{\substack{j=1 \\ j\neq i}}^{N} \phi_{0(j)}\right), \\ \dot{e}_{pv(i)} = \frac{1}{C_{PV(i)}}\left(\phi_{PV(i)} - \phi_{L(i)}u_{(i)}\right) - \dot{v}^*_{pv(i)}, \\ 0\dot{v}^*_{pv(i)} = \phi_{PV(i)} - n_{p(i)}\gamma_{(i)}I_{rs(i)}v^*_{pv(i)}e^{\gamma_{(i)}v^*_{pv(i)}}. \end{cases} \tag{8.32}$$

Here, choose $z_{i1} = \frac{\phi_{0(i)}}{\phi_{L(i)}}, z_{i2} = \phi_{L(i)}, z_{i3} = v_{pv(i)}, z_{i4} = \frac{\phi_{pv(i)}}{e_{pv(i)}}, z_{i5} = \frac{\dot{v}^*_{pv(i)}}{v^*_{pv(i)}}$ and $z_{i6} = e^{\gamma v^*_{pv(i)}}$ as the fuzzy premise variables, and define $x_i(t) = [\phi_{L(i)}\ v_{0(i)}\ e_{pv(i)}\ v^*_{pv(i)}]^T$, then the MPPT of multi-PV generator is given by the following T-S model,

Plant Rule $\mathscr{R}_i^l$: IF z_{1i} is $\mathscr{F}_{i1}^l$, z_{2i} is $\mathscr{F}_{i2}^l$, $\cdots$, $z_{i6}(t)$ is $\mathscr{F}_{i6}^l$, **THEN**

$$E\dot{x}_i(t) = A_{iil}x_i(t) + B_{il}u_i(t) + \sum_{\substack{j=1 \\ j\neq i}}^{N} A_{ijl}x_j(t) + \omega_i(t), l \in \mathscr{L}_i := \{1,2,\ldots,r_i\} \tag{8.33}$$

where $\mathscr{R}_i^l$ denotes the l-th fuzzy inference rule; r_i is the number of inference rules; $z_i(t) \triangleq [z_{i1}, z_{i2}, \cdots, z_{i6}]$ are the measurable variables; $\{A_{iil}, B_{il}, A_{ijl}\}$ is the l-th local model as below:

$$A_{iil} = \begin{bmatrix} \frac{R_{0(i)}\left(\mathscr{F}_{i1}^l - 1 - R_{L(i)}\right)}{L} & \frac{-R_{0(i)}}{L} & 0 & 0 \\ \frac{1}{C_{0(i)}} & \frac{-1}{R_{load}C_{0(i)}} & 0 & 0 \\ 0 & 0 & \frac{\mathscr{F}_{i4}^l}{C_{PV(i)}} & -\mathscr{F}_{i5}^l \\ 0 & 0 & \mathscr{F}_{i4}^l & -n_{p(i)}\gamma_{(i)}I_{rs(i)}\mathscr{F}_{i6}^l \end{bmatrix},$$

$$A_{ijl} = \begin{bmatrix} 0 & 0 & 0 & 0 \\ \frac{\mathscr{F}_{j1}^l}{C_{0(i)}} & 0 & 0 & 0 \\ 0 & 0 & 0 & 0 \\ 0 & 0 & 0 & 0 \end{bmatrix}, B_{il} = \begin{bmatrix} \frac{V_{D(i)} + \mathscr{F}_{i3}^l - R_{M(i)}\mathscr{F}_{i2}^l}{L} \\ 0 \\ -\frac{\mathscr{F}_{i2}^l}{C_{PV(i)}} \\ 0 \end{bmatrix},$$

$$E = \begin{bmatrix} 1 & 0 & 0 & 0 \\ 0 & 1 & 0 & 0 \\ 0 & 0 & 1 & 0 \\ 0 & 0 & 0 & 0 \end{bmatrix}, D_i = \begin{bmatrix} -1 \\ 0 \\ 0 \\ 0 \end{bmatrix}, \omega_i(t) = \frac{V_{D(i)}}{L_{(i)}}. \tag{8.34}$$

By fuzzy blending,

$$E\dot{x}_i(t) = A_{ii}(\mu_i)x_i(t) + B_i(\mu_i)u_i(t) + \sum_{\substack{j=1 \\ j\neq i}}^{N} A_{ij}(\mu_j)x_j(t) + \omega_i(t), \tag{8.35}$$

where $A_i(\mu_i) := \sum_{l=1}^{r_i} \mu_{il}A_{il}, B_i(\mu_i) := \sum_{l=1}^{r_i} \mu_{il}B_{il}, A_{ij}(\mu_j) := \sum_{l=1}^{r_j} \mu_{jl}A_{ijl}$.

Consider a decentralized fuzzy controller, which shares the same premise variables in (8.33), as follows:

Controller Rule $\mathscr{R}_i^l$: **IF** z_{1i} is $\mathscr{F}_{i1}^l$, z_{2i} is $\mathscr{F}_{i2}^l$, $\cdots$, $z_{i6}(t)$ is $\mathscr{F}_{i6}^l$, **THEN**

$$u_i(t) = K_{iil}x_i(t), l \in \mathscr{L} \tag{8.36}$$

where $K_{iil} \in \Re^{n_{ui}\times n_{xi}}$ are the controller gains to be designed.

Likewise, the global T-S fuzzy controller is given by

$$u(t) = K_{ii}(\mu_i)x(t), \tag{8.37}$$

where $K_{ii}(\mu_i) := \sum_{l=1}^{r_i} \mu_{il}K_{iil}$.

By submitting the controller into the fuzzy system (8.33), we get

$$E\dot{x}_i(t) = \bar{A}_{ii}(\mu_i)x_i(t) + \sum_{\substack{j=1 \\ j\neq i}}^{N} A_{ij}(\mu_j)x_j(t) + \omega_i(t), \tag{8.38}$$

where $\bar{A}_{ii}(\mu_i) = A_{ii}(\mu_i)x_i(t) + B_i(\mu_i)K_{ii}(\mu_i)$.

Then, we define $\omega(t) = \begin{bmatrix} \omega_1^T(t) & \omega_2^T(t) & \cdots & \omega_N^T(t) \end{bmatrix}^T$, and consider, without loss of generality, only the class of norm-bounded square integrable disturbance that acts on the output voltage v_{dc} over the time interval $[t_1, t_2]$, which is defined as below:

$$\mathbb{W}_{[t_1,t_2],\delta} \triangleq \left\{ \omega \in L_2[t_1,t_2] : \int_{t_1}^{t_2} \omega_i^T(s)\,\omega_i(s)\,ds \leq \delta_i \right\}, \tag{8.39}$$

where is a known positive scalar, and $\delta = \sum_{i=1}^{N} \delta_i$.

The following theorem presents sufficient conditions for guaranteeing the FTB of the closed-loop system in (8.38) as below:

Theorem 8.2: Finite-Time Decentralized Controller Design

Consider the positive scalars $\{b_1, b_2, T, \phi, \sigma_i, \beta\}$, the positive-definite symmetric matrices $\{R, M_i\}$, and the matrices $\{\bar{K}_{ii}, L_i\}$. The closed-loop system in (8.38) is the

FTB with respect to $(b_1, b_2, [0,T], R, \mathbb{W}_{[0,T],\phi})$, if the following LMIs for all p hold:

$$\Phi_{llp} < 0, 1 \leq l \leq r \tag{8.40}$$

$$\Phi_{ls} + \Phi_{slp} < 0, 1 \leq l < s \leq r \tag{8.41}$$

where

$$\Phi_{lsp} = \begin{bmatrix} \text{Sym}(A_{iil}X_i + B_{il}\bar{K}_{iis}) + A_{ijp}M_iA_{ijp}^T - \eta EX_i & I & X_i^T \\ \star & -\eta I & 0 \\ \star & \star & -\frac{M_0}{N-1} \end{bmatrix}. \tag{8.42}$$

Furthermore, the corresponding controller gains are given by

$$K_{iil} = \bar{K}_{iil}X_i^{-1}. \tag{8.43}$$

■

Proof. Consider the following Lyapunov functional

$$V(t) = \sum_{i=1}^{N} x_i^T(t)E^T P_i x_i(t), \forall t \in [0, T^*]. \tag{8.44}$$

It is easy to see from (8.50) that $E^T P_i = P_i^T E \geq 0$.

Along the trajectory of the system (8.38), one gets

$$\begin{aligned} \dot{V}(t) &= \sum_{i=1}^{N} 2\left[E\dot{x}_i(t)\right]^T P_i x_i(t) \\ &= \sum_{i=1}^{N} 2\left[\bar{A}_{ii}(\mu_i)x_i(t) + \sum_{\substack{j=1 \\ j \neq i}}^{N} A_{ij}(\mu_j)x_j(t) + \omega_i(t)\right]^T P_i x_i(t). \end{aligned} \tag{8.45}$$

Note that

$$2\bar{x}^T\bar{y} \leq \kappa^{-1}\bar{x}^T\bar{x} + \kappa\bar{y}^T\bar{y}, \tag{8.46}$$

where $\{\bar{x}, \bar{y}\} \in \Re^n$ and the scalar $\kappa > 0$. Define the matrix $0 < M_i = M_i^T$,

$$\begin{aligned} &\sum_{i=1}^{N} \sum_{j=1, i \neq j}^{N} 2x_j^T(t)A_{ij}^T(\mu_j)P_i x_i(t) \\ &\leq \sum_{i=1}^{N} \sum_{j=1, i \neq j}^{N} x_j^T(t)M_i^{-1}x_j(t) + \sum_{i=1}^{N} \sum_{j=1, i \neq j}^{N} x_i^T(t)P_i^T A_{ij}(\mu_j)M_iA_{ij}^T(\mu_j)P_i x_i(t) \\ &= \sum_{i=1}^{N} \sum_{j=1, i \neq j}^{N} x_i^T(t)M_j^{-1}x_i(t) + \sum_{i=1}^{N} \sum_{j=1, i \neq j}^{N} x_i^T(t)P_i^T A_{ij}(\mu_j)M_iA_{ij}^T(\mu_j)P_i x_i(t). \end{aligned} \tag{8.47}$$

An auxiliary function is introduced as below:

$$J(t) = \dot{V}(t) - \eta V(t) - \eta \omega^T(t)\omega(t), \tag{8.48}$$

where η is a positive scalar.

Define $\chi_i(t) = \begin{bmatrix} x_i^T(t) & \omega_i^T(t) \end{bmatrix}^T$, and it follows from (8.44)-(8.48) that

$$\begin{aligned} J(t) \leq & \sum_{i=1}^{N} 2\left[\bar{A}_{ii}(\mu_i)x_i(t) + \omega_i(t)\right]^T P_i x_i(t) + \sum_{i=1}^{N} \sum_{j=1, i\neq j}^{N} x_i^T(t) M_j^{-1} x_i(t) \\ & + \sum_{i=1}^{N} \sum_{j=1, i\neq j}^{N} x_i^T(t) P_i^T A_{ij}(\mu_j) M_i A_{ij}^T(\mu_j) P_i x_i(t) \\ & - \eta \sum_{i=1}^{N} x_i^T(t) E^T P_i x_i(t) - \eta \sum_{i=1}^{N} \omega_i^T(t)\omega_i(t) \\ & = \chi_i^T(t)\Phi_i(\mu_i)\chi_i(t), \end{aligned} \tag{8.49}$$

where $\Phi_i(\mu_i) = \begin{bmatrix} \Phi_{i(1)}(\mu_i) & P_i^T \\ \star & -\eta I \end{bmatrix}$, $\Phi_{i(1)}(\mu_i) = \text{Sym}\left(P_i^T \bar{A}_{ii}(\mu_i)\right) + P_i^T A_{ij}(\mu_j) M_i(\star) + \sum\limits_{j=1, i\neq j}^{N} M_j^{-1} - \eta P_i^T E$.

It is easy to see that $\Phi_i(\mu_i) < 0$ implies $J(t) < 0$. Now, define

$$P_i = \begin{bmatrix} P_{i1} & 0 \\ P_{i2} & P_{i3} \end{bmatrix}, \tag{8.50}$$

where $0 < P_{i1} = P_{i1}^T \in \Re^{(n_{xi}-1)\times(n_{xi}-1)}$, $P_{i2} \in \Re^{1\times(n_{xi}-1)}$, P_{i3} is a scalar. It is easy to see that the inequality $E^T P_i = P_i^T E \geq 0$ holds.

In order to derive an LMI-based result, we define $X_i = P_i^{-1}$, that is

$$X_i = \begin{bmatrix} X_{i1} & 0 \\ X_{i2} & X_{i3} \end{bmatrix}, \tag{8.51}$$

where $0 < X_{i1} = X_{i1}^T \in \Re^{(n_{xi}-1)\times(n_{xi}-1)}$, $X_{i2} \in \Re^{1\times(n_{xi}-1)}$, X_{i3} is a scalar.

By performing a congruence transformation to $\Phi_i(\mu_i) < 0$ by $\Gamma_i = \text{diag}\left\{X_i^T, I\right\}$, $X_i = P_i^{-1}$, we have

$$\begin{bmatrix} \text{Sym}\left(\bar{A}_{ii}(\mu_i)X_i\right) + A_{ij}(\mu_j)M_i A_{ij}^T(\mu_j) + \sum\limits_{j=1, i\neq j}^{N} X_i^T M_j^{-1} X_i - \eta E X_i & I \\ \star & -\eta I \end{bmatrix} < 0. \tag{8.52}$$

Define $\bar{K}_{ii}(\mu_i) = K_{ii}(\mu_i)X_i$ and $M_0 \leq M_i$; then $\sum\limits_{j=1, i\neq j}^{N} M_j^{-1} \leq (N-1)M_0^{-1}$. The following matrix inequality implies (8.52),

$$\begin{bmatrix} \text{Sym}\left(\bar{A}_{ii}(\mu_i)X_i\right) + A_{ij}(\mu_j)M_i A_{ij}^T(\mu_j) - \eta E X_i & I & X_i^T \\ \star & -\eta I & 0 \\ \star & \star & -\frac{M_0}{N-1} \end{bmatrix} < 0. \tag{8.53}$$

By extracting the fuzzy premise variables, the inequality in (8.40) and (8.41) can be obtained.

Due to $J(t) < 0$,

$$\dot{V}(t) < \eta V(t) + \eta \omega^T(t)\omega(t). \tag{8.54}$$

By multiplying both sides of (8.54) by $e^{-\eta t}$ and integrating the resulting inequality from 0 to t with $t \in [0, T^*]$, it is easy to see that

$$\begin{aligned} e^{-\eta t}V(t) &< V(0) + \eta \int_0^t e^{-\eta s}\omega^T(s)\omega(s)ds \\ &\leq x^T(0)\mathscr{E}^T\mathscr{P}x(0) + \eta\delta, \end{aligned} \tag{8.55}$$

where $\mathscr{E} = \text{diag}\underbrace{\{E, \cdots, E\}}_{N}$, $\mathscr{P} = \text{diag}\underbrace{\{P_1, \cdots, P_N\}}_{N}$, $x(t) = \left[x_1^T(t)\; x_2^T(t)\; \cdots\; x_N^T(t)\right]^T$.

On the other hand, it follows from (8.55) that

$$e^{-\eta t}V(t) \geq e^{-\eta T^*}x^T(t)\mathscr{E}^T\mathscr{P}x(t), \tag{8.56}$$

which implies that

$$e^{-\eta T^*}x^T(t)\mathscr{E}^T\mathscr{P}x(t) < x^T(0)\mathscr{E}^T\mathscr{P}x(0) + \eta\delta. \tag{8.57}$$

Now, we partition $x_i(t)$ as

$$x_i(t) = \begin{bmatrix} \bar{x}_i(t) \\ x_{i3}(t) \end{bmatrix}, \tag{8.58}$$

where $\bar{x}_i(t) = \begin{bmatrix} x_{i1}(t) \\ x_{i2}(t) \end{bmatrix}$.

Define $\bar{x}(t) = \begin{bmatrix} \bar{x}_1^T(t) & \bar{x}_2^T(t) & \cdots & \bar{x}_N^T(t) \end{bmatrix}^T$ and $\mathscr{P}_1 = \text{diag}\underbrace{\{P_{11}, \cdots, P_{N1}\}}_{N}$. It follows from (8.56)-(8.58) that

$$e^{-\eta T^*}\bar{x}^T(t)\mathscr{P}_1\bar{x}(t) < \bar{x}^T(0)\mathscr{P}_1\bar{x}(0) + \eta\delta. \tag{8.59}$$

We introduce the matrix $0 < \mathscr{R}_1^T = \mathscr{R}_1 \in \Re^{Nn_2 \times Nn_2}$, and further define

$$\begin{aligned} &c_1 = \bar{x}^T(0)\mathscr{R}_1\bar{x}(0), \\ &\bar{\sigma}_{\mathscr{P}1} = \lambda_{\max}\left(\mathscr{R}_1^{-\frac{1}{2}}\mathscr{P}_1\mathscr{R}_1^{-\frac{1}{2}}\right), \underline{\sigma}_{\mathscr{P}1} = \lambda_{\min}\left(\mathscr{R}_1^{-\frac{1}{2}}\mathscr{P}_1\mathscr{R}_1^{-\frac{1}{2}}\right). \end{aligned} \tag{8.60}$$

Based on the relationships (8.59) and (8.60), we have

$$\bar{x}^T(t)\mathscr{R}_1\bar{x}(t) < \frac{\bar{\sigma}_{\mathscr{P}1}c_1 + \eta\delta}{e^{-\eta T^*}\underline{\sigma}_{\mathscr{P}1}}. \tag{8.61}$$

From Definition 8.1, the descriptor fuzzy system in (8.38) is the FTB. This completes the proof.

It follows from (8.60) and (8.61) that

$$\|\bar{x}(t)\| < \frac{\bar{\sigma}_{\mathscr{P}1}c_1 + \eta\delta}{e^{-\eta T^*}\underline{\sigma}_{\mathscr{P}1}\lambda_{\min}(\mathscr{R}_1)}, t \in [0, T^*]. \tag{8.62}$$

8.2.3 OPTIMAL ALGORITHM FOR MULTI-MODE OPERATION

Consider these three operation modes for the deficit power mode, the floating power mode, and the excess power mode described in Chapter 7. By using a three-switch operation, the system dynamics can be considered as below:

$$\dot{x}(t) = A_i(\mu_i)x(t) + B_i(\mu_i)u(t) + \omega_i(t), i \in \mathscr{I} := \{1,2,3\} \tag{8.63}$$

where $x(t) \in \Re^{n_x}$ and $u(t) \in \Re^{n_u}$ denote the system state and control input, respectively; $\mathscr{I}$ is the set of switched subsystems; $A_i(\mu_i) := \sum_{l=1}^{r_i} \mu_{il}[z_i(t)]A_{il}, B_i(\mu_i) := \sum_{l=1}^{r_i} \mu_{il}[z_i(t)]B_{il}, \mu_{il}[z_i(t)] := \frac{\Pi_{\phi=1}^{g}\mu_{il\phi}[z_{i\phi}(t)]}{\Sigma_{\varsigma=1}^{r_i}\Pi_{\phi=1}^{g}\mu_{i\varsigma\phi}[z_{i\phi}(t)]} \geq 0, \sum_{l=1}^{r_i} \mu_{il}[z_i(t)] = 1$. For simplicity, we define $\mu_{il} = \mu_{il}[z_i(t)]$.

Consider a switching fuzzy controller as below [15, 16]:

$$u_i(t) = K_i(\mu_i)x(t), i \in \mathscr{I} \tag{8.64}$$

where $K_i(\mu_i) := \sum_{l=1}^{r_i} \mu_{il}K_{il}$, $K_{il} \in \Re^{n_{ui} \times n_{xi}}$ is the controller gain matrix to be determined.

Submitting the controller (8.64) into the system (8.63), the i-th closed-loop system can be represented as

$$\dot{x}(t) = \bar{A}_i(\mu_i)x(t) + \omega_i(t), i \in \mathscr{I} \tag{8.65}$$

where $\bar{A}_i(\mu_i) = A_i(\mu_i) + B_i(\mu_i)K_i(\mu_i)$.

Then, we consider, without loss of generality, only the class of norm-bounded square integrable disturbance that acts on the output voltage v_{dc} over the time interval $[t_1, t_2]$, which is defined as below:

$$\mathbb{W}_{[t_1,t_2],\delta} \triangleq \left\{\omega_i \in L_2[t_1,t_2] : \int_{t_1}^{t_2} \omega_i^T(s)\omega_i(s)ds \leq \delta\right\}, \tag{8.66}$$

where δ is a positive scalar.

The aim is to design the fuzzy controller for the switching system in (8.63) such that the resulting closed-loop control system is the finite time boundedness (FTB) with respect to the specified parameters $(c_1, c_2, [0,T], R, \mathbb{W}_{[0,T],\phi})$. First, the definition of the FTB for nonlinear systems generalized from [13] is recalled as below:

Definition 8.2 [13]. Consider the following nonlinear dynamic system:

$$\dot{x}(t) = f(x,u,\omega).$$

Consider a time interval $[0,T]$, two positive scalars a_1, a_2, with $a_1 < a_2$, and a weighted matrix $R > 0$. The system in (8.63) with $u(t) = 0$ is said to be the FTB with respect to $(a_1, a_2, [0,T], R, \mathbb{W}_{[0,T],\phi})$, if

$$x^T(0)Rx(0) \leq a_1 \Longrightarrow x^T(t)Rx(t) < a_2, \forall t \in [0,T],$$

for all $\omega(t) \in \mathbb{W}_{[0,T],\phi}$.

The following theorem presents sufficient conditions for guaranteeing the FTB of the closed-loop system in (8.65), which is proposed as below:

Theorem 8.3: Finite-Time Switched Controller Design

Consider the positive scalars $\{b_1, b_2, T, \phi, \sigma_i, \beta\}$, and the positive-definite symmetric matrices $\{R, M_i\}$, and the matrices $\{\bar{K}_{ii}, L_i\}$. The fuzzy system in (8.63) is the FTB with respect to $(b_1, b_2, [0,T], R, \mathbb{W}_{[0,T],\phi})$, if the following LMIs hold:

$$\Phi_{ll} < 0, 1 \le l \le r \tag{8.67}$$

$$\Phi_{ls} + \Phi_{sl} < 0, 1 \le l < s \le r \tag{8.68}$$

where

$$\Phi_{ls} = \begin{bmatrix} \mathrm{Sym}\left(A_{il}X + B_{il}\bar{K}_{is}\right) - \eta X & I \\ \star & -\eta I \end{bmatrix}. \tag{8.69}$$

Then, the closed-loop system in (8.65) is the FTB with respect to (b_1, b_2, $[0,T]$, R, $\mathbb{W}_{[0,T],\phi}$). Furthermore, the corresponding controller gains are given by

$$K_{il} = \bar{K}_{il} X^{-1}. \tag{8.70}$$

■

Proof. Consider the following Lyapunov functional

$$V(t) = x^T(t) P x(t), \forall t \in [0, T^*] \tag{8.71}$$

where P is a symmetric positive definite matrix.

Along the trajectory of the system (8.65), we have

$$\dot{V}(t) = 2x^T(t) P \left[\bar{A}_i(\mu_i) x(t) + \omega_i(t)\right], i \in \mathscr{I} \tag{8.72}$$

An auxiliary function is introduced as below:

$$J(t) = \dot{V}(t) - \eta V(t) - \eta \omega_i^T(t) \omega_i(t), \tag{8.73}$$

where η is a positive scalar.

It follows from (8.72) and (8.73) that

$$\begin{aligned} J(t) &\le 2x^T(t) P \left[\bar{A}_i(\mu_i) x(t) + \omega_i(t)\right] - \eta V(t) - \eta \omega_i^T(t) \omega_i(t) \\ &= \chi_i^T(t) \Phi_i(\mu_i) \chi_i(t), \end{aligned} \tag{8.74}$$

where $\chi_i(t) = \begin{bmatrix} x^T(t) & \omega_i^T(t) \end{bmatrix}^T$, and $\Phi_i(\mu_i) = \begin{bmatrix} \mathrm{Sym}\left(P\bar{A}_i(\mu_i)\right) - \eta P & P \\ \star & -\eta I \end{bmatrix}$.

It is easy to see that $\Phi_i(\mu_i) < 0$ implies $J(t) < 0$. In order to derive an LMI-based result, we define $X = P^{-1}$. By performing a congruence transformation to $\Phi_i(\mu_i) < 0$ by $\Gamma = \text{diag}\{X, I\}$, we have

$$\begin{bmatrix} \text{Sym}\left(\bar{A}_i(\mu_i)X\right) - \eta X & I \\ \star & -\eta I \end{bmatrix} < 0. \tag{8.75}$$

Define $\bar{K}_i(\mu_i) = K_i(\mu_i)X$ and by extracting the fuzzy premise variables, the inequalities in (8.67) and (8.68) can be obtained.

Due to $J(t) < 0$,

$$\dot{V}(t) < \eta V(t) + \eta \omega_i^T(t)\omega_i(t). \tag{8.76}$$

By multiplying both sides of (8.76) by $e^{-\eta t}$ and integrating the resulting inequality from 0 to t with $t \in [0, T^*]$, it is easy to see that

$$\begin{aligned} e^{-\eta t}V(t) &< V(0) + \eta \int_0^t e^{-\eta s}\omega_i^T(s)\omega_i(s)ds \\ &\leq x^T(0)Px(0) + \eta\delta. \end{aligned} \tag{8.77}$$

On the other hand, it follows from (8.71) that

$$e^{-\eta t}V(t) \geq e^{-\eta T^*}x^T(t)Px(t), \tag{8.78}$$

which implies that

$$e^{-\eta T^*}x^T(t)Px(t) < x^T(0)Px(0) + \eta\delta. \tag{8.79}$$

We introduce the matrix $0 < R^T = R \in \Re^{n_x \times n_x}$, and further define

$$\begin{aligned} &c_1 = x^T(0)Rx(0), \\ &\bar{\sigma}_P = \lambda_{\max}\left(R^{-\frac{1}{2}}PR^{-\frac{1}{2}}\right), \underline{\sigma}_P = \lambda_{\min}\left(R^{-\frac{1}{2}}PR^{-\frac{1}{2}}\right). \end{aligned} \tag{8.80}$$

Based on the relationships (8.79) and (8.80), we have

$$x^T(t)Rx(t) < \frac{\bar{\sigma}_{P1}c_1 + \eta\delta}{e^{-\eta T^*}\underline{\sigma}_P}. \tag{8.81}$$

From Definition 8.2, the switching fuzzy system in (8.65) is the FTB. This completes the proof.

It follows from (8.80) and (8.81) that

$$\|x(t)\| < \frac{\bar{\sigma}_{P1}c_1 + \eta\delta}{e^{-\eta T^*}\underline{\sigma}_P\lambda_{\min}(R)}, t \in [0, T^*]. \tag{8.82}$$

8.3 STEADY-STATE PERFORMANCE ANALYSIS

8.3.1 MPPT OPTIMAL ALGORITHM FOR SINGLE GENERATOR

Recall the MPPT of multi-PV generator in (8.33) as below:

$$E\dot{x}(t) = A(\mu)x(t) + B(\mu)u(t) + D\omega(t), \tag{8.83}$$

where $A(\mu) := \sum_{l=1}^{r} \mu_l A_l, B(\mu) := \sum_{l=1}^{r} \mu_l B_l$, and the local model parameters $\{E, A_l, B_l, D\}$ are defined in (8.1).

Then, we consider, without loss of generality, only the class of bounded disturbance that acts on the output voltage v_{dc}, which is defined as below:

$$\omega^T(t)\,\omega(t) \le \bar{\omega}^2, \tag{8.84}$$

where $\bar{\omega}$ is a positive scalar.

Consider a T-S fuzzy controller, which shares the same premise variables in the fuzzy system (8.83), is proposed as follows:

$$u(t) = K(\mu)x(t), \tag{8.85}$$

where $K(\mu) := \sum_{l=1}^{r} \mu_l K_l, K_l \in \Re^{n_u \times n_x}$ are the controller gains to be designed.

By submitting the controller (8.85) into the fuzzy system (8.83),

$$E\dot{x}(t) = \bar{A}(\mu)x(t) + D\omega(t), \tag{8.86}$$

where $\bar{A}(\mu) = A(\mu) + B(\mu)K(\mu)$.

The aim is to design the fuzzy controller in (8.85), such that all of the system states are bounded by the following reachable set:

$$\mathbb{S} \triangleq \{x(t) \in \Re^{n_x}\, |x(t) \text{ and } \omega(t) \text{ satisfy } (8.83) \text{ and } (8.84), \text{ respectively}, t \ge 0\}. \tag{8.87}$$

An ellipsoid that bounds the reachable set of the closed-loop system in (8.86) is given by

$$\mathbb{E} \triangleq \{x \mid x^T P x < 1,\; x \in \Re^{n_x}\}. \tag{8.88}$$

In the following theorem, we derive a sufficient condition for the bounding of the reachable set of the closed-loop system (8.86).

Theorem 8.4: Controller Design of Reachable Set Estimation

Consider the fuzzy law (8.85), which guarantees that the reachable set of the closed-loop system (8.86) is bounded by the intersection of ellipsoid in (8.88), if there exist matrices $X = \begin{bmatrix} X_{(1)} & 0 \\ X_{(2)} & X_{(3)} \end{bmatrix}$, $0 < X_{(1)} = X_{(1)}^T \in \Re^{(n_x-1)\times(n_x-1)}, X_{(2)} \in \Re^{1\times(n_x-1)}$, $X_{(3)}$ is a scalar, and $\bar{K}_p \in \Re^{n_u \times n_x}$, and positive scalars $\{\eta, \bar{\omega}\}$, such that the following LMIs hold:

$$\Phi_{ll} < 0, 1 \le l \le r \tag{8.89}$$

$$\Phi_{lp} + \Phi_{pl} < 0, 1 \le l < p \le r \tag{8.90}$$

where

$$\Phi_{lp} = \begin{bmatrix} \mathrm{Sym}(A_l X + B_l \bar{K}_p) - \eta E X & D \\ \star & -\frac{\eta}{\bar{\omega}^2} I \end{bmatrix}. \tag{8.91}$$

Furthermore, the controller gains can be calculated by

$$\bar{K}_p = \bar{K}_p X^{-1}. \tag{8.92}$$

■

Proof. Consider the following Lyapunov functional

$$V(t) = x^T(t) E^T P x(t). \tag{8.93}$$

It is easy to see from (8.97) that $E^T P = P^T E \ge 0$.

Along the trajectory of the system (8.86), one gets

$$\begin{aligned} \dot{V}(t) &= 2\left[E\dot{x}(t)\right]^T P x(t) \\ &= 2\left[\bar{A}(\mu)x(t) + D\omega(t)\right]^T P x(t). \end{aligned} \tag{8.94}$$

An auxiliary function is introduced as below:

$$J(t) = \dot{V}(t) + \eta V(t) - \frac{\eta}{\bar{\omega}^2} \omega^T(t) \omega(t), \tag{8.95}$$

where η is a positive scalar.

It follows from (8.93)-(8.95) that

$$\begin{aligned} J(t) &= 2\left[\bar{A}(\mu)x(t) + D\omega(t)\right]^T P x(t) - \eta x^T(t) E^T P x(t) - \frac{\eta}{\bar{\omega}^2} \omega^2(t) \\ &= \chi^T(t) \Phi(\mu) \chi(t), \end{aligned} \tag{8.96}$$

where $\chi(t) = \left[x^T(t) \quad \omega^T(t)\right]^T$, and $\Phi(\mu) = \begin{bmatrix} \mathrm{Sym}\left(P^T \bar{A}(\mu)\right) - \eta P^T E & P^T D \\ \star & -\frac{\eta}{\bar{\omega}^2} I \end{bmatrix}$.

It is easy to see that $\Phi(\mu) < 0$ implies $J(t) < 0$.

Now, define

$$P = \begin{bmatrix} P_{(1)} & 0 \\ P_{(2)} & P_{(3)} \end{bmatrix}, \tag{8.97}$$

where $0 < P_{(1)} = P_{(1)}^T \in \Re^{(n_x-1)\times(n_x-1)}, P_{(2)} \in \Re^{1\times(n_x-1)}$, $P_{(3)}$ is a scalar. Thus, the inequality $E^T P = P^T E \ge 0$ holds.

In order to derive an LMI-based result, we define $X = P^{-1}$, that is

$$X = \begin{bmatrix} X_{(1)} & 0 \\ X_{(2)} & X_{(3)} \end{bmatrix}, \tag{8.98}$$

where $0 < X_{(1)} = X_{(1)}^T \in \Re^{(n_x-1)\times(n_x-1)}, X_{(2)} \in \Re^{1\times(n_x-1)}$, $X_{(3)}$ is a scalar.

By performing a congruence transformation to $\Phi(\mu) < 0$ by $\Gamma = \text{diag}\{X^T, I\}$, $X = P^{-1}$, we have

$$\begin{bmatrix} \text{Sym}\left(\bar{A}(\mu)X\right) - \eta EX & D \\ \star & -\frac{\eta}{\bar{\omega}^2} I \end{bmatrix} < 0. \tag{8.99}$$

Define $\bar{K}(\mu) = K(\mu)X$, and by extracting the fuzzy premises, the inequalities in (8.89) and (8.90) can be obtained.

Due to $J(t) < 0$, it implies that $\dot{V}(t) + \eta V(t) - \frac{\eta}{\bar{\omega}^2}\omega^T(t)\omega(t) < 0$. Then, by multiplying both its sides with $e^{\eta t}$, it yields

$$\begin{aligned} \frac{d\left(e^{\eta t}V(t)\right)}{dt} &= e^{\eta t}\dot{V}(t) + e^{\eta t}aV(t) \\ &< e^{\eta t}\frac{\eta}{\bar{\omega}^2}\omega^T(t)\omega(t). \end{aligned} \tag{8.100}$$

Now, by performing the integral of (8.100) from 0 to $T > 0$, it yields

$$\begin{aligned} e^{\eta T}V(T) &< \int_0^T e^{\eta t}\frac{\eta}{\bar{\omega}^2}\omega^T(t)\omega(t)dt \\ &< e^{\eta T} - 1. \end{aligned} \tag{8.101}$$

Thus, for any time $T > 0$, one can obtain $V(T) < 1$.

Now, we partition $x(t)$ as

$$x(t) = \begin{bmatrix} \bar{x}(t) \\ x_3(t) \end{bmatrix}, \tag{8.102}$$

where $\bar{x}(t) = \begin{bmatrix} x_1(t) \\ x_2(t) \end{bmatrix}$.

Since $V(T) < 1$, it shows

$$\begin{aligned} 1 &> V(t) \\ &= \bar{x}^T(t)P_{(1)}\bar{x}(t). \end{aligned} \tag{8.103}$$

Thus completing the proof to the reachable set in (8.88).

Here, the aim is to design the controller in the form of (8.85) such that the "smallest" bound for the reachable set in (8.88) can be obtained. To do so, a simple optimization algorithm is pointed out in [14], i.e. maximize δ subject to $\delta\mathbf{I} < P_{(1)}$. By using Schur complement, and performing a congruence transformation by $\Gamma = \text{diag}\{I, X_{(1)}\}$, one can easily solve the optimization problem as below:

Algorithm 8.1.

$$\text{Minimize } \bar{\delta}, \text{ subject to } \begin{bmatrix} \bar{\delta}\mathbf{I} & X_{(1)} \\ \star & X_{(1)} \end{bmatrix} > 0, (8.89) \text{ and } (8.90),$$

where $\bar{\delta} = \delta^{-1}$.

8.3.2 MPPT OPTIMAL ALGORITHM FOR MULTI-MACHINE GENERATORS

Recall the MPPT model of the multi-PV generator in (8.33) as below:

$$E\dot{x}_i(t) = A_{ii}(\mu_i)x_i(t) + B_i(\mu_i)u_i(t) + \sum_{\substack{j=1 \\ j\neq i}}^{N} A_{ij}(\mu_j)x_j(t) + \omega_i(t), \tag{8.104}$$

where $A_i(\mu_i) := \sum_{l=1}^{r_i} \mu_{il}A_{il}, B_i(\mu_i) := \sum_{l=1}^{r_i} \mu_{il}B_{il}, A_{ij}(\mu_j) := \sum_{l=1}^{r_j} \mu_{jl}A_{ijl}$, and the local parameters $\{A_{il}, B_{il}, A_{ijl}\}$ are defined in (8.32).

Consider a decentralized T-S fuzzy controller, which shares the same premise variables in (8.104), as follows:

$$u(t) = K_{ii}(\mu_i)x(t), \tag{8.105}$$

where $K_{ii}(\mu_i) := \sum_{l=1}^{r_i} \mu_{il}K_{iil}, K_{iil} \in \Re^{n_{ui}\times n_{xi}}$ are controller gains to be designed.

By submitting the fuzzy controller (8.105) into the fuzzy system (8.104),

$$E\dot{x}_i(t) = \bar{A}_{ii}(\mu_i)x_i(t) + \sum_{\substack{j=1 \\ j\neq i}}^{N} A_{ij}(\mu_j)x_j(t) + \omega_i(t), \tag{8.106}$$

where $\bar{A}_{ii}(\mu_i) = A_{ii}(\mu_i)x_i(t) + B_i(\mu_i)K_{ii}(\mu_i)$.

Define $\omega(t) = \left[\omega_1^T(t)\ \omega_2^T(t)\ \cdots\ \omega_N^T(t)\right]^T$ and $x(t) = \left[x_1^T(t)\ x_2^T(t)\ \cdots\ x_N^T(t)\right]^T$. Assuming that the disturbance is bounded, the following conditions can be satisfied:

$$\omega^T(t)\,\omega(t) \leq \bar{\omega}^2, \tag{8.107}$$

where $\bar{\omega}$ is a positive scalar.

The aim is to design the decentralized fuzzy controller in (8.105), such that all of the system states are bounded by the following reachable set:

$$\begin{aligned} \mathbb{S} \triangleq \{x(t) \in \Re^{n_x}\,|x(t) \text{ and } \omega(t) \text{ satisfy} \\ (8.106) \text{ and } (8.107), \text{ respectively}, t \geq 0\}. \end{aligned} \tag{8.108}$$

An ellipsoid that bounds the reachable set of the closed-loop system in (8.106) is given by

$$\mathbb{E} \triangleq \left\{x \,|\, x^T P x < 1,\ x \in \Re^{n_x}\right\}. \tag{8.109}$$

In the following theorem, we derive a sufficient condition for the bounding of the reachable set of the closed-loop system in (8.106).

Theorem 8.5: Decentralized Controller Design of Reachable Set

Consider the fuzzy law (8.105), which guarantees that the reachable set of the closed-loop system (8.106) is bounded by the intersection of ellipsoid in (8.109), if there exist matrices $X_i = \begin{bmatrix} X_{i1} & 0 \\ X_{i2} & X_{i3} \end{bmatrix}, 0 < X_{i1} = X_{i1}^T \in \Re^{(n_{xi}-1)\times(n_{xi}-1)}, X_{i2} \in \Re^{1\times(n_{xi}-1)}$, X_{i3} is a scalar, and $\bar{K}_{iis} \in \Re^{n_{ui}\times n_{xi}}$, and positive definite symmetric matrix M_i, and positive scalars $\{\eta, \bar{\omega}\}$, such that for all p the following LMIs hold:

$$\Phi_{llp} < 0, 1 \leq l \leq r \tag{8.110}$$

$$\Phi_{lsp} + \Phi_{slp} < 0, 1 \leq l < s \leq r \tag{8.111}$$

where

$$\Phi_{lsp} = \begin{bmatrix} \mathrm{Sym}\left(A_{iil}X_i + B_{il}\bar{K}_{iis}\right) + A_{ijp}M_iA_{ijp}^T - \eta EX_i & I & X_i^T \\ \star & -\eta I & 0 \\ \star & \star & -\frac{M_0}{N-1} \end{bmatrix}. \tag{8.112}$$

Furthermore, the corresponding controller gains are given by

$$K_{iil} = \bar{K}_{iil}X_i^{-1}. \tag{8.113}$$

■

Proof. Consider the following Lyapunov functional

$$V(t) = \sum_{i=1}^{N} x_i^T(t)E^T P_i x_i(t), \forall t \in [0, T^*]. \tag{8.114}$$

It is easy to see from (8.121) that $E^T P_i = P_i^T E \geq 0$.

Along the trajectory of system (8.106), we have

$$\begin{aligned} \dot{V}(t) &= \sum_{i=1}^{N} 2\left[E\dot{x}_i(t)\right]^T P_i x_i(t) \\ &= \sum_{i=1}^{N} 2\left[\bar{A}_{ii}(\mu_i)x_i(t) + \sum_{\substack{j=1 \\ j\neq i}}^{N} A_{ij}(\mu_j)x_j(t) + \omega_i(t)\right]^T P_i x_i(t). \end{aligned} \tag{8.115}$$

Note that

$$2\bar{x}^T\bar{y} \leq \kappa^{-1}\bar{x}^T\bar{x} + \kappa\bar{y}^T\bar{y}, \tag{8.116}$$

where $\bar{x}, \bar{y} \in \Re^n$ and scalar $\kappa > 0$.

Define the matrix $0 < M_i = M_i^T$,

$$\sum_{i=1}^{N}\sum_{j=1,i\neq j}^{N} 2x_j^T(t)A_{ij}^T(\mu_j)P_ix_i(t)$$
$$\leq \sum_{i=1}^{N}\sum_{j=1,i\neq j}^{N} x_j^T(t)M_i^{-1}x_j(t) + \sum_{i=1}^{N}\sum_{j=1,i\neq j}^{N} x_i^T(t)P_i^TA_{ij}(\mu_j)M_iA_{ij}^T(\mu_j)P_ix_i(t)$$
$$= \sum_{i=1}^{N}\sum_{j=1,i\neq j}^{N} x_i^T(t)M_j^{-1}x_i(t) + \sum_{i=1}^{N}\sum_{j=1,i\neq j}^{N} x_i^T(t)P_i^TA_{ij}(\mu_j)M_iA_{ij}^T(\mu_j)P_ix_i(t). \quad (8.117)$$

An auxiliary function is introduced as below:

$$J(t) = \dot{V}(t) + \eta V(t) - \frac{\eta}{\bar{\omega}^2}\omega^T(t)\,\omega(t), \quad (8.118)$$

where η is a positive scalar.

It follows from (8.115)-(8.118) that

$$J(t) \leq \sum_{i=1}^{N} 2\left[\bar{A}_{ii}(\mu_i)x_i(t) + \omega_i(t)\right]^T P_ix_i(t) + \sum_{i=1}^{N}\sum_{j=1,i\neq j}^{N} x_i^T(t)M_j^{-1}x_i(t)$$
$$+ \sum_{i=1}^{N}\sum_{j=1,i\neq j}^{N} x_i^T(t)P_i^TA_{ij}(\mu_j)M_iA_{ij}^T(\mu_j)P_ix_i(t)$$
$$-\eta\sum_{i=1}^{N} x_i^T(t)E^TP_ix_i(t) - \frac{\eta}{\bar{\omega}^2}\sum_{i=1}^{N}\omega_i^T(t)\omega_i(t)$$
$$= \chi_i^T(t)\,\Phi_i(\mu_i)\chi_i(t), \quad (8.119)$$

where $\chi_i(t) = \left[\begin{array}{cc} x_i^T(t) & \omega_i^T(t), \end{array}\right]^T$, and

$$\Phi_i(\mu_i) = \left[\begin{array}{cc} \mathrm{Sym}\left(P_i^T\bar{A}_{ii}(\mu_i)\right) + P_i^TA_{ij}(\mu_j)M_i(\star) + \sum_{j=1,i\neq j}^{N} M_j^{-1} - \eta P_i^TE & P_i^T \\ \star & -\frac{\eta}{\bar{\omega}^2}I \end{array}\right]. \quad (8.120)$$

It is easy to see that $\Phi_i(\mu_i) < 0$ implies $J(t) < 0$. Now, define

$$P_i = \left[\begin{array}{cc} P_{i1} & 0 \\ P_{i2} & P_{i3} \end{array}\right], \quad (8.121)$$

where $0 < P_{i1} = P_{i1}^T \in \Re^{(n_{xi}-1)\times(n_{xi}-1)}, P_{i2} \in \Re^{1\times(n_{xi}-1)}$, P_{i3} is a scalar. It is easy to see that the inequality $E^TP_i = P_i^TE \geq 0$ holds.

In order to derive an LMI-based result, we define $X_i = P_i^{-1}$, that is

$$X_i = \left[\begin{array}{cc} X_{i1} & 0 \\ X_{i2} & X_{i3} \end{array}\right], \quad (8.122)$$

where $0 < X_{i1} = X_{i1}^T \in \Re^{(n_{xi}-1)\times(n_{xi}-1)}, X_{i2} \in \Re^{1\times(n_{xi}-1)}$, X_{i3} is a scalar.

By performing a congruence transformation to $\Phi_i(\mu_i) < 0$ by $\Gamma_i = \text{diag}\left\{X_i^T, I\right\}$, $X_i = P_i^{-1}$, one gets

$$\begin{bmatrix} \text{Sym}\left(\bar{A}_{ii}(\mu_i)X_i\right) + A_{ij}(\mu_j)M_iA_{ij}^T(\mu_j) + \sum_{j=1,i\neq j}^{N} X_i^T M_j^{-1} X_i - \eta E X_i & I \\ \star & -\frac{\eta}{\bar{\omega}^2} I \end{bmatrix} < 0. \tag{8.123}$$

Define $\bar{K}_{ii}(\mu_i) = K_{ii}(\mu_i)X_i$ and $M_0 \leq M_i$, it has $\sum_{j=1,i\neq j}^{N} M_j^{-1} \leq (N-1)M_0^{-1}$. The following matrix inequality implies (8.123),

$$\begin{bmatrix} \text{Sym}\left(\bar{A}_{ii}(\mu_i)X_i\right) + A_{ij}(\mu_j)M_iA_{ij}^T(\mu_j) - \eta E X_i & I & X_i^T \\ \star & -\frac{\eta}{\bar{\omega}^2} I & 0 \\ \star & \star & -\frac{M_0}{N-1} \end{bmatrix} < 0. \tag{8.124}$$

By extracting the fuzzy premises, the inequalities in (8.110) and (8.111) can be obtained.

Due to $J(t) < 0$, it implies that $\dot{V}(t) + \eta V(t) - \frac{\eta}{\bar{\omega}^2}\omega^T(t)\omega(t) < 0$. Then, by multiplying both its sides with $e^{\eta t}$, it yields

$$\begin{aligned} \frac{d\left(e^{\eta t}V(t)\right)}{dt} &= e^{\eta t}\dot{V}(t) + e^{\eta t}\eta V(t) \\ &< e^{\eta t}\frac{\eta}{\bar{\omega}^2}\omega^T(t)\omega(t). \end{aligned} \tag{8.125}$$

Now, by performing the integral of (8.125) from 0 to $T > 0$, it yields

$$\begin{aligned} e^{\eta T}V(T) &< \int_0^T e^{\eta t}\frac{\eta}{\bar{\omega}^2}\omega^T(t)\omega(t)dt \\ &< e^{\eta T} - 1. \end{aligned} \tag{8.126}$$

Thus, for any time $T > 0$, one can obtain $V(T) < 1$.

We further partition $x_i(t)$ as

$$x_i(t) = \begin{bmatrix} \bar{x}_i(t) \\ x_{i3}(t) \end{bmatrix}, \tag{8.127}$$

where $\bar{x}_i(t) = \begin{bmatrix} x_{i1}(t) \\ x_{i2}(t) \end{bmatrix}$.

Since $V(T) < 1$, it shows

$$\begin{aligned} 1 &> V(t) \\ &= \sum_{i=1}^{N} \bar{x}_i^T(t) P_{i1} \bar{x}_i(t) \\ &= \bar{x}^T(t)\mathscr{P}_1\bar{x}(t), \end{aligned} \tag{8.128}$$

where $\bar{x}(t) = \begin{bmatrix} \bar{x}_1^T(t) & \bar{x}_2^T(t) & \cdots & \bar{x}_N^T(t) \end{bmatrix}^T$ and $\mathscr{P}_1 = \text{diag}\underbrace{\{P_{11}, \cdots, P_{N1}\}}_{N}$. Thus completing the proof of reachable set in (8.109).

Here, the aim is to design the controller in the form of (8.105) such that the "smallest" bound for the reachable set in (8.128) can be obtained. To do so, a simple optimization algorithm is pointed out in [14], i.e. maximize δ subject to $\delta\mathbf{I} < \mathscr{P}_1$. By using Schur complement, and performing a congruence transformation by $\Gamma = \text{diag}\{I, \mathscr{X}_1\}$, where $\mathscr{X}_1 = \text{diag}\underbrace{\{X_{11}, \cdots, X_{N1}\}}_{N}$, one can easily solve the optimization problem as below:

Algorithm 8.2.

$$\text{Minimize } \bar{\delta}, \text{ subject to } \begin{bmatrix} \bar{\delta}\mathbf{I} & \mathscr{X}_1 \\ \star & \mathscr{X}_1 \end{bmatrix} > 0, (8.110) \text{ and } (8.111),$$

where $\bar{\delta} = \delta^{-1}$.

8.3.3 OPTIMAL ALGORITHM FOR MULTI-MODE OPERATION

Recall the system dynamics with a three-switch operation as shown in (8.63),

$$\dot{x}(t) = A_i(\mu_i)x(t) + B_i(\mu_i)u(t) + \omega_i(t), i \in \mathscr{I} := \{1,2,3\} \tag{8.129}$$

where $x(t) \in \Re^{n_x}$ and $u(t) \in \Re^{n_u}$ denote the system state and control input, respectively; $\mathscr{I}$ is the set of switched subsystems; $A_i(\mu_i) := \sum_{l=1}^{r_i} \mu_{il}[z_i(t)]A_{il}, B_i(\mu_i) := \sum_{l=1}^{r_i} \mu_{il}[z_i(t)]B_{il}, \mu_{il}[z_i(t)] := \frac{\Pi_{\phi=1}^{g}\mu_{il\phi}[z_{i\phi}(t)]}{\Sigma_{\varsigma=1}^{r_i}\Pi_{\phi=1}^{g}\mu_{i\varsigma\phi}[z_{i\phi}(t)]} \geq 0, \sum_{l=1}^{r_i} \mu_{il}[z_i(t)] = 1$. For simplicity, we define $\mu_{il} = \mu_{il}[z_i(t)]$.

Consider a switching fuzzy controller as below:

$$u(t) = K_i(\mu_i)x(t), i \in \mathscr{I} \tag{8.130}$$

where $K_i(\mu_i) := \sum_{l=1}^{r_i} \mu_{il}K_{il}$, $K_{il} \in \Re^{n_u \times n_x}$ is the controller gain matrix.

Submitting the controller (8.130) into the system (8.129), the i-th closed-loop control system can be represented as

$$\dot{x}(t) = \bar{A}_i(\mu_i)x(t) + \omega_i(t), i \in \mathscr{I} \tag{8.131}$$

where $\bar{A}_i(\mu_i) = A_i(\mu_i) + B_i(\mu_i)K_i(\mu_i)$.

Assuming that the disturbance is bounded, the following conditions can be satisfied:

$$\omega_i^T(t)\omega_i(t) \leq \bar{\omega}^2, i \in \mathscr{I} \tag{8.132}$$

where $\bar{\omega}$ is a positive scalar.

The aim is to design the switching fuzzy controller in (8.130), such that all of the system states are bounded by the following reachable set:

$$\mathbb{S} \triangleq \{x(t) \in \Re^{n_x} \,|x(t) \text{ and } \omega_i(t) \text{ satisfy } \\ (8.131) \text{ and } (8.132), \text{ respectively}, t \geq 0\}. \tag{8.133}$$

An ellipsoid that bounds the reachable set of the closed-loop control system in (8.131) is given by

$$\mathbb{E} \triangleq \left\{x \,|\, x^T P x < 1,\, x \in \Re^{n_x}\right\}. \tag{8.134}$$

In the following theorem, we derive a sufficient condition for the bounding of the reachable set of closed-loop control system (8.131).

Theorem 8.6: Switched Controller Design of Reachable Set Estimation

Consider the switching fuzzy law (8.130), which guarantees that the reachable set of the closed-loop control system (8.131) is bounded by the intersection of ellipsoid in (8.134), if there exist the matrices $X \in \Re^{n_x \times n_x}$, and $\bar{K}_{is} \in \Re^{n_u \times n_x}$, and the positive scalars $\{\eta, \bar{\omega}\}$, such that the following LMIs hold:

$$\Phi_{ll} < 0, 1 \leq l \leq r \tag{8.135}$$

$$\Phi_{ls} + \Phi_{sl} < 0, 1 \leq l < s \leq r \tag{8.136}$$

where

$$\Phi_{ls} = \begin{bmatrix} \text{Sym}(A_{il}X + B_{il}\bar{K}_{is}) - \eta X & I \\ \star & -\frac{\eta}{\bar{\omega}^2} I \end{bmatrix}. \tag{8.137}$$

Furthermore, the corresponding controller gains are given by

$$K_{il} = \bar{K}_{il} X^{-1}. \tag{8.138}$$

■

Proof. Consider the following Lyapunov functional

$$V(t) = x^T(t) P x(t), \tag{8.139}$$

where P is a symmetric positive definite matrix.

Along the trajectory of system (8.131), one gets

$$\dot{V}(t) = 2x^T(t) P \left[\bar{A}_i(\mu_i) x(t) + \omega_i(t)\right], i \in \mathscr{I}. \tag{8.140}$$

An auxiliary function is introduced as below:

$$J(t) = \dot{V}(t) - \eta V(t) - \frac{\eta}{\bar{\omega}^2} \omega_i^T(t) \omega_i(t), \tag{8.141}$$

where η is a positive scalar.

It follows from (8.140) and (8.141) that

$$J(t) \leq 2x^T(t)P\left[\bar{A}_i(\mu_i)x(t) + \omega_i(t)\right] - \eta V(t) - \frac{\eta}{\bar{\omega}^2}\omega_i^T(t)\omega_i(t)$$
$$= \chi_i^T(t)\Phi_i(\mu_i)\chi_i(t), \tag{8.142}$$

where $\chi_i(t) = \begin{bmatrix} x^T(t) & \omega_i^T(t) \end{bmatrix}^T$, and $\Phi_i(\mu_i) = \begin{bmatrix} \text{Sym}\left(P\bar{A}_i(\mu_i)\right) - \eta P & P \\ \star & -\frac{\eta}{\bar{\omega}^2}I \end{bmatrix}$.

It is easy to see that $\Phi_i(\mu_i) < 0$ implies $J(t) < 0$. In order to derive an LMI-based result, we define $X = P^{-1}$. By using the Schur complement and performing a congruence transformation by $\Gamma = \text{diag}\{X, I\}$, one gets

$$\begin{bmatrix} \text{Sym}\left(\bar{A}_i(\mu_i)X\right) - \eta X & I \\ \star & -\frac{\eta}{\bar{\omega}^2}I \end{bmatrix} < 0. \tag{8.143}$$

Define $\bar{K}_i(\mu_i) = K_i(\mu_i)X$ and by extracting the fuzzy premise variables, the inequalities in (8.135) and (8.136) can be obtained.

Due to $J(t) < 0$, it implies that $\dot{V}(t) + \eta V(t) - \frac{\eta}{\bar{\omega}^2}\omega_i^T(t)\omega_i(t) < 0$. Then, by multiplying both its sides with $e^{\eta t}$, it yields

$$\frac{d\left(e^{\eta t}V(t)\right)}{dt} = e^{\eta t}\dot{V}(t) + e^{\eta t}\eta V(t)$$
$$< e^{\eta t}\frac{\eta}{\bar{\omega}^2}\omega_i^T(t)\omega_i(t). \tag{8.144}$$

Now, by performing the integral of (8.144) from 0 to $T > 0$, it yields

$$e^{\eta T}V(T) < \int_0^T e^{\eta t}\frac{\eta}{\bar{\omega}^2}\omega_i^T(t)\omega_i(t)dt$$
$$< e^{\eta T} - 1. \tag{8.145}$$

Thus, for any time $T > 0$, one can obtain $V(T) < 1$.

Since $V(T) < 1$, it shows

$$1 > V(t)$$
$$= x^T(t)Px(t). \tag{8.146}$$

This completes the proof of the reachable set in (8.134).

Here, the aim is to design the switched fuzzy controller in the form of (8.130) such that the "smallest" bound for the reachable set in (8.134) can be obtained. To do so, a simple optimization algorithm is pointed out in [14], i.e. maximize δ subject to $\delta\mathbf{I} < P$. By using Schur complement, and performing a congruence transformation by $\Gamma = \text{diag}\{I, X\}$, one can easily solve the optimization problem as below:

Algorithm 8.3.

$$\text{Minimize } \bar{\delta}, \text{ subject to } \begin{bmatrix} \bar{\delta}\mathbf{I} & \mathbf{I} \\ \star & X \end{bmatrix} > 0, (8.135) \text{ and } (8.136).$$

where $\bar{\delta} = \delta^{-1}$.

8.4 SIMULATION STUDIES

In order to confirm the effectiveness of the proposed MPPT control method, we consider a solar PV system, and its dynamic model can be described as shown in (8.5). The parameters are chosen as below: $L = 150\mu\text{H}, C_0 = 1000\mu\text{F}, n_p = 36, \gamma = 38.63, I_{rs} = 4\text{A}, T = 300\text{K}$. Now, the proposed MPPT algorithm can be implemented as below:

i) Use the descriptor system approach to represent the MPPT control problem of the PV system, that is

$$\begin{cases} \dot{\phi}_{pv} = -\frac{1}{L}(1-u)v_{dc} + \frac{1}{L}v_{pv}, \\ \dot{v}_{dc} = \frac{1}{C_0}(1-u)\phi_{pv} - \frac{1}{C_0}\phi_0, \\ 0 \cdot \dot{\varepsilon}_{pv} = \phi_{pv} - n_p\gamma I_{rs}e^{\gamma v^*_{pv}}\varepsilon_{pv} - n_p\gamma I_{rs}v_{pv}e^{\gamma v^*_{pv}}. \end{cases}$$

ii) Choose $z_1 = \frac{v_{pv}}{\phi_{pv}}$, $z_2 = \frac{\phi_0}{v_{dc}}$, $z_3 = e^{\gamma v^*_{pv}}$, $z_4 = \frac{v_{pv}}{\phi_{pv}}e^{\gamma v^*_{pv}}$, $z_5 = v_{dc}$, and $z_6 = \phi_{pv}$ as fuzzy premise variables, and linearize the above mentioned nonlinear system around the operation points listed in Table 8.1.

Table 8.1
Linearization of operation points

Parameters	Values	
$\frac{v_{pv}}{\phi_{pv}}$	3	5
$\frac{\phi_0}{v_{dc}}$	0.3	0.5
$e^{\gamma v^*_{pv}}$	70	103
$\frac{v_{pv}}{\phi_{pv}}e^{\gamma v^*_{pv}}$	335	350
v_{dc}	11	12
ϕ_{pv}	2	4

In that case, the nonlinear descriptor system is represented by the following T-S fuzzy model:

$$E\dot{x}(t) = A(\mu)x(t) + B(\mu)u(t) + D\omega(t),$$

where $A(\mu) := \sum_{l=1}^{r} \mu_l A_l$, $B(\mu) := \sum_{l=1}^{r} \mu_l B_l$, $D = \begin{bmatrix} 0 & 1 & 0 \end{bmatrix}^T$, $r = 64$. Due to space limitations, the model parameters are omitted.

iii) According to Theorem 8.1, the problem of MPPT control with reachable set estimation can be solved, and the controller gains are as below:

$$\begin{aligned} K_{1\sim16} &= \begin{bmatrix} -175.8929 & 49.2130 & -0.0334 \end{bmatrix}, \\ K_{17\sim32} &= \begin{bmatrix} -138.1634 & 47.3412 & -0.0334 \end{bmatrix}, \\ K_{33\sim48} &= \begin{bmatrix} -164.7768 & 55.4520 & -0.0334 \end{bmatrix}, \\ K_{49\sim64} &= \begin{bmatrix} -192.2057 & 76.2387 & -0.0334 \end{bmatrix}. \end{aligned}$$

iv) Use (8.30) to calculate the bounding for the MPPT error $\varepsilon_{pv} = 31.84$.

Using the above solution and choosing the initial state $\bar{x}(0) = \begin{bmatrix} 1 \\ 15 \end{bmatrix}$, the responses of system states by proposed control strategy are shown in Figure 8.1. It is easy to see that the proposed finite time controller can force the tracking error to converge around zero in a pre-specified finite time T.

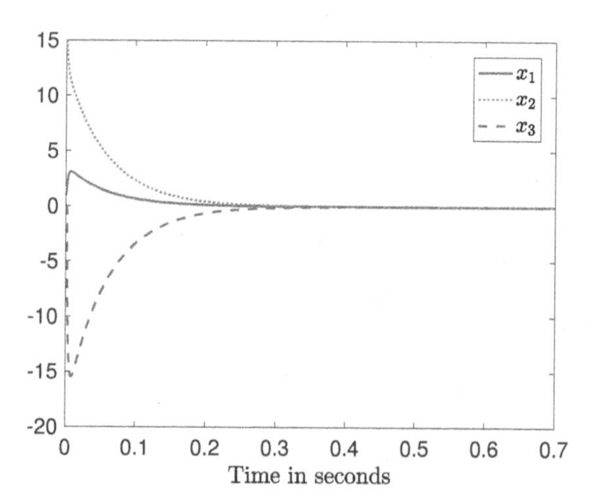

Figure 8.1 State responses for closed-loop control system.

8.5 REFERENCES

1. Jiang, L. L., Srivatsan, R., and Maskell, D. L. (2018). Computational intelligence techniques for maximum power point tracking in PV systems: A review. Renewable and Sustainable Energy Reviews, 85, 14-45.
2. Justo, J. J., Mwasilu, F., Lee, J., and Jung, J. W. (2013). AC-microgrids versus DC-microgrids with distributed energy resources: A review. Renewable and Sustainable Energy Reviews, 24, 387-405.
3. Li, X., Li, Y., and Seem, J. E. (2013). Maximum power point tracking for photovoltaic system using adaptive extremum seeking control. IEEE Transactions on Control Systems Technology, 21(6), 2315-2322.
4. Salas, V., E. Ollas, Barrado, A., and A. Lázaro. (2006). Review of the maximum power point tracking algorithms for stand-alone photovoltaic systems. Solar Energy Materials and Solar Cells, 90(11), 1555-1578.
5. Ishaque, K., Salam, Z., Amjad, M., and Mekhilef, S. (2012). An improved particle swarm optimization (PSO) C-based MPPT for PV with reduced steady-state oscillation. IEEE Transactions on Power Electronics, 27(8), 3627-3638.
6. Guo, L., Meng, Z., Sun, Y., and Wang, L. (2018). A modified CAT swarm optimization based maximum power point tracking method for photovoltaic system under partially shaded condition. Energy, 144, 501-514.
7. Samangkool, K. and Premrudeepreechacharn, S. (2005). Maximum power point tracking using neural networks for grid-connected photovoltaic system. International Conference on Future Power Systems. In Proceedings of the 2005 International Conference on Future Power Systems, 1-4.

8. Kumar, N., Hussain, I., Singh, B., and Panigrahi, B. (2017). Rapid MPPT for uniformly and partial shaded PV system by using JayaDE algorithm in highly fluctuating atmospheric conditions. IEEE Transactions on Industrial Informatics, 13(5), 2406-2416.
9. Merida, J., Aguilar, L. T., and Davila, J. (2014). Analysis and synthesis of sliding mode control for large scale variable speed wind turbine for power optimization. Renewable Energy, 71(11), 715-728.
10. Yang, J., Ding, Z., Li, S., and Zhang, C. (2017). Continuous finite-time output regulation of nonlinear systems with unmatched time-varying disturbances. IEEE Control Systems Letters, 2(1), 97-102.
11. Basin, M. V., Panathula, C. B., Shtessel, Y. B., and Ramrez, P. C. R. (2016). Continuous finite-time higher order output regulators for systems with unmatched unbounded disturbances. IEEE Transactions on Industrial Electronics, 63(8), 5036-5043.
12. Song, J., Niu, Y., and Zou, Y. (2017). Finite-time stabilization via sliding mode control. IEEE Transactions on Automatic Control, 62(3), 1478-1483.
13. ELBSAT, Mohammad, N., YAZ, and Edwin, E. (2013). Robust and resilient finite-time bounded control of discrete-time uncertain nonlinear systems. Automatica, 49(7), 2292-2296.
14. Fridman, E. and Shaked, U. (2003). On reachable sets for linear systems with delay and bounded peak inputs . Automatica, 39(11), 2005-2010.
15. Zhang, L. and Gao, H. (2010). Technical communique: Asynchronously switched control of switched linear systems with average dwell time. Automatica, 46(5), 953-958.
16. Zhang, L., Shi, P., and Basin, M. (2008). Robust stability and stabilisation of uncertain switched linear discrete time-delay systems. IET Control Theory and Applications, 2(7), 606-614.

Part IV

Cyber-Physical Control Framework for Microgrids

Preview

Cyber-physical systems CPSs integrate cyber and physical subsystems. CPSs perform realtime monitoring and feedback control via adopting a mixture of sensors, actuators and available computation and communication modules. Due to the increasing developments in distributed sensing, action and networking technologies, CPSs that function within large-scale networks are called networked CPSs (NCPSs).

Power systems are recognized as large-scale nonlinear systems. Traditionally, simplified linear models have been generally utilized in synchronous generators for a long time. Due to the inherent simplicity of analysis and design on implementation of linear control methodologies, they were well developed in utilities. However, such linear control results only provide asymptotic stability in a small region of the equilibrium and attenuate the impacts on small disturbances. During the past several decades, as an effective means of representing nonlinear systems, Takagi-Sugeno (T-S) model-based approach has been intensively applied. As a smooth nonlinear system could be characterized by using the T-S fuzzy model, the central features of the approach are twofold: i) The T-S fuzzy model has the ability of approximating the nonlinear system with an arbitrary precision; ii) The extensive linear control methodologies can be used to resolve the corresponding control problems of nonlinear systems.

Due to the openness of system operation and communication protocol, CPSs are extremely vulnerable to cyber attacks that can result in catastrophic consequences. Thus, security is nowadays a major challenge for implementing CPSs. Many researchers have extensively investigated security issues from different perspectives. One of the main issues is maintaining system security by analyzing the vulnerabilities of CPSs to cyber attacks. A great deal of the literature focuses on the effects of cyber attacks such as false data injection attacks, denial-of-service (DoS) attack, time delay switch (TDS) attacks, deception attack, against particular systems. More recently, NCPSs have been used to monitor and control distributed systems and revealed many important research perspectives. Modern power systems depend on computers to monitor and control distributed systems over wide areas. The complexity of such computer systems makes them highly vulnerable to attacks.

This part is organized as follows. Chapter 9 investigates network-induced delays in local and interconnected power systems. Chapter 10 examines the centralized event-triggered control, decentralized event-triggered control, and distributed event-triggered control for networked cyber-physical power systems. Chapter 11 focuses on nonlinear power networks subject to TDS attacks and proposes an effective method to countervail the negative impacts of such attacks.

9 Fuzzy Control with Network-Induced Delay

Network-induced delay often happens in networked cyber-physical systems. It is well known that such delays may lead to unsatisfactory performance, or instability [1, 2]. To derive less conservative stability conditions, various approaches have been investigated for time-delay systems (TDSs). The work of [3] introduced the augmented Lyapunov functional with triple-integral terms. The work of [13] relaxed the condition that all of Lyapunov matrices were necessarily positive definite. The work of [5] investigated more free-weighting matrices by using redundant variables. All the aforementioned references derived delay-dependent criteria for time-delay systems based on the direct Lyapunov-based approach. It is noted that the stability analysis for systems with time-varying delays can be reformulated as the robust control problem subject to input-output (IO) stability; the technique is known as the indirect Lyapunov-based method [6], consisting of model transformation approach and the SSG theorem. It has been recently developed for a broad class of systems with time-varying delays [7, 8].

In this chapter, the renewable energy sources are considered as the distributed connection to the common bus in a microgrid. They communicate with each other by using networks. Network-induced delays in local and interconnection are investigated, and some results for solving the stability analysis and control synthesis problems are proposed.

9.1 NETWORK-INDUCED DELAYS IN LOCAL SUBSYSTEMS

9.1.1 DECENTRALIZED CONTROL PROBLEMS

Consider a networked environment consisting of a multi-PV or multi-wind generator utilizing N nonlinear subsystems with interconnections and time-varying delays. Then, the i-th nonlinear subsystem is described by the following T-S fuzzy model:

Plant Rule $\mathscr{R}_i^l$: **IF** ζ_{1i} is $\mathscr{F}_{i1}^l$, ζ_{2i} is $\mathscr{F}_{i2}^l$, $\cdots$, $\zeta_{ig}(t)$ is $\mathscr{F}_{ig}^l$, **THEN**

$$\begin{cases} \dot{x}_i(t) = A_{il}x_i(t) + A_{dil}x_i(t-d_i(t)) + B_{il}u_i(t) + \sum\limits_{\substack{k=1\\k\neq i}}^{N} \bar{A}_{ikl}x_k(t) + B_{\omega il}\omega_i(t) \\ y_i(t) = C_{il}x_i(t) + C_{dil}x_i(t-d_i(t)) + D_{\omega il}\omega_i(t) \\ z_i(t) = L_{il}x_i(t) + L_{dil}x_i(t-d_i(t)) + F_{il}u_i(t) + F_{\omega il}\omega_i(t) \\ x_i(t) = \varphi_i(t), -d_{i2} \le t \le 0, l \in \mathscr{L}_i := \{1,2,\ldots,r_i\} \end{cases} \tag{9.1}$$

where $i \in \mathscr{N} := \{1,2,\ldots,N\}$, N is the number of subsystems. For the i-th subsystem, $\mathscr{R}_i^l$ is the l-th fuzzy inference rule; r_i is the number of inference rules;$\mathscr{F}_{i\phi}^l$ $(\phi = 1,2,\ldots,g)$ are the fuzzy sets; $x_i(t) \in \Re^{n_{xi}}, u_i(t) \in \Re^{n_{ui}}, y_i(t) \in \Re^{n_{yi}}$, and

$z_i(t) \in \Re^{n_{zi}}$ are the system state, the control input, the measured output, and the regulated output, respectively; $\omega_i(t) \in \Re^{n_{\omega i}}$ is the disturbance input, which belongs to $L_2[0,\infty)$; $\zeta_i(t) := [\zeta_{i1}(t), \zeta_{i2}(t), \ldots, \zeta_{ig}(t)]$ are some measurable variables; $(A_{il}, A_{dil}, B_{il}, B_{\omega il}, C_{il}, C_{dil}, D_{\omega il}, L_{il}, L_{dil}, F_{il}, F_{\omega il})$ denotes the l-th local model; $\bar{A}_{ikl}$ is the interconnection matrix of the i-th and k-th subsystems for the l-th local model; $d_i(t)$ is a continuous function, which satisfies $0 < d_{i1} \leq d_i(t) \leq d_{i2}$ and $\dot{d}_i(t) \leq \tau_i < \infty$, where d_{i1} and d_{i2} are the lower and upper bounds of $d_i(t)$, respectively; $\varphi_i(t)$ is a continuous real-valued initial function in the interval $[-d_{i2}\ 0]$.

Define $\mu_{il}[\zeta_i(t)]$ as the normalized membership function of the inferred fuzzy set $\mathscr{F}_i^l := \prod_{\phi=1}^{g} \mathscr{F}_{i\phi}^l$, and

$$\mu_{il}[\zeta_i(t)] := \frac{\prod_{\phi=1}^{g} \mu_{il\phi}[\zeta_{i\phi}(t)]}{\sum_{\varsigma=1}^{r_i} \prod_{\phi=1}^{g} \mu_{i\varsigma\phi}[\zeta_{i\phi}(t)]} \geq 0, \sum_{l=1}^{r_i} \mu_{il}[\zeta_i(t)] = 1. \tag{9.2}$$

In this section, we will denote $\mu_{il} := \mu_{il}[\zeta_i(t)]$ for brevity.

By fuzzy blending, the i-th global T-S fuzzy dynamic model is given by

$$\begin{cases} \dot{x}_i(t) = A_i(\mu_i)x_i(t) + A_{di}(\mu_i)x_i(t - d_i(t)) + B_i(\mu_i)u_i(t) \\ \qquad + \sum\limits_{\substack{k=1 \\ k \neq i}}^{N} \bar{A}_{ik}(\mu_i)x_k(t) + B_{\omega i}(\mu_i)\omega_i(t) \\ y_i(t) = C_i(\mu_i)x_i(t) + C_{di}(\mu_i)x_i(t - d_i(t)) + D_{\omega i}(\mu_i)\omega_i(t) \\ z_i(t) = L_i(\mu_i)x_i(t) + L_{di}(\mu_i)x_i(t - d_i(t)) + F_i(\mu_i)u_i(t) + F_{\omega i}(\mu_i)\omega_i(t) \\ x_i(t) = \varphi_i(t), -d_{i2} \leq t \leq 0, i \in \mathscr{N} \end{cases} \tag{9.3}$$

where

$$\begin{aligned} A_i(\mu_i) &:= \sum_{l=1}^{r_i} \mu_{il}A_{il}, A_{di}(\mu_i) := \sum_{l=1}^{r_i} \mu_{il}A_{dil}, B_i(\mu_i) := \sum_{l=1}^{r_i} \mu_{il}B_{il}, \\ B_{\omega i}(\mu_i) &:= \sum_{l=1}^{r_i} \mu_{il}B_{\omega il}, C_i(\mu_i) := \sum_{l=1}^{r_i} \mu_{il}C_{il}, C_{di}(\mu_i) := \sum_{l=1}^{r_i} \mu_{il}C_{dil}, \\ D_{\omega i}(\mu_i) &:= \sum_{l=1}^{r_i} \mu_{il}D_{\omega il}, \bar{A}_{ik}(\mu_i) := \sum_{l=1}^{r_i} \mu_{il}\bar{A}_{ikl}, L_i(\mu_i) := \sum_{l=1}^{r_i} \mu_{il}L_{il}, \\ L_{di}(\mu_i) &:= \sum_{l=1}^{r_i} \mu_{il}L_{dil}, F_i(\mu_i) := \sum_{l=1}^{r_i} \mu_{il}F_{il}, F_{\omega i}(\mu_i) := \sum_{l=1}^{r_i} \mu_{il}F_{\omega il}. \end{aligned} \tag{9.4}$$

Assume that the bounds of the time-varying delay $d_i(t)$ are known a priori. Our objective in this section is to design a decentralized memory fuzzy dynamic output feedback (DOF) controller as below:

$$\begin{cases} \dot{\hat{x}}_i(t) = A_{c0i}(\mu_i)\hat{x}_i(t) + A_{c1i}(\mu_i)\hat{x}_i(t - d_{i1}) \\ \qquad + A_{c2i}(\mu_i)\hat{x}_i(t - d_{i2}) + B_{ci}(\mu_i)y_i(t) \\ u_i(t) = C_{c0i}(\mu_i)\hat{x}_i(t) + C_{c1i}(\mu_i)\hat{x}_i(t - d_{i1}) \\ \qquad + C_{c2i}(\mu_i)\hat{x}_i(t - d_{i2}) + D_{ci}y_i(t), i \in \mathscr{N} \end{cases} \tag{9.5}$$

where $\hat{x}_i(t) \in \Re^{n_{xi}}$ is the i-th controller state, and

$$A_{c0i}(\mu_i) := \sum_{l=1}^{r_i}\sum_{j=1}^{r_i} \mu_{il}\mu_{ij}A_{c0ilj}, A_{c1i}(\mu_i) := \sum_{l=1}^{r_i}\sum_{j=1}^{r_i} \mu_{il}\mu_{ij}A_{c1ilj},$$
$$A_{c2i}(\mu_i) := \sum_{l=1}^{r_i}\sum_{j=1}^{r_i} \mu_{il}\mu_{ij}A_{c2ilj}, C_{c0i}(\mu_i) := \sum_{l=1}^{r_i} \mu_{il}C_{c0il},$$
$$C_{c1i}(\mu_i) := \sum_{l=1}^{r_i} \mu_{il}C_{c1il}, C_{c2i}(\mu_i) := \sum_{l=1}^{r_i} \mu_{il}C_{c2il}, B_{ci}(\mu_i) := \sum_{l=1}^{r_i} \mu_{il}B_{cil}, \tag{9.6}$$

and $\{A_{c0ilj}, A_{c1ilj}, A_{c2ilj}, C_{c0il}, C_{c1il}, C_{c2il}, B_{cil}, D_{ci}\}, (l,j) \in \mathscr{L}_i, i \in \mathscr{N}$ are the controller gains to be determined with compatible dimensions.

Define $\bar{x}_i(t) = \begin{bmatrix} x_i^T(t) & \hat{x}_i^T(t) \end{bmatrix}^T$, it follows from (9.3) and (9.5) that the closed-loop fuzzy control system is given by

$$\begin{cases} \dot{\bar{x}}_i(t) = \bar{A}_i(\mu_i)\bar{\xi}_i(t) + A_{di}(\mu_i)\bar{x}_i(t - d_i(t)) \\ \qquad + E\sum\limits_{k=1, k\neq i}^{N} \bar{A}_{ik}(\mu_i)x_k(t) + B_{\omega i}(\mu_i)w_i(t) \\ z_i(t) = \bar{C}_i(\mu_i)\bar{\xi}_i(t) + C_{di}(\mu_i)\bar{x}_i(t - d_i(t)) + D_{\omega i}(\mu_i)\omega_i(t) \\ \bar{x}_i(t) = \begin{bmatrix} \varphi_i^T(t) & 0 \end{bmatrix}^T, -d_{i2} \le t \le 0, i \in \mathscr{N} \end{cases} \tag{9.7}$$

where

$$\begin{aligned}
&\bar{\xi}_i(t) = \begin{bmatrix} \bar{x}_i^T(t) & \bar{x}_i^T(t-d_{i1}) & \bar{x}_i^T(t-d_{i2}) \end{bmatrix}^T, \\
&\bar{A}_i(\mu_i) = \begin{bmatrix} A_i(\mu_i) & A_{d1i}(\mu_i) & A_{d2i}(\mu_i) \end{bmatrix}, \\
&A_i(\mu_i) = \begin{bmatrix} A_i(\mu_i) + B_i(\mu_i)D_{ci}C_i(\mu_i) & B_i(\mu_i)C_{c0i}(\mu_i) \\ B_{ci}(\mu_i)C_i(\mu_i) & A_{c0i}(\mu_i) \end{bmatrix}, \\
&A_{d1i}(\mu_i) = \begin{bmatrix} 0 & B_i(\mu_i)C_{c1i}(\mu_i) \\ 0 & A_{c1i}(\mu_i) \end{bmatrix}, A_{d2i}(\mu_i) = \begin{bmatrix} 0 & B_i(\mu_i)C_{c2i}(\mu_i) \\ 0 & A_{c2i}(\mu_i) \end{bmatrix}, \\
&A_{di}(\mu_i) = \begin{bmatrix} A_{di}(\mu_i) + B_i(\mu_i)D_{ci}C_{di}(\mu_i) & 0 \\ B_{ci}(\mu_i)C_{di}(\mu_i) & 0 \end{bmatrix}, E = \begin{bmatrix} \mathbf{I} \\ 0 \end{bmatrix}, \\
&B_{\omega i}(\mu_i) = \begin{bmatrix} B_i(\mu_i)D_{ci}D_{wi}(\mu_i) + B_{wi}(\mu_i) \\ B_{ci}(\mu_i)D_{wi}(\mu_i) \end{bmatrix}, \bar{C}_i(\mu_i) = \begin{bmatrix} C_i(\mu_i) & C_{d1i}(\mu_i) & C_{d2i}(\mu_i) \end{bmatrix}, \\
&C_i(\mu_i) = \begin{bmatrix} L_i(\mu_i) + F_i(\mu_i)D_{ci}C_i(\mu_i) & F_i(\mu_i)C_{c0i}(\mu_i) \end{bmatrix}, \\
&C_{d1i}(\mu_i) = \begin{bmatrix} 0 & F_i(\mu_i)C_{c1i}(\mu_i) \end{bmatrix}, C_{d2i}(\mu_i) = \begin{bmatrix} 0 & F_i(\mu_i)C_{c2i}(\mu_i) \end{bmatrix}, \\
&C_{di}(\mu_i) = \begin{bmatrix} L_{di}(\mu_i) + F_i(\mu_i)D_{ci}C_{di}(\mu_i) & 0 \end{bmatrix}, D_{\omega i}(\mu_i) = F_i(\mu_i)D_{ci}D_{\omega i}(\mu_i) + F_{\omega i}(\mu_i).
\end{aligned} \tag{9.8}$$

By defining $\tilde{z}(t) = \begin{bmatrix} z_1^T(t) \; z_2^T(t) \; \cdots \; z_N^T(t) \end{bmatrix}^T$ and $\tilde{w}(t) = \begin{bmatrix} \omega_1^T(t) \; \omega_2^T(t) \; \cdots \; \omega_N^T(t) \end{bmatrix}^T$, we can formulate the problem of the decentralized dynamic output feedback $\mathscr{H}_\infty$ control as follows:

Given the large-scale T-S fuzzy system in (9.3) and a disturbance attenuation level $\gamma > 0$, the purpose of this section is to design a decentralized memory fuzzy DOF controller in the form of (9.5) to satisfy the following two requirements simultaneously:

1) When $w_i(t) = 0$, the closed-loop fuzzy control system in (9.7) is asymptotically stable;

2) The induced L_2 norm of the operator from $\tilde{\omega}$ to $\tilde{z}$ is less than γ under zero initial conditions

$$\int_0^\infty \tilde{z}^T(t)\tilde{z}(t)dt < \gamma^2 \int_0^\infty \tilde{\omega}^T(t)\tilde{w}(t)dt \tag{9.9}$$

for any nonzero $\tilde{\omega} \in L_2[0\ \infty)$.

9.1.2 MODEL TRANSFORMATION

In this subsection, a model transformation will be proposed to convert the closed-loop fuzzy control system (9.7) into an interconnected formulation, such that the problem of decentralized DOF $\mathscr{H}_\infty$ control can be tackled by using IO stability. The IO approach proposed in this section benefits from the application of the scaled small gain (SSG) theorem. Interested readers can refer to the SSG theorem proposed in [6] for more details. Here, we just review some essential ideas on the SSG theorem. Consider an interconnected system with input and output as below:

$$\mathscr{S}_1 : \xi_d(t) = \mathbf{G}\eta_d(t), \mathscr{S}_2 : \eta_d(t) = \Delta\xi_d(t), \tag{9.10}$$

where the operator $\mathbf{G}$ is known for the forward subsystem $\mathscr{L}_1$; the operator Δ is unknown for the feedback system $\mathscr{S}_2$. Assume that the operator $\Delta \in \mathscr{W} := \{\Delta : \|\Delta\|_\infty \leq 1\}$, then, a sufficient condition for the stability of the interconnected system is directly followed by using the following SSG theorem [6].

Lemma 9.1: Model Transformation

Assume that $\mathscr{S}_1$ is internally stable, then the interconnected system with two subsystems $\mathscr{S}_1$ and $\mathscr{S}_2$ in (9.10) is robustly stable, if there exist nonsingular matrices $T_\eta \in \Re^{n_\eta \times n_\eta}$ and $T_\xi \in \Re^{n_\xi \times n_\xi}$, such that for all $\Delta \in \mathscr{W}$, the condition $\left\|T_\xi \mathbf{G} T_\eta^{-1}\right\|_\infty < 1$ holds. ■

In this section, we will firstly reformulate the closed-loop fuzzy control system (9.7) into an interconnected structure with extra input and output. Then, we introduce an LKF and use the SSG theorem to determine sufficient conditions for the decentralized DOF $\mathscr{H}_\infty$ controller performance analysis and design of the large-scale T-S fuzzy system in (9.3).

Here, by using a two-term approximation method [7], the time-varying delay term $\bar{x}_i(t - d_i(t))$ will be approximated by $\bar{x}_i(t - d_{i1})$ and $\bar{x}_i(t - d_{i2})$. The approximation error is given by

$$\begin{aligned}\frac{\bar{d}_i}{2}\eta_{di}(t) &= \bar{x}_i(t-d_i(t)) - \frac{1}{2}\left[\bar{x}_i(t-d_{i1})+\bar{x}_i(t-d_{i2})\right] \\ &= \frac{1}{2}\int_{-d_{i2}}^{-d_i(t)} \dot{\bar{x}}_i(t+\alpha)\,d\alpha - \frac{1}{2}\int_{-d_i(t)}^{-d_{i1}} \dot{\bar{x}}_i(t+\alpha)\,d\alpha \\ &= \frac{1}{2}\int_{-d_{i2}}^{-d_{i1}} \rho_i(\alpha)\,\xi_{di}(t+\alpha)\,d\alpha, \end{aligned} \tag{9.11}$$

where $\bar{d}_i = d_{i2} - d_{i1}, \xi_{di}(t) = \dot{\bar{x}}_i(t)$ and

$$\rho_i(\alpha) = \begin{cases} 1, & \text{if } \alpha \leq -d_i(t), \\ -1, & \text{if } \alpha > -d_i(t). \end{cases} \tag{9.12}$$

By substituting (9.11) into the system (9.7), and putting the delay uncertainty $\eta_{di}(t)$ into a feedback interconnection, the closed-loop fuzzy control system in (9.7) is rewritten as

$$\begin{aligned} &\mathscr{L}_{i1}: \begin{cases} \dot{\bar{x}}_i(t) = \tilde{A}_i(\mu_i)\bar{\xi}_i(t) + \frac{\bar{d}_i}{2}A_{di}(\mu_i)\eta_{di}(t) + E\sum\limits_{\substack{k=1\\k\neq i}}^{N}\bar{A}_{ik}(\mu_i)x_k(t) + B_{wi}(\mu_i)w_i(t) \\ z_i(t) = \tilde{C}_i(\mu_i)\bar{\xi}_i(t) + \frac{\bar{d}_i}{2}C_{di}(\mu_i)\eta_{di}(t) + D_{wi}(\mu_i)w_i(t) \\ \xi_{di}(t) = \dot{\bar{x}}_i(t), \end{cases} \\ &\mathscr{L}_{i2}: \eta_{di}(t) = \Delta_i\xi_{di}(t) \end{aligned} \tag{9.13}$$

with $\bar{x}_i(t) = \begin{bmatrix} \varphi_i^T(t) & 0 \end{bmatrix}^T, -d_{i2} \leq t \leq 0, i \in \mathscr{N}$, and Δ_i is an uncertain operator in $\mathscr{S}_{i2}$, and

$$\begin{cases} \tilde{A}_i(\mu_i) = \begin{bmatrix} A_i(\mu_i) & A_{d1i}(\mu_i) + \frac{1}{2}A_{di}(\mu_i) & A_{d2i}(\mu_i) + \frac{1}{2}A_{di}(\mu_i) \end{bmatrix}, \\ \tilde{C}_i(\mu_i) = \begin{bmatrix} C_i(\mu_i) & C_{d1i}(\mu_i) + \frac{1}{2}C_{di}(\mu_i) & C_{d2i}(\mu_i) + \frac{1}{2}C_{di}(\mu_i) \end{bmatrix}. \end{cases} \tag{9.14}$$

Now, it can be shown that the system in (9.13) is a feedback interconnection with two subsystems $\mathscr{S}_{i1}$ and $\mathscr{S}_{i2}$. Based on the interconnected system in (9.13), we present the following lemma:

Lemma 9.2: SSG Analysis of Interconnected System

Given the interconnected system with two subsystems $\mathscr{S}_{i1}$ and $\mathscr{S}_{i2}$ in (9.13), there exist nonsingular matrices $X_i, i \in \mathscr{N}$ such that the operator $\Delta_i : \xi_{di}(t) \longmapsto \eta_{di}(t)$ satisfies the property $\left\|X_i\Delta_iX_i^{-1}\right\|_\infty \leq 1$. ■

Proof. Consider zero initial conditions, and by using the relation in (9.11) and Jensen's inequality, it yields

$$\begin{aligned}
&\int_0^t \eta_{di}^T(\alpha)X_i^T X_i \eta_{di}(\alpha)d\alpha \\
&= \frac{1}{\bar{d}_i^2}\int_0^t \left[\int_{-d_{i2}}^{-d_{i1}} \rho_i(\beta)\xi_{di}(\alpha+\beta)d\beta\right]^T X_i^T X_i \left[\int_{-d_{i2}}^{-d_{i1}} \rho_i(\beta)\xi_{di}(\alpha+\beta)d\beta\right] d\alpha \\
&\le \frac{1}{\bar{d}_i^2}\int_0^t \left[\bar{d}_i \int_{-d_{i2}}^{-d_{i1}} \rho_i^2(\beta)\xi_{di}^T(\alpha+\beta)X_i^T X_i \xi_{di}(\alpha+\beta)d\beta\right] d\alpha \\
&= \frac{1}{\bar{d}_i}\int_{-d_{i2}}^{-d_{i1}} \left[\int_0^t \xi_{di}^T(\alpha+\beta)X_i^T X_i \xi_{di}(\alpha+\beta)d\alpha\right] d\beta \\
&= \frac{1}{\bar{d}_i}\int_{-d_{i2}}^{-d_{i1}} \left[\int_\beta^{t+\beta} \xi_{di}^T(\alpha)X_i^T X_i \xi_{di}(\alpha)d\alpha\right] d\beta \\
&\le \frac{1}{\bar{d}_i}\int_{-d_{i2}}^{-d_{i1}} \left[\int_0^t \xi_{di}^T(\alpha)X_i^T X_i \xi_{di}(\alpha)d\alpha\right] d\beta \\
&= \int_0^t \xi_{di}^T(\alpha)X_i^T X_i \xi_{di}(\alpha)d\alpha. \qquad (9.15)
\end{aligned}$$

Then, substituting $\eta_{di}(t) = \Delta_i X_i^{-1} X_i \xi_{di}(t)$ into the inequality (9.15), we have

$$\int_0^t \xi_{di}(t)X_i^T X_i^{-T}\Delta_i^T X_i^T X_i \Delta_i X_i^{-1} X_i \xi_{di}(t)d\alpha \le \int_0^t \xi_{di}^T(\alpha)X_i^T X_i \xi_{di}(\alpha)d\alpha, \quad (9.16)$$

which implies that the property $\left\|X_i\Delta_i X_i^{-1}\right\|_\infty \le 1$ holds, thus completing this proof.

Note: In this section we just consider the case when a single time-varying state delay exists in the system (9.3). However, the method proposed can be easily extended to the system (9.3) with multiple time-varying delays.

Note: In Lemma 9.2, the feedback subsystem $\mathscr{S}_{i2}$ satisfies the property $\left\|X_i\Delta_i X_i^{-1}\right\|_\infty \le 1$, which implies that $\left\|X\Delta X^{-1}\right\|_\infty \le 1$ holds, where $X = \text{diag}\underbrace{\{X_1\cdots X_i\cdots X_N\}}_N$, $\Delta = \text{diag}\underbrace{\{\Delta_1\cdots\Delta_i\cdots\Delta_N\}}_N$. Since the closed-loop fuzzy control system in (9.7) is equivalent to the interconnected system in (9.13), the system (9.7) is asymptotically stable by using Lemma 9.1, if the forward system $\mathscr{S}_{i1}$ is internally stable and the condition $\left\|X\mathbf{G}X^{-1}\right\|_\infty < 1$ holds.

9.1.3 DESIGN OF DECENTRALIZED DYNAMIC OUTPUT FEEDBACK CONTROL

In this subchapter, we present an $\mathscr{H}_\infty$ performance analysis result for the closed-loop fuzzy control system in (9.7). Consider the following Lyapunov-Krasovskii functional (LKF):

$$\begin{aligned} V(t) &= \sum_{i=1}^{N} V_i(t) \\ &= \sum_{i=1}^{N} [V_{i1}(t) + V_{i2}(t) + V_{i3}(t)] \end{aligned} \tag{9.17}$$

with

$$\begin{cases} V_{i1}(t) = \bar{x}_i^T(t) P_{i1} \bar{x}_i(t) + \int_{t-d_{i1}}^{t} \bar{x}_i^T(\alpha) Q_{i1} \bar{x}_i(\alpha) d\alpha \\ \quad + d_{i1} \int_{-d_{i1}}^{0} \int_{t+\beta}^{t} \dot{\bar{x}}_i^T(\alpha) Z_{i1} \dot{\bar{x}}_i(\alpha) d\alpha d\beta, \\ V_{i2}(t) = \bar{x}_i^T(t) P_{i2} \bar{x}_i(t) + \int_{t-d_{i2}}^{t} \bar{x}_i^T(\alpha) Q_{i2} \bar{x}_i(\alpha) d\alpha \\ \quad + d_{i2} \int_{-d_{i2}}^{0} \int_{t+\beta}^{t} \dot{\bar{x}}_i^T(\alpha) Z_{i2} \dot{\bar{x}}_i(\alpha) d\alpha d\beta, \\ V_{i3}(t) = \int_{t-d_i(t)}^{t} \bar{x}_i^T(\alpha) Q_{i3} \bar{x}_i(\alpha) d\alpha, \end{cases} \tag{9.18}$$

where $\{P_{i1}, P_{i2}, Q_{i3}, Z_{i1}, Z_{i2}\} \in \Re^{2n_{xi} \times 2n_{xi}}, i \in \mathscr{N}$, are positive definite symmetric matrices, and $\{Q_{i1}, Q_{i2}\} \in \Re^{2n_{xi} \times 2n_{xi}}, i \in \mathscr{N}$, are symmetric matrices.

Inspired by [13], we do not require the condition of $Q_{i1} > 0$ and $Q_{i2} > 0$ in (9.18). To ensure the positive property of $V(t)$, we define $\bar{x}(t) = \begin{bmatrix} \bar{x}_1^T(t) & \bar{x}_2^T(t) & \cdots & \bar{x}_N^T(t) \end{bmatrix}^T$ and give the following lemma.

Lemma 9.3: Novel Lyapunov-Krasovskii Functional

Consider the Lyapunov–Krasovskii functional (LKF) in (9.17), then $V(t) \geq \varepsilon \|\bar{x}(t)\|^2$ if there exist the positive definite symmetric matrices $\{P_{i1}, P_{i2}, Q_{i3}, Z_{i1}, Z_{i2}\} \in \Re^{2n_{xi} \times 2n_{xi}}$, and symmetric matrices $\{Q_{i1}, Q_{i2}\} \in \Re^{2n_{xi} \times 2n_{xi}}$, and the scalar $\varepsilon > 0$, such that for all $i \in \mathscr{N}$ the following inequalities hold:

$$\begin{bmatrix} \frac{P_{i1}}{d_{i1}} + Z_{i1} & -Z_{i1} \\ \star & Q_{i1} + Q_{i3} + Z_{i1} \end{bmatrix} > 0, \tag{9.19}$$

$$\begin{bmatrix} \frac{P_{i2}}{d_{i2}} + Z_{i2} & -Z_{i2} \\ \star & Q_{i2} + Z_{i2} \end{bmatrix} > 0. \tag{9.20}$$

■

Proof. Firstly, by using Jensen's inequality,

$$\begin{aligned}
d_{i1} \int_{-d_{i1}}^{0} \int_{t+\beta}^{t} \dot{\bar{x}}_i^T(\alpha) Z_{i1} \dot{\bar{x}}_i(\alpha)\, d\alpha d\beta \\
&\geq d_{i1} \int_{-d_{i1}}^{0} \frac{-1}{\beta} \left[\int_{t+\beta}^{t} \dot{\bar{x}}_i^T(\alpha)\, d\alpha\right]^T Z_{i1} \left[\int_{t+\beta}^{t} \dot{\bar{x}}_i(\alpha)\, d\alpha\right] d\beta \\
&= d_{i1} \int_{-d_{i1}}^{0} \frac{-1}{\beta} [\bar{x}_i(t) - \bar{x}_i(t+\beta)]^T Z_{i1} [\bar{x}_i(t) - \bar{x}_i(t+\beta)]\, d\beta \\
&= d_{i1} \int_{0}^{d_{i1}} \frac{1}{s} [\bar{x}_i(t) - \bar{x}_i(t-s)]^T Z_{i1} [\bar{x}_i(t) - \bar{x}_i(t-s)]\, ds \\
&\geq \int_{0}^{d_{i1}} [\bar{x}_i(t) - \bar{x}_i(t-s)]^T Z_{i1} [\bar{x}_i(t) - \bar{x}_i(t-s)]\, ds \\
&= \int_{t-d_{i1}}^{t} [\bar{x}_i(t) - \bar{x}_i(\alpha)]^T Z_{i1} [\bar{x}_i(t) - \bar{x}_i(\alpha)]\, d\alpha. \qquad (9.21)
\end{aligned}$$

Since the matrices $Q_{i3} > 0$, it yields

$$\int_{t-d_i(t)}^{t} \bar{x}_i^T(\alpha) Q_{i3} \bar{x}_i(\alpha)\, d\alpha \geq \int_{t-d_{i1}}^{t} \bar{x}_i^T(\alpha) Q_{i3} \bar{x}_i(\alpha)\, d\alpha. \qquad (9.22)$$

It follows from (9.18), (9.21), and (9.22) that

$$V_{i1}(t) + V_{i3}(t) \geq \int_{t-d_{i1}}^{t} \begin{bmatrix} \bar{x}_i(t) \\ \bar{x}_i(\alpha) \end{bmatrix}^T \begin{bmatrix} \frac{P_{i1}}{d_{i1}} + Z_{i1} & -Z_{i1} \\ \star & Q_{i1} + Q_{i3} + Z_{i1} \end{bmatrix} \begin{bmatrix} \bar{x}_i(t) \\ \bar{x}_i(\alpha) \end{bmatrix} d\alpha. \qquad (9.23)$$

Similarly, we also have

$$\begin{aligned}
d_{i2} \int_{-d_{i2}}^{0} \int_{t+\beta}^{t} \dot{\bar{x}}_i^T(\alpha) Z_{i2} \dot{\bar{x}}_i(\alpha)\, d\alpha d\beta \\
&\geq \int_{t-d_{i2}}^{t} [\bar{x}_i(t) - \bar{x}_i(\alpha)]^T Z_{i2} [\bar{x}_i(t) - \bar{x}_i(\alpha)]\, d\alpha, \qquad (9.24)
\end{aligned}$$

which implies that

$$V_{i2}(t) \geq \int_{t-d_{i2}}^{t} \begin{bmatrix} \bar{x}_i(t) \\ \bar{x}_i(\alpha) \end{bmatrix}^T \begin{bmatrix} \frac{P_{i2}}{d_{i2}} + Z_{i2} & -Z_{i2} \\ \star & Q_{i2} + Z_{i2} \end{bmatrix} \begin{bmatrix} \bar{x}_i(t) \\ \bar{x}_i(\alpha) \end{bmatrix} d\alpha. \qquad (9.25)$$

Based on the relations in (9.23) and (9.25), it is easy to see that the property $V(t) \geq \varepsilon \|\bar{x}(t)\|^2$ can be verified if the inequalities in (9.19) and (9.20) hold, thus completing this proof.

Note: The LKFs introduced in [9] require that all Lyapunov matrices are positive definite symmetric. We relax the restricted condition in the sense that matrices Q_{i1} and Q_{i2} may be negative definite symmetric. It is thus expected to reduce the design conservatism, and will be validated in the simulation examples.

Based on the interconnected fuzzy system in (9.13) and the LKF in (9.17), we have the following $\mathscr{H}_\infty$ performance analysis result.

Lemma 9.4: $\mathscr{H}_\infty$ Performance Analysis

Given the large-scale T-S fuzzy system in (9.3) and decentralized fuzzy output feedback controller in (9.5), then the closed-loop fuzzy control system in (9.7) is asymptotically stable with an $\mathscr{H}_\infty$ disturbance attenuation level γ, if there exist the positive definite symmetric matrices $\{P_{i1}, P_{i2}, Q_{i3}, Z_{i1}, Z_{i2}, \bar{X}_i\} \in \Re^{2n_{xi}\times 2n_{xi}}$, symmetric matrices $\{Q_{i1}, Q_{i2}\} \in \Re^{2n_{xi}\times 2n_{xi}}$, matrix multipliers $\mathscr{G}_i \in \Re^{(10n_{xi}+n_{wi})\times 2n_{xi}}$, and the scalars $0 < \varepsilon_i(\mu_i) \leq \varepsilon_0$, such that for all $i \in \mathscr{N}$ the following matrix inequalities hold:

$$\begin{bmatrix} \frac{P_{i1}}{d_{i1}} + Z_{i1} & -Z_{i1} \\ \star & Q_{i1} + Q_{i3} + Z_{i1} \end{bmatrix} > 0, \tag{9.26}$$

$$\begin{bmatrix} \frac{P_{i2}}{d_{i2}} + Z_{i2} & -Z_{i2} \\ \star & Q_{i2} + Z_{i2} \end{bmatrix} > 0, \tag{9.27}$$

$$\begin{bmatrix} -\mathbf{I} & \mathbb{C}_i(\mu_i) & 0 \\ \star & \Theta_i + \mathrm{Sym}\{\mathscr{G}_i \mathbb{A}_i(\mu_i)\} & \mathscr{G}_i E A_{ik,k\neq i}(\mu_i) \\ \star & \star & -E_{ik,k\neq i}(\mu_i)\mathbf{I} \end{bmatrix} < 0, \tag{9.28}$$

where

$$\Theta_i = \begin{bmatrix} \Theta_i^{(1)} & P_{i1}+P_{i2} & 0 & 0 & 0 & 0 \\ \star & \Theta_i^{(2)} & Z_{i1} & Z_{i2} & 0 & 0 \\ \star & \star & \Theta_i^{(3)} & -\frac{1-\tau_i}{4}Q_{i3} & -\frac{(1-\tau_i)\bar{d}_i}{4}Q_{i3} & 0 \\ \star & \star & \star & -Q_{i2}-Z_{i2}-\frac{1-\tau_i}{4}Q_{i3} & -\frac{(1-\tau_i)\bar{d}_i}{4}Q_{i3} & 0 \\ \star & \star & \star & \star & -\bar{X}_i - \frac{(1-\tau_i)\bar{d}_i^2}{4}Q_{i3} & 0 \\ \star & \star & \star & \star & \star & -\gamma^2\mathbf{I} \end{bmatrix},$$

$$\Theta_i^{(1)} = d_{i1}^2 Z_{i1} + d_{i2}^2 Z_{i2} + \bar{X}_i, \Theta_i^{(3)} = -Q_{i1} - Z_{i1} - \frac{1-\tau_i}{4}Q_{i3},$$

$$\Theta_i^{(2)} = Q_{i1} + Q_{i2} + Q_{i3} - Z_{i1} - Z_{i2} + \varepsilon_0(N-1)EE^T, E = \begin{bmatrix} \mathbf{I} & 0 \end{bmatrix}^T,$$

$$A_{ik,k\neq i}(\mu_i) = \underbrace{\begin{bmatrix} \bar{A}_{i1}(\mu_i) & \cdots & \bar{A}_{ik,k\neq i}(\mu_i) & \cdots & \bar{A}_{iN}(\mu_i) \end{bmatrix}}_{N-1},$$

$$E_{ik,k\neq i}(\mu_i) = diag\underbrace{\{ \varepsilon_{i1}(\mu_i)\mathbf{I}_{n_{xi}} \ \cdots \ \varepsilon_{ik,k\neq i}(\mu_i)\mathbf{I}_{n_{xi}} \ \cdots \ \varepsilon_{iN}(\mu_i)\mathbf{I}_{n_{xi}} \}}_{N-1},$$

$$\mathbb{A}_i(\mu_i) = \left[-\mathbf{I} \; A_i(\mu_i) \; A_{d1i}(\mu_i) + \frac{1}{2}A_{di}(\mu_i) \; A_{d2i}(\mu_i) + \frac{1}{2}A_{di}(\mu_i) \; \frac{\bar{d}_i}{2}A_{di}(\mu_i) \; B_{\omega i}(\mu_i) \right],$$

$$\mathbb{C}_i(\mu_i) = \left[0 \; C_i(\mu_i) \; C_{d1i}(\mu_i) + \frac{1}{2}C_{di}(\mu_i) \; C_{d2i}(\mu_i) + \frac{1}{2}C_{di}(\mu_i) \; \frac{\bar{d}_i}{2}C_{di}(\mu_i) \; D_{\omega i}(\mu_i) \right]. \tag{9.29}$$

■

Proof. By taking the derivative of $V_i(t)$ along the trajectory of the forward subsystem $\mathscr{S}_{i1}$ in (9.13), it yields

$$\begin{aligned}\dot{V}_i(t) &= \dot{V}_{i1}(t)+\dot{V}_{i2}(t)+\dot{V}_{i3}(t)\\ &\leq 2\bar{x}_i^T(t)(P_{i1}+P_{i2})\dot{\bar{x}}_i(t)+\bar{x}_i^T(t)Q_{i3}\bar{x}_i(t)\\ &\quad -(1-\tau_i)\bar{x}_i^T(t-d_i(t))Q_{i3}\bar{x}_i(t-d_i(t))\\ &\quad +\bar{x}_i^T(t)Q_{i1}\bar{x}_i(t)-\bar{x}_i^T(t-d_{i1})Q_{i1}\bar{x}_i(t-d_{i1})\\ &\quad +\bar{x}_i^T(t)Q_{i2}\bar{x}_i(t)-\bar{x}_i^T(t-d_{i2})Q_{i2}\bar{x}_i(t-d_{i2})\\ &\quad +d_{i1}^2\dot{\bar{x}}_i^T(t)Z_{i1}\dot{\bar{x}}_i(t)-d_{i1}\int_{t-d_{i1}}^{t}\dot{\bar{x}}_i^T(\alpha)Z_{i1}\dot{\bar{x}}_i(\alpha)\,d\alpha\\ &\quad +d_{i2}^2\dot{\bar{x}}_i^T(t)Z_{i2}\dot{\bar{x}}_i(t)-d_{i2}\int_{t-d_{i2}}^{t}\dot{\bar{x}}_i^T(\alpha)Z_{i2}\dot{\bar{x}}_i(\alpha)\,d\alpha. \end{aligned} \tag{9.30}$$

Define $\bar{\chi}_i(t)=\begin{bmatrix}\dot{\bar{x}}_i^T(t) & \bar{x}_i^T(t) & \bar{x}_i^T(t-d_{i1}) & \bar{x}_i^T(t-d_{i2}) & \eta_{di}^T(t) & w_i^T(t)\end{bmatrix}^T$ and the matrix multipliers $\mathscr{G}_i\in\Re^{(10n_{xi}+n_{wi})\times 2n_{xi}}$, and then it follows from the forward subsystem $\mathscr{S}_{i1}$ in (9.13) that

$$\begin{aligned} 0 &= 2\sum_{i=1}^{N}\bar{\chi}_i^T(t)\mathscr{G}_i\left[-\dot{\bar{x}}_i(t)+\tilde{A}_i(\mu_i)\bar{\xi}_i(t)+\frac{\bar{d}_i}{2}A_{di}(\mu_i)\eta_{di}(t)+B_{wi}(\mu_i)w_i(t)\right]\\ &\quad +2\sum_{i=1}^{N}\bar{\chi}_i^T(t)\mathscr{G}_iE\sum_{\substack{k=1\\k\neq i}}^{N}\bar{A}_{ik}(\mu_i)x_k(t). \end{aligned} \tag{9.31}$$

Note that

$$2\bar{x}^T\bar{y}\leq\kappa^{-1}\bar{x}^T\bar{x}+\kappa\bar{y}^T\bar{y}, \tag{9.32}$$

where $\bar{x},\bar{y}\in\Re^n$ and scalar $\kappa>0$, and

$$\sum_{i=1}^{N}\sum_{\substack{k=1\\k\neq i}}^{N}x_k^T(t)x_k(t)=(N-1)\sum_{i=1}^{N}x_i^T(t)x_i(t). \tag{9.33}$$

Define the scalar parameters $0<\varepsilon_{ik}(\mu_i)\leq\varepsilon_0$, where $\varepsilon_{ik}(\mu_i):=\sum_{l=1}^{r_i}\mu_{il}\varepsilon_{ikl}, i\in\mathscr{N}$, and by using the relations in (9.32) and (9.33), we get

$$
\begin{aligned}
0 &= 2\sum_{i=1}^{N} \bar{\chi}_i^T(t)\mathcal{G}_i\left[-\dot{\bar{x}}_i(t) + \tilde{A}_i(\mu_i)\bar{\xi}_i(t) + \frac{\bar{d}_i}{2}A_{di}(\mu_i)\eta_{di}(t) + B_{wi}(\mu_i)w_i(t)\right] \\
&\quad + 2\sum_{i=1}^{N} \bar{\chi}_i^T(t)\mathcal{G}_i E \sum_{\substack{k=1 \\ k\neq i}}^{N} \bar{A}_{ik}(\mu_i)x_k(t) \\
&= 2\sum_{i=1}^{N} \bar{\chi}_i^T(t)\mathcal{G}_i\left[-\dot{\bar{x}}_i(t) + \tilde{A}_i(\mu_i)\bar{\xi}_i(t) + \frac{\bar{d}_i}{2}A_{di}(\mu_i)\eta_{di}(t) + B_{wi}(\mu_i)w_i(t)\right] \\
&\quad + 2\sum_{i=1}^{N}\sum_{\substack{k=1 \\ k\neq i}}^{N} \bar{\chi}_i^T(t)\mathcal{G}_i E\bar{A}_{ik}(\mu_i)x_k(t) \\
&\leq 2\sum_{i=1}^{N} \bar{\chi}_i^T(t)\mathcal{G}_i\left[-\dot{\bar{x}}_i(t) + \tilde{A}_i(\mu_i)\bar{\xi}_i(t) + \frac{\bar{d}_i}{2}A_{di}(\mu_i)\eta_{di}(t) + B_{wi}(\mu_i)w_i(t)\right] \\
&\quad + \sum_{i=1}^{N}\sum_{\substack{k=1 \\ k\neq i}}^{N} \varepsilon_{ik}^{-1}(\mu_i)\bar{\chi}_i^T(t)\mathcal{G}_i E\bar{A}_{ik}(\mu_i)\bar{A}_{ik}^T(\mu_i)E^T\mathcal{G}_i^T\bar{\chi}_i(t) \\
&\quad + \sum_{i=1}^{N} \varepsilon_0(N-1)x_i^T(t)x_i(t).
\end{aligned} \tag{9.34}
$$

Then, by using Jensen's inequality,

$$
\begin{aligned}
-d_{i1}\int_{t-d_{i1}}^{t} \dot{\bar{x}}_i^T(\alpha)Z_{i1}\dot{\bar{x}}_i(\alpha)d\alpha &\leq -\left[\int_{t-d_{i1}}^{t} \dot{\bar{x}}_i(\alpha)d\alpha\right]^T Z_{i1}\left[\int_{t-d_{i1}}^{t} \dot{\bar{x}}_i(\alpha)d\alpha\right] \\
&= -\left(\bar{x}_i(t) - \bar{x}_i(t-d_{i1})\right)^T Z_{i1}\left(\bar{x}_i(t) - \bar{x}_i(t-d_{i1})\right).
\end{aligned} \tag{9.35}
$$

Similarly,

$$
\begin{aligned}
-d_{i2}\int_{t-d_{i2}}^{t} \dot{\bar{x}}_i^T(a)Z_{i2}\dot{\bar{x}}_i(\alpha)d\alpha &\leq -\left[\int_{t-d_{i2}}^{t} \dot{\bar{x}}_i(\alpha)d\alpha\right]^T Z_{i2}\left[\int_{t-d_{i2}}^{t} \dot{\bar{x}}_i(\alpha)d\alpha\right] \\
&= -\left(\bar{x}_i(t) - \bar{x}_i(t-d_{i2})\right)^T Z_{i2}\left(\bar{x}_i(t) - \bar{x}_i(t-d_{i2})\right).
\end{aligned} \tag{9.36}
$$

Denote $\bar{X}_i = X_i^T X_i, i \in \mathcal{N}$, and we consider the following index

$$
\begin{aligned}
J(t) &= \sum_{i=1}^{N} J_i(t) \\
&= \sum_{i=1}^{N}\int_0^{\infty} \left\{\xi_{di}^T(t)\bar{X}_i\xi_{di}(t) - \eta_{di}^T(t)\bar{X}_i\eta_{di}(t) + z_i^T(t)z_i(t) - \gamma^2\omega_i^T(t)\omega_i(t)\right\}dt.
\end{aligned} \tag{9.37}
$$

Under zero initial conditions, it can be known that $V(0) = 0$ and $V(\infty) \geq 0$. Then, it follows from (9.11), (9.30), and (9.34)-(9.37) that

$$\begin{aligned} J(t) &\leq J(t) + V(\infty) - V(0) \\ &= \sum_{i=1}^{N} \int_{0}^{\infty} \{\dot{V}_i(t) + \xi_{di}^T(t)\bar{X}_i\xi_{di}(t) - \eta_{di}^T(t)\bar{X}_i\eta_{di}(t) + z_i^T(t)z_i(t) - \gamma^2\omega_i^T(t)\omega_i(t)\}\, dt \\ &\leq \sum_{i=1}^{N} \int_{0}^{\infty} \bar{\chi}_i^T(t)\Sigma_i(\mu_i)\bar{\chi}_i(t)\, dt, \end{aligned} \tag{9.38}$$

with

$$\Sigma_i(\mu_i) = \Theta_i + \mathrm{Sym}\{\mathcal{G}_i\mathbb{A}_i(\mu_i)\} + \sum_{\substack{k=1\\k\neq i}}^{N} \varepsilon_{ik}^{-1}(\mu_i)\mathcal{G}_iE\bar{A}_{ik}(\mu_i)\bar{A}_{ik}^T(\mu_i)E^T\mathcal{G}_i^T + \mathbb{C}_i^T(\mu_i)\mathbb{C}_i(\mu_i), \tag{9.39}$$

where $\Theta_i, \mathbb{A}_i(\mu_i)$ and $\mathbb{C}_i(\mu_i)$ are defined in (9.29).

By using the Schur complement to (9.28), it is easy to see from (9.38) that the inequality in (9.28) implies $\dot{V}(t) < 0$ and $\|\mathbf{X}\mathbf{G}X^{-1}\|_{\infty} < 1$, when $w_i(t) = 0$. Then, based on Lemmas 9.1 and 9.2, it can be known that the closed-loop fuzzy control system (9.7) is asymptotically stable. In addition, it follows from (9.28) and (9.38) that $J(t) < 0$. Together with the relation (9.15), it yields $\int_0^{\infty} \tilde{z}^T(t)\tilde{z}(t)dt < \gamma^2 \int_0^{\infty} \tilde{w}^T(t)\tilde{w}(t)dt$, thus completing this proof.

It is noted that the $\mathcal{H}_{\infty}$ performance analysis result given in (9.28) is nonlinear when the controller gains are unknown. In the following we will address the decentralized DOF $\mathcal{H}_{\infty}$ controller design for the large-scale T-S fuzzy system in (9.3). By specifying the matrix multipliers $\mathcal{G}_i$ and using the matrix decomposition technique, the nonlinear matrix inequalities in (9.28) are formulated into the linear ones. The corresponding result is given by the following theorem.

Theorem 9.1: $\mathcal{H}_{\infty}$ DOF Controller Design

Consider a large-scale T-S fuzzy system in (9.3) and two positive scalar parameters ε_0 and ε. Then, a decentralized dynamic output feedback controller in the form (9.5) exists, such that the closed-loop fuzzy control system (9.7) is asymptotically stable with an $\mathcal{H}_{\infty}$ disturbance attenuation level γ, if there exist the positive definite symmetric matrices $\{\bar{P}_{i1}, \bar{P}_{i2}, \bar{Q}_{i3}, \bar{Z}_{i1}, \bar{Z}_{i2}, \tilde{X}_i\} \in \Re^{2n_{xi}\times 2n_{xi}}$, the symmetric matrices $\{\bar{Q}_{i1}, \bar{Q}_{i2}\} \in \Re^{2n_{xi}\times 2n_{xi}}$, and matrices $\{V_{i1}, U_{i1}, M_i, \bar{A}_{c0ilj}, \bar{A}_{c1ilj}, \bar{A}_{c2ilj}\} \in \Re^{n_{xi}\times n_{xi}}$, $\bar{B}_{cil} \in \Re^{n_{xi}\times n_{yi}}$, $\{\bar{C}_{c0il}, \bar{C}_{c1il}, \bar{C}_{c2il}\} \in \Re^{n_{ui}\times n_{xi}}, D_{ci} \in \Re^{n_{ui}\times n_{yi}}$, and the positive scalars $\varepsilon_{il} \leq \varepsilon_0$, such that for all $i \in \mathcal{N}$ the following LMIs hold:

$$\begin{bmatrix} \frac{\bar{P}_{i1}}{d_{i1}} + \bar{Z}_{i1} & -\bar{Z}_{i1} \\ \star & \bar{Q}_{i1} + \bar{Q}_{i3} + \bar{Z}_{i1} \end{bmatrix} > 0, \tag{9.40}$$

$$\begin{bmatrix} \frac{\bar{P}_{i2}}{d_{i2}}+\bar{Z}_{i2} & -\bar{Z}_{i2} \\ \star & \bar{Q}_{i2}+\bar{Z}_{i2} \end{bmatrix} > 0, \tag{9.41}$$

and

$$\bar{\Sigma}_{ill} < 0, 1 \leq l \leq r_i \tag{9.42}$$
$$\bar{\Sigma}_{ilj} + \bar{\Sigma}_{ijl} < 0, 1 \leq l < j \leq r_i \tag{9.43}$$

where

$$\bar{\Sigma}_{ilj} = \begin{bmatrix} -\varepsilon_0^{-1}(N-1)^{-1}\mathbf{I} & 0 & \bar{\mathscr{E}} & 0 & 0 & 0 \\ \star & -\mathbf{I} & \bar{\Xi}_{ilj} & 0 & \frac{\bar{d}_i}{2}F_{il}D_{ci}C_{dij} & 0 \\ \star & \star & \tilde{\Theta}_i + Sym\{\bar{\Pi}_{ilj}\} & \mathscr{G}_i A_{ikl,k\neq i} & \Upsilon_{ilj} & \varepsilon \mathscr{U}_i^T \\ \star & \star & \star & -E_{ikl,k\neq i}\mathbf{I} & 0 & 0 \\ \star & \star & \star & \star & -\varepsilon\mathbf{I} & 0 \\ \star & \star & \star & \star & \star & -\varepsilon\mathbf{I} \end{bmatrix} \tag{9.44}$$

with

$$\tilde{\Theta}_i = \begin{bmatrix} \tilde{\Theta}_i^{(1)} & \bar{P}_{i1}+\bar{P}_{i2} & 0 & 0 & 0 & 0 \\ \star & \tilde{\Theta}_i^{(2)} & \bar{Z}_{i1} & \bar{Z}_{i2} & 0 & 0 \\ \star & \star & \tilde{\Theta}_i^{(3)} & -\frac{1-\tau_i}{4}\bar{Q}_{i3} & -\frac{(1-\tau_i)\bar{d}_i}{4}\bar{Q}_{i3} & 0 \\ \star & \star & \star & -\bar{Q}_{i2}-\bar{Z}_{i2}-\frac{1-\tau_i}{4}\bar{Q}_{i3} & -\frac{(1-\tau_i)\bar{d}_i}{4}\bar{Q}_{i3} & 0 \\ \star & \star & \star & \star & -\tilde{X}_i - \frac{(1-\tau)\bar{d}_i^2}{4}\bar{Q}_{i3} & 0 \\ \star & \star & \star & \star & \star & -\gamma^2\mathbf{I} \end{bmatrix},$$

$$\tilde{\Theta}_i^{(1)} = d_{i1}^2\bar{Z}_{i1} + d_{i2}^2\bar{Z}_{i2} + \tilde{X}_i, \tilde{\Theta}_i^{(2)} = \bar{Q}_{i1}+\bar{Q}_{i2}+\bar{Q}_{i3}-\bar{Z}_{i1}-\bar{Z}_{i2}, \tilde{\Theta}_i^{(3)}$$
$$= -\bar{Q}_{i1} - \bar{Z}_{i1} - \frac{1-\tau_i}{4}\bar{Q}_{i3},$$

$$\bar{\Pi}_{ilj} = \begin{bmatrix} -\bar{M}_i & \Pi_{ilj}^{(1)} & \Pi_{ilj}^{(2)} & \Pi_{ilj}^{(3)} & \Pi_{ilj}^{(4)} & \Pi_{ilj}^{(5)} \\ -\bar{M}_i & \Pi_{ilj}^{(1)} & \Pi_{ilj}^{(2)} & \Pi_{ilj}^{(3)} & \Pi_{ilj}^{(4)} & \Pi_{ilj}^{(5)} \\ 0_{(6n_{xi}+n_{wi})\times 2n_{xi}} & 0 & 0 & 0 & 0 & 0 \end{bmatrix},$$

$$\Pi_{ilj}^{(1)} = \begin{bmatrix} A_{il}U_{i1}+B_{il}\bar{C}_{c0ij} & A_{il}+B_{il}D_{ci}C_{ij} \\ \bar{A}_{c0ilj} & V_{i1}^T A_{il}+\bar{B}_{cil}C_{ij} \end{bmatrix}, \bar{M}_i = \begin{bmatrix} U_{i1} & \mathbf{I} \\ M_i & V_{i1}^T \end{bmatrix},$$

$$\Pi_{ilj}^{(2)} = \begin{bmatrix} \frac{1}{2}A_{dil}U_{i1}+B_{il}\bar{C}_{c1ij} & \frac{1}{2}A_{dil}+\frac{1}{2}B_{il}D_{ci}C_{dij} \\ \bar{A}_{c1ilj} & \frac{1}{2}V_{i1}^T A_{dil}+\frac{1}{2}\bar{B}_{cil}C_{dij} \end{bmatrix},$$

$$\Pi_{ilj}^{(3)} = \begin{bmatrix} \frac{1}{2}A_{dil}U_{i1}+B_{il}\bar{C}_{c2ij} & \frac{1}{2}A_{dil}+\frac{1}{2}B_{il}D_{ci}C_{dij} \\ \bar{A}_{c2ilj} & \frac{1}{2}V_{i1}^T A_{dil}+\frac{1}{2}\bar{B}_{cil}C_{dij} \end{bmatrix},$$

$$\Pi_{ilj}^{(4)} = \frac{\bar{d}_i}{2}\begin{bmatrix} A_{dil}U_{i1} & A_{dil}+B_{il}D_{ci}C_{dij} \\ 0 & V_{i1}^T A_{dil}+\bar{B}_{cil}C_{dij} \end{bmatrix}, \Pi_{ilj}^{(5)} = \begin{bmatrix} B_{il}D_{ci}D_{wij}+B_{wil} \\ \bar{B}_{cil}D_{wij}+V_{i1}^T B_{wil} \end{bmatrix}, \tag{9.45}$$

and

$$\begin{aligned}
&\bar{\Xi}_{ilj} = \left[\begin{array}{cccccc} 0 & \Xi_{ilj}^{(1)} & \Xi_{ilj}^{(2)} & \Xi_{ilj}^{(3)} & \Xi_{ilj}^{(4)} & F_{il}D_{ci}D_{wij}+F_{wil} \end{array}\right],\\
&\Xi_{ilj}^{(1)} = \left[\begin{array}{cc} L_{il}U_{i1}+F_{il}\bar{C}_{c0ij} & L_{il}+F_{il}D_{ci}C_{ij} \end{array}\right],\\
&\Xi_{ilj}^{(2)} = \left[\begin{array}{cc} \frac{1}{2}L_{dil}U_{i1}+F_{il}\bar{C}_{c1ij} & \frac{1}{2}L_{dil}+\frac{1}{2}F_{il}D_{ci}C_{dij} \end{array}\right],\\
&\Xi_{ilj}^{(3)} = \left[\begin{array}{cc} \frac{1}{2}L_{dil}U_{i1}+F_{il}\bar{C}_{c2ij} & \frac{1}{2}L_{dil}+\frac{1}{2}F_{il}D_{ci}C_{dij} \end{array}\right],\\
&\Xi_{ilj}^{(4)} = \frac{\bar{d}_i}{2}\left[\begin{array}{cc} L_{dil}U_{i1} & L_{dil}+F_{il}D_{ci}C_{dij} \end{array}\right],\\
&\Upsilon_{ilj} = \left[\begin{array}{c} \frac{\bar{d}_i}{2}B_{il}D_{ci}C_{dij} \\ \frac{\bar{d}_i}{2}V_{i1}^{T}A_{dil}+\frac{\bar{d}_i}{2}\bar{B}_{cil}C_{dij} \\ \frac{\bar{d}_i}{2}B_{il}D_{ci}C_{dij} \\ \frac{\bar{d}_i}{2}V_{i1}^{T}A_{dil}+\frac{\bar{d}_i}{2}\bar{B}_{cil}C_{dij} \\ 0_{(6n_{xi}+n_{wi})\times n_{xi}} \end{array}\right], \mathscr{G}_i = \left[\begin{array}{c} \left[\begin{array}{c} \mathbf{I} \\ V_{i1}^{T} \end{array}\right] \\ \left[\begin{array}{c} \mathbf{I} \\ V_{i1}^{T} \end{array}\right] \\ 0_{(6n_{xi}+n_{wi})\times n_{xi}} \end{array}\right],\\
&\mathscr{U}_i = \left[\begin{array}{cccccc} 0 & 0 & 0 & 0 & \left[\begin{array}{cc} U_{i1} & 0 \end{array}\right] & 0 \end{array}\right], \bar{\mathscr{E}} = \left[\begin{array}{cccccc} 0 & \left[\begin{array}{cc} U_{i1} & \mathbf{I} \end{array}\right] & 0 & 0 & 0 & 0 \end{array}\right],\\
&A_{ikl,k\neq i} = \underbrace{\left[\begin{array}{ccccc} \bar{A}_{i1l} & \cdots & \bar{A}_{ikl,k\neq i} & \cdots & \bar{A}_{iNl} \end{array}\right]}_{N-1},\\
&E_{ikl,k\neq i} = \mathrm{diag}\underbrace{\left\{\begin{array}{ccccc} \varepsilon_{i1l}\mathbf{I}_{n_{xi}} & \cdots & \varepsilon_{ikl,k\neq i}\mathbf{I}_{n_{xi}} & \cdots & \varepsilon_{iNl}\mathbf{I}_{n_{xi}} \end{array}\right\}}_{N-1}.
\end{aligned} \tag{9.46}$$

Furthermore, a decentralized dynamic output feedback controller in the form of (9.5) is given by

$$\begin{aligned}
&D_{ci}, C_{c0il} = \left(\bar{C}_{c0il}-D_{ci}C_{il}U_{i1}\right)U_{i2}^{-1}, C_{c1il} = \left(\bar{C}_{c1il}-\frac{1}{2}D_{ci}C_{dil}U_{i1}\right)U_{i2}^{-1},\\
&C_{c2il} = \left(\bar{C}_{c2il}-\frac{1}{2}D_{ci}C_{dil}U_{i1}\right)U_{i2}^{-1}, B_{cil} = V_{i2}^{-T}\left(\bar{B}_{cil}-V_{i1}^{T}B_{il}D_{ci}\right),\\
&A_{c0ilj} = V_{i2}^{-T}\left\{\bar{A}_{c0ilj}-V_{i1}^{T}A_{il}U_{i1}-V_{i1}^{T}B_{il}D_{ci}C_{ij}U_{i1}-V_{i2}^{T}B_{cil}C_{ij}U_{i1}-V_{i1}^{T}B_{il}C_{c0ij}U_{i2}\right\}U_{i2}^{-1},\\
&A_{c1ilj} = V_{i2}^{-T}\left\{\bar{A}_{c1ilj}-V_{i1}^{T}B_{il}C_{c1ij}U_{i2}-\frac{1}{2}\bar{V}_{ilj}\right\}U_{i2}^{-1},\\
&A_{c2ilj} = V_{2i}^{-T}\left\{\bar{A}_{c2ilj}-V_{i1}^{T}B_{il}C_{c2ij}U_{i2}-\frac{1}{2}\bar{V}_{ilj}\right\}U_{i2}^{-1},\\
&\bar{V}_{ilj} = \left(V_{i1}^{T}A_{dil}+V_{i1}^{T}B_{il}D_{ci}C_{dij}+V_{i2}^{T}B_{cil}C_{dij}\right)U_{i1},
\end{aligned} \tag{9.47}$$

where $U_{i2} = V_{i2}^{-T}M_i - V_{i2}^{-T}V_{i1}^{T}U_{i1}, (l,j)\in L_i, i\in\mathscr{N}$. ■

Proof. Firstly, by using Schur complement, the inequality in (9.28) can be rewritten as

$$\begin{bmatrix} -\varepsilon_0^{-1}(N-1)^{-1}\mathbf{I} & 0 & \mathscr{E} & 0 \\ \star & -\mathbf{I} & \mathbb{C}_i(\mu_i) & 0 \\ \star & \star & \bar{\Theta}_i+\mathrm{Sym}\{\mathscr{G}_i\mathbb{A}_i(\mu_i)\} & \mathscr{G}_iEA_{ik,k\neq i}(\mu_i) \\ \star & \star & \star & -E_{ik,k\neq i}(\mu_i)\mathbf{I} \end{bmatrix}<0, \quad (9.48)$$

where $\mathbb{A}_i(\mu_i),\mathbb{C}_i(\mu_i),A_{ik,k\neq i}(\mu_i)$, and $E_{ik,k\neq i}(\mu_i)$ are defined in (9.29), and

$$\bar{\Theta}_i=\begin{bmatrix} \Theta_i^{(1)} & P_{i1}+P_{i2} & 0 & 0 & 0 & 0 \\ \star & \bar{\Theta}_i^{(2)} & Z_{i1} & Z_{i2} & 0 & 0 \\ \star & \star & \Theta_i^{(3)} & -\frac{1-\tau_i}{4}Q_{i3} & -\frac{(1-\tau_i)\bar{d}_i}{4}Q_{i3} & 0 \\ \star & \star & \star & -Q_{i2}-Z_{i2}-\frac{1-\tau_i}{4}Q_{i3} & -\frac{(1-\tau_i)\bar{d}_i}{4}Q_{i3} & 0 \\ \star & \star & \star & \star & -\bar{X}_i-\frac{(1-\tau)\bar{d}_i^2}{4}Q_{i3} & 0 \\ \star & \star & \star & \star & \star & -\gamma^2\mathbf{I} \end{bmatrix},$$

$$\Theta_i^{(1)}=d_{i1}^2Z_{i1}+d_{i2}^2Z_{i2}+\bar{X}_i,\bar{\Theta}_i^{(2)}=Q_{i1}+Q_{i2}+Q_{i3}-Z_{i1}-Z_{i2},$$
$$\Theta_i^{(3)}=-Q_{i1}-Z_{i1}-\frac{1-\tau_i}{4}Q_{i3},\mathscr{E}=\begin{bmatrix} 0 & E^T & 0 & 0 & 0 & 0 \end{bmatrix},E=\begin{bmatrix} \mathbf{I} & 0 \end{bmatrix}^T. \quad (9.49)$$

For the simplification of controller design, we directly specify the matrix multipliers $\mathscr{G}_i$ as

$$\mathscr{G}_i=\begin{bmatrix} G_i & G_i & 0_{2n_{xi}\times(6n_{xi}+n_{wi})} \end{bmatrix}^T, i\in\mathscr{N} \quad (9.50)$$

where matrices $G_i\in\Re^{2n_{xi}\times 2n_{xi}}$ are nonsingular.

Now, we further define [10]

$$W_i=\begin{bmatrix} U_{i1} & \mathbf{I} \\ U_{i2} & 0 \end{bmatrix},G_i=\begin{bmatrix} V_{i1} & \bullet \\ V_{i2} & \bullet \end{bmatrix},G_i^{-1}=\begin{bmatrix} U_{i1} & \bullet \\ U_{i2} & \bullet \end{bmatrix},$$
$$\Gamma_1:=\mathrm{diag}\{\mathbf{I}_{n_{xi}} \quad \mathbf{I}_{n_{zi}} \quad W_i \quad W_i \quad W_i \quad W_i \quad W_i \quad \mathbf{I}_{n_{wi}} \quad E\},$$
$$E=\mathrm{diag}\underbrace{\{\mathbf{I}_{n_{xi}} \quad \cdots \quad \mathbf{I}_{n_{xi}}\}}_{N-1},\bar{P}_{i1}=W_i^TP_{i1}W_i,\bar{P}_{i2}=W_i^TP_{i2}W_i,$$
$$\bar{Q}_{i1}=W_i^TQ_{i1}W_i,\bar{Q}_{i2}=W_i^TQ_{i2}W_i,\bar{Q}_{i3}=W_i^TQ_{i3}W_i,\bar{Z}_{i1}=W_i^TZ_{i1}W_i,$$
$$\bar{Z}_{i2}=W_i^TZ_{i2}W_i,\tilde{X}_i=W_i^T\bar{X}_iW_i,M_i=V_{i1}^TU_{i1}+V_{i2}^TU_{i2},$$
$$\bar{A}_{c0ilj}=V_{i1}^TA_{il}U_{i1}+V_{i1}^TB_{il}D_{ci}C_{ij}U_{i1}+V_{i2}^TB_{cil}C_{ij}U_{i1}+V_{i1}^TB_{il}C_{c0ij}U_{i2}+V_{i2}^TA_{c0ilj}U_{i2},$$
$$\bar{A}_{c1ilj}=V_{i1}^TB_{il}C_{c1ij}U_{i2}+V_{i2}^TA_{c1ilj}U_{i2}+\frac{1}{2}\left[V_{i1}^TA_{dil}+V_{i1}^TB_{il}D_{ci}C_{dij}+V_{i2}^TB_{cil}C_{dij}\right]U_{i1},$$
$$\bar{A}_{c2ilj}=V_{i1}^TB_{il}C_{c2ij}U_{i2}+V_{i2}^TA_{c2ilj}U_{i2}+\frac{1}{2}\left[V_{i1}^TA_{dil}+V_{i1}^TB_{il}D_{ci}C_{dij}+V_{i2}^TB_{cil}C_{dij}\right]U_{i1},$$
$$\bar{C}_{c0il}=D_{ci}C_{il}U_{i1}+C_{c0il}U_{i2},\bar{C}_{c1il}=C_{c1il}U_{i2}+\frac{1}{2}D_{ci}C_{dil}U_{i1},$$
$$\bar{C}_{c2il}=C_{c2il}U_{i2}+\frac{1}{2}D_{ci}C_{dil}U_{i1},\bar{B}_{cil}=V_{i1}^TB_{il}D_{ci}+V_{i2}^TB_{cil}, \quad (9.51)$$

where matrices $W_i \in \Re^{2n_{xi}\times 2n_{xi}}$ are nonsingular; • denote the elements satisfying $G_iG_i^{-1} = \mathbf{I}$.

By substituting the matrix multipliers $\mathscr{G}_i$ defined in (9.50) into (9.48), and performing a congruence transformation by Γ_1, we can extract the fuzzy basic functions to yield

$$\sum_{l=1}^{r_i} \mu_{il}^2 \tilde{\Sigma}_{ill} + \sum_{l=1}^{r_i-1} \sum_{j=l+1}^{r_i} \mu_{il}\mu_{ij} \left[\tilde{\Sigma}_{ilj} + \tilde{\Sigma}_{ijl}\right] < 0, \tag{9.52}$$

with

$$\tilde{\Sigma}_{ilj} = \begin{bmatrix} -\varepsilon_0^{-1}(N-1)^{-1}\mathbf{I} & 0 & \bar{\mathscr{E}}_i & 0 \\ \star & -\mathbf{I} & \bar{\Xi}_{ilj} + \frac{\bar{d}_i}{2} F_{il} D_{ci} C_{dij} \mathscr{U}_i & 0 \\ \star & \star & \tilde{\Theta}_i + \mathrm{Sym}\{\bar{\Pi}_{ilj} + \Upsilon_{ilj}\mathscr{U}_i\} & \bar{\mathscr{G}}_i A_{ikl,k\neq i} \\ \star & \star & \star & -E_{ikl,k\neq i}\mathbf{I} \end{bmatrix}, \tag{9.53}$$

where $\{\tilde{\Theta}_i, \bar{\Pi}_{ilj}\}$ and $\{A_{ikl,k\neq i}, E_{ikl,k\neq i}, \bar{\mathscr{E}}_i, \bar{\Xi}_{ilj}, \Upsilon_{ilj}, \mathscr{U}_i, \bar{\mathscr{G}}_i\}$ are defined in (9.45) and (9.46), respectively.

The terms $\Upsilon_{ilj}\mathscr{U}_i$ and $\frac{\bar{d}_i}{2} F_{il} D_{ci} C_{dij} \mathscr{U}_i$ in (9.53) are nonlinear. In order to cast $\tilde{\Sigma}_{ilj} < 0$ in (9.52) into the linear matrix conditions, by using the relation in (9.32),

$$\mathrm{Sym}\left\{ \begin{array}{cc} 0 & \frac{\bar{d}_i}{2} F_{il} D_{ci} C_{dij} \mathscr{U}_i \\ \star & \Upsilon_{ilj}\mathscr{U}_i \end{array} \right\} \leq \varepsilon^{-1} \begin{bmatrix} \frac{\bar{d}_i}{2} F_{il} D_{ci} C_{dij} \\ \Upsilon_{ilj} \end{bmatrix} (\star)^T + \varepsilon(\star)^T \begin{bmatrix} 0 & \mathscr{U}_i \end{bmatrix}, \tag{9.54}$$

where scalar $\varepsilon > 0$.

By using Schur complement to (9.42) and (9.43), and taking the relation in (9.54), it is easy to see that the inequalities in (9.42) and (9.43) imply (9.52). Then, by performing congruence transformations to (9.26) and (9.27) by $\Gamma_2 := \mathrm{diag}\{W_i, W_i\}$, respectively, the inequalities (9.40) and (9.41) can be obtained.

In addition, it follows from the conditions in (9.42) and (9.43) that

$$d_{i1}^2 \bar{Z}_{i1} + d_{i2}^2 \bar{Z}_{i2} + \tilde{X}_i - \mathrm{Sym}\{\bar{M}_i\} < 0, i \in \mathscr{N}, \tag{9.55}$$

which means $\bar{M}_i + \bar{M}_i^T > 0$. In this way, it can be directly inferred that $M_i - V_{i1}U_{i1}^T$ is nonsingular and together with relation $M_i = V_{i1}^T U_{i1} + V_{i2}^T U_{i2}$, we can see that both matrices V_{i2} and U_{i2} are nonsingular. Thus, the controller gains can be obtained by (9.47). On the other hand, since matrices $\bar{M}_i$ are nonsingular, there always exist nonsingular matrices W_i and G_i satisfying $\bar{M}_i = W_i^T G_i W_i$, therefore completing this proof.

Note: A structural constraint is imposed on the matrix multipliers $\mathscr{G}_i$ given in (9.50) for obtaining the LMI-based conditions, which will result in conservative design. By unfolding $\dot{x}_i(t)$ in (9.30), we can remove the matrix multipliers $\mathscr{G}_i$ introducing in (9.34). However, in this case, many tuning scalars will be introduced due to the coupling term $\sum_{k=1, k\neq i}^{N} \bar{A}_{ik}(\mu_i)x_k(t)$. In this case, it is very difficult to find appropriate values of those scalars such that the obtained LMIs have a feasible solution.

Note: The conditions in Theorem 9.1 are strict LMIs only when two tuning scalars ε_0 and ε are known. A natural issue arises as to how to find appropriate values of these two scalars such that the obtained LMIs have a feasible solution. A simple way solving the tuning issue is to use the trial-and-error method. Alternatively, a number of optimization-search algorithms, such as the program fminsearch in the optimization toolbox of MATLAB, or genetic algorithm, can be applied to the tuning issue. These optimization techniques present effective means for solving the LMI-based parameter-tuning problem [11].

9.2 NETWORK-INDUCED DELAY IN INTERCONNECTED SYSTEMS

In a networked environment, we consider a multi-PV generator or/and a multi-machine WG, where nonlinearities are involved in interconnections to other subsystems. The T-S fuzzy model is used to represent the nonlinear subsystem as below:

$$\dot{x}(t) = A(t)x(t) + B(t)u(t) + \omega(t), \tag{9.56}$$

where

$$x(t) = \begin{bmatrix} x_1^T(t) & x_2^T(t) & \cdots & x_N^T(t) \end{bmatrix}^T,$$

$$A(t) = \begin{bmatrix} A_{11}(t) & A_{12}(t) & \cdots & A_{1N}(t) \\ A_{21}(t) & A_{22}(t) & \cdots & A_{2N}(t) \\ \vdots & \vdots & \ddots & \vdots \\ A_{N1}(t) & A_{N2}(t) & \cdots & A_{NN}(t) \end{bmatrix}, u(t) = \begin{bmatrix} u_1(t) \\ u_2(t) \\ \vdots \\ u_N(t) \end{bmatrix},$$

$$B(t) = \begin{bmatrix} B_1(t) & 0 & \cdots & 0 \\ 0 & B_2(t) & \cdots & 0 \\ \vdots & \vdots & \ddots & \vdots \\ 0 & 0 & \cdots & B_N(t) \end{bmatrix}, \omega(t) = \begin{bmatrix} \omega_1(t) \\ \omega_2(t) \\ \vdots \\ \omega_N(t) \end{bmatrix}. \tag{9.57}$$

Choose $\{z_1(t), \cdots, z_g(t)\}$ as fuzzy premise variables. Thus, the nonlinear system in (4.5) can be represented by the following fuzzy system,

Plant Rule $\mathscr{R}^l$: **IF** $z_1(t)$ is $\mathscr{F}_1^l$ and $\cdots$ and $z_g(t)$ is $\mathscr{F}_g^l$, **THEN**

$$\dot{x}(t) = A_l x(t) + B_l u(t), l \in \mathscr{L} := \{1, 2, \ldots, r\} \tag{9.58}$$

where $\mathscr{R}^l$ denotes the l-th fuzzy inference rule; r is the number of inference rules; $z(t) \triangleq [z_1, \cdots, z_g]$ are the measurable variables; $\{A_l, B_l\}$ is the l-th local model.

Denote as $\mathscr{F}^l := \prod_{\phi=1}^N \mathscr{F}_\phi^l$ the inferred fuzzy set, and $\mu_l[z(t)]$ as the normalized membership function,

$$\mu_l[z(t)] := \frac{\prod_{\phi=1}^g \mu_{l\phi}[z_\phi(t)]}{\sum_{\varsigma=1}^r \prod_{\phi=1}^g \mu_{\varsigma\phi}[z_\phi(t)]} \geq 0, \sum_{l=1}^r \mu_l[z(t)] = 1. \tag{9.59}$$

We denote $\mu_l := \mu_l[z(t)]$ for simplicity.

After fuzzy blending, the global T-S fuzzy dynamic model is given by

$$\dot{x}(t) = A(\mu)x(t) + B(\mu)u(t), \tag{9.60}$$

where $A(\mu) := \sum_{l=1}^{r} \mu_l A_l, B(\mu) := \sum_{l=1}^{r} \mu_l B_l$. Consider the decentralized state-feedback fuzzy controller as below:

Controller Rule $\mathscr{R}_i^l$: **IF** $\zeta_{i1}(t)$ is $\mathscr{F}_{i1}^l$ and $\zeta_{i2}(t)$ is $\mathscr{F}_{i2}^l$ and $\cdots$ and $\zeta_{ig}(t)$ is $\mathscr{F}_{ig}^l$, **THEN**

$$u_i(t) = K_{il} x_i(t), i \in \mathscr{N} \tag{9.61}$$

where $K_{il} \in \Re^{n_{ui} \times n_{xi}}$ is the controller gain matrix.

Thus, the overall fuzzy controller is inferred as

$$u_i(t) = K_i(\mu_i) x_i(t), i \in \mathscr{N} \tag{9.62}$$

where $K_i(\mu_i) := \sum_{l=1}^{r_i} \mu_{il} K_{il}$.

Submitting the controller (9.62) into the system (9.60), the i-th closed-loop system can be represented as

$$\begin{cases} \dot{x}_i(t) = (A_i(\mu_i) + B_i(\mu_i)K_i(\mu_i))x_i(t) \\ \qquad + D_i(\mu_i)\omega_i(t) + \sum_{\substack{k=1 \\ k \neq i}}^{N} \bar{A}_{dik}(\mu_i)x_k(t - d_{ik}(t)), \\ x_i(t) \equiv 0, -\bar{d} \le t \le 0, i \in \mathscr{N}. \end{cases} \tag{9.63}$$

Define $x(t) = \left[x_1^T(t)\ x_2^T(t)\ \cdots\ x_N^T(t)\right]^T$, $\omega(t) = \left[\omega_1^T(t)\ \omega_2^T(t)\ \cdots\ \omega_N^T(t)\right]^T$, and $u(t) = \left[u_1^T(t)\ u_2^T(t)\ \cdots\ u_N^T(t)\right]^T$, and assume that the disturbance and input signal are bounded, the following conditions can be satisfied:

$$\omega^T(t)\omega(t) \le \bar{\omega}^2, \tag{9.64}$$

and

$$u^T(t)u(t) \le \bar{u}^2, \tag{9.65}$$

where $\bar{\omega}$ and $\bar{u}$ are two positive scalars.

The aim is to design the fuzzy controller in (9.62), such that all of the system states are bounded by the following reachable set:

$$\begin{aligned} \mathbb{S} \triangleq \{x(t) \in \Re^{n_x} \,|\, & x(t), \omega(t) \text{ and } u(t) \text{ satisfy} \\ & (9.63), (9.64) \text{ and } (9.65), \text{respectively}, t \ge 0\}. \end{aligned} \tag{9.66}$$

An ellipsoid that bounds the reachable set of the closed-loop system in (9.64) is given by

$$\mathbb{E} \triangleq \left\{x \,|\, x^T P x < 1, x \in \Re^{n_x}\right\}, \tag{9.67}$$

where $P = \text{diag}\underbrace{\{P_1 \cdots P_i \cdots P_N\}}_{N}, P_i = P_i^T > 0, i \in \mathscr{N}$.

This section will propose a model transformation that reformulates the closed-loop system (9.63) as several feedback interconnections with extra inputs and outputs. The problem of reachable set estimation and synthesis is thus subject to the input-output (IO) stability. Before moving on, consider the following lemmas, which will be utilized to derive the main results [6].

Lemma 9.5: SSG Method

Consider an interconnected system $\mathscr{S}_1 : \xi(t) = \mathbf{G}\eta(t), \mathscr{S}_2 : \eta(t) = \Delta\xi(t)$. The system is said to be robustly stable for all Δ, if $\|\Delta\|_\infty \leq 1$, $\|\mathbf{G}\|_\infty < 1$, and the subsystem $\mathscr{S}_1$ is internally stable. ■

Lemma 9.6: Relaxing Technique

Given the interconnected matrix $\bar{A}_{dikl}$ in the fuzzy system (9.56), and the symmetric positive definite matrix $H_i \in \Re^{n_{xi} \times n_{xi}}$, the following inequality holds:

$$\sum_{i=1}^{N}\sum_{\substack{k=1\\k\neq i}}^{N} \bar{A}_{dik}(\mu_i)H_i\bar{A}_{dik}^T(\mu_i) \leq \sum_{i=1}^{N}\sum_{\substack{k=1\\k\neq i}}^{N}\sum_{l=1}^{r_i} \mu_{il}\bar{A}_{dikl}H_i\bar{A}_{dikl}^T. \tag{9.68}$$

■

Proof. Note that

$$\left[\bar{A}_{dikl} - \bar{A}_{dikf}\right] H_i \left[\bar{A}_{dikl} - \bar{A}_{dikf}\right]^T \geq 0, \tag{9.69}$$

which implies that

$$\bar{A}_{dikl}H_i\bar{A}_{dikl}^T + \bar{A}_{dikf}H_i\bar{A}_{dikf}^T \geq \bar{A}_{dikl}H_i\bar{A}_{dikf}^T + \bar{A}_{dikf}H_i\bar{A}_{dikl}^T. \tag{9.70}$$

By taking the relationships in (9.60) and (9.70),

$$\begin{aligned}
&\sum_{i=1}^{N}\sum_{\substack{k=1\\k\neq i}}^{N}\bar{A}_{dik}(\mu_i)H_i\bar{A}_{dik}^{T}(\mu_i)\\
&=\sum_{i=1}^{N}\sum_{\substack{k=1\\k\neq i}}^{N}\sum_{l=1}^{r_i}\sum_{f=1}^{r_i}\mu_{il}\mu_{if}\bar{A}_{dikl}H_i\bar{A}_{dikf}^{T}\\
&=\frac{1}{2}\sum_{i=1}^{N}\sum_{\substack{k=1\\k\neq i}}^{N}\sum_{l=1}^{r_i}\sum_{f=1}^{r_i}\mu_{il}\mu_{if}\left[\bar{A}_{dikl}H_i\bar{A}_{dikf}^{T}+\bar{A}_{dikf}H_i\bar{A}_{dikl}^{T}\right]\\
&\leq\frac{1}{2}\sum_{i=1}^{N}\sum_{\substack{k=1\\k\neq i}}^{N}\sum_{l=1}^{r_i}\sum_{f=1}^{r_i}\mu_{il}\mu_{if}\left[\bar{A}_{dikl}H_i\bar{A}_{dikl}^{T}+\bar{A}_{dikf}H_i\bar{A}_{dikf}^{T}\right]\\
&=\frac{1}{2}\sum_{i=1}^{N}\sum_{\substack{k=1\\k\neq i}}^{N}\sum_{l=1}^{r_i}\mu_{il}\bar{A}_{dikl}H_i\bar{A}_{dikl}^{T}+\frac{1}{2}\sum_{i=1}^{N}\sum_{\substack{k=1\\k\neq i}}^{N}\sum_{f=1}^{r_i}\mu_{if}\bar{A}_{dikf}H_i\bar{A}_{dikf}^{T}\\
&=\sum_{i=1}^{N}\sum_{\substack{k=1\\k\neq i}}^{N}\sum_{l=1}^{r_i}\mu_{il}\bar{A}_{dikl}H_i\bar{A}_{dikl}^{T}.
\end{aligned}\tag{9.71}$$

This completes the proof.

9.2.1 MODEL TRANSFORMATION

This subsection firstly proposes a model transformation, which reformulates the closed-loop system (9.63) as several feedback interconnections. Then, based on a combined application of the LKF and the SSG theorems, the IO approach is developed to the reachable set estimation and synthesis.

Inspired by [6], one approximates the uncertain term $x_k(t-d_{ik}(t))$ in (9.63) by $x_k(t)$ and $x_k\left(t-\bar{d}\right)$, and the approximation error is given by

$$\begin{aligned}
\frac{\bar{d}}{2}\eta_{ik}(t)&=x_k(t-d_{ik}(t))-\frac{1}{2}\left[x_k(t)+x_k\left(t-\bar{d}\right)\right]\\
&=\frac{1}{2}\int_{-\bar{d}}^{-d_{ik}(t)}\dot{x}_k(t+\alpha)\,d\alpha-\frac{1}{2}\int_{-d_{ik}(t)}^{0}\dot{x}_k(t+\alpha)\,d\alpha\\
&=\frac{1}{2}\int_{-\bar{d}}^{0}\rho_{ik}(\alpha)\,\xi_k(t+\alpha)\,d\alpha,
\end{aligned}\tag{9.72}$$

where $\xi_k(t)=\dot{x}_k(t)$ and

$$\rho_{ik}(\alpha)=\begin{cases}1, & \text{if } \alpha\leq -d_{ik}(t),\\ -1, & \text{if } \alpha> -d_{ik}(t).\end{cases}\tag{9.73}$$

By substituting (9.72) into (9.63), we eliminate the delay uncertainty $\eta_{ik}(t)$ and utilize several feedback formulations. Thus, the closed-loop system (9.63) can be rewritten as

$$\mathscr{S}_{i1}: \begin{cases} \dot{x}_i(t) = (A_i(\mu_i) + B_i(\mu_i)K_i(\mu_i))x_i(t) + \sum_{\substack{k=1\\k\neq i}}^{N} \frac{1}{2}\bar{A}_{dik}(\mu_i)x_k(t) \\ \qquad + \sum_{\substack{k=1\\k\neq i}}^{N} \frac{1}{2}\bar{A}_{dik}(\mu_i)x_k(t-\bar{d}) + \sum_{\substack{k=1\\k\neq i}}^{N} \frac{\bar{d}}{2}\bar{A}_{dik}(\mu_i)\eta_{ik}(t) + D_i(\mu_i)w_i(t), \\ \bar{\xi}_i(t) = \dot{\bar{x}}_i(t), \end{cases}$$

$$\mathscr{S}_{i2}: \bar{\eta}_i(t) = \bar{\Delta}_i\bar{\xi}_i(t), i \in \mathscr{N} \tag{9.74}$$

where $x_i(t) \equiv 0, -\bar{d} \leq t \leq 0$, and $\bar{\Delta}_i$ denotes the relationship from system input to output, and

$$\begin{cases} \bar{\xi}_i(t) = \left[\xi_1^T(t) \cdots \xi_{k,k\neq i}^T(t) \cdots \xi_N^T(t)\right]^T, \dot{\bar{x}}_i(t) = \left[\dot{x}_1^T(t) \cdots \dot{x}_{k,k\neq i}^T(t) \cdots \dot{x}_N^T(t)\right]^T, \\ \bar{\eta}_i(t) = \left[\eta_{i1}^T(t) \cdots \eta_{ik,k\neq i}^T(t) \cdots \eta_{iN}^T(t)\right]^T, \bar{\Delta}_i = \text{diag}\left\{\Delta_{i1} \cdots \Delta_{ik,k\neq i} \cdots \Delta_{iN}\right\}. \end{cases} \tag{9.75}$$

Lemma 9.7: SSG Analysis

Given the interconnected system in (9.74), the operator $\bar{\Delta}_i : \bar{\xi}_i(t) \longmapsto \bar{\eta}_i(t)$ satisfies the SSG condition $\left\|\bar{\Delta}_i\right\|_\infty \leq 1$. ■

Proof. Consider zero initial conditions, and by taking (9.72) and using Jensen's inequality, one can obtain

$$\begin{aligned} &\int_0^t \eta_{ik}^T(s)\eta_{ik}(s)ds \\ &= \frac{1}{\bar{d}^2}\int_0^t \left[\int_{-\bar{d}}^0 \rho_{ik}(\beta)\xi_k(s+\beta)d\beta\right]^T (\star)ds \\ &\leq \frac{1}{\bar{d}^2}\int_0^t \left[\bar{d}\int_{-\bar{d}}^0 \rho_{ik}^2(\beta)\xi_k^T(s+\beta)\xi_k(s+\beta)d\beta\right]ds \\ &= \frac{1}{\bar{d}}\int_{-\bar{d}}^0 \left[\int_0^t \xi_k^T(s+\beta)\xi_k(s+\beta)ds\right]d\beta \\ &= \frac{1}{\bar{d}}\int_{-\bar{d}}^0 \left[\int_\beta^{t+\beta} \xi_k^T(s)\xi_k(s)ds\right]d\beta \\ &\leq \frac{1}{\bar{d}}\int_{-\bar{d}}^0 \left[\int_0^t \xi_k^T(s)\xi_k(s)ds\right]d\beta \\ &= \int_0^t \xi_k^T(s)\xi_k(s)ds. \end{aligned} \tag{9.76}$$

Substituting $\eta_{ik}(s) = \Delta_{ik}\xi_k(s)$ into (9.76), it shows

$$\int_0^t [\Delta_{ik}\xi_k(s)]^T \Delta_{ik}\xi_k(s)\,ds \leq \int_0^t \xi_k^T(s)\,\xi_k(s)\,ds, \tag{9.77}$$

which means that $\|\Delta_{ik}\|_\infty \leq 1$.

It follows from (9.77) that

$$\sum_{\substack{k=1\\k\neq i}}^{N} \int_0^t [\Delta_{ik}\xi_k(s)]^T \Delta_{ik}\xi_k(s)\,ds \leq \sum_{\substack{k=1\\k\neq i}}^{N} \int_0^t \xi_k^T(s)\,\xi_k(s)\,ds, \tag{9.78}$$

which implies that $\left\|\bar{\Delta}_i\right\|_\infty \leq 1$, thus completing this proof.

Note: In this section, only the time-varying delays involving in interconnections are considered. However, the proposed method can be easily developed for the condition that the local time-varying delay term $x_i(t - d_i(t))$ appears in the system (9.63).

Note: In Lemma 9.7, the feedback subsystem $\mathscr{S}_{i2}$ satisfies the property $\left\|\bar{\Delta}_i\right\|_\infty \leq 1$, which implies that $\left\|\tilde{\Delta}\right\|_\infty \leq 1$ holds, where $\tilde{\Delta} = \text{diag}\underbrace{\left\{\bar{\Delta}_1 \cdots \bar{\Delta}_i \cdots \bar{\Delta}_N\right\}}_{N}$. Due to the fact that the system in (9.63) and the system in (9.74) are equivalent, the system in (9.63) is asymptotically stable by using Lemma 9.7, if the system in (9.74) is internally stable and the condition $\left\|\tilde{\mathbf{G}}\right\|_\infty < 1$ holds, where $\tilde{\mathbf{G}} = \text{diag}\underbrace{\{\mathbf{G}_1 \cdots \mathbf{G}_i \cdots \mathbf{G}_N\}}_{N}$.

9.2.2 DESIGN OF DECENTRALIZED CONTROL OF REACHABLE SET

This section presents a condition sufficient to guarantee that the reachable set of the closed-loop control system (9.63) with input constraint (9.65) is bounded by the intersection of ellipsoids in (9.67).

Theorem 9.2: Reachable Set Analysis

Consider the large-scale T-S fuzzy system in (9.60). A decentralized fuzzy controller in (9.62) can guarantee that the reachable set of the closed-loop system (9.63) with input constraint (9.65) is bounded by the intersection of ellipsoid in (9.67), if there exist the positive definite symmetric matrices $\{P_i, Q_{ik}, Z_i, M_i, W_i, H_{01}, H_{02}, H_{ik1}, H_{ik2}, H_{ik3}\} \in \Re^{n_{xi}\times n_{xi}}$, $H_{01} \leq H_{ik1}, H_{02} \leq H_{ik2}$, and the matrices $K_i(\mu_i) \in \Re^{n_{ui}\times n_{xi}}$, and the multipliers $\mathscr{G}_i \in \Re^{((2+N)n_{xi}+n_{wi})\times n_{xi}}$, such that for all $i \in \mathscr{N}$, the following matrix inequalities hold:

$$\begin{bmatrix} \Theta_i(\mu_i) & \mathscr{E}_1 & \mathscr{E}_2 & \mathscr{E}_{i3} \\ \star & -\frac{H_{01}}{N-1} & 0 & 0 \\ \star & \star & -\frac{H_{02}}{N-1} & 0 \\ \star & \star & \star & -\mathscr{H}_{i3} \end{bmatrix} < 0, \tag{9.79}$$

$$\begin{bmatrix} -P_i & \star \\ K_i(\mu_i) & -\bar{u}^2\mathbf{I} \end{bmatrix} < 0, \tag{9.80}$$

where

$$\Theta_i(\mu_i) = \Theta_i^{(1)} - \sum_{\substack{k=1\\k\neq i}}^{N} (1-\tau_{ki})\, e^{-a\bar{d}}\Pi_{ik}^T Q_{ik}\Pi_{ik} + \mathrm{Sym}\{\mathscr{G}_i\mathbb{A}_i(\mu_i)\} + \sum_{\substack{k=1\\k\neq i}}^{N}\Theta_{ik}^{(2)},$$

$$\Theta_i^{(1)} = \begin{bmatrix} \Theta_i^{(11)} & P_i & 0 & 0 & 0 \\ \star & \Theta_i^{(12)} & e^{-a\bar{d}}Z_i & 0 & 0 \\ \star & \star & \Theta_i^{(13)} & 0 & 0 \\ \star & \star & \star & -\mathscr{M}_i & 0 \\ \star & \star & \star & \star & -\frac{a}{\bar{w}^2}\mathbf{I} \end{bmatrix},$$

$$\Theta_i^{(11)} = \bar{d}^2 Z_i + (N-1)M_i, \Theta_i^{(12)} = aP_i + \sum_{\substack{k=1\\k\neq i}}^{N} Q_{ik} + W_i - e^{-a\bar{d}}Z_i,$$

$$\Theta_i^{(13)} = -e^{-a\bar{d}}W_i - e^{-a\bar{d}}Z_i, \Theta_{ik}^{(2)} = \left[\mathscr{G}_i\bar{A}_{dik}(\mu_i)\right]\Theta_{ik}^{(21)}\left[\mathscr{G}_i\bar{A}_{dik}(\mu_i)\right]^T,$$

$$\Theta_{ik}^{(21)} = \left[\frac{1}{4}H_{ik1} + \frac{1}{4}H_{ik2} + \frac{\bar{d}^2}{4}H_{ik3}\right], \tag{9.81}$$

and

$$\mathbb{A}_i(\mu_i) = \begin{bmatrix} -\mathbf{I} & \mathscr{A}_i(\mu_i) & 0 & 0 & D_i(\mu_i) \end{bmatrix}, \mathscr{A}_i(\mu_i) = A_i(\mu_i) + B_i(\mu_i)K_i(\mu_i),$$

$$\mathscr{E}_1 = \begin{bmatrix} 0 & \mathbf{I} & 0 & 0 & 0 \end{bmatrix}^T, \mathscr{E}_2 = \begin{bmatrix} 0 & 0 & \mathbf{I} & 0 & 0 \end{bmatrix}^T, \mathscr{E}_{i3} = \underbrace{\left[\mathscr{E}_{1i3}\cdots\mathscr{E}_{ki3,k\neq i}\cdots\mathscr{E}_{Ni3}\right]}_{N-1},$$

$$\mathscr{E}_{ki3} = \begin{bmatrix} 0 & 0 & 0 & \underbrace{\overbrace{[0\cdots\mathbf{I}\cdots 0]}^{k,k\neq i}}_{N-1} & 0 \end{bmatrix}^T, \Pi_{ik} = \begin{bmatrix} 0 & \frac{1}{2}\mathbf{I} & \frac{1}{2}\mathbf{I} & \frac{\bar{d}}{2}\underbrace{\overbrace{[0\cdots\mathbf{I}\cdots 0]}^{k,k\neq i}}_{N-1} & 0 \end{bmatrix},$$

$$\mathscr{H}_{i3} = \mathrm{diag}\underbrace{\{H_{1i3}\cdots H_{ki3,k\neq i}\cdots H_{Ni3}\}}_{N-1}, \mathscr{M}_i = \mathrm{diag}\underbrace{\{M_i\cdots M_i\cdots M_i\}}_{N-1}. \tag{9.82}$$

■

Proof. The proof is divided into three steps. Step 1 will show the stability of the resulting closed-loop control system in (9.63), the reachable set estimation will be given in Step 2, Step 3 will handle the input constraint in (9.65) [16].

Step 1. Consider the following LKF:

$$V(t) = \sum_{i=1}^{N} V_i(t) = \sum_{i=1}^{N} \left[V_{i1}(t) + V_{i2}(t) + V_{i3}(t) + V_{i4}(t)\right] \tag{9.83}$$

with

$$\begin{cases} V_{i1}(t) = x_i^T(t) P_i x_i(t), \\ V_{i2}(t) = \int_{t-\bar{d}}^{t} e^{a(s-t)} x_i^T(s) W_i x_i(s) ds, \\ V_{i3}(t) = \sum_{\substack{k=1 \\ k \neq i}}^{N} \int_{t-d_{ik}(t)}^{t} e^{a(s-t)} x_k^T(s) Q_{ki} x_k(s) ds, \\ V_{i4}(t) = \bar{d} \int_{-\bar{d}}^{0} \int_{t+\beta}^{t} e^{a(s-t)} \dot{x}_i^T(s) Z_i \dot{x}_i(s) ds d\beta, \end{cases} \tag{9.84}$$

where $\{P_i, W_i, Q_{ki}, Z_i\} \in \Re^{n_{xi} \times n_{xi}}, i \in \mathcal{N}$ are positive definite symmetric matrices. By taking the derivative of $V(t)$, it yields

$$\dot{V}_{i1}(t) = 2x_i^T(t) P_i \dot{x}_i(t), \tag{9.85}$$

$$\dot{V}_{i2}(t) = -aV_{i2}(t) + x_i^T(t) W_i x_i(t) - e^{-a\bar{d}} x_i^T\left(t-\bar{d}\right) W_i x_i\left(t-\bar{d}\right), \tag{9.86}$$

$$\begin{aligned} \dot{V}_{i3}(t) &= -aV_{i3}(t) + \sum_{\substack{k=1 \\ k \neq i}}^{N} x_k^T(t) Q_{ki} x_k(t) \\ &\quad - \sum_{\substack{k=1 \\ k \neq i}}^{N} (1 - \dot{d}_{ik}(t)) e^{-ad_{ik}(t)} x_k^T(t - d_{ik}(t)) \times Q_{ki} x_k(t - d_{ik}(t)) \\ &\leq -aV_{i3}(t) + \sum_{\substack{k=1 \\ k \neq i}}^{N} x_k^T(t) Q_{ki} x_k(t) \\ &\quad - \sum_{\substack{k=1 \\ k \neq i}}^{N} (1 - \tau_{ik}) e^{-a\bar{d}} x_k^T(t - d_{ik}(t)) \times Q_{ki} x_k(t - d_{ik}(t)), \end{aligned} \tag{9.87}$$

$$\begin{aligned} \dot{V}_{i4}(t) &= -aV_{i4}(t) + \bar{d}^2 \dot{x}_i^T(t) Z_i \dot{x}_i(t) \\ &\quad - \bar{d} e^{-a\bar{d}} \int_{t-\bar{d}}^{t} \dot{x}_i^T(s) Z_i \dot{x}_i(s) ds. \end{aligned} \tag{9.88}$$

Note that

$$\sum_{i=1}^{N} \sum_{\substack{k=1 \\ k \neq i}}^{N} x_k^T(t) Q_{ki} x_k(t) = \sum_{i=1}^{N} \sum_{\substack{k=1 \\ k \neq i}}^{N} x_i^T(t) Q_{ik} x_i(t). \tag{9.89}$$

It follows from (9.89) that

$$\sum_{i=1}^{N}\sum_{\substack{k=1\\k\neq i}}^{N}(1-\tau_{ik})e^{-a\bar{d}}x_k^T(t-d_{ik}(t))Q_{ki}x_k(t-d_{ik}(t))$$
$$=\sum_{i=1}^{N}\sum_{\substack{k=1\\k\neq i}}^{N}(1-\tau_{ki})e^{-a\bar{d}}\phi_{ki}^T(t)Q_{ik}\phi_{ki}(t), \tag{9.90}$$

where $\phi_{ki}(t)=\left[\frac{1}{2}x_i(t)+\frac{1}{2}x_i(t-\bar{d})+\frac{\bar{d}}{2}\eta_{ki}(t)\right]$.

By using Jensen's inequality,

$$-\bar{d}e^{-a\bar{d}}\int_{t-\bar{d}}^{t}\dot{x}_i^T(s)Z_i\dot{x}_i(s)ds$$
$$\leq -e^{-a\bar{d}}\left[\int_{t-\bar{d}}^{t}\dot{x}_i(s)ds\right]^T Z_i\left[\int_{t-\bar{d}}^{t}\dot{x}_i(s)ds\right]$$
$$=-e^{-a\bar{d}}\left[x_i(t)-x_i(t-\bar{d})\right]^T Z_i\left[x_i(t)-x_i(t-\bar{d})\right]. \tag{9.91}$$

Define the multipliers $\mathscr{G}_i\in\Re^{((2+N)n_{xi}+n_{wi})\times n_{xi}}, i\in\mathscr{N}$, and it follows from (9.74) that

$$\begin{aligned}0=&2\sum_{i=1}^{N}\chi_i^T(t)\mathscr{G}_i[A_i(\mu_i)+B_i(\mu_i)K_i(\mu_i)]x_i(t)\\&+2\sum_{i=1}^{N}\chi_i^T(t)\mathscr{G}_i[-\dot{x}_i(t)+D_i(\mu_i)w_i(t)]\\&+2\sum_{i=1}^{N}\chi_i^T(t)\mathscr{G}_i\sum_{\substack{k=1\\k\neq i}}^{N}\frac{1}{2}\bar{A}_{dik}(\mu_i)x_k(t)\\&+2\sum_{i=1}^{N}\chi_i^T(t)\mathscr{G}_i\sum_{\substack{k=1\\k\neq i}}^{N}\frac{1}{2}\bar{A}_{dik}(\mu_i)x_k(t-\bar{d})\\&+2\sum_{i=1}^{N}\chi_i^T(t)\mathscr{G}_i\sum_{\substack{k=1\\k\neq i}}^{N}\frac{\bar{d}}{2}\bar{A}_{dik}(\mu_i)\eta_{ik}(t),\end{aligned} \tag{9.92}$$

where

$$\begin{aligned}\chi_i(t)&=\left[\begin{array}{ccccc}\dot{x}_i^T(t) & x_i^T(t) & x_i^T(t-\bar{d}) & \tilde{\eta}_i^T(t) & w_i^T(t)\end{array}\right]^T,\\ \tilde{\eta}_i(t)&=\left[\eta_{1i}^T(t)\cdots\eta_{ki,i\neq k}^T(t)\cdots\eta_{Ni}^T(t)\right]^T.\end{aligned} \tag{9.93}$$

Note that

$$2\bar{x}^T\bar{y}\leq\kappa^{-1}\bar{x}^T\bar{x}+\kappa\bar{y}^T\bar{y}, \tag{9.94}$$

where $\bar{x},\bar{y}\in\Re^n$ and scalar $\kappa>0$.

Define positive definite symmetric matrices $\{H_{01}, H_{02}, H_{ik1}, H_{ik2}, H_{ik3}\} \in \Re^{n_{xi} \times n_{xi}}$, where $H_{01} \leq H_{ik1}$, $H_{02} \leq H_{ik2}, i \in \mathcal{N}$, and by relationship of (9.94),

$$\begin{aligned} 2\sum_{i=1}^{N} \chi_i^T(t)\mathscr{G}_i \sum_{\substack{k=1\\k\neq i}}^{N} \frac{1}{2}\bar{A}_{dik}(\mu_i)x_k(t) &\leq \sum_{i=1}^{N}\sum_{\substack{k=1\\k\neq i}}^{N} \frac{1}{4}\chi_i^T(t)\Xi_{ik}(\mu_i)H_{ik1}\Xi_{ik}^T(\mu_i)\chi_i(t) + \sum_{i=1}^{N}\sum_{\substack{k=1\\k\neq i}}^{N} x_k^T(t)H_{ik1}^{-1}x_k(t) \\ &\leq \sum_{i=1}^{N}\sum_{\substack{k=1\\k\neq i}}^{N} \frac{1}{4}\chi_i^T(t)\Xi_{ik}(\mu_i)H_{ik1}\Xi_{ik}^T(\mu_i)\chi_i(t) + \sum_{i=1}^{N}(N-1)x_i^T(t)H_{01}^{-1}x_i(t), \end{aligned} \tag{9.95}$$

$$\begin{aligned} 2\sum_{i=1}^{N} \chi_i^T(t)\mathscr{G}_i \sum_{\substack{k=1\\k\neq i}}^{N} \frac{1}{2}\bar{A}_{dik}(\mu_i)x_k(t-\bar{d}) &\leq \sum_{i=1}^{N}\sum_{\substack{k=1\\k\neq i}}^{N} \frac{1}{4}\chi_i^T(t)\Xi_{ik}(\mu_i)H_{ik2}\Xi_{ik}^T(\mu_i)\chi_i(t) + \sum_{i=1}^{N}\sum_{\substack{k=1\\k\neq i}}^{N} x_k^T(t-\bar{d})H_{ik2}^{-1}x_k(t-\bar{d}) \\ &\leq \sum_{i=1}^{N}\sum_{\substack{k=1\\k\neq i}}^{N} \frac{1}{4}\chi_i^T(t)\Xi_{ik}(\mu_i)H_{ik2}\Xi_{ik}^T(\mu_i)\chi_i(t) + \sum_{i=1}^{N}(N-1)x_i^T(t-\bar{d})H_{02}^{-1}x_i(t-\bar{d}), \end{aligned} \tag{9.96}$$

$$\begin{aligned} 2\sum_{i=1}^{N} \chi_i^T(t)\mathscr{G}_i \sum_{\substack{k=1\\k\neq i}}^{N} \frac{\bar{d}}{2}\bar{A}_{dik}(\mu_i)\eta_{ik}(t) &\leq \sum_{i=1}^{N}\sum_{\substack{k=1\\k\neq i}}^{N} \frac{\bar{d}^2}{4}\chi_i^T(t)\Xi_{ik}(\mu_i)H_{ik3}\Xi_{ik}^T(\mu_i)\chi_i(t) + \sum_{i=1}^{N}\sum_{\substack{k=1\\k\neq i}}^{N} \eta_{ik}^T(t)H_{ik3}^{-1}\eta_{ik}(t) \\ &= \sum_{i=1}^{N}\sum_{\substack{k=1\\k\neq i}}^{N} \frac{\bar{d}^2}{4}\chi_i^T(t)\Xi_{ik}(\mu_i)H_{ik3}\Xi_{ik}^T(\mu_i)\chi_i(t) + \sum_{i=1}^{N}\sum_{\substack{k=1\\k\neq i}}^{N} \eta_{ki}^T(t)H_{ki3}^{-1}\eta_{ki}(t), \end{aligned} \tag{9.97}$$

where $\Xi_{ik}(\mu_i) = \mathscr{G}_i\bar{A}_{dik}(\mu_i)$.

It follows from (9.83)-(9.92) and (9.95)-(9.97) that

$$
\begin{aligned}
&\dot{V}(t)+aV(t)-\frac{a}{\bar{\omega}^2}w^T(t)\,\omega(t) \\
&\quad=\sum_{i=1}^{N}\left[\dot{V}_i(t)+aV_i(t)-\frac{a}{\bar{\omega}^2}\omega_i^T(t)\,\omega_i(t)\right] \\
&\quad\le\sum_{i=1}^{N}\left[ax_i^T(t)P_ix_i(t)+2x_i^T(t)P_i\dot{x}_i(t)\right]+\sum_{i=1}^{N}x_i^T(t)W_ix_i(t) \\
&\qquad-\sum_{i=1}^{N}e^{-a\bar{d}}x_i^T\left(t-\bar{d}\right)W_ix_i\left(t-\bar{d}\right)+\sum_{i=1}^{N}\sum_{\substack{k=1\\k\neq i}}^{N}x_i^T(t)\,Q_{ik}x_i(t) \\
&\qquad-\sum_{i=1}^{N}\sum_{\substack{k=1\\k\neq i}}^{N}(1-\tau_{ki})\,e^{-a\bar{d}}\phi_{ki}^T(t)\,Q_{ik}\phi_{ki}(t)+\sum_{i=1}^{N}\bar{d}^2\dot{x}_i^T(t)\,Z_i\dot{x}_i(t) \\
&\qquad-\sum_{i=1}^{N}e^{-a\bar{d}}\left(x_i(t)-x_i\left(t-\bar{d}\right)\right)^TZ_i\left(x_i(t)-x_i\left(t-\bar{d}\right)\right) \\
&\qquad+\sum_{i=1}^{N}2\chi_i^T(t)\,\mathscr{G}_i\{-\dot{x}_i(t)+\mathscr{A}_i(\mu_i)x_i(t)+D_i(\mu_i)w_i(t)\} \\
&\qquad+\sum_{i=1}^{N}\sum_{\substack{k=1\\k\neq i}}^{N}\frac{1}{4}\chi_i^T(t)\,\Xi_{ik}(\mu_i)H_{ik1}\Xi_{ik}^T(\mu_i)\chi_i(t) \\
&\qquad+\sum_{i=1}^{N}(N-1)x_i^T(t)H_{01}^{-1}x_i(t)+\sum_{i=1}^{N}\sum_{\substack{k=1\\k\neq i}}^{N}\frac{1}{4}\chi_i^T(t)\,\Xi_{ik}(\mu_i)H_{ik2}\Xi_{ik}^T(\mu_i)\chi_i(t) \\
&\qquad+\sum_{i=1}^{N}(N-1)x_i^T(t-\bar{d})H_{02}^{-1}x_i(t-\bar{d})+\sum_{i=1}^{N}\sum_{\substack{k=1\\k\neq i}}^{N}\frac{\bar{d}^2}{4}\chi_i^T(t)\,\Xi_{ik}(\mu_i)H_{ik3}\Xi_{ik}^T(\mu_i)\chi_i(t) \\
&\qquad+\sum_{i=1}^{N}\sum_{\substack{k=1\\k\neq i}}^{N}\frac{\bar{d}^2}{4}\eta_{ki}^T(t)H_{ki3}^{-1}\eta_{ki}(t)-\sum_{i=1}^{N}\frac{a}{\bar{\omega}^2}\omega_i^T(t)\,\omega_i(t),
\end{aligned}
\tag{9.98}
$$

where $\phi_{ki}(t)=\left[\frac{1}{2}x_i(t)+\frac{1}{2}x_i\left(t-\bar{d}\right)+\frac{\bar{d}}{2}\eta_{ki}(t)\right]$, $\mathscr{A}_i(\mu_i)=A_i(\mu_i)+B_i(\mu_i)K_i(\mu_i)$.

Let $\mathscr{M}_i = \text{diag}\underbrace{\{M_1 \cdots M_{k,k\neq i} \cdots M_N\}}_{N-1}$, where $0 < M_i = M_i^T \in \Re^{n_{xi}\times n_{xi}}$, to yield

$$\begin{aligned}
&\sum_{i=1}^{N} \left[\bar{\xi}_i^T(t)\mathscr{M}_i\bar{\xi}_i(t) - \bar{\eta}_i^T(t)\mathscr{M}_i\bar{\eta}_i(t)\right] \\
&\quad = \sum_{i=1}^{N}\sum_{\substack{k=1\\k\neq i}}^{N} \left[\xi_k^T(t)M_k\xi_k(t) - \eta_{ik}^T(t)M_k\eta_{ik}(t)\right] \\
&\quad = (N-1)\sum_{i=1}^{N} \xi_i^T(t)M_i\xi_i(t) - \sum_{i=1}^{N}\sum_{\substack{k=1\\k\neq i}}^{N} \eta_{ki}^T(t)M_i\eta_{ki}(t),
\end{aligned} \tag{9.99}$$

where $\bar{\xi}_i(t)$ and $\bar{\eta}_i(t)$ are defined in (9.75).

Consider the index

$$J(t) = \int_0^\infty \sum_{i=1}^{N} \left\{\bar{\xi}_i^T(t)\mathscr{M}_i\bar{\xi}_i(t) - \bar{\eta}_i^T(t)\mathscr{M}_i\bar{\eta}_i(t) - \frac{a}{\bar{w}^2}w_i^T(t)w_i(t)\right\}dt. \tag{9.100}$$

Under zero initial conditions, one can obtain $V(0) = 0$ and $V(\infty) \geq 0$, thus it yields

$$\begin{aligned}
J(t) &\leq J(t) + V(\infty) - V(0) \\
&< J(t) + \int_0^\infty \{\dot{V}(t) + aV(t)\}dt \\
&= \int_0^\infty \sum_{i=1}^{N} \{(N-1)\xi_i^T(t)M_i\xi_i(t) + \dot{V}_i(t) \\
&\quad + aV_i(t) - \frac{a}{\bar{w}^2}w_i^T(t)w_i(t)\}dt - \int_0^\infty \sum_{i=1}^{N}\sum_{\substack{k=1\\k\neq i}}^{N} \eta_{ki}^T(t)M_i\eta_{ki}(t)dt \\
&\leq \int_0^\infty \sum_{i=1}^{N} \{\chi_i^T(t)\Sigma_i(\mu_i)\chi_i(t)\}dt
\end{aligned} \tag{9.101}$$

with

$$\begin{aligned}
\Sigma_i(\mu_i) &= \Theta_i^{(1)} - \sum_{\substack{k=1\\k\neq i}}^{N} (1-\tau_{ki})e^{-a\bar{d}}\Pi_{ik}^T Q_{ik}\Pi_{ik} + \text{Sym}\{\mathscr{G}_i\mathbb{A}_i(\mu_i)\} \\
&\quad + \mathscr{E}_1 H_{01}^{-1}\mathscr{E}_1^T + \mathscr{E}_2 H_{02}^{-1}\mathscr{E}_2^T + \sum_{\substack{k=1\\k\neq i}}^{N} \mathscr{E}_{ki3}H_{ki3}^{-1}\mathscr{E}_{ki3}^T + \sum_{\substack{k=1\\k\neq i}}^{N} \Theta_{ik}^{(2)},
\end{aligned} \tag{9.102}$$

where $\Theta_i^{(1)}, \Theta_{ik}^{(2)}, \Pi_{ik}, \mathbb{A}_i(\mu_i), \mathscr{E}_1, \mathscr{E}_2$ and $\mathscr{E}_{ki3}$ are defined in (9.81) and (9.82).

By using Schur complement to (9.79), it shows $\Sigma_i(\mu_i) < 0, i \in \mathscr{N}$. When $w(t) = 0$, the inequality $\Sigma_i(\mu_i) < 0$ implies that $\dot{V}(t) < 0$ and $J(t) < 0$. By using Lemmas 9.5 and 9.7, the stability of the closed-loop system in (9.63) can be verified.

Step 2. Since $\Sigma_i(\mu_i) < 0$, one can obtain $\dot{V}(t) + aV(t) - \frac{a}{\bar{w}^2} w^T(t) w(t) < 0$. Then, by multiplying both its sides with e^{at}, it yields

$$\begin{aligned} \frac{d\left(e^{at}V(t)\right)}{dt} &= e^{at}\dot{V}(t) + e^{at}aV(t) \\ &< e^{at}\frac{a}{\bar{w}^2} w^T(t) w(t). \end{aligned} \tag{9.103}$$

Now, by performing the integral of (9.103) from 0 to $T > 0$,

$$\begin{aligned} e^{aT}V(T) &< \int_0^T e^{at}\frac{a}{\bar{\omega}^2}\omega^T(t)\omega(t)\,dt \\ &< e^{aT} - 1. \end{aligned} \tag{9.104}$$

Thus, for any time $T > 0$, one can obtain $V(T) < 1$.

Since $V(T) < 1$, it shows

$$\begin{aligned} 1 &> V(t) \\ &> \sum_{i=1}^{N} x_i^T(t) P_i x_i(t) \\ &= x^T(t)\mathscr{P}x(t), \end{aligned} \tag{9.105}$$

where $x(t) = \begin{bmatrix} x_1^T(t) & x_2^T(t) & \cdots & x_N^T(t) \end{bmatrix}^T, P = \text{diag}\{P_1 \cdots \underset{N-1}{P_i} \cdots P_N\}$. This completes the proof of reachable set in (9.67).

Step 3. It follows from (9.65) that

$$\sum_{i=1}^{N} \frac{u_i(t)^T u_i(t)}{\bar{u}^2} \leq 1. \tag{9.106}$$

It can be seen that the following inequality implies (9.106):

$$\sum_{i=1}^{N} \frac{u_i(t)^T u_i(t)}{\bar{u}^2} < \sum_{i=1}^{N} x_i^T(t) P_i x_i(t). \tag{9.107}$$

Substituting $u_i(t) = K_i(\mu_i)x_i(t)$ into (9.107), and using Schur complement, the inequality in (9.80) implies (9.107), thus completing the overall proof.

Here, the aim is to design the controller in the form of (9.62) such that the "smallest" bound for the reachable set in (9.66) can be obtained. To do so, a simple optimization algorithm is pointed out in [12], i.e. maximize δ subject to $\delta\mathbf{I} < \mathscr{P}$. By using Schur complement, one can easily solve the optimization problem as below:

Algorithm 9.1.

$$\text{Minimize } \bar{\delta}, \text{ subject to } \begin{bmatrix} \bar{\delta}\mathbf{I} & \mathbf{I} \\ \star & \mathscr{P} \end{bmatrix} > 0, (9.79) \text{ and } (9.80), i \in \mathscr{N}$$

where $\bar{\delta} = \delta^{-1}$.

Note: From the IO perspective, the smaller $\bar{\Delta}_i$ in (9.74) is, the less conservative the stability criteria are. It is easy to obtain the property $\|\bar{\Delta}_i\|_\infty \leq \frac{\bar{d}}{2}$ when the approximation error $\frac{\bar{d}}{2}\eta_{ik}(t)$ in (9.74) is used instead of $\eta_{ik}(t)$. When considering the different approximation approaches in [13] and [14], the system input-output properties are $\|\bar{\Delta}_i\|_\infty \leq \bar{d}$ and $\|\bar{\Delta}_i\|_\infty \leq \frac{\bar{d}}{\sqrt{2}}$, respectively.

Note: The Lyapunov function $V_{i2}(t)$ in (9.84) considers the delay derivative τ_{ik}, which reduces the conservatism to reachable set estimation when the delay d_{ik} is slow-varying case, i.e. $\tau_{ik} < 1$. When considering the condition $\tau_{ik} \geq 1$ or the unknown delay derivative, one can set $Q_{ki} \equiv 0$ in Theorem 9.2 for that case.

It is easy to see from (9.79) that the result in Theorem 9.2 is nonconvex. In the following, we derive the LMI-based conditions for a decentralized fuzzy controller that ensures the reachable set bounding in (9.66).

Theorem 9.3: Decentralized Controller Design of Reachable Set

Consider the large-scale T-S fuzzy system in ((9.60). A decentralized fuzzy controller in (9.62) can guarantee that the reachable set of the closed-loop system (9.63) with input constraint (9.65) bounded by the intersection of ellipsoid in (9.67), if there exist the positive definite symmetric matrices $\{\bar{P}_i, \bar{Q}_{ik}, \bar{Z}_i, \bar{M}_i, \bar{W}_i, H_{01}, H_{02}, H_{ik1}, H_{ik2}, H_{ik3}\} \in \Re^{n_{xi}\times n_{xi}}$, $H_{01} \leq H_{ik1}, H_{02} \leq H_{ik2}$, and the matrices $\bar{K}_{il} \in \Re^{n_{ui}\times n_{xi}}, G_i \in \Re^{n_{xi}\times n_{xi}}$, and the scalar $\bar{d}$, such that for all $i \in \mathcal{N}$, the following LMIs hold:

$$\begin{bmatrix} -\bar{P}_i & \star \\ \bar{K}_{il} & -\bar{u}^2\mathbf{I} \end{bmatrix} < 0, 1 \leq l \leq r_i \tag{9.108}$$

$$\bar{\Sigma}_{ill} < 0, 1 \leq l \leq r_i \tag{9.109}$$

$$\bar{\Sigma}_{ilj} + \bar{\Sigma}_{ijl} < 0, 1 \leq l < j \leq r_i \tag{9.110}$$

with

$$\bar{\Sigma}_{ilj} = \begin{bmatrix} \bar{\Theta}_{ilj} & \bar{\mathscr{E}}_{i1} & \bar{\mathscr{E}}_{i2} & \bar{E}_{i3} \\ \star & -\frac{H_{01}}{N-1} & 0 & 0 \\ \star & \star & -\frac{H_{02}}{N-1} & 0 \\ \star & \star & \star & -H_{i3} \end{bmatrix}, \tag{9.111}$$

where

$$\left\{\begin{array}{l}\bar{\Theta}_{ilj}=\bar{\Theta}_{i}^{(1)}-\sum\limits_{\substack{k=1\\k\neq i}}^{N}(1-\tau_{ki})e^{-a\bar{d}}\Pi_{ik}^{T}\bar{Q}_{ik}\Pi_{ik}+\mathrm{Sym}\{\mathscr{E}_{i}\bar{\mathbb{A}}_{ilj}\}+\sum\limits_{\substack{k=1\\k\neq i}}^{N}\bar{\Theta}_{ik}^{(2)},\\ \bar{\Theta}_{i}^{(1)}=\begin{bmatrix}\bar{\Theta}_{i}^{(11)} & \bar{P}_{i} & 0 & 0 & 0\\ \star & \bar{\Theta}_{i}^{(12)} & e^{-a\bar{d}}\bar{Z}_{i} & 0 & 0\\ \star & \star & \bar{\Theta}_{i}^{(13)} & 0 & 0\\ \star & \star & \star & -\bar{M}_{i} & 0\\ \star & \star & \star & \star & -\frac{a}{\bar{w}^{2}}\mathbf{I}\end{bmatrix},\\ \bar{\Theta}_{ik}^{(2)}=\left[\mathscr{E}_{i}\bar{A}_{dikl}\right]\bar{\Theta}_{ik}^{(21)}\left[\mathscr{E}_{i}\bar{A}_{dikl}\right]^{T},\bar{\Theta}_{i}^{(11)}=\bar{d}^{2}\bar{Z}_{i}+(N-1)\bar{M}_{i},\\ \bar{\Theta}_{i}^{(12)}=a\bar{P}_{i}+\sum\limits_{\substack{k=1\\k\neq i}}^{N}\bar{Q}_{ik}+\bar{W}_{i}-e^{-a\bar{d}}\bar{Z}_{i},\\ \bar{\Theta}_{i}^{(13)}=-e^{-a\bar{d}}\bar{W}_{i}-e^{-a\bar{d}}\bar{Z}_{i},\bar{\Theta}_{ik}^{(21)}=\frac{1}{4}H_{ik1}+\frac{1}{4}H_{ik2}+\frac{\bar{d}^{2}}{4}H_{ik3},\end{array}\right. \tag{9.112}$$

and

$$\left\{\begin{array}{l}\bar{\mathbb{A}}_{ilj}=\begin{bmatrix}-G_{i} & A_{il}G_{i}+B_{il}\bar{K}_{ij} & 0 & 0 & D_{il}\end{bmatrix},\\ \mathscr{E}_{i}=\begin{bmatrix}\mathbf{I} & \mathbf{I} & 0 & 0 & 0\end{bmatrix}^{T},\bar{\mathscr{E}}_{i1}=\begin{bmatrix}0 & G_{i} & 0 & 0 & 0\end{bmatrix}^{T},\\ \bar{\mathscr{E}}_{i2}=\begin{bmatrix}0 & 0 & G_{i} & 0 & 0\end{bmatrix}^{T},\bar{E}_{i3}=\underbrace{\left[\bar{\mathscr{E}}_{1i3}\cdots\bar{\mathscr{E}}_{ki3,k\neq i}\cdots\bar{\mathscr{E}}_{Ni3}\right]}_{N-1},\\ \bar{\mathscr{E}}_{ki3}=\begin{bmatrix}0 & 0 & 0 & \underbrace{\overbrace{[0\cdots G_{i}\cdots 0]}^{k,k\neq i}}_{N-1} & 0\end{bmatrix}^{T},\\ \Pi_{ik}=\begin{bmatrix}0 & \frac{1}{2}\mathbf{I} & \frac{1}{2}\mathbf{I} & \frac{\bar{d}}{2}\underbrace{\overbrace{[0\cdots\mathbf{I}\cdots 0]}^{k,k\neq i}}_{N-1} & 0\end{bmatrix},\\ H_{i3}=\mathrm{diag}\underbrace{\{H_{1i3}\cdots H_{ki3,k\neq i}\cdots H_{Ni3}\}}_{N-1},\bar{M}_{i}=\mathrm{diag}\underbrace{\{\bar{M}_{i}\cdots\bar{M}_{i}\cdots\bar{M}_{i}\}}_{N-1}.\end{array}\right. \tag{9.113}$$

Moreover, the controller gain matrix in (9.62) is obtained by calculating

$$K_{il}=\bar{K}_{il}G_{i}^{-1},l\in\mathscr{L}_{i},i\in\mathscr{N}. \tag{9.114}$$

■

Proof. It follows from (9.109) and (9.110) that

$$\bar{d}^{2}\bar{Z}_{i}+(N-1)\bar{M}_{i}-\mathrm{Sym}\{G_{i}\}<0,i\in\mathscr{N} \tag{9.115}$$

which implies that $G_i, i\in\mathscr{N}$ are nonsingular matrices.

Define

$$\left\{\begin{array}{l} \mathscr{G}_i = \left[\begin{array}{ccccc} G_i^{-1} & G_i^{-1} & 0 & 0 & 0 \end{array}\right]^T, \bar{P}_i = G_i^T P_i G_i, \\ \bar{Q}_{ik} = G_i^T Q_{ik} G_i, \bar{W}_i = G_i^T W_i G_i, \bar{Z}_i = G_i^T Z_i G_i, \\ \bar{M}_i = G_i^T M_i G_i, \bar{K}_{il} = K_{il} G_i, \mathcal{G}_i = \mathrm{diag}\underbrace{\{G_i \cdots G_i \cdots G_i\}}_{N-1}, \\ E = \mathrm{diag}\underbrace{\{\mathbf{I} \cdots \mathbf{I} \cdots \mathbf{I}\}}_{N-1}, \Gamma_1 := \mathrm{diag}\left\{\begin{array}{cccccccc} G_i & G_i & \mathcal{G}_i & G_i & \mathbf{I} & \mathbf{I} & \mathbf{I} & E \end{array}\right\}, i \in \mathscr{N}. \end{array}\right. \tag{9.116}$$

In addition, by using Lemma 9.6,

$$\begin{aligned} &\sum_{\substack{k=1 \\ k \neq i}}^{N} \bar{A}_{dik}(\mu_i) \left[\frac{1}{4} H_{ik1} + \frac{1}{4} H_{ik2} + \frac{d^2}{4} H_{ik3}\right] \bar{A}_{dik}^T(\mu_i) \\ &\quad \leq \sum_{\substack{k=1 \\ k \neq i}}^{N} \sum_{l=1}^{r_i} \mu_{il} \bar{A}_{dikl} \left[\frac{1}{4} H_{ik1} + \frac{1}{4} H_{ik2} + \frac{d^2}{4} H_{ik3}\right] \bar{A}_{dikl}^T. \end{aligned} \tag{9.117}$$

We now perform a congruence transformation to (9.79) by Γ_1 and consider the relationships in (9.117). Then, by extracting the fuzzy basis functions, we have

$$\sum_{l=1}^{r_i} \mu_{il}^2 \bar{\Sigma}_{ill} + \sum_{l=1}^{r_i - 1} \sum_{j=l+1}^{r_i} \mu_{il} \mu_{ij} \left[\bar{\Sigma}_{ilj} + \bar{\Sigma}_{ijl}\right] < 0, \tag{9.118}$$

where $\bar{\Sigma}_{ilj}$ is defined in (9.111).

We define $\Gamma_2 := \mathrm{diag}\{G_i, \mathbf{I}\}$, and perform a congruence transformation to (9.80) by Γ_2. After extracting the fuzzy basis functions, the inequality in (9.108) is directly obtained, thus completing this proof.

To obtain the "smallest" bound for the reachable set in (9.67), one can define $\Gamma_3 := \mathrm{diag}\{\mathbf{I}, \bar{G}\}$, where $\bar{G} = \mathrm{diag}\underbrace{\{G_1 \cdots G_i \cdots G_N\}}_{N}$. Then, by performing a congruence transformation to the inequality in Algorithm 9.1, the following LMI-based algorithm can be obtained:

Algorithm 9.2.

$$\text{Minimize } \bar{\delta}, \text{ subject to } \left[\begin{array}{cc} \bar{\delta}\mathbf{I} & \bar{G} \\ \star & \bar{P} \end{array}\right] > 0, \text{and } (9.108) - (9.110), i \in \mathscr{N}$$

where $\bar{P} = \mathrm{diag}\underbrace{\{\bar{P}_1 \cdots \bar{P}_i \cdots \bar{P}_N\}}$.

Note: In order to obtain the controller design in the form of LMIs, one casts the multipliers $\mathscr{G}_i$ as $\left[\begin{array}{ccccc} G_i^{-1} & G_i^{-1} & 0 & 0 & 0 \end{array}\right]^T$ in (9.116). The structural constraint on the multipliers $\mathscr{G}_i$ might lead to conservatism to some extent. One would specify the multipliers $\mathscr{G}_i$ as $\left[\begin{array}{ccccc} G_i^{-1} & \varepsilon_i G_i^{-1} & 0 & 0 & 0 \end{array}\right]^T$ with scalar ε_i to be searched or manually prescribed. In that case, the conservatism can be further reduced.

To further reduce the conservatism, inspired by [13], a relaxing LKF, where not all Lyapunov matrices are necessarily positive definite, is introduced. The following lemma ensures the positive property of $V(t)$.

Lemma 9.8: Relaxing LKF

Consider the relaxing LKF $V(t) = \sum_{i=1}^{N} [V_{i1}(t) + V_{i2}(t) + V_{i3}(t) + V_{i4}(t)]$, where $V_{i1}(t) = x_i^T(t)(P_i + R_i)x_i(t)$, $\{V_{i2}(t), V_{i3}(t), V_{i4}(t)\}$ is defined in (9.83). Then $V(t) \geq \varepsilon \|x(t)\|^2$, where $\varepsilon > 0$, $x(t) = \begin{bmatrix} x_1^T(t) & x_2^T(t) & \cdots & x_N^T(t) \end{bmatrix}^T$, if there exist symmetric positive definite matrices $\{P_i, Q_{ik}, Z_i\} \in \Re^{n_{xi} \times n_{xi}}$, and symmetric matrices $\{R_i, W_i\} \in \Re^{n_{xi} \times n_{xi}}$, such that for all $i \in \mathscr{N}$, the following LMIs hold:

$$\begin{bmatrix} \frac{a}{1-e^{-a\bar{d}}} R_i + Z_i & -Z_i \\ \star & W_i + Z_i \end{bmatrix} > 0. \tag{9.119}$$

■

Proof. Firstly, by applying Jensen's inequality [6], it follows from (9.88) that

$$\begin{aligned} V_{i4}(t) &= \bar{d} \int_{-\bar{d}}^{0} \int_{t+\beta}^{t} e^{a(s-t)} \dot{x}_i^T(s) Z_i \dot{x}_i(s) ds d\beta \\ &\geq \bar{d} \int_{-\bar{d}}^{0} \frac{-1}{\beta} \left[\int_{t+\beta}^{t} e^{0.5a(s-t)} \dot{x}_i^T(s) ds \right] Z_i(\star) d\beta \\ &\geq \bar{d} \int_{-\bar{d}}^{0} \frac{-e^{a\beta}}{\beta} \left[\int_{t+\beta}^{t} \dot{x}_i^T(s) ds \right] Z_i(\star) d\beta \\ &\geq \int_{-\bar{d}}^{0} e^{a\beta} [x_i(t) - x_i(t+\beta)]^T Z_i(\star) d\beta \\ &= \int_{t-\bar{d}}^{t} e^{a(s-t)} [x_i(t) - x_i(s)]^T Z_i [x_i(t) - x_i(s)] ds. \end{aligned} \tag{9.120}$$

The inequality in (9.83) implies that the Lyapunov matrix P_i is positive definite. To partly relax this constraint, an extra matrix variable R_i is introduced. Then, based on the relaxing LKF and by taking the relationship in (9.120), it shows

$$
\begin{aligned}
&V_{i1}(t)+V_{i2}(t)+V_{i4}(t)\\
&=x_i^T(t)(P_i+R_i)x_i(t)+\int_{t-\bar{d}}^{t}e^{a(s-t)}x_i^T(s)W_ix_i(s)ds\\
&+\bar{d}\int_{-\bar{d}}^{0}\int_{t+\beta}^{t}e^{a(s-t)}\dot{x}_i^T(s)Z_i\dot{x}_i(s)dsd\beta\\
&\geq x_i^T(t)P_ix_i(t)+x_i^T(t)\left(\frac{a}{1-e^{-a\bar{d}}}R_i\right)x_i(t)\int_{t-\bar{d}}^{t}e^{a(s-t)}ds\\
&+\int_{t-\bar{d}}^{t}e^{a(s-t)}\begin{bmatrix}x_i(t)\\x_i(s)\end{bmatrix}^T\begin{bmatrix}Z_i & -Z_i\\ \star & W_i+Z_i\end{bmatrix}\begin{bmatrix}x_i(t)\\x_i(s)\end{bmatrix}ds\\
&=x_i^T(t)P_ix_i(t)\\
&+\int_{t-\bar{d}}^{t}e^{a(s-t)}\begin{bmatrix}x_i(t)\\x_i(s)\end{bmatrix}^T\begin{bmatrix}\frac{a}{1-e^{-a\bar{d}}}R_i+Z_i & -Z_i\\ \star & W_i+Z_i\end{bmatrix}\begin{bmatrix}x_i(t)\\x_i(s)\end{bmatrix}ds, \qquad (9.121)
\end{aligned}
$$

where $\{R_i,W_i\}\in\Re^{n_{xi}\times n_{xi}}$ are symmetric matrices, and $\{P_i,Z_i\}\in\Re^{n_{xi}\times n_{xi}}$ are symmetric positive definite matrices.

It is easy to see that there always exists a positive scalar ε such that the inequality $V(t)\geq\varepsilon\|x(t)\|^2$ holds if the matrices $P_i>0,Q_{ik}>0,Z_i>0$ and the inequality in (64) holds, thus completing this proof.

Based on the relaxing LKF in Lemma 9.8, an LMI-based condition for existence of a decentralized controller that ensures the reachable set bounding in (9.67) can be summarized as below.

Theorem 9.4: Relaxing Decentralized Controller Design of Reachable Set

Consider the large-scale T-S fuzzy system in (9.60). A decentralized fuzzy controller in (9.62) can guarantee that the reachable set of the closed-loop system (9.63) with input constraint (9.65) is bounded by the intersection of ellipsoid in (9.67), if there exist the positive definite symmetric matrices $\{\bar{P}_i,\bar{R}_i,\bar{Z}_i,\bar{M}_i,\bar{W}_i,H_{01},H_{02},H_{ik1},H_{ik2},H_{ik3}\}\in\Re^{n_{xi}\times n_{xi}}$, $H_{01}\leq H_{ik1},H_{02}\leq H_{ik2}$, and the matrices $\bar{K}_{il}\in\Re^{n_{ui}\times n_{xi}},G_i\in\Re^{n_{xi}\times n_{xi}}$, and the scalar $\bar{d}$, such that for all $i\in\mathscr{N}$, the following LMIs hold:

$$\begin{bmatrix}\frac{a}{1-e^{-ad}}\bar{R}_i+\bar{Z}_i & -\bar{Z}_i\\ \star & \bar{W}_i+\bar{Z}_i\end{bmatrix}>0, \tag{9.122}$$

$$\begin{bmatrix}-\bar{P}_i & \star\\ \bar{K}_{il} & -\bar{u}^2\mathbf{I}\end{bmatrix}<0, 1\leq l\leq r_i \tag{9.123}$$

$$\tilde{\Sigma}_{ill}<0, 1\leq l\leq r_i \tag{9.124}$$

$$\tilde{\Sigma}_{ilj}+\tilde{\Sigma}_{ijl}<0, 1\leq l<j\leq r_i \tag{9.125}$$

with

$$\tilde{\Sigma}_{ilj} = \begin{bmatrix} \tilde{\Theta}_{ilj} & \bar{\mathscr{E}}_{i1} & \bar{\mathscr{E}}_{i2} & \bar{E}_{i3} \\ \star & -\frac{H_{01}}{N-1} & 0 & 0 \\ \star & \star & -\frac{H_{02}}{N-1} & 0 \\ \star & \star & \star & -H_{i3} \end{bmatrix}, \tag{9.126}$$

where

$$\begin{cases} \tilde{\Theta}_{ilj} = \tilde{\Theta}_i^{(1)} + \mathrm{Sym}\{\mathscr{E}_i \bar{\mathbb{A}}_{ilj}\} + \sum_{k=1, k\neq i}^{N} \tilde{\Theta}_{ik}^{(2)}, \\ \tilde{\Theta}_i^{(1)} = \begin{bmatrix} \tilde{\Theta}_i^{(11)} & \bar{P}_i + \bar{R}_i & 0 & 0 & 0 \\ \star & \tilde{\Theta}_i^{(12)} & e^{-a\bar{d}}\bar{Z}_i & 0 & 0 \\ \star & \star & \bar{\Theta}_i^{(13)} & 0 & 0 \\ \star & \star & \star & -\bar{M}_i & 0 \\ \star & \star & \star & \star & -\frac{a}{\bar{w}^2}\mathbf{I} \end{bmatrix}, \\ \tilde{\Theta}_i^{(11)} = \bar{d}^2\bar{Z}_i + (N-1)\bar{M}_i, \tilde{\Theta}_i^{(12)} = a(\bar{P}_i + \bar{R}_i) + \bar{W}_i - e^{-a\bar{d}}\bar{Z}_i, \\ \bar{\Theta}_i^{(13)} = -e^{-a\bar{d}}\bar{W}_i - e^{-a\bar{d}}\bar{Z}_i, \tilde{\Theta}_{ik}^{(2)} = [\mathscr{E}_i \bar{A}_{dikl}]\, \tilde{\Theta}_{ik}^{(21)}\, [\mathscr{E}_i \bar{A}_{dikl}]^T, \\ \tilde{\Theta}_{ik}^{(21)} = \left[\frac{1}{4}H_{ik1} + \frac{1}{4}H_{ik2} + \frac{\bar{d}^2}{4}H_{ik3}\right], \\ \bar{\mathbb{A}}_{ilj} = \begin{bmatrix} -G_i & A_{il}G_i + B_{il}\bar{K}_{ij} & 0 & 0 & D_{il} \end{bmatrix}, \mathscr{E}_i = \begin{bmatrix} \mathbf{I} & \mathbf{I} & 0 & 0 & 0 \end{bmatrix}^T, \\ \bar{\mathscr{E}}_{i1} = \begin{bmatrix} 0 & G_i & 0 & 0 & 0 \end{bmatrix}^T, \bar{\mathscr{E}}_{i2} = \begin{bmatrix} 0 & 0 & G_i & 0 & 0 \end{bmatrix}^T, \\ \bar{E}_{i3} = \underbrace{[\bar{\mathscr{E}}_{1i3} \cdots \bar{\mathscr{E}}_{ki3,k\neq i} \cdots \bar{\mathscr{E}}_{Ni3}]}_{N-1}, \bar{\mathscr{E}}_{ki3} = \begin{bmatrix} 0 & 0 & 0 & \underbrace{\overbrace{[0 \cdots G_i \cdots 0]}^{k,k\neq i}}_{N-1} & 0 \end{bmatrix}^T, \\ \Pi_{ik} = \begin{bmatrix} 0 & \frac{1}{2}\mathbf{I} & \frac{1}{2}\mathbf{I} & \frac{\bar{d}}{2}\underbrace{\overbrace{[0 \cdots \mathbf{I} \cdots 0]}^{k,k\neq i}}_{N-1} & 0 \end{bmatrix}, \\ H_{i3} = \mathrm{diag}\underbrace{\{H_{1i3} \cdots H_{ki3,k\neq i} \cdots H_{Ni3}\}}_{N-1}, \bar{M}_i = \mathrm{diag}\underbrace{\{\bar{M}_i \cdots \bar{M}_i \cdots \bar{M}_i\}}_{N-1}. \end{cases} \tag{9.127}$$

Moreover, the controller gain matrix in (9.62) is obtained by calculating

$$K_{il} = \bar{K}_{il}G_i^{-1}, l \in \mathscr{L}_i, i \in \mathscr{N}. \tag{9.128}$$

■

Similar to Algorithm 9.2, the LMI-based algorithm to the “smallest” bound for the reachable set in (9.67) can also be given by:

Algorithm 9.3.

Minimize $\bar{\delta}$, subject to $\begin{bmatrix} \bar{\delta}\mathbf{I} & \bar{G} \\ \star & \bar{P} \end{bmatrix} > 0$, and (9.122) – (9.125), $i \in \mathscr{N}$

where $\bar{P} = \mathrm{diag}\underbrace{\{\bar{P}_1 \cdots \bar{P}_i \cdots \bar{P}_N\}}$.

Note: Based on the model transformation in (9.74) and the SSG method, the results proposed in this section avoid the use of the bounding inequalities in Lemma 1 of [15], and (16) of [9]. Moreover, the Lyapunov matrices R_i and W_i in Lemma 9.8 are not necessarily positive definite. It is expected that these results lead to less conservatism than the ones in [15, 9].

9.3 SIMULATION STUDIES

Consider a microgrid containing three nonlinear subsystems as below:

Plant Rule $\mathscr{R}_i^l$: IF $x_{i1}(t)$ is $\mathscr{F}_{i1}^l$, THEN

$$\begin{cases} \dot{x}_i(t) = A_{il}x_i(t) + B_{il}u_i(t) + \sum_{\substack{k=1 \\ k\neq i}}^{N} \bar{A}_{dikl}x_k(t - d_{ik}(t)) + D_{il}w_i(t), \\ x_i(t) \equiv 0, -\bar{d} \le t \le 0, i = \{1,2,3\}, l = \{1,2\} \end{cases}$$

where the system parameters are given by

$$A_{11} = \begin{bmatrix} 0.21 & -0.22 \\ 1.02 & 0 \end{bmatrix}, A_{12} = \begin{bmatrix} 0.18 & -0.26 \\ 1.23 & 0 \end{bmatrix},$$
$$\bar{A}_{d121} = \begin{bmatrix} -0.15 & -0.05 \\ 0.40 & 0 \end{bmatrix}, \bar{A}_{d131} = \begin{bmatrix} 0.18 & -0.26 \\ 1.23 & 0 \end{bmatrix},$$
$$\bar{A}_{d122} = \begin{bmatrix} -0.20 & 0 \\ 0.30 & 0 \end{bmatrix}, \bar{A}_{d132} = \begin{bmatrix} -0.20 & 0 \\ 0.40 & 0 \end{bmatrix},$$
$$B_{12} = \begin{bmatrix} 1.17 \\ 0.44 \end{bmatrix}, D_{12} = \begin{bmatrix} 0.82 \\ 0.25 \end{bmatrix}, B_{11} = \begin{bmatrix} 0.91 \\ 0.22 \end{bmatrix}, D_{11} = \begin{bmatrix} 0.91 \\ 0.17 \end{bmatrix}$$

for the first subsystem, and

$$A_{21} = \begin{bmatrix} 0.28 & -0.14 \\ 1.16 & 0 \end{bmatrix}, A_{22} = \begin{bmatrix} 0.15 & -0.26 \\ 1.08 & 0 \end{bmatrix},$$
$$\bar{A}_{d211} = \begin{bmatrix} -0.30 & -0.10 \\ 0.40 & 0 \end{bmatrix}, \bar{A}_{d231} = \begin{bmatrix} 0.25 & -0.05 \\ 0.30 & 0 \end{bmatrix},$$
$$\bar{A}_{d212} = \begin{bmatrix} -0.20 & 0 \\ 0.25 & 0 \end{bmatrix}, \bar{A}_{d232} = \begin{bmatrix} -0.35 & -0.15 \\ 0.45 & 0 \end{bmatrix},$$
$$B_{21} = \begin{bmatrix} 1.13 \\ 0.28 \end{bmatrix}, D_{21} = \begin{bmatrix} 0.75 \\ 0.42 \end{bmatrix}, B_{22} = \begin{bmatrix} 0.84 \\ 0.19 \end{bmatrix}, D_{22} = \begin{bmatrix} 0.84 \\ 0.51 \end{bmatrix}$$

for the second subsystem, and

$$A_{31} = \begin{bmatrix} 0.21 & -0.16 \\ 1.21 & 0 \end{bmatrix}, A_{32} = \begin{bmatrix} 0.24 & -0.28 \\ 1.53 & 0 \end{bmatrix},$$
$$\bar{A}_{d311} = \begin{bmatrix} -0.25 & -0.05 \\ 0.40 & 0 \end{bmatrix}, \bar{A}_{d321} = \begin{bmatrix} -0.25 & -0.15 \\ 0.40 & 0 \end{bmatrix},$$
$$\bar{A}_{d312} = \begin{bmatrix} -0.20 & -0.10 \\ 0.30 & 0 \end{bmatrix}, \bar{A}_{d322} = \begin{bmatrix} -0.40 & -0.10 \\ 0.35 & 0 \end{bmatrix},$$
$$B_{31} = \begin{bmatrix} 1.16 \\ 0.22 \end{bmatrix}, D_{31} = \begin{bmatrix} 0.87 \\ 0.38 \end{bmatrix}, B_{32} = \begin{bmatrix} 0.68 \\ 0.17 \end{bmatrix}, D_{32} = \begin{bmatrix} 0.76 \\ 0.49 \end{bmatrix}$$

for the third subsystem.

For the open-loop system, the reachable set is unbounded since it is unstable. The aim is to design a controller in the form of (9.62) such that the reachable set of the resulting closed-loop system with input constraint (9.66) is bounded by an intersection of ellipsoids. Assume that $\bar{d} = 0.45, a = 0.11, \bar{w} = 1.1, \bar{u} = 8, \tau_{ik} = 0.9$, and choose $\mathscr{G}_i = \begin{bmatrix} G_{i1}^T & G_{i2}^T & 0 & 0 & 0 \end{bmatrix}^T, i = \{1,2,3\}$, where

$$\begin{aligned} G_{11} &= \begin{bmatrix} 0.0884 & 0.0183 \\ 0.0485 & 0.0912 \end{bmatrix}, G_{12} = \begin{bmatrix} 0.1254 & 0.0377 \\ 0.0718 & 0.1349 \end{bmatrix}, \\ G_{21} &= \begin{bmatrix} 0.0753 & -0.0098 \\ 0.0403 & 0.0547 \end{bmatrix}, G_{22} = \begin{bmatrix} 0.0969 & -0.0185 \\ 0.0522 & 0.0708 \end{bmatrix}, \\ G_{31} &= \begin{bmatrix} 0.1030 & 0.0024 \\ 0.0525 & 0.0531 \end{bmatrix}, G_{32} = \begin{bmatrix} 0.1475 & 0.0110 \\ 0.0767 & 0.0745 \end{bmatrix}. \end{aligned}$$

By applying Algorithm 9.1, one can obtain $\bar{\delta}_{\min} = 17.0779$. When applying Algorithm 9.2 and Algorithm 9.3, $\bar{\delta}_{\min} = 25.0885$ and $\bar{\delta}_{\min} = 8.7187$ are obtained, respectively. The corresponding controller gains are calculated as

$$\begin{aligned} K_{11} &= \begin{bmatrix} -4.5142 & -2.9842 \end{bmatrix}, K_{12} = \begin{bmatrix} -3.1057 & -1.8298 \end{bmatrix}, \\ K_{21} &= \begin{bmatrix} -2.1912 & -1.1897 \end{bmatrix}, K_{22} = \begin{bmatrix} -3.4934 & -2.0992 \end{bmatrix}, \\ K_{31} &= \begin{bmatrix} -3.2320 & -1.4896 \end{bmatrix}, K_{32} = \begin{bmatrix} -4.2721 & -1.7366 \end{bmatrix}, \end{aligned}$$

for Algorithm 9.1, and

$$\begin{aligned} K_{11} &= \begin{bmatrix} -4.1435 & -2.6725 \end{bmatrix}, K_{12} = \begin{bmatrix} -2.8929 & -1.6232 \end{bmatrix}, \\ K_{21} &= \begin{bmatrix} -3.8094 & -2.3317 \end{bmatrix}, K_{22} = \begin{bmatrix} -2.6989 & -1.4697 \end{bmatrix}, \\ K_{31} &= \begin{bmatrix} -5.3441 & -3.0617 \end{bmatrix}, K_{32} = \begin{bmatrix} -3.7377 & -1.9859 \end{bmatrix}, \end{aligned}$$

for Algorithm 9.2, and

$$\begin{aligned} K_{11} &= \begin{bmatrix} -5.6150 & -4.9634 \end{bmatrix}, K_{12} = \begin{bmatrix} -4.1363 & -3.5400 \end{bmatrix}, \\ K_{21} &= \begin{bmatrix} -4.9866 & -4.7645 \end{bmatrix}, K_{22} = \begin{bmatrix} -3.7001 & -3.4291 \end{bmatrix}, \\ K_{31} &= \begin{bmatrix} -6.0989 & -5.4938 \end{bmatrix}, K_{32} = \begin{bmatrix} -4.4989 & -3.9681 \end{bmatrix}, \end{aligned}$$

for Algorithm 9.3.

Table 9.1
Comparison of minimum $\bar{\delta}$ for different methods

with $a = 0.11, \bar{w} = 1.1, \bar{u} = 8, \tau_{ik} = 0.9$ in Example 1

Methods	$\bar{d} = 0.45$	$\bar{d} = 0.46$	$\bar{d} = 0.50$
[15]	∞	∞	∞
[9]	∞	∞	∞
Algorithm 9.1 with given $\mathscr{G}_i$	17.0779	17.7853	∞
Algorithm 9.2	25.0885	∞	∞
Algorithm 9.3	8.7187	8.8214	9.2925

For different $\bar{d}$, one calculates respectively the minimum $\bar{\delta}$ by using the different methods. The detailed comparison is listed in Table 9.1. It can be seen that the direct Lyapunov design method proposed in [9] fails to find bounding reachable sets for this case. The results in Algorithm 9.1 could be less conservative than those in Algorithm 9.2, because imposing the constraints on the matrix multipliers $\mathscr{G}_i$ brings conservatism to Algorithm 9.2. Due to the relaxing condition that the Lyapunov matrices R_i and W_i in (9.119) are not necessarily positive definite, the minimum $\bar{\delta}$ obtained by Algorithm 9.3 is much better than those obtained in Algorithm 9.1 or in Algorithm 9.2.

9.4 REFERENCES

1. Zhang, L., Gao, H., and Kaynak, O. (2012). Network-induced constraints in networked control systems: A survey. IEEE Transactions on Industrial Informatics, 9(1), 403-416.
2. Basin, M., Perez, J., and Martinez-Zuniga, R. (2006). Alternative optimal filter for linear state delay systems. International Journal of Adaptive Control and Signal Processing, 20(20), 509-517.
3. Sun, J., Liu, G. P., Chen, J., and Rees, D. (2010). Improved delay-range-dependent stability criteria for linear systems with time-varying delays. Automatica, 46(2), 466-470.
4. Xu, S., Lam, J., Zhang, B., and Zou, Y. (2015). New insight into delay-dependent stability of time-delay systems. International Journal of Robust and Nonlinear Control, 25(7), 961-970.
5. Kwon, O. M., Park, M. J., Park, J. H., Lee, S. M., and Cha, E. J. (2013). Stability and stabilization for discrete-time systems with time-varying delays via augmented Lyapunov-Krasovskii functional. Journal of the Franklin Institute, 350(3), 521-540.
6. K. Gu, V. Kharitonov, and J. Chen, Stability of Time-Delay Systems. Boston: Birkhauser, 2003.
7. Gu, K., Zhang, Y., and Xu, S. (2011). Small gain problem in coupled differential-difference equations, time-varying delays, and direct Lyapunov method. International Journal of Robust and Nonlinear Control, 21(4), 429-451.
8. Li, X. and Gao, H. (2011). A new model transformation of discrete-time systems with time-varying delay and its application to stability analysis. IEEE Transactions on Automatic Control, 56(9), 2172-2178.

9. Zhang, H., Yu, G., Zhou, C., and Dang, C. (2013). Delay-dependent decentralised $\mathscr{H}_\infty$ filtering for fuzzy interconnected systems with time-varying delay based on Takagi-Sugeno fuzzy model. IET Control Theory and Applications, 7(5), 720-729.
10. Xiang, W., Jian, X., and Iqbal, M. N. (2013). $\mathscr{H}_\infty$ control for switched fuzzy systems via dynamic output feedback: hybrid and switched approaches. Communications in Nonlinear Science and Numerical Simulation, 18(6), 1499-1514.
11. Xie, L., Lu, L., Zhang, D., and Zhang, H. (2004). Improved robust $\mathscr{H}_2$ and $\mathscr{H}_\infty$ filtering for uncertain discrete-time systems. Automatica. 40(5), 873-880.
12. Fridman, E. and Shaked, U. (2003). On reachable sets for linear systems with delay and bounded peak inputs. Automatica, 39(11), 2005-2010.
13. Kao, C. Y. and Lincoln, B. (2004). Simple stability criteria for systems with time-varying delays. Automatica, 40(8), 1429-1434.
14. Fridman, E. and Shaked, U. (2006). Input-output approach to stability and $\mathscr{L}_2$-gain analysis of systems with time-varying delays. Systems and Control Letters, 55(12), 1041-1053.
15. Zhang, Z., Lin, C., and Chen, B. (2015). New decentralized $\mathscr{H}_\infty$ filter design for nonlinear interconnected systems based on Takagi-Sugeno fuzzy models. IEEE Transactions on Cybernetics, 45(12).
16. Zhong, Z., Wai, R. J., Shao, Z., and Xu, M. (2017). Reachable set estimation and decentralized controller design for large-scale nonlinear systems with time-varying delay and input constraint. IEEE Transactions on Fuzzy Systems, 25(6), 1629-1643.

10 Event-Triggered Fuzzy Control

In many digital implementations of NCSs, computers are often required to execute control tasks comprising of sampling, quantizing, transmitting the output of the plant, and computing, implementing the control input [1]. In the execution of control tasks, the conventional principle is based on time-triggered control in the sense that the control task is executed in a periodic manner, and it will bring collision or channel congestion or larger time delays in the network due to the limited communication bandwidth. Recently, interest is shown in the event-triggered control aiming at reduction in data transmissions. The working principle based on event-triggered control is to decide whether or not to transmit control signals in term of a given threshold [2, 3, 4, 5, 6, 7]. In other words, the control signals are not always implemented in every sampling period. The idea of event-based control has appeared under a variety of names, such as event-triggered feedback [2, 3], interrupt-based feedback [5], self-triggered feedback [6], state-triggered feedback [7].

In this chapter, the renewable energy sources are considered as the distributed connection to the common bus in a microgrid. They communicate with each other by using networks. In order to reduce data communication, centralized event-triggered control, decentralized event-triggered control, and distributed event-triggered control are investigated, respectively, and some new results for solving the stability analysis and control synthesis problems are proposed.

10.1 CENTRALIZED EVENT-TRIGGERED FUZZY CONTROL

10.1.1 PROBLEM FORMULATION

In a networked environment, we consider a multi-PV generator or/and a multi-machine WG, which consists of several nonlinear subsystems with interconnections. Then, T-S fuzzy model is used to represent the nonlinear system as below:

$$\dot{x}(t) = A(t)x(t) + B(t)u(t) + \omega(t), \tag{10.1}$$

where

$$x(t) = \begin{bmatrix} x_1^T(t) & x_2^T(t) & \cdots & x_N^T(t) \end{bmatrix}^T,$$

$$A(t) = \begin{bmatrix} A_{11}(t) & A_{12}(t) & \cdots & A_{1N}(t) \\ A_{21}(t) & A_{22}(t) & \cdots & A_{2N}(t) \\ \vdots & \vdots & \ddots & \vdots \\ A_{N1}(t) & A_{N2}(t) & \cdots & A_{NN}(t) \end{bmatrix}, u(t) = \begin{bmatrix} u_1(t) \\ u_2(t) \\ \vdots \\ u_N(t) \end{bmatrix},$$

$$B(t) = \begin{bmatrix} B_1(t) & 0 & \cdots & 0 \\ 0 & B_2(t) & \cdots & 0 \\ \vdots & \vdots & \ddots & \vdots \\ 0 & 0 & \cdots & B_N(t) \end{bmatrix}, \omega(t) = \begin{bmatrix} \omega_1(t) \\ \omega_2(t) \\ \vdots \\ \omega_N(t) \end{bmatrix}. \tag{10.2}$$

Choose $z(t) = [z_{11}, \cdots, z_{N1}; z_{12}, \cdots, z_{N2}; \cdots; z_{1g}, \cdots, z_{Ng}]$ as fuzzy premise variables. Thus, the nonlinear system in (10.1) can be represented by the following fuzzy system,

Plant Rule $\mathscr{R}^l$: **IF** $z_{11}(t)$ is $\mathscr{F}_{11}^l$ and $\cdots$ and $z_{N1}(t)$ is $\mathscr{F}_{N1}^l; z_{12}(t)$ is $\mathscr{F}_{12}^l$ and $\cdots$ and $z_{N2}(t)$ is $\mathscr{F}_{N2}^l$, $\cdots$, $z_{1g}(t)$ is $\mathscr{F}_{1g}^l$ and $\cdots$ and $z_{Ng}(t)$ is $\mathscr{F}_{Ng}^l$, **THEN**

$$\dot{x}(t) = A_l x(t) + B_l u(t) + \omega(t), l \in \mathscr{L} := \{1, 2, \ldots, r\} \tag{10.3}$$

where $\mathscr{R}^l$ denotes the l-th fuzzy inference rule; r is the number of inference rules; $\{A_l, B_l\}$ is the l-th local model.

Denote as $\mathscr{F}^l := \prod_{\phi=1}^{gN} \mathscr{F}_\phi^l$ the inferred fuzzy set, and $\mu_l[z(t)]$ as the normalized membership function, it yields

$$\mu_l[z(t)] := \frac{\prod_{\phi=1}^{g} \mu_{l\phi}[z_\phi(t)]}{\sum_{\varsigma=1}^{r} \prod_{\phi=1}^{g} \mu_{\varsigma\phi}[z_\phi(t)]} \geq 0, \sum_{l=1}^{r} \mu_l[z(t)] = 1. \tag{10.4}$$

We denote $\mu_l := \mu_l[z(t)]$ for simplicity.

After fuzzy blending, the global T-S fuzzy dynamic model is given by

$$\dot{x}(t) = A(\mu)x(t) + B(\mu)u(t) + \omega(t), \tag{10.5}$$

where $A(\mu) := \sum_{l=1}^{r} \mu_l A_l, B(\mu) := \sum_{l=1}^{r} \mu_l B_l$.

Before moving on, the following assumptions are firstly required [8].

Assumption 10.1. All samplers are time-driven. Let $\bar{s}$ denote constant sampling intervals, $\bar{s} > 0$, that is

$$s_{k+1} - s_k = \bar{s}, k \in \mathbb{N}.$$

Assumption 10.2. The sampling signals are transmitted via the network with a constant delay τ, and communication data are transmitted and received without disorder.

Assumption 10.3. The ZOH is time-driven and updated at instants z_k. Thus, it has

$$\underline{z} \leq z_{k+1} - z_k \leq \bar{z}, k \in \mathbb{N},$$

where $z > 0, \bar{z} > 0$.

Motivated by [8], we define the time elapsed for the sampler as $\rho^s(t) = t - s_k$. According to Assumption 10.1, it yields

$$0 \leq \rho^s(t) < \bar{s}, t \in [s_k, s_{k+1}). \tag{10.6}$$

Then the time elapsed for th ZOH is defined as $\rho^z(t) = t - z_k$. Based on the Assumption 10.2, it yields

$$0 \leq \rho^z(t) < \bar{z}, t \in [z_k, z_{k+1}). \tag{10.7}$$

Finally, based on the relations in (10.6) and (10.7) and defining as $\bar{\rho}^{sz}(t)$ as the total elapsed time from the sampling instant to the updating one, it yields

$$\eta(t) = \rho^s(t) + \tau + \rho^z(t), \tag{10.8}$$

which implies $\tau \leq \eta(t) < \bar{\eta}$, where $\bar{\eta} = \bar{s} + \tau + \bar{z}$.

It is noted that in the context of networked control systems, the traditionally time-triggered implementation is undesirable due to the existence of the limit communication bandwidth. Here, in order to reduce data transmissions, inspired by [2], we will propose an event-triggering mechanism (ETM) in the sense that it determines when information should be transmitted to the controller. Assume that the premise variable $z(t)$ and the system state $x(t)$ are measurable, in that case both $z(t)$ and $x(t)$ are involved in the sampled-data measurement, event-triggered control, and network-induced delay. Now, without loss of generality, we further assume that both $z(t)$ and $x(t)$ are packed, transmitted, and updated at the same time. Then, a centralized event-triggered state-feedback fuzzy controller can be given by

Controller Rule $\mathscr{R}^s$: **IF** $\hat{z}_1(z_k)$ is $\mathscr{F}^s$ and $\hat{z}_2(z_k)$ is $\mathscr{F}^s$ and $\cdots$ and $\hat{z}_g(z_k)$ is $\mathscr{F}^s$, **THEN**

$$u(t) = K_s\hat{x}(z_k), t \in [z_k, z_{k+1}) \tag{10.9}$$

where $K_s \in \Re^{n_u \times n_x}, s \in \mathscr{L}$ are controller gains to be determined; $\hat{z}(z_k) := [\hat{z}_1(z_k), \hat{z}_2(z_k), \cdots, \hat{z}_g(z_k)]$; $\hat{z}(z_k)$ and $\hat{x}(z_k)$ denote the updating signals in the fuzzy controller.

Similarly, the overall event-triggered state-feedback fuzzy controller is

$$u(t) = K(\hat{\mu})\hat{x}(z_k), t \in [z_k, z_{k+1}) \tag{10.10}$$

where

$$K(\hat{\mu}) := \sum_{s=1}^{r} \hat{\mu}_s[\hat{z}(z_k)] K_s, \sum_{s=1}^{r} \hat{\mu}_s[\hat{z}(z_k)] = 1, \hat{\mu}_s[\hat{z}(z_k)] := \frac{\prod_{\phi=1}^{g} \hat{\mu}_{s\phi}\left[\hat{z}_\phi(z_k)\right]}{\sum_{\varsigma=1}^{r}\prod_{\phi=1}^{g} \hat{\mu}_{\varsigma\phi}\left[\hat{\zeta}_\phi(z_k)\right]} \geq 0. \tag{10.11}$$

In the following, we will denote $\hat{\mu}_s := \hat{\mu}_s[\hat{z}(z_k)]$ for brevity.

Note: The centralized event-triggered fuzzy controller reduces to a PDC when $\mu_l = \hat{\mu}_l$. However, the premise variables of the fuzzy controller (10.10) undergo sampled-data measurement, event-triggered control, and network-induced delay. In such circumstances, the asynchronous variables between μ_l and $\hat{\mu}_l$ are more realistic. As pointed out in [9], when the knowledge between μ_l and $\hat{\mu}_l$ is unavailable, the condition $\mu_l \neq \hat{\mu}_l$ generally leads to a linear controller instead of a fuzzy one, which degrades the stabilization ability of the controller. When the knowledge on μ_l and $\hat{\mu}_l$ is available, the design conservatism can be improved, and we obtain the corresponding fuzzy controller.

In order to implement the event-triggered fuzzy controller given by (10.10), we assume that each subsystem transmits its measurements through a networked channel, and propose an event-triggered solution, where SP, BF and ETM are the sampler, buffer and event-triggering mechanism, respectively. For each subsystem, a smart sensor consists of an BF that is to store $\hat{x}(s_{k-1})$, which represents the latest measurement datum transmitted successfully to the controller, and an ETM that determines whether or not to transmit both $x(s_k)$ and $z(s_k)$ to the controller. Hence, in every sample period both $x(t)$ and $z(t)$ are firstly sampled by the SP. Then, they are transmitted to the controller and are executed, only when a prescribed event is violated. This leads to a reduction of data transmissions.

To formalize the described solution, the ETM in the sensor can operate as

$$\text{ETM: Both } x(s_k) \text{ and } z(s_k) \text{ are sent} \Leftrightarrow \|x(s_k) - \hat{x}(s_{k-1})\| > \sigma \|x(s_k)\|, \quad (10.12)$$

where $\sigma \geq 0$ is a suitably chosen design parameter.

Based on the operating condition given in (10.12), an event-triggered strategy is formulated as follows:

$$\hat{x}(s_k) = \begin{cases} x(s_k), & \text{when } \|x(s_k) - \hat{x}(s_{k-1})\| > \sigma \|x(s_k)\|, \\ \hat{x}(s_{k-1}), & \text{when } \|x(s_k) - \hat{x}(s_{k-1})\| \leq \sigma \|x(s_k)\|, \end{cases} \quad (10.13)$$

$$\hat{z}(s_k) = \begin{cases} z(s_k), & \text{when } x(s_k) \text{ is sent,} \\ \hat{z}(s_{k-1}), & \text{when } x(s_k) \text{ is not sent.} \end{cases} \quad (10.14)$$

In the case, the closed-loop fuzzy control system is given by

$$\dot{x}(t) = A(\mu)x(t) + B(\mu)K(\hat{\mu})\hat{x}(z_k). \quad (10.15)$$

Note: The event-triggered strategy $\|x(t) - x(r_k)\| > \sigma \|x(r_k)\|$ proposed in [6] is required to examine the triggered condition, continuously. However, the event-triggered scheme given in (10.13) and (10.14) only verifies the triggered condition at each sampling instant.

Note: A state-feedback fuzzy controller generally depends on premise variables and system states. The event-triggered scheme given in (10.13) and (10.14) shows that at the instant t_k both the premise variable $z(t_k)$ and system state $x(t_k)$ are not always transmitted to the fuzzy controller only when a prescribed threshold based on the system state is violated. Thus, the proposed triggered scheme significantly reduces data transmissions.

Based on the input delay approach [10], the sampled-data controller in (10.10) is reformulated as a delayed controller as follows:

$$u(t) = K(\hat{\mu})\hat{x}(t - \eta(t)), t \in [z_k, z_{k+1}) \quad (10.16)$$

where $\eta(t) = t - z_k$.

Combined with the fuzzy system in (10.5) and the delayed controller in (10.16), the closed-loop fuzzy event-triggered control system is given by

$$\dot{x}(t) = A(\mu)x(t) + B(\mu)K(\hat{\mu})\hat{x}(t-\eta(t)). \tag{10.17}$$

Here, we model the event-triggered counterpart as a disturbance [2],

$$\begin{aligned} e(t) &= \hat{x}(t-\eta(t)) - x(t-\eta(t)), \\ x(v) &= x(t-\eta(t)) - x(t), t \in [z_k, z_{k+1}). \end{aligned} \tag{10.18}$$

Then, by substituting (10.18) into (10.17), the closed-loop fuzzy control system in (10.17) can be rewritten as

$$\dot{x}(t) = A(\mu)x(t) + B(\mu)K(\hat{\mu})\left(x(t-\tau) + x(v) + e(t)\right). \tag{10.19}$$

10.1.2 DESIGN OF CENTRALIZED EVENT-TRIGGERED CONTROL

Now, we introduce the following LKF by utilizing Wirtinger's inequality [11]:

$$V(t) = V_1(t) + V_2(t), t \in [z_k, z_{k+1}) \tag{10.20}$$

with

$$\begin{cases} V_1(t) = x^T(t)Px(t) + \int_{t-\tau}^{t} x^T(\alpha)Qx(\alpha)d\alpha \\ \qquad + \tau\int_{-\tau}^{0}\int_{t+\beta}^{t}\dot{x}^T(\alpha)Z\dot{x}(\alpha)d\alpha d\beta, \\ V_2(t) = \bar{\eta}^2\int_{z_k}^{t}\dot{x}^T(\alpha)W\dot{x}(\alpha)d\alpha \\ \qquad - \frac{\pi^2}{4}\int_{z_k}^{t}[x(\alpha)-x(z_k)]^T W [x(\alpha)-x(z_k)]d\alpha, \end{cases} \tag{10.21}$$

where $\{P, Q, Z, W\} \in \Re^{n_x \times n_x}$, are symmetric positive definite matrices.

Based on the LKF in (10.20), a sufficient condition for the stability of the closed-loop fuzzy control system in (10.19) is given by the following theorem.

Theorem 10.1: $\mathscr{H}_\infty$ Performance Analysis of Centralized Event-Triggering

Consider the fuzzy system in (10.5), and a centralized event-triggered controller in the form of (10.10), the closed-loop fuzzy control system in (10.15) is asymptotically stable with $\mathscr{H}_\infty$ performance index γ, if there exist the symmetric positive definite matrices $\{P, Z, U, W\} \in \Re^{Nn_x \times Nn_x}$, and symmetric matrix $Q \in \Re^{Nn_x \times Nn_x}$, and positive scalars $\{\bar{\eta}^2, \tau, \varepsilon, \sigma\}$, such that the following matrix inequalities hold:

$$\Theta_1 + \mathrm{Sym}(\mathbb{P}\mathbb{A}(\mu, \hat{\mu})) < 0, \tag{10.22}$$

where

$$\Theta_1 = \begin{bmatrix} \Theta_{11} & 2P & 0 & 0 & 0 & 0 \\ \star & \Theta_{22} & Z & 0 & 0 & 0 \\ \star & \star & -Q-Z+\sigma^2 U & \sigma^2 U & 0 & 0 \\ \star & \star & \star & -\frac{\pi^2}{4}W+\sigma^2 U & 0 & 0 \\ \star & \star & \star & \star & -U & 0 \\ \star & \star & \star & \star & \star & -\gamma^2 I \end{bmatrix},$$

$$\Theta_{11} = \tau^2 Z + \bar{\eta}^2 W, \Theta_{22} = Q - Z + C^T C, \mathbb{P} = \begin{bmatrix} P & \varepsilon P & 0 & 0 & 0 & 0 \end{bmatrix}^T,$$
$$\mathbb{A}(\mu,\hat{\mu}) = \begin{bmatrix} -I & A(\mu) & B(\mu)K(\hat{\mu}) & B(\mu)K(\hat{\mu}) & B(\mu)K(\hat{\mu}) & I \end{bmatrix}. \quad (10.23)$$

■

Proof. Consider the Lyapunov function in (10.20), and by taking the time derivative of $V(t)$. Based on Jensen's inequality [12], one has

$$\begin{aligned} \dot{V}_1(t) &\leq 2x^T(t)P\dot{x}(t) + x^T(t)Qx(t) - x^T(t-\tau)Qx(t-\tau) \\ &\quad + \tau^2\dot{x}^T(t)Z\dot{x}(t) - \tau\int_{t-\tau}^{t}\dot{x}^T(\alpha)Z\dot{x}(\alpha)d\alpha \\ &\leq 2x^T(t)P\dot{x}(t) + x^T(t)Qx(t) - x^T(t-\tau)Qx(t-\tau) \\ &\quad + \tau^2\dot{x}^T(t)Z\dot{x}(t) - (x(t)-x(t-\tau))^T Z(x(t)-x(t-\tau)), \end{aligned} \quad (10.24)$$

$$\dot{V}_2(t) \leq \bar{\eta}^2\dot{x}^T(t)W\dot{x}(t) - \frac{\pi^2}{4}x^T(v)Wx(v). \quad (10.25)$$

Consider the following performance index,

$$J = \dot{V}(t) + y^T(t)y(t) - \gamma^2\omega^T(t)\omega(t), \quad (10.26)$$

where $y(t) = Cx(t)$ denotes the regulated output.

It is well-known that $J < 0$ implies the closed-loop control system is asymptotically stable with $\mathscr{H}_\infty$ performance.

It follows from (10.19) that

$$\begin{aligned} 0 &= 2\left[\dot{x}^T(t)P + x^T(t)\varepsilon P\right] \\ &\quad \times\left[-\dot{x}(t) + A(\mu)x(t) + B(\mu)K(\hat{\mu})(x(t-\tau)+x(v)+e(t)) + \omega(t)\right]. \end{aligned} \quad (10.27)$$

Define

$$\mathbb{A}(\mu,\hat{\mu}) = \begin{bmatrix} -I & A(\mu) & B(\mu)K(\hat{\mu}) & B(\mu)K(\hat{\mu}) & B(\mu)K(\hat{\mu}) & I \end{bmatrix},$$
$$\chi(t) = \begin{bmatrix} \dot{x}^T(t) & x^T(t) & x^T(t-\tau) & x^T(v) & e^T(t) & \omega^T(t) \end{bmatrix}^T. \quad (10.28)$$

It follows from (10.24)-(10.28) that

$$\begin{aligned}
J &\leq 2x^T(t)P\dot{x}(t) + x^T(t)Qx(t) - x^T(t-\tau)Qx(t-\tau) \\
&+ \tau^2\dot{x}^T(t)Z\dot{x}(t) - (x(t) - x(t-\tau))^T Z(x(t) - x(t-\tau)) \\
&+ \bar{\eta}^2\dot{x}^T(t)W\dot{x}(t) - \frac{\pi^2}{4}x^T(v)Wx(v) \\
&+ x^T(t)C^TCx(t) - \gamma^2\omega^T(t)\omega(t) \\
&+ [x(v) + x(t-\tau)]^T U[x(v) + x(t-\tau)] - e^T(t)Ue(t) \\
&+ 2\left[\dot{x}^T(t)P + x^T(t)\varepsilon P\right]\mathbb{A}(\mu,\hat{\mu})\chi(t) \\
&= \chi^T(t)\Theta(\mu,\hat{\mu})\chi(t),
\end{aligned} \tag{10.29}$$

where $\Theta(\mu,\hat{\mu}) = \Theta_1 + \mathrm{Sym}(\mathbb{P}\mathbb{A}(\mu,\hat{\mu}))$, and $\{\Theta_1, \mathbb{P}, \mathbb{A}(\mu,\hat{\mu})\}$ is defined in (10.23). It is easy to see that $\Theta(\mu,\hat{\mu}) < 0$ which implies $J < 0$,thus completing this proof.

10.1.3 RELAXING DESIGN OF CENTRALIZED EVENT-TRIGGERED CONTROL

Inspired by [13], we do not require that the matrix Q in (10.20) is necessarily positive definite. To ensure the positive property of $V(t)$, we give the following lemma:

Lemma 10.1: Relaxing Lyapunov-Krasovskii function

Consider the Lyapunov-Krasovskii function (LKF) in (10.20), then $V(t) \geq \varepsilon\|x(t)\|^2$, where $\varepsilon > 0$, if there exist the symmetric positive definite matrices $\{P, Z, W\} \in \Re^{n_x \times n_x}$, and symmetric matrix $Q \in \Re^{n_x \times n_x}$, such that the following inequalities hold:

$$\begin{bmatrix} \frac{1}{\tau}P + Z & -Z \\ \star & Q + Z \end{bmatrix} > 0. \tag{10.30}$$

■

Proof. Firstly, by using Jensen's inequality, we have

$$\begin{aligned}
\tau_i \int_{-\tau}^{0}\int_{t+\beta}^{t} x^T(\alpha) Z x(\alpha) d\alpha d\beta \\
&\geq \tau \int_{-\tau}^{0} \frac{-1}{\beta}\left[\int_{t+\beta}^{t} \dot{x}^T(\alpha) d\alpha\right] Z \left[\int_{t+\beta}^{t} \dot{x}(\alpha) d\alpha\right] d\beta \\
&= \tau \int_{-\tau}^{0} \frac{-1}{\beta}[x(t)-x(t+\beta)]^T Z [x(t)-x(t+\beta)] d\beta \\
&= \tau \int_{0}^{\tau} \frac{1}{\beta}[x(t)-x(t-\beta)]^T Z [x(t)-x(t-\beta)] d\beta \\
&\geq \int_{0}^{\tau} [x(t)-x(t-\beta)]^T Z [x(t)-x(t-\beta)] d\beta \\
&= \int_{t-\tau}^{t} [x(t)-x(\alpha)]^T Z [x(t)-x(\alpha)] d\alpha.
\end{aligned} \tag{10.31}$$

It follows from (10.20) and (10.31) that

$$\begin{aligned}
V_1(t) =& x^T(t) P x(t) + \int_{t-\tau}^{t} x^T(\alpha) Q x(\alpha) d\alpha + \tau \int_{-\tau}^{0}\int_{t+\beta}^{t} \dot{x}^T(\alpha) Z \dot{x}(\alpha) d\alpha d\beta \\
\geq& \int_{t-\tau}^{t} \begin{bmatrix} x(t) \\ x(\alpha) \end{bmatrix}^T \begin{bmatrix} \frac{1}{\tau}P+Z & -Z \\ \star & Q+Z \end{bmatrix} \begin{bmatrix} x(t) \\ x(\alpha) \end{bmatrix} d\alpha.
\end{aligned} \tag{10.32}$$

For $V_2(t)$ given in (10.20), we have $x(\alpha)-x(z_k)=0$ when $\alpha = z_k$. By using Wirtinger's inequality in [11], it is easy to see that $V_2(t) \geq 0$. Therefore, there always exists a positive scalar ε such that the inequality $V(t) \geq \varepsilon \|x(t)\|^2$ holds if the inequality in (10.3) holds. This completes the proof.

Based on the LKF in (10.20), a sufficient condition for the stability of the closed-loop fuzzy control system in (10.19) is directly given by the following theorem.

Theorem 10.2: $\mathscr{H}_\infty$ Performance Analysis of Centralized Event-Triggering

Consider the fuzzy system in (10.5), and a centralized event-triggered controller in the form of (10.10), the closed-loop fuzzy control system in (10.15) is asymptotically stable with $\mathscr{H}_\infty$ performance index γ, if there exist the symmetric positive definite matrices $\{P, Z, U, W\} \in \Re^{Nn_x \times Nn_x}$, and symmetric matrix $Q \in \Re^{Nn_x \times Nn_x}$, and positive scalars $\{\bar{\eta}^2, \tau, \varepsilon, \sigma\}$, such that the following matrix inequalities hold:

$$\begin{bmatrix} \frac{1}{\tau}P+Z & -Z \\ \star & Q+Z \end{bmatrix} > 0, \tag{10.33}$$

$$\Theta_1 + \mathrm{Sym}(\mathbb{P}\mathbb{A}(\mu,\hat{\mu})) < 0, \tag{10.34}$$

where

$$\Theta_1 = \begin{bmatrix} \Theta_{11} & 2P & 0 & 0 & 0 & 0 \\ \star & \Theta_{22} & Z & 0 & 0 & 0 \\ \star & \star & -Q-Z+\sigma^2 U & \sigma^2 U & 0 & 0 \\ \star & \star & \star & -\frac{\pi^2}{4}W+\sigma^2 U & 0 & 0 \\ \star & \star & \star & \star & -U & 0 \\ \star & \star & \star & \star & \star & -\gamma^2 I \end{bmatrix},$$

$$\Theta_{11} = \tau^2 Z + \bar{\eta}^2 W, \Theta_{22} = Q - Z + C^T C, \mathbb{P} = \begin{bmatrix} P & \varepsilon P & 0 & 0 & 0 & 0 \end{bmatrix}^T,$$
$$\mathbb{A}(\mu,\hat{\mu}) = \begin{bmatrix} -I & A(\mu) & B(\mu)K(\hat{\mu}) & B(\mu)K(\hat{\mu}) & B(\mu)K(\hat{\mu}) & I \end{bmatrix}. \quad (10.35)$$

■

It is noted that the results on Theorem 10.2 are not LMIs. It is also noted that the existing relaxation technique $\sum_{l=1}^{r} [\mu_l]^2 \Sigma_{ll} + \sum_{l=1}^{r} \sum_{l<s\leq r}^{r} \mu_l \mu_s \Sigma_{ls} < 0$ is no longer applicable to fuzzy controller synthesis, since $\mu_s \neq \hat{\mu}_s$. Here, assume that the asynchronized information is known, and it is subject to

$$\underline{\rho}_l \leq \frac{\hat{\mu}_l}{\mu_l} \leq \bar{\rho}_l, \quad (10.36)$$

where $\underline{\rho}_l$ and $\bar{\rho}_l$ are positive scalars.

It follows (10.34) and (10.36), and the asynchronized method proposed in [9] that the design result on the centralized event-triggered fuzzy controller can be summarized as below:

Theorem 10.3: Design of $\mathscr{H}_\infty$ Centralized Event-Triggered Using Asynchronized Method

Consider the fuzzy system in (10.5), and a centralized event-triggered controller in the form of (10.10), the closed-loop fuzzy control system (10.15) with the asynchronized relationship 10.36 is asymptotically stable with $\mathscr{H}_\infty$ performance γ, if there exist the symmetric positive definite matrices $\{P, \bar{Z}, \bar{W}, \bar{U}, X\} \in \Re^{Nn_x \times Nn_x}$, and matrices $M_{ls} = M_{sl}^T \in \Re^{N(4n_x+n_\omega+n_y)\times N(4n_x+n_\omega+n_y)}$, $\bar{Q} = \bar{Q}^T \in \Re^{Nn_x \times Nn_x}$, $\bar{K}_s \in \Re^{Nn_u \times Nn_x}$, and positive scalars $\left\{\bar{\eta}, \tau, \bar{\rho}_l, \underline{\rho}_l, \varepsilon, \sigma\right\}$, such that for all $(l,s) \in \mathscr{L}$, the following

LMIs hold:

$$\begin{bmatrix} \frac{1}{\tau}X+\bar{Z} & -\bar{Z} \\ \star & \bar{Q}+\bar{Z} \end{bmatrix} > 0, \tag{10.37}$$

$$\bar{\rho}_l \Sigma_{ll} + M_{ll} < 0, \tag{10.38}$$

$$\underline{\rho}_l \Sigma_{ll} + M_{ll} < 0, \tag{10.39}$$

$$\bar{\rho}_s \Sigma_{ls} + \bar{\rho}_l \Sigma_{sl} + M_{ls} + M_{sl} < 0, \tag{10.40}$$

$$\underline{\rho}_s \Sigma_{ls} + \underline{\rho}_l \Sigma_{sl} + M_{ls} + M_{sl} < 0, \tag{10.41}$$

$$\underline{\rho}_s \Sigma_{ls} + \bar{\rho}_l \Sigma_{sl} + M_{ls} + M_{sl} < 0, \tag{10.42}$$

$$\bar{\rho}_s \Sigma_{ls} + \underline{\rho}_l \Sigma_{sl} + M_{ls} + M_{sl} < 0, \tag{10.43}$$

$$\begin{bmatrix} M_{11} & \cdots & M_{1r} \\ \vdots & \ddots & \vdots \\ M_{r1} & \cdots & M_{rr} \end{bmatrix} > 0, \tag{10.44}$$

where

$$\begin{aligned}
\Sigma_{ls} &= \begin{bmatrix} \tilde{\Theta}_1 + Sym\left(\mathbb{I}\bar{\mathbb{A}}_{is}\right) & \mathbb{X} \\ \star & -I \end{bmatrix}, \mathbb{X} = \begin{bmatrix} 0 & CX & 0 & 0 & 0 & 0 \end{bmatrix}^T, \\
\tilde{\Theta}_1 &= \begin{bmatrix} \tilde{\Theta}_{11} & 2X & 0 & 0 & 0 & 0 \\ \star & \bar{Q}-\bar{Z} & \bar{Z} & 0 & 0 & 0 \\ \star & \star & -\bar{Q}-\bar{Z}+\sigma^2\bar{U} & \sigma^2\bar{U} & 0 & 0 \\ \star & \star & \star & -\frac{\pi^2}{4}\bar{W}+\sigma^2\bar{U} & 0 & 0 \\ \star & \star & \star & \star & -\bar{U} & 0 \\ \star & \star & \star & \star & \star & -\gamma^2 I \end{bmatrix}, \\
\tilde{\Theta}_{11} &= \tau^2\bar{Z} + \bar{\eta}^2\bar{W}, \mathbb{I} = \begin{bmatrix} I & \varepsilon I & 0 & 0 & 0 & 0 \end{bmatrix}^T, \\
\bar{\mathbb{A}}_{is} &= \begin{bmatrix} -X & A_l X & B_l\bar{K}_s & B_l\bar{K}_s & B_l\bar{K}_s & X \end{bmatrix}.
\end{aligned} \tag{10.45}$$

In that case, the proposed fuzzy controller gains can be calculated by

$$K_s = \bar{K}_s X^{-1}, s \in \mathscr{L}. \tag{10.46}$$

■

Proof. Define Γ_1 =diag$\{X,X,X,X,X,I\}$, and Γ_2 =diag$\{X,X\}$, where $X = P^{-1}$, and define $\bar{W} = XWX, \bar{Z} = XZX, \bar{Q} = XQX, \bar{U} = XUX, \bar{K}(\hat{\mu}) = K(\hat{\mu})X$. By performing the congruence transformation to (10.34) by Γ_1,

$$\bar{\Theta}_1 + Sym\left(\mathbb{I}\bar{\mathbb{A}}(\mu,\hat{\mu})\right) < 0, \tag{10.47}$$

where

$$\bar{\Theta}_1 = \begin{bmatrix} \bar{\Theta}_{11} & 2X & 0 & 0 & 0 & 0 \\ \star & \bar{Q}-\bar{Z}+XC^TCX & \bar{Z} & 0 & 0 & 0 \\ \star & \star & -\bar{Q}-\bar{Z}+\bar{U} & \bar{U} & 0 & 0 \\ \star & \star & \star & -\frac{\pi^2}{4}\bar{W}+\bar{U} & 0 & 0 \\ \star & \star & \star & \star & -\bar{U} & 0 \\ \star & \star & \star & \star & \star & -\gamma^2 I \end{bmatrix},$$

$$\bar{\Theta}_{11} = \tau^2\bar{Z}+\bar{\eta}^2\bar{W}, \mathbb{I}=\begin{bmatrix} I & \varepsilon I & 0 & 0 & 0 & 0 \end{bmatrix}^T,$$
$$\bar{\mathbb{A}}(\mu,\hat{\mu}) = \begin{bmatrix} -X & A(\mu)X & B(\mu)\bar{K}(\hat{\mu}) & B(\mu)\bar{K}(\hat{\mu}) & B(\mu)\bar{K}(\hat{\mu}) & X \end{bmatrix}. \quad (10.48)$$

By performing the congruence transformation to (10.33) by Γ_2, the inequality in (10.37) can be directly obtained. By applying Schur complement lemma to (10.47), and using the asynchronized method proposed in [9], the LMI-based results on (10.38)-(10.44) are obtained, thus completing this proof.

It is worth pointing output that the number of LMIs on 10.3 is large. It is also noted that the existing relaxation technique $\sum_{l=1}^{r}[\mu_l]^2\Sigma_{ll}+\sum_{l=1}^{r}\sum_{l<s\leq r}^{r}\mu_l\mu_s\Sigma_{ls}<0$ is no longer applicable to fuzzy controller synthesis, since $\mu_s \neq \hat{\mu}_s$. Similarly to the asynchronous relaxation technique in [14], it is assumed that

$$|\mu_l - \hat{\mu}_l| \leq \delta_l, l \in \mathscr{L}, \quad (10.49)$$

where δ_l is a positive scalar. If $\Sigma_{ls}+M_l \geq 0$, where M_l is a symmetric matrix, one obtains

$$\begin{aligned} &\sum_{l=1}^{r}\sum_{s=1}^{r}\mu_l\hat{\mu}_s\Sigma_{ls} \\ &= \sum_{l=1}^{r}\sum_{s=1}^{r}\mu_l\mu_s\Sigma_{ls} + \sum_{l=1}^{r}\sum_{s=1}^{r}\mu_l(\hat{\mu}_s-\mu_s)(\Sigma_{ls}+M_l) \\ &\leq \sum_{l=1}^{r}\sum_{s=1}^{r}\mu_l\mu_s\left[\Sigma_{ls}+\sum_{s=1}^{r}\delta_s(\Sigma_{ls}+M_l)\right]. \end{aligned} \quad (10.50)$$

Based on the synchronized relationship (10.50), the design of a centralized fuzzy controller is given as below.

Theorem 10.4: Design of $\mathscr{H}_\infty$ Centralized Event-Triggered Controller Using Synchronized Method

For the fuzzy system in (10.5), a centralized event-triggered controller in the form of (10.10) can be used to stabilize its closed-loop control system with $\mathscr{H}_\infty$ performance γ subject to the asynchronized relationship (10.36), if there exist the symmetric positive definite matrices $\{P,\bar{Z},\bar{W},X,\bar{U}\} \in \Re^{Nn_x\times Nn_x}$, and matrices $M_l = M_l^T \in \Re^{N(4n_x+n_\omega)\times N(4n_x+n_\omega)}$, $\bar{Q}=\bar{Q}^T \in \Re^{Nn_x\times Nn_x}$, $\bar{K}_s \in \Re^{Nn_u\times Nn_x}$, and positive scalars

$\{\bar{s}, \bar{z}, \tau, \delta_l, \sigma\}$, such that for all $(l,s) \in \mathscr{L}$, the following LMIs hold:

$$\begin{bmatrix} \frac{1}{\tau}X + \bar{Z} & -\bar{Z} \\ \star & \bar{Q} + \bar{Z} \end{bmatrix} > 0, \tag{10.51}$$

$$\Sigma_{ll} < 0, l \in \mathscr{L} \tag{10.52}$$

$$\Sigma_{ls} + \Sigma_{sl} < 0, 1 \leq l < s \leq r \tag{10.53}$$

where

$$\Sigma_{ls} = \Phi_{ls} + \sum\nolimits_{s=1}^{r} \delta_s (\Phi_{ls} + M_l), \Phi_{ls} = \begin{bmatrix} \tilde{\Theta}_1 + Sym(\mathbb{I}\bar{\mathbb{A}}_{is}) & \mathbb{X} \\ \star & -I \end{bmatrix},$$

$$\tilde{\Theta}_1 = \begin{bmatrix} \tilde{\Theta}_{11} & 2X & 0 & 0 & 0 & 0 \\ \star & \bar{Q} - \bar{Z} & \bar{Z} & 0 & 0 & 0 \\ \star & \star & -\bar{Q} - \bar{Z} + \sigma^2 \bar{U} & \sigma^2 \bar{U} & 0 & 0 \\ \star & \star & \star & -\frac{\pi^2}{4}\bar{W} + \sigma^2 \bar{U} & 0 & 0 \\ \star & \star & \star & \star & -\bar{U} & 0 \\ \star & \star & \star & \star & \star & -\gamma^2 I \end{bmatrix},$$

$$\tilde{\Theta}_{11} = \tau^2 \bar{Z} + (\bar{z} + \bar{s})^2 \bar{W}, \mathbb{I} = \begin{bmatrix} I & \varepsilon I & 0 & 0 & 0 & 0 \end{bmatrix}^T,$$

$$\bar{\mathbb{A}}_{is} = \begin{bmatrix} -X & A_l X & B_l \bar{K}_s & B_l \bar{K}_s & B_l \bar{K}_s & X \end{bmatrix}. \tag{10.54}$$

In that case, the controller gains can be calculated by

$$K_l = \bar{K}_l X^{-1}. \tag{10.55}$$

■

Note that when the asynchronized information of μ_l and $\hat{\mu}_l$ is unknown, the design result on a centralized event-triggered linear controller can be directly derived as below:

Theorem 10.5: Design of $\mathscr{H}_\infty$ Centralized Event-Triggered Linear Controller

For the fuzzy system in (10.5), a centralized event-triggered controller $u(t) = K\hat{x}(z_k)$ can be used to stabilize its closed-loop control system with $\mathscr{H}_\infty$ performance γ, if there exist the symmetric positive definite matrices $\{P, \bar{Z}, \bar{W}, U, X\} \in \Re^{Nn_x \times Nn_x}$, and matrices $\bar{Q} = \bar{Q}^T \in \Re^{Nn_x \times Nn_x}$, $\bar{K} \in \Re^{Nn_u \times Nn_x}$, and positive scalars $\{\bar{s}, \bar{z}, \tau, \varepsilon, \delta_l\}$, such that for all $l \in \mathscr{L}$, the following LMIs hold:

$$\begin{bmatrix} \frac{1}{\tau}X + \bar{Z} & -\bar{Z} \\ \star & \bar{Q} + \bar{Z} \end{bmatrix} > 0, \tag{10.56}$$

$$\begin{bmatrix} \tilde{\Theta}_1 + Sym(\mathbb{I}\bar{\mathbb{A}}_i) & \mathbb{X} \\ \star & -I \end{bmatrix} < 0, \tag{10.57}$$

where

$$\tilde{\Theta}_1 = \begin{bmatrix} \tilde{\Theta}_{11} & 2X & 0 & 0 & 0 & 0 \\ \star & \bar{Q}-\bar{Z} & \bar{Z} & 0 & 0 & 0 \\ \star & \star & -\bar{Q}-\bar{Z}+\sigma^2\bar{U} & \sigma^2\bar{U} & 0 & 0 \\ \star & \star & \star & -\frac{\pi^2}{4}\bar{W}+\sigma^2\bar{U} & 0 & 0 \\ \star & \star & \star & \star & -\bar{U} & 0 \\ \star & \star & \star & \star & \star & -\gamma^2 I \end{bmatrix},$$

$$\tilde{\Theta}_{11} = \tau^2\bar{Z}+(\bar{z}+\bar{s})^2\bar{W}, \mathbb{I}=\begin{bmatrix} I & \varepsilon I & 0 & 0 & 0 & 0 \end{bmatrix}^T,$$
$$\bar{\mathbb{A}}_i = \begin{bmatrix} -X & A_l X & B_l\bar{K} & B_l\bar{K} & B_l\bar{K} & X \end{bmatrix}. \tag{10.58}$$

In that case, the proposed linear controller gains can be calculated by

$$K = \bar{K}X^{-1}. \tag{10.59}$$

■

Note: When designing an event-triggered linear controller in Theorem 10.5, the premise variables are no longer required to transmit through communication networks. Compared with the event-triggered fuzzy controller in (10.10), the linear one reduces the requirements for extra hardware and software and raises the design conservatism.

10.2 DECENTRALIZED EVENT-TRIGGERED FUZZY CONTROL

10.2.1 PROBLEM FORMULATION

In a networked environment, we consider a multi-PV generator or/and a multi-wind generator, which consists of several nonlinear subsystems with interconnections. Then, the i-th nonlinear subsystem is represented by the following T-S fuzzy model:

Plant Rule $\mathscr{R}_i^l$: **IF** $z_{i1}(t)$ is $\mathscr{F}_{i1}^l$ and $z_{i2}(t)$ is $\mathscr{F}_{i2}^l$ and $\cdots$ and $z_{ig}(t)$ is $\mathscr{F}_{ig}^l$, **THEN**

$$\dot{x}_i(t) = A_{il}x_i(t) + B_{il}u_i(t) + \sum_{\substack{j=1 \\ j\neq i}}^{N} \bar{A}_{ijl}x_j(t), \tag{10.60}$$

where $l \in \mathscr{L}_i := \{1,2,\ldots,r_i\}$, $i \in \mathscr{N} := \{1,2,\ldots,N\}, N$ is the number of the subsystems. For the i-th subsystem, $\mathscr{R}_i^l$ is the l-th fuzzy inference rule; r_i is the number of inference rules;$\mathscr{F}_{i\phi}^l$ $(\phi = 1,2,\ldots,g)$ are the fuzzy sets; $x_i(t) \in \Re^{n_{xi}}$ and $u_i(t) \in \Re^{n_{ui}}$ denote the system state and control input, respectively; $z_i(t) := [z_{i1}(t), z_{i2}(t), \ldots, z_{ig}(t)]$ are the measurable variables; A_{il} and B_{il} are the l-th local model; $\bar{A}_{ijl}$ denotes the interconnected matrix of the i-th and j-th subsystems for the l-th local model.

Define the inferred fuzzy set $\mathscr{F}_i^l := \prod_{\phi=1}^{g}\mathscr{F}_{i\phi}^l$ and normalized membership function $\mu_{il}[z_i(t)]$,

$$\mu_{il}[z_i(t)] := \frac{\prod_{\phi=1}^{g}\mu_{il\phi}[z_{i\phi}(t)]}{\sum_{\varsigma=1}^{r_i}\prod_{\phi=1}^{g}\mu_{i\varsigma\phi}[z_{i\phi}(t)]} \geq 0, \sum_{l=1}^{r_i}\mu_{il}[z_i(t)] = 1. \tag{10.61}$$

In the following, we will denote $\mu_{il} := \mu_{il}[z_i(t)]$ for brevity.

By fuzzy blending, the i-th global T-S fuzzy dynamic model is obtained by

$$\dot{x}_i(t) = A_i(\mu_i)x_i(t) + B_i(\mu_i)u_i(t) + \sum_{\substack{j=1 \\ j\neq i}}^{N} \bar{A}_{ij}(\mu_i)x_j(t), \tag{10.62}$$

where $A_i(\mu_i) := \sum_{l=1}^{r_i} \mu_{il}A_{il}, B_i(\mu_i) := \sum_{l=1}^{r_i} \mu_{il}B_{il}, \bar{A}_{ij}(\mu_i) := \sum_{l=1}^{r_i} \mu_{il}\bar{A}_{ijl}$.

Note: Instead of a special class of large-scale fuzzy systems with linear interconnection matrix $\bar{A}_{ij}$ in [15], this section considers a general class of large-scale fuzzy systems in (10.62), where nonlinearities appear in interconnections to other subsystems. Control problems for the large-scale fuzzy system (10.62) with nonlinear interconnections are more complex and challenging than those with linear interconnections in [15].

Before moving on, the following assumptions are required [20].

Assumption 10.4. The sampler in each subsystem is clock-driven. Let h_i denote the upper bound of sampling intervals, we have

$$t_{k+1}^i - t_k^i \leq h_i, k \in \mathbb{N}$$

where $h_i > 0$.

Assumption 10.5. Assume that each subsystem in the large-scale system (1) is closed by a communication channel. The sampled signals at the instant t_k^i are transmitted over the communication network inducing a constant time delay $\tau_i \geq 0$.

Assumption 10.6. The zero-order-hold (ZOH) is event-driven, and it uses the latest sampled-data signals and holds them until the next transmitted data are received.

In the context of networked control systems, the traditionally time-triggered implementation is undesirable due to the existence of the limit communication bandwidth. Here, in order to reduce data transmissions, inspired by [2], we will propose an event-triggering mechanism (ETM) in the sense that it determines when information should be transmitted to the controller. Assume that the premise variable $z_i(t)$ and the system state $x_i(t)$ are measurable, in that case both $z_i(t)$ and $x_i(t)$ are involved in the sampled-data measurement, event-triggered control, and network-induced delay. Now, without loss of generality, we further assume that both $z_i(t)$ and $x_i(t)$ are packed, transmitted, and updated at the same time. Then, a decentralized event-triggered state-feedback fuzzy controller can be given by

Controller Rule $\mathscr{R}_i^s$: **IF** $\hat{z}_{i1}(t_k^i - \tau_i)$ is $\mathscr{F}_{i1}^s$ and $\hat{z}_{i2}(t_k^i - \tau_i)$ is $\mathscr{F}_{i2}^s$ and $\cdots$ and $\hat{z}_{ig}(t_k^i - \tau_i)$ is $\mathscr{F}_{ig}^s$, **THEN**

$$u_i(t) = K_{is}\hat{x}_i(t_k^i - \tau_i), t \in [t_k^i, t_{k+1}^i) \tag{10.63}$$

where $K_{is} \in \Re^{n_u \times n_x}, s \in \mathscr{L}_i, i \in \mathscr{N}$ are controller gains to be determined; $\hat{z}_i(t_k^i - \tau_i) := [\hat{z}_{i1}(t_k^i - \tau_i), \hat{z}_{i2}(t_k^i - \tau_i), \ldots, \hat{z}_{ig}(t_k^i - \tau_i)]$; $\hat{z}_i(t_k^i - \tau_i)$ and $\hat{x}_i(t_k^i - \tau_i)$ denote the updating signals in the fuzzy controller.

Similarly, the overall event-triggered state-feedback fuzzy controller is

$$u_i(t) = K_i(\hat{\mu}_i)\hat{x}_i(t_k^i - \tau_i), t \in [t_k^i, t_{k+1}^i) \tag{10.64}$$

where

$$\begin{cases} K_i(\hat{\mu}_i) := \sum_{s=1}^{r_i} \hat{\mu}_{is}\left[\hat{z}_i(t_k^i - \tau_i)\right] K_{is}, \sum_{s=1}^{r_i} \hat{\mu}_{is}\left[\hat{z}_i(t_k^i - \tau_i)\right] = 1, \\ \hat{\mu}_{is}\left[\hat{z}_i(t_k^i - \tau_i)\right] := \frac{\Pi_{\phi=1}^{g} \hat{\mu}_{is\phi}\left[\hat{z}_{i\phi}(t_k^i - \tau_i)\right]}{\Sigma_{\varsigma=1}^{r_i} \Pi_{\phi=1}^{g} \hat{\mu}_{i\varsigma\phi}\left[\hat{z}_{i\phi}(t_k^i - \tau_i)\right]} \geq 0. \end{cases} \tag{10.65}$$

In the following, we will denote $\hat{\mu}_{is} := \hat{\mu}_{is}\left[\hat{z}_i(t_k^i - \tau_i)\right]$ for brevity.

Note: The decentralized event-triggered fuzzy controller reduces to a PDC when $\mu_{il} = \hat{\mu}_{il}$. However, the premise variables of the fuzzy controller (7) undergo sampled-data measurement, event-triggered control, and network-induced delay. In such circumstances, the asynchronous variables between μ_{il} and $\hat{\mu}_{il}$ are more realistic. As pointed out in [9] , when the knowledge between μ_{il} and $\hat{\mu}_{il}$ is unavailable, the condition $\mu_{il} \neq \hat{\mu}_{il}$ generally leads to a linear controller instead of a fuzzy one, which degrades the stabilization ability of the controller. When the knowledge on μ_{il} and $\hat{\mu}_{il}$ is available, the design conservatism can be improved, and we obtain the corresponding fuzzy controller.

In order to implement the event-triggered fuzzy controller given by (10.64), we assume that each subsystem transmits its measurements through a networked channel, and propose a solution, where SP, BF and ETM are the sampler, buffer and event-triggering mechanism, respectively. For each subsystem, a smart sensor consists of an BF that is to store $\hat{x}_i\left(t_{k-1}^i\right)$, which represents the latest measurement datum transmitted successfully to the controller, and an ETM that determines whether or not to transmit both $x_i\left(t_k^i\right)$ and $z_i(t_k^i)$ to the controller. Hence, in every sample period both $x_i(t)$ and $z_i(t)$ are firstly sampled by the SP. Then, they are transmitted to the controller and are executed, only when a prescribed event is violated. This leads to a reduction of data transmissions.

To formalize the described solution, the ETM in the sensor can operate as

$$\text{ETM: Both } x_i\left(t_k^i\right) \text{ and } z_i(t_k) \text{ are sent} \Leftrightarrow \left\|x_i\left(t_k^i\right) - \hat{x}_i\left(t_{k-1}^i\right)\right\| > \sigma_i \left\|x_i\left(t_k^i\right)\right\|, \tag{10.66}$$

where $\sigma_i \geq 0$ is a suitably chosen design parameter.

Based on the operating condition given in (10.66), an event-triggered strategy is formulated as follows:

$$\hat{x}_i\left(t_k^i\right) = \begin{cases} x_i\left(t_k^i\right), & \text{when } \left\|x_i\left(t_k^i\right) - \hat{x}_i\left(t_{k-1}^i\right)\right\| > \sigma_i\left\|x_i\left(t_k^i\right)\right\|, \\ \hat{x}_i\left(t_{k-1}^i\right), & \text{when } \left\|x_i\left(t_k^i\right) - \hat{x}_i\left(t_{k-1}^i\right)\right\| \leq \sigma_i\left\|x_i\left(t_k^i\right)\right\|, \end{cases} \tag{10.67}$$

$$\hat{z}_i\left(t_k^i\right) = \begin{cases} z_i\left(t_k^i\right), & \text{when } x_i\left(t_k^i\right) \text{ is sent,} \\ \hat{z}_i\left(t_{k-1}^i\right), & \text{when } x_i\left(t_k^i\right) \text{ is not sent.} \end{cases} \tag{10.68}$$

In the case, the i-th closed-loop fuzzy control system is given by

$$\dot{x}_i(t) = A_i(\mu_i)x_i(t) + B_i(\mu_i)K_i(\hat{\mu}_i)\hat{x}_i(t_k^i - \tau_i) + \sum_{\substack{j=1 \\ j\neq i}}^{N} \bar{A}_{ij}(\mu_i)x_j(t), i \in \mathcal{N}. \tag{10.69}$$

Note: The event-triggered strategy $\|x_i(t) - x_i(r_k^i)\| > \sigma_i \|x_i(r_k^i)\|$ proposed in [6] is required to examine the triggered condition, continuously. However, the event-triggered scheme given in (10.67) and (10.68) only verifies the triggered condition at each sampling instant.

Note: A state-feedback fuzzy controller generally depends on premise variables and system states. The event-triggered scheme given in (10.67) and (10.68) shows that at the instant t_k^i both the premise variable $z_i(t_k^i)$ and system state $x_i(t_k^i)$ are not always transmitted to the fuzzy controller; that occurs only when a prescribed threshold based on the system state is violated. Thus, the proposed triggered scheme significantly reduces data transmissions.

Before ending this section, we give the following lemma, which will be used to obtain the main results.

Lemma 10.2: Relaxing Inequality

Given the interconnected matrix $\bar{A}_{ijl}$ in the system (10.62), and the symmetric positive definite matrix $W_i \in \Re^{n_{xi} \times n_{xi}}$, the following inequality holds:

$$\sum_{i=1}^{N} \sum_{\substack{j=1 \\ j \neq i}}^{N} \bar{A}_{ij}(\mu_i) W_i \bar{A}_{ij}^T(\mu_i) \leq \sum_{i=1}^{N} \sum_{\substack{j=1 \\ j \neq i}}^{N} \sum_{l=1}^{r_i} \mu_{il} \bar{A}_{ijl} W_i \bar{A}_{ijl}^T. \tag{10.70}$$

■

Proof. Note that for $(i, j) \in \mathcal{N}, j \neq i, l \in \mathcal{L}_i$

$$\left[\bar{A}_{ijl} - \bar{A}_{ijf}\right] W_i \left[\bar{A}_{ijl} - \bar{A}_{ijf}\right]^T \geq 0, \tag{10.71}$$

which implies that

$$\bar{A}_{ijl} W_i \bar{A}_{ijl}^T + \bar{A}_{ijf} W_i \bar{A}_{ijf}^T \geq \bar{A}_{ijl} W_i \bar{A}_{ijf}^T + \bar{A}_{ijf} W_i \bar{A}_{ijl}^T. \tag{10.72}$$

By taking the relationship (10.72), we have

$$\begin{aligned}\sum_{i=1}^{N}\sum_{\substack{j=1\\j\neq i}}^{N}\bar{A}_{ij}(\mu_i)W_i\bar{A}_{ij}^T(\mu_i) &= \sum_{i=1}^{N}\sum_{\substack{j=1\\j\neq i}}^{N}\sum_{l=1}^{r_i}\sum_{f=1}^{r_i}\mu_{il}\mu_{if}\bar{A}_{ijl}W_i\bar{A}_{ijf}^T \\ &= \frac{1}{2}\sum_{i=1}^{N}\sum_{\substack{j=1\\j\neq i}}^{N}\sum_{l=1}^{r_i}\sum_{f=1}^{r_i}\mu_{il}\mu_{if}\left[\bar{A}_{ijl}W_i\bar{A}_{ijf}^T+\bar{A}_{ijf}W_i\bar{A}_{ijl}^T\right] \\ &\le \frac{1}{2}\sum_{i=1}^{N}\sum_{\substack{j=1\\j\neq i}}^{N}\sum_{l=1}^{r_i}\sum_{f=1}^{r_i}\mu_{il}\mu_{if}\left[\bar{A}_{ijl}W_i\bar{A}_{ijl}^T+\bar{A}_{ijf}W_i\bar{A}_{ijf}^T\right] \\ &= \frac{1}{2}\sum_{i=1}^{N}\sum_{\substack{j=1\\j\neq i}}^{N}\sum_{l=1}^{r_i}\mu_{il}\bar{A}_{ijl}W_i\bar{A}_{ijl}^T+\frac{1}{2}\sum_{i=1}^{N}\sum_{\substack{j=1\\j\neq i}}^{N}\sum_{f=1}^{r_i}\mu_{is}\bar{A}_{ijf}W_i\bar{A}_{ijf}^T \\ &= \sum_{i=1}^{N}\sum_{\substack{j=1\\j\neq i}}^{N}\sum_{l=1}^{r_i}\mu_{il}\bar{A}_{ijl}W_i\bar{A}_{ijl}^T. \end{aligned} \tag{10.73}$$

This completes the proof.

10.2.2 CO-DESIGN OF DECENTRALIZED EVENT-TRIGGERED CONTROL

This subsection will firstly reformulate the closed-loop fuzzy control system (10.62) into a continuous-time system with time-varying delay and extra disturbance by using the input-delay and perturbed system approaches. Then, based on an LKF along with Wirtinger's inequality, we will present stability analysis and controller synthesis for the large-scale networked fuzzy system in (10.62), respectively. It will be shown that the co-design result consisting of the controller gain, sampled period, network delay, and event-triggered parameter is derived in terms of a set of LMIs.

Based on the input delay approach [10], the sampled-data controller in (10.64) is reformulated as a delayed controller as follows:

$$u_i(t) = K_i(\hat{\mu}_i)\hat{x}_i(t-\eta_i(t)), t \in [t_k^i, t_{k+1}^i) \tag{10.74}$$

where $\eta_i(t) = t - t_k^i + \tau_i$. It follows from the Assumptions 10.4-10.6 that

$$\tau_i \le \eta_i(t) < \bar{\eta}_i, \bar{\eta}_i = \tau_i + h_i, t \in [t_k^i, t_{k+1}^i), k \in \mathbb{N}. \tag{10.75}$$

Combined with the large-scale fuzzy system in (10.62) and the delayed controller in (10.74), the closed-loop fuzzy event-triggered control system is given by

$$\dot{x}_i(t) = A_i(\mu_i)x_i(t) + B_i(\mu_i)K_i(\hat{\mu}_i)\hat{x}_i(t-\eta_i(t)) + \sum_{\substack{j=1\\j\neq i}}^{N}\bar{A}_{ij}(\mu_i)x_j(t), i \in \mathscr{N}. \tag{10.76}$$

Here, we model the event-triggered counterpart as a disturbance [2],

$$\begin{aligned} e_i(t-\eta_i(t)) &= \hat{x}_i(t-\eta_i(t)) - x_i(t-\eta_i(t)), \\ x_i(v) &= x_i(t-\eta_i(t)) - x_i(t-\tau_i), t \in [t_k^i, t_{k+1}^i). \end{aligned} \tag{10.77}$$

Then, by substituting (10.77) into (10.76), the closed-loop fuzzy control system in (10.76) can be rewritten as

$$\begin{aligned} \dot{x}_i(t) =& A_i(\mu_i)x_i(t) + B_i(\mu_i)K_i(\hat{\mu}_i)\left(x_i(t-\tau_i) + x_i(v) + e_i(t-\eta_i(t))\right) \\ &+ \sum_{\substack{j=1 \\ j\neq i}}^{N} \bar{A}_{ij}(\mu_i)x_j(t), i \in \mathcal{N}. \end{aligned} \tag{10.78}$$

Now, we introduce the following LKF utilizing Wirtinger's inequality:

$$V(t) = \sum_{i=1}^{N} \left[V_{i1}(t) + V_{i2}(t)\right], t \in [t_k^i, t_{k+1}^i) \tag{10.79}$$

with

$$\begin{cases} V_{i1}(t) = x_i^T(t) P_i x_i(t) + \int_{t-\tau_i}^{t} x_i^T(\alpha) Q_i x_i(\alpha) d\alpha \\ \qquad + \tau_i \int_{-\tau_i}^{0} \int_{t+\beta}^{t} \dot{x}_i^T(\alpha) Z_i \dot{x}_i(\alpha) d\alpha d\beta, \\ V_{i2}(t) = (\bar{\eta}_i - \tau_i)^2 \int_{t_k^i-\tau_i}^{t} \dot{x}_i^T(\alpha) W_i \dot{x}_i(\alpha) d\alpha \\ \qquad - \frac{\pi^2}{4} \int_{t_k^i-\tau_i}^{t-\tau_i} \left[x_i(\alpha) - x_i(t_k^i - \tau_i)\right]^T W_i \left[x_i(\alpha) - x_i(t_k^i - \tau_i)\right] d\alpha, \end{cases} \tag{10.80}$$

where $\{P_i, Q_i, Z_i, W_i\} \in \Re^{n_{xi} \times n_{xi}}, i \in \mathcal{N}$ are symmetric matrices, and $P_i > 0, Z_i > 0, W_i > 0$.

Inspired by [13], we do not require that the matrix Q_i in (10.79) is necessarily positive definite. To ensure the positive property of $V(t)$, we give the following lemma:

Lemma 10.3: Relaxing Lyapunov–Krasovskii Function

Consider the Lyapunov–Krasovskii function (LKF) in (10.79), then $V(t) \geq \varepsilon \|x(t)\|^2$, where $\varepsilon > 0$, $x(t) = \begin{bmatrix} x_1^T(t) & x_2^T(t) & \cdots & x_N^T(t) \end{bmatrix}^T$, if there exist the symmetric positive definite matrices $\{P_i, Z_i, W_i\} \in \Re^{n_{xi} \times n_{xi}}$, and symmetric matrix $Q_i \in \Re^{n_{xi} \times n_{xi}}$, such that for all $i \in \mathcal{N}$ the following inequalities hold:

$$\begin{bmatrix} \frac{1}{\tau_i} P_i + Z_i & -Z_i \\ \star & Q_i + Z_i \end{bmatrix} > 0. \tag{10.81}$$

■

Proof. Firstly, by using Jensen's inequality, we have

$$\begin{aligned}
&\tau_i \int_{-\tau_i}^{0} \int_{t+\beta}^{t} x_i^T(\alpha) Z_i x_i(\alpha)\, d\alpha d\beta \\
&\geq \tau_i \int_{-\tau_i}^{0} \frac{-1}{\beta} \left[\int_{t+\beta}^{t} \dot{x}_i^T(\alpha)\, d\alpha\right] Z_i \left[\int_{t+\beta}^{t} \dot{x}_i(\alpha)\, d\alpha\right] d\beta \\
&= \tau_i \int_{-\tau_i}^{0} \frac{-1}{\beta} \left[x_i(t) - x_i(t+\beta)\right]^T Z_i \left[x_i(t) - x_i(t+\beta)\right] d\beta \\
&= \tau_i \int_{0}^{\tau_i} \frac{1}{\beta} \left[x_i(t) - x_i(t-\beta)\right]^T Z_i \left[x_i(t) - x_i(t-\beta)\right] d\beta \\
&\geq \int_{0}^{\tau_i} \left[x_i(t) - x_i(t-\beta)\right]^T Z_i \left[x_i(t) - x_i(t-\beta)\right] d\beta \\
&= \int_{t-\tau_i}^{t} \left[x_i(t) - x_i(\alpha)\right]^T Z_i \left[x_i(t) - x_i(\alpha)\right] d\alpha. \qquad (10.82)
\end{aligned}$$

It follows from (10.80)-(10.82) that

$$\begin{aligned}
V_{i1}(t) =& x_i^T(t) P_i x_i(t) + \int_{t-\tau_i}^{t} x_i^T(\alpha) Q_i x_i(\alpha)\, d\alpha + \tau_i \int_{-\tau_i}^{0} \int_{t+\beta}^{t} \dot{x}_i^T(\alpha) Z_i \dot{x}_i(\alpha)\, d\alpha d\beta \\
\geq& \int_{t-\tau_i}^{t} \begin{bmatrix} x_i(t) \\ x_i(\alpha) \end{bmatrix}^T \begin{bmatrix} \frac{1}{\tau_i} P_i + Z_i & -Z_i \\ \star & Q_i + Z_i \end{bmatrix} \begin{bmatrix} x_i(t) \\ x_i(\alpha) \end{bmatrix} d\alpha. \qquad (10.83)
\end{aligned}$$

For $V_{i2}(t)$ given in (10.80), we have $x_i(\alpha) - x_i(t_k^i - \tau_i) = 0$ when $\alpha = t_k^i - \tau_i$. By using [11], it is easy to see that $V_{i2}(t) \geq 0$. Therefore, there always exists a positive scalar ε such that the inequality $V(t) \geq \varepsilon \|x(t)\|^2$ holds if the inequality in (10.81) holds. Thus, completing this proof.

Based on the LKF in (10.79), a sufficient condition for the stability of the closed-loop fuzzy control system in (10.69) is given by the following theorem.

Theorem 10.6: Performance Analysis of Decentralized Event-Triggering

Given the large-scale T-S fuzzy system in (10.62) and a decentralized event-triggered fuzzy controller in the form of (10.64), the closed-loop fuzzy control system in (10.78) is asymptotically stable, if there exist the symmetric positive definite matrices $\{P_i, Z_i, W_i, M_i, U_i\} \in \Re^{n_{xi} \times n_{xi}}$, and symmetric matrix $Q_i \in \Re^{n_{xi} \times n_{xi}}$, matrix multipliers $\mathscr{G}_i \in \Re^{5n_{xi} \times n_{xi}}$, and positive scalars $\{\bar{\eta}_i, \tau_i, \sigma_i\}$, such that for all $i \in \mathscr{N}$ the following matrix inequalities hold:

$$\begin{bmatrix} \frac{1}{\tau_i} P_i + Z_i & -Z_i \\ \star & Q_i + Z_i \end{bmatrix} > 0, \qquad (10.84)$$

$$\begin{bmatrix} \Theta_i + \mathrm{Sym}\{\mathscr{G}_i \mathbb{A}_i(\mu_i, \hat{\mu}_i)\} & \mathscr{G}_i A_{ij}(\mu_i) \\ \star & -M_i \end{bmatrix} < 0, \qquad (10.85)$$

where

$$\Theta_i = \begin{bmatrix} \Theta_i^{(1)} & P_i & 0 & 0 & 0 \\ \star & \Theta_i^{(2)} & Z_i & 0 & 0 \\ \star & \star & \Theta_i^{(3)} & \sigma_i^2 U_i & 0 \\ \star & \star & \star & \Theta_i^{(4)} & 0 \\ \star & \star & \star & \star & -U_i \end{bmatrix}, \Theta_i^{(1)} = \tau_i^2 Z_i + (\bar{\eta}_i - \tau_i)^2 W_i,$$

$$\Theta_i^{(2)} = Q_i - Z_i + \sum_{\substack{j=1 \\ j\neq i}}^{N} M_j, \Theta_i^{(3)} = -Q_i - Z_i + \sigma_i^2 U_i, \Theta_i^{(4)} = -\frac{\pi^2}{4} W_i + \sigma_i^2 U_i,$$

$$\mathbb{A}_i(\mu_i, \hat{\mu}_i) = \begin{bmatrix} -\mathbf{I} & A_i(\mu_i) & B_i(\mu_i)K_i(\hat{\mu}_i) & B_i(\mu_i)K_i(\hat{\mu}_i) & B_i(\mu_i)K_i(\hat{\mu}_i) \end{bmatrix},$$
$$A_{ij}(\mu_i) = \underbrace{\left[\bar{A}_{i1}(\mu_i) \cdots \bar{A}_{ij, j\neq i}(\mu_i) \cdots \bar{A}_{iN}(\mu_i)\right]}_{N-1}, M_i = \operatorname{diag}\underbrace{\{M_i \cdots M_i \cdots M_i\}}_{N-1}. \quad (10.86)$$

■

Proof. By taking the time derivative of $V(t)$, one has

$$\begin{aligned} \dot{V}_{i1}(t) \leq{} & 2x_i^T(t) P_i \dot{x}_i(t) + x_i^T(t) Q_i x_i(t) \\ & - x_i^T(t-\tau_i) Q_i x_i(t-\tau_i) + \tau_i^2 \dot{x}_i^T(t) Z_i \dot{x}_i(t) \\ & - \tau_i \int_{t-\tau_i}^{t} \dot{x}_i^T(\alpha) Z_i \dot{x}_i(\alpha) d\alpha, \end{aligned} \quad (10.87)$$

$$\dot{V}_{i2}(t) \leq (\bar{\eta}_i - \tau_i)^2 \dot{x}_i^T(t) W_i \dot{x}_i(t) - \frac{\pi^2}{4} x_i^T(v) W_i x_i(v). \quad (10.88)$$

Based on Jensen's inequality, we have

$$\begin{aligned} & -\tau_i \int_{t-\tau_i}^{t} \dot{x}_i^T(\alpha) Z_i \dot{x}_i(\alpha) d\alpha \\ & \leq -\left[\int_{t-\tau_i}^{t} \dot{x}_i(\alpha) d\alpha\right]^T Z_i \left[\int_{t-\tau_i}^{t} \dot{x}_i(\alpha) d\alpha\right] \\ & = -(x_i(t) - x_i(t-\tau_i))^T Z_i (x_i(t) - x_i(t-\tau_i)). \end{aligned} \quad (10.89)$$

Define the matrix multipliers $\mathscr{G}_i \in \Re^{5n_{xi} \times n_{xi}}, i \in \mathscr{N}$, and it follows from (10.78) that

$$0 = \sum_{i=1}^{N} 2\chi_i^T(t) \mathscr{G}_i \mathbb{A}_i(\mu_i, \hat{\mu}_i) \chi_i(t) + \sum_{i=1}^{N} 2\chi_i^T(t) \mathscr{G}_i \sum_{\substack{j=1 \\ j\neq i}}^{N} \bar{A}_{ij}(\mu_i) x_j(t), \quad (10.90)$$

where

$$\chi_i(t) = \begin{bmatrix} \dot{x}_i^T(t) & x_i^T(t) & x_i^T(t-\tau_i) & x_i^T(v) & e_i^T(t-\eta_i(t)) \end{bmatrix}^T,$$
$$\mathbb{A}_i(\mu_i, \hat{\mu}_i) = \begin{bmatrix} -\mathbf{I} & A_i(\mu_i) & B_i(\mu_i)K_i(\hat{\mu}_i) & B_i(\mu_i)K_i(\hat{\mu}_i) & B_i(\mu_i)K_i(\hat{\mu}_i) \end{bmatrix}. \quad (10.91)$$

Note that

$$2\bar{x}^T\bar{y} \leq \bar{x}^T M^{-1}\bar{x} + \bar{y}^T M\bar{y}, \tag{10.92}$$

where $\{\bar{x}, \bar{y}\} \in \Re^n$ and symmetric matrix $M > 0$.

Define $M_i = M_i^T > 0$, and by using the relation of (10.92), we have

$$\begin{aligned}
&\sum_{i=1}^{N} 2\chi_i^T(t)\mathscr{G}_i \sum_{\substack{j=1 \\ j\neq i}}^{N} \bar{A}_{ij}(\mu_i)x_j(t) \\
&\quad \leq \sum_{i=1}^{N}\sum_{\substack{j=1 \\ j\neq i}}^{N} \chi_i^T(t)\mathscr{G}_i\bar{A}_{ij}(\mu_i)M_i^{-1}\bar{A}_{ij}^T(\mu_i)\mathscr{G}_i^T\chi_i(t) + \sum_{i=1}^{N}\sum_{\substack{j=1 \\ j\neq i}}^{N} x_j^T(t)M_ix_j(t) \\
&\quad \leq \sum_{i=1}^{N}\sum_{\substack{j=1 \\ j\neq i}}^{N} \chi_i^T(t)\mathscr{G}_i\bar{A}_{ij}(\mu_i)M_i^{-1}\bar{A}_{ij}^T(\mu_i)\mathscr{G}_i^T\chi_i(t) + \sum_{i=1}^{N}\sum_{\substack{j=1 \\ j\neq i}}^{N} x_i^T(t)M_jx_i(t).
\end{aligned} \tag{10.93}$$

In addition, it follows from (10.67), (10.68) and (10.77) that

$$\begin{aligned}
\|e_i(t-\eta_i(t))\| &= \|\hat{x}_i(t-\eta_i(t)) - x_i(t-\eta_i(t))\| \\
&\leq \sigma_i \|x_i(t-\eta_i(t))\| \\
&= \sigma_i \|x_i(t-\tau_i) + x_i(v)\|.
\end{aligned} \tag{10.94}$$

Based on the relation in (10.94), we define the symmetric positive definite matrices U_i,

$$\begin{aligned}
&e_i^T(t-\eta_i(t))U_ie_i(t-\eta_i(t)) \\
&\quad \leq \sigma_i^2 \begin{bmatrix} x_i(t-\tau_i) \\ x_i(v) \end{bmatrix}^T \begin{bmatrix} U_i & U_i \\ \star & U_i \end{bmatrix} \begin{bmatrix} x_i(t-\tau_i) \\ x_i(v) \end{bmatrix}.
\end{aligned} \tag{10.95}$$

It follows from (10.87)-(10.95) that

$$
\begin{aligned}
\dot{V}(t) \leq & \sum_{i=1}^{N} \left[2x_i^T(t) P_i \dot{x}_i(t) + x_i^T(t) Q_i x_i(t)\right] \\
& - \sum_{i=1}^{N} x_i^T(t-\tau_i) Q_i x_i(t-\tau_i) + \sum_{i=1}^{N} \tau_i^2 \dot{x}_i^T(t) Z_i \dot{x}_i(t) \\
& - \sum_{i=1}^{N} (x_i(t) - x_i(t-\tau_i))^T Z_i (x_i(t) - x_i(t-\tau_i)) \\
& + \sum_{i=1}^{N} (\bar{\eta}_i - \tau_i)^2 \dot{x}_i^T(t) W_i \dot{x}_i(t) - \sum_{i=1}^{N} \frac{\pi^2}{4} x_i^T(v) W_i x_i(v) \\
& + \sum_{i=1}^{N} 2\chi_i^T(t) \mathscr{G}_i \mathbb{A}_i(\mu_i, \hat{\mu}_i) \chi_i(t) + \sum_{i=1}^{N} \sum_{\substack{j=1 \\ j \neq i}}^{N} \chi_i^T(t) \mathscr{G}_i \bar{A}_{ij}(\mu_i) M_i^{-1} \bar{A}_{ij}^T(\mu_i) \mathscr{G}_i^T \chi_i(t) \\
& + \sum_{i=1}^{N} \sum_{\substack{j=1 \\ j \neq i}}^{N} x_i^T(t) M_j x_i(t) - \sum_{i=1}^{N} \left\{e_i^T(t-\eta_i(t)) U_i e_i(t-\eta_i(t))\right\} \\
& + \sum_{i=1}^{N} \sigma_i^2 \begin{bmatrix} x_i(t-\tau_i) \\ x_i(v) \end{bmatrix}^T \begin{bmatrix} U_i & U_i \\ \star & U_i \end{bmatrix} \begin{bmatrix} x_i(t-\tau_i) \\ x_i(v) \end{bmatrix} \\
= & \sum_{i=1}^{N} \chi_i^T(t) \left[\Theta_i + \mathrm{Sym}\left\{\mathscr{G}_i \mathbb{A}_i(\mu_i, \hat{\mu}_i)\right\}\right] \chi_i(t) \\
& + \sum_{i=1}^{N} \sum_{\substack{j=1 \\ j \neq i}}^{N} \chi_i^T(t) \mathscr{G}_i \bar{A}_{ij}(\mu_i) M_i^{-1} \bar{A}_{ij}^T(\mu_i) \mathscr{G}_i^T \chi_i(t),
\end{aligned}
\tag{10.96}
$$

where Θ_i and $\mathbb{A}_i(\mu_i, \hat{\mu}_i)$ are defined in (10.86).

By applying Schur complement lemma to (10.85), it can be seen that the inequality in (10.85) implies $\dot{V}(t) < 0$, thus completing this proof.

The conditions given in (10.85) are nonlinear matrix inequalities when the controller gains are unknown. Furthermore, when the information between μ_{il} and $\hat{\mu}_{il}$ is unavailable, the condition $\mu_{il} \neq \hat{\mu}_{il}$ generally leads to a linear controller instead of a fuzzy one, which induces the design conservatism. From a practical perspective, it is possible to obtain a priori knowledge between μ_{il} and $\hat{\mu}_{il}$. Thus, we assume

$$
\underline{\rho}_{il} \leq \frac{\hat{\mu}_{il}}{\mu_{il}} \leq \bar{\rho}_{il},
\tag{10.97}
$$

where $\underline{\rho}_{il}$ and $\bar{\rho}_{il}$ are known positive scalars.

Based on Theorem 10.6 and (10.97), we will present the co-design result consisting of the fuzzy controller gains, sampled period, network delay, and event-triggered parameter in terms of a set of LMIs; the result is summarized as follows:

Theorem 10.7: Co-Design of Decentralized Event-Triggered Fuzzy Control Using Asynchronous Method

Given the large-scale T-S fuzzy system in (10.62) and a decentralized event-triggered fuzzy controller in the form of (10.64), the closed-loop fuzzy control system (10.78) with the asynchronous condition (10.97) is asymptotically stable, if there exist the symmetric positive definite matrices $\{\bar{P}_i, \bar{W}_i, \bar{Z}_i, \bar{U}_i, V_i, V_0\} \in \Re^{n_{xi} \times n_{xi}}$, $V_0 \leq V_i$, and symmetric matrices $\bar{Q}_i \in \Re^{n_{xi} \times n_{xi}}$ and matrices $X_{ils} = X_{isl}^T$, $G_i \in \Re^{n_{xi} \times n_{xi}}$, $\bar{K}_{is} \in \Re^{n_{ui} \times n_{xi}}$, and positive scalars $\left\{\bar{\eta}_i, \tau_i, \sigma_i, \bar{\rho}_{il}, \underline{\rho}_{il}\right\}$, such that for all $(l,s) \in \mathscr{L}_i, i \in \mathscr{N}$ the following LMIs hold:

$$\begin{bmatrix} \frac{1}{\tau_i}\bar{P}_i + \bar{Z}_i & -\bar{Z}_i \\ \star & \bar{Q}_i + \bar{Z}_i \end{bmatrix} > 0, \tag{10.98}$$

$$\bar{\rho}_{il}\Sigma_{ill} + X_{ill} < 0, \tag{10.99}$$

$$\underline{\rho}_{il}\Sigma_{ill} + X_{ill} < 0, \tag{10.100}$$

$$\bar{\rho}_{is}\Sigma_{ils} + \bar{\rho}_{il}\Sigma_{isl} + X_{ils} + X_{isl} < 0, \tag{10.101}$$

$$\underline{\rho}_{is}\Sigma_{ils} + \underline{\rho}_{il}\Sigma_{isl} + X_{ils} + X_{isl} < 0, \tag{10.102}$$

$$\underline{\rho}_{is}\Sigma_{ils} + \bar{\rho}_{il}\Sigma_{isl} + X_{ils} + X_{isl} < 0, \tag{10.103}$$

$$\bar{\rho}_{is}\Sigma_{ils} + \underline{\rho}_{il}\Sigma_{isl} + X_{ils} + X_{isl} < 0, \tag{10.104}$$

$$\begin{bmatrix} X_{i11} & \cdots & X_{i1r_i} \\ \vdots & \ddots & \vdots \\ X_{ir_i1} & \cdots & X_{ir_ir_i} \end{bmatrix} > 0, \tag{10.105}$$

where

$$\Sigma_{ils} = \begin{bmatrix} \Sigma_{ils}^{(1)} & \mathscr{E}_{(1)}G_i^T \\ \star & -(N-1)^{-1}V_0 \end{bmatrix}, \Sigma_{ils}^{(1)} = \bar{\Theta}_i + \mathrm{Sym}\left\{\mathscr{E}_{(2)}\bar{\mathbb{A}}_{ils}\right\} + \sum_{\substack{j=1 \\ j\neq i}}^{N} \mathscr{E}_{(2)}\bar{A}_{ijl}V_i\bar{A}_{ijl}^T\mathscr{E}_{(2)}^T,$$

$$\bar{\Theta}_i = \begin{bmatrix} \bar{\Theta}_i^{(1)} & \bar{P}_i & 0 & 0 & 0 \\ \star & \bar{Q}_i - \bar{Z}_i & \bar{Z}_i & 0 & 0 \\ \star & \star & \bar{\Theta}_i^{(3)} & \sigma_i^2\bar{U}_i & 0 \\ \star & \star & \star & -\frac{\pi^2}{4}\bar{W}_i + \sigma_i^2\bar{U}_i & 0 \\ \star & \star & \star & \star & -\bar{U}_i \end{bmatrix},$$

$$\bar{\Theta}_i^{(1)} = \tau_i^2\bar{Z}_i + (\bar{\eta}_i - \tau_i)^2\bar{W}_i, \bar{\Theta}_i^{(3)} = -\bar{Q}_i - \bar{Z}_i + \sigma_i^2\bar{U}_i,$$

$$\bar{\mathbb{A}}_{ils} = \begin{bmatrix} -G_i & A_{il}G_i & B_{il}\bar{K}_{is} & B_{il}\bar{K}_{is} & B_{il}\bar{K}_{is} \end{bmatrix},$$

$$\mathscr{E}_{(1)} = \begin{bmatrix} 0 & \mathbf{I} & 0 & 0 & 0 \end{bmatrix}^T, \mathscr{E}_{(2)} = \begin{bmatrix} \mathbf{I} & \mathbf{I} & 0 & 0 & 0 \end{bmatrix}^T. \tag{10.106}$$

Furthermore, a decentralized event-triggered fuzzy controller in the form of (10.63) is given by

$$K_{is} = \bar{K}_{is} G_i^{-1}, s \in \mathscr{L}_i, i \in \mathscr{N}. \tag{10.107}$$

■

Proof. For matrix inequality linearization purpose, define $M_i = V_i^{-1}, V_0 \leq V_i, i \in \mathscr{N}$,

$$\sum_{\substack{j=1 \\ j\neq i}}^{N} M_j \leq (N-1) V_0^{-1}. \tag{10.108}$$

Now, by substituting (10.108) into (10.85), and applying Schur complement lemma, the following inequality implies (10.85),

$$\begin{bmatrix} \Sigma_{ils}^{(1)} & \mathscr{E}_{(1)} \\ \star & -(N-1)^{-1} V_0 \end{bmatrix} < 0, \tag{10.109}$$

where

$$\Sigma_{ils}^{(1)} = \bar{\Theta}_i + \mathrm{Sym}\{\mathscr{G}_i \mathbb{A}_i(\mu_i, \hat{\mu}_i)\} + \sum_{\substack{j=1 \\ j\neq i}}^{N} \mathscr{G}_i \bar{A}_{ij}(\mu_i) V_i \bar{A}_{ij}^T(\mu_i) \mathscr{G}_i^T,$$

$$\bar{\Theta}_i = \begin{bmatrix} \Theta_i^{(1)} & P_i & 0 & 0 & 0 \\ \star & Q_i - Z_i & Z_i & 0 & 0 \\ \star & \star & \Theta_i^{(3)} & \sigma_i^2 U_i & 0 \\ \star & \star & \star & \Theta_i^{(4)} & 0 \\ \star & \star & \star & \star & -U_i \end{bmatrix},$$

$$\Theta_i^{(1)} = \tau_i^2 Z_i + (\bar{\eta}_i - \tau_i)^2 W_i, \Theta_i^{(3)} = -Q_i - Z_i + \sigma_i^2 U_i,$$

$$\Theta_i^{(4)} = -\frac{\pi^2}{4} W_i + \sigma_i^2 U_i, \mathscr{E}_{(1)} = \begin{bmatrix} 0 & \mathbf{I} & 0 & 0 & 0 \end{bmatrix}^T,$$

$$\mathbb{A}_i(\mu_i, \hat{\mu}_i) = \begin{bmatrix} -\mathbf{I} & A_i(\mu_i) & B_i(\mu_i) K_i(\hat{\mu}_i) & B_i(\mu_i) K_i(\hat{\mu}_i) & B_i(\mu_i) K_i(\hat{\mu}_i) \end{bmatrix}. \tag{10.110}$$

It follows from (10.99) that

$$\tau_i^2 \bar{Z}_i + (\bar{\eta}_i - \tau_i)^2 \bar{W}_i - \mathrm{Sym}\{G_i\} < 0, i \in \mathscr{N} \tag{10.111}$$

which implies that $G_i, i \in \mathscr{N}$ are nonsingular matrices.

We further define

$$\begin{cases} \mathscr{G}_i = \begin{bmatrix} G_i^{-1} & G_i^{-1} & 0 & 0 & 0 \end{bmatrix}^T, \\ \Gamma_1 := \mathrm{diag}\{ G_i \quad G_i \quad G_i \quad G_i \quad G_i \quad \mathbf{I} \}, \\ \bar{P}_i = G_i^T P_i G_i, \bar{Q}_i = G_i^T Q_i G_i, \\ \bar{Z}_i = G_i^T Z_i G_i, \bar{U}_i = G_i^T U_i G_i, \bar{W}_i = G_i^T W_i G_i. \end{cases} \tag{10.112}$$

By substituting (10.112) into (10.109), and performing a congruence transformation by Γ_1, and extracting the fuzzy membership functions, we have

$$\sum_{l=1}^{r_i}\sum_{f=1}^{r_i}\sum_{s=1}^{r_i}\mu_{il}\mu_{if}\hat{\mu}_{is}\Sigma_{ilfs}<0, \tag{10.113}$$

where

$$\Sigma_{ilfs}=\begin{bmatrix}\Sigma_{ilfs}^{(1)} & \mathscr{E}_{(1)}G_i^T\\ \star & -(N-1)^{-1}V_0\end{bmatrix}, \tag{10.114}$$

and $\Sigma_{ilfs}^{(1)}, \bar{\Theta}_i, \mathscr{E}_{(2)}$ and $\bar{\mathrm{A}}_{ils}$ are defined in (10.106).

Then, by using the relaxing inequality in Lemma10.2, the following inequality implies (10.113),

$$\sum_{l=1}^{r_i}\sum_{s=1}^{r_i}\mu_{il}\hat{\mu}_{is}\Sigma_{ils}<0, \tag{10.115}$$

where Σ_{ils} is defined in (10.106).

By using the asynchronous method of [14], the inequality in (10.115) holds if the inequalities (10.99)-(10.105) hold. Then, by performing congruence transformations to (10.84) by Γ_2, where $\Gamma_2:=\mathrm{diag}\{G_i,G_i\}$, the inequalities in (10.98) can be obtained, thus completing this proof.

It should be noted that the existing relaxation technique $\sum_{l=1}^{r_i}[\mu_{il}]^2\Sigma_{ill}+\sum_{l=1}^{r_i}\sum_{l<s\le r_i}\mu_{il}\mu_{is}\Sigma_{ils}<0$ is no longer applicable to fuzzy controller synthesis, since $\mu_{il}\neq\hat{\mu}_{il}$. Similarly to the asynchronous relaxation technique in [14], assume that $|\mu_{il}-\hat{\mu}_{il}|\le\delta_{il}, l\in\mathscr{L}_i, i\in\mathscr{N}$, where δ_{il} is a positive scalar. If $\Sigma_{ils}+X_{il}\ge 0$, where X_{il} is a symmetric matrix,

$$\begin{aligned}\sum_{l=1}^{r_i}\sum_{s=1}^{r_i}\mu_i^l\hat{\mu}_i^s\Sigma_{ils}&=\sum_{l=1}^{r_i}\sum_{s=1}^{r_i}\mu_{il}\mu_{is}\Sigma_{ils}+\sum_{l=1}^{r_i}\sum_{s=1}^{r_i}\mu_{il}(\hat{\mu}_{is}-\mu_{is})(\Sigma_{ils}+X_{il})\\&\le\sum_{l=1}^{r_i}\sum_{s=1}^{r_i}\mu_{il}\mu_{is}\left[\Sigma_{ils}+\sum_{s=1}^{r_i}\delta_{is}(\Sigma_{ils}+X_{il})\right].\end{aligned} \tag{10.116}$$

Therefore, upon defining $\bar{\Sigma}_{ils}=\Sigma_{ils}+\sum_{s=1}^{r_i}\delta_{is}(\Sigma_{ils}+X_{il})$, the existing relaxation technique from [17] can be applied to (10.115). Thus, the synchronized condition in (10.116) can be derived from the asynchronous condition in (10.115). Based on this property, we will present the co-design result with the reduced number of LMIs, which is summarized as follows:

Theorem 10.8: Co-Design of Decentralized Event-Triggered Fuzzy Controller Using Synchronized Method

Given the large-scale T-S fuzzy system in (10.62) and a decentralized event-triggered fuzzy controller in the form of (10.64), the closed-loop fuzzy control system in

(10.78) is asymptotically stable, if there exist the symmetric positive definite matrices $\{\bar{P}_i, \bar{W}_i, \bar{Z}_i, \bar{U}_i, V_i, V_0\} \in \Re^{n_{xi} \times n_{xi}}$, $V_0 \leq V_i$, and symmetric matrices $\bar{Q}_i \in \Re^{n_{xi} \times n_{xi}}$ and matrices $G_i \in \Re^{n_{xi} \times n_{xi}}$, $\bar{K}_{is} \in \Re^{n_{ui} \times n_{xi}}$, and positive scalars $\{\bar{\eta}_i, \tau_i, \sigma_i\}$, such that for all $(l,s) \in \mathscr{L}_i, i \in \mathscr{N}$ the following LMIs hold:

$$\begin{bmatrix} \frac{1}{\tau_i}\bar{P}_i + \bar{Z}_i & -\bar{Z}_i \\ \star & \bar{Q}_i + \bar{Z}_i \end{bmatrix} > 0, \tag{10.117}$$

$$\Sigma_{ill} < 0, l \in \mathscr{L} \tag{10.118}$$

$$\Sigma_{ils} + \Sigma_{isl} < 0, 1 \leq l < s \leq r \tag{10.119}$$

where

$$\Sigma_{ils} = \begin{bmatrix} \Sigma_{ils}^{(1)} & \mathscr{E}_{(1)} G_i^T \\ \star & -(N-1)^{-1} V_0 \end{bmatrix}, \Sigma_{ils}^{(1)} = \bar{\Theta}_i + \mathrm{Sym}\{\mathscr{E}_{(2)} \bar{\mathbb{A}}_{ils}\} + \sum_{\substack{j=1 \\ j \neq i}}^{N} \mathscr{E}_{(2)} \bar{A}_{ijl} V_i \bar{A}_{ijl}^T \mathscr{E}_{(2)}^T,$$

$$\bar{\Theta}_i = \begin{bmatrix} \bar{\Theta}_i^{(1)} & \bar{P}_i & 0 & 0 & 0 \\ \star & \bar{Q}_i - \bar{Z}_i & \bar{Z}_i & 0 & 0 \\ \star & \star & \bar{\Theta}_i^{(3)} & \sigma_i^2 \bar{U}_i & 0 \\ \star & \star & \star & -\frac{\pi^2}{4}\bar{W}_i + \sigma_i^2 \bar{U}_i & 0 \\ \star & \star & \star & \star & -\bar{U}_i \end{bmatrix},$$

$$\bar{\Theta}_i^{(1)} = \tau_i^2 \bar{Z}_i + (\bar{\eta}_i - \tau_i)^2 \bar{W}_i, \bar{\Theta}_i^{(3)} = -\bar{Q}_i - \bar{Z}_i + \sigma_i^2 \bar{U}_i,$$
$$\bar{\mathbb{A}}_{ils} = \begin{bmatrix} -G_i & A_{il} G_i & B_{il} \bar{K}_{is} & B_{il} \bar{K}_{is} & B_{il} \bar{K}_{is} \end{bmatrix},$$
$$\mathscr{E}_{(1)} = \begin{bmatrix} 0 & \mathbf{I} & 0 & 0 & 0 \end{bmatrix}^T, \mathscr{E}_{(2)} = \begin{bmatrix} \mathbf{I} & \mathbf{I} & 0 & 0 & 0 \end{bmatrix}^T. \tag{10.120}$$

Furthermore, a decentralized event-triggered fuzzy controller in the form of (10.64) is given by

$$K_{is} = \bar{K}_{is} G_i^{-1}, s \in \mathscr{L}_i, i \in \mathscr{N}. \tag{10.121}$$

■

For the case where the information between μ_{il} and $\hat{\mu}_{il}$ is unavailable, the corresponding result on decentralized event-triggered linear controller design can be obtained as follows:

Lemma 10.4: Co-Design of Decentralized Event-Triggered Linear Controller

Given the large-scale T-S fuzzy system in (10.62) and a decentralized event-triggered linear controller $u_i(t) = K_i \hat{x}_i(t - \eta_i(t))$, the resulting closed-loop fuzzy control system is asymptotically stable, if there exist the symmetric positive definite matrices $\{\bar{P}_i, \bar{W}_i, \bar{Z}_i, \bar{U}_i, V_i, V_0\} \in \Re^{n_{xi} \times n_{xi}}$, $V_0 \leq V_i$, and symmetric matrices $\bar{Q}_i \in \Re^{n_{xi} \times n_{xi}}$, and matrices $G_i \in \Re^{n_{xi} \times n_{xi}}$, $\bar{K}_i \in \Re^{n_{ui} \times n_{xi}}$, and positive scalars $\{\bar{\eta}_i, \tau_i, \sigma_i\}$, such that for all $l \in \mathscr{L}_i, i \in \mathscr{N}$ the following LMIs hold:

$$\begin{bmatrix} \frac{1}{\tau_i}\bar{P}_i + \bar{Z}_i & -\bar{Z}_i \\ \star & \bar{Q}_i + \bar{Z}_i \end{bmatrix} > 0, \tag{10.122}$$

$$\begin{bmatrix} \Sigma_{ilfs}^{(1)} & \mathscr{E}_{(1)} G_i^T \\ \star & -(N-1)^{-1} V_0 \end{bmatrix} < 0, \tag{10.123}$$

where

$$\Sigma_{ilfs}^{(1)} = \bar{\Theta}_i + \mathrm{Sym}\left\{\mathscr{E}_{(2)} \bar{\mathbb{A}}_{il}\right\} + \sum_{\substack{j=1 \\ j\neq i}}^{N} \mathscr{E}_{(2)} \bar{A}_{ijl} V_i \bar{A}_{ijl}^T \mathscr{E}_{(2)}^T,$$

$$\bar{\Theta}_i = \begin{bmatrix} \bar{\Theta}_i^{(1)} & \bar{P}_i & 0 & 0 & 0 \\ \star & \bar{Q}_i - \bar{Z}_i & \bar{Z}_i & 0 & 0 \\ \star & \star & \bar{\Theta}_i^{(3)} & \sigma_i^2 \bar{U}_i & 0 \\ \star & \star & \star & \bar{\Theta}_i^{(4)} & 0 \\ \star & \star & \star & \star & -\bar{U}_i \end{bmatrix}, \bar{\Theta}_i^{(1)} = \tau_i^2 \bar{Z}_i + (\bar{\eta}_i - \tau_i)^2 \bar{W}_i,$$

$$\begin{aligned}
&\bar{\Theta}_i^{(3)} = -\bar{Q}_i - \bar{Z}_i + \sigma_i^2 \bar{U}_i, \bar{\Theta}_i^{(4)} = -\frac{\pi^2}{4} \bar{W}_i + \sigma_i^2 \bar{U}_i, \\
&\bar{\mathbb{A}}_{il} = \begin{bmatrix} -G_i & A_{il} G_i & B_{il} \bar{K}_i & B_{il} \bar{K}_i & B_{il} \bar{K}_i \end{bmatrix}, \\
&\mathscr{E}_{(1)} = \begin{bmatrix} 0 & \mathbf{I} & 0 & 0 & 0 \end{bmatrix}^T, \mathscr{E}_{(2)} = \begin{bmatrix} \mathbf{I} & \mathbf{I} & 0 & 0 & 0 \end{bmatrix}^T.
\end{aligned} \tag{10.124}$$

■

Note: When designing an event-triggered linear controller $u_i(t) = K_i \hat{x}_i(t - \eta_i(t))$, the premise variables are no longer required to transmit through communication networks. Compared with the event-triggered fuzzy controller in (10.64), the linear one reduces the requirements for extra hardware and software and raises the design conservatism.

10.3 DISTRIBUTED EVENT-TRIGGERED FUZZY CONTROL

In the context of networked systems, the standard time-triggering implementation is unachievable due to limited communication bandwidth. Inspired by [2], an event-triggering scheme, which makes a decision on the transmission of system outputs, its objective is to save the communication resources and to guarantee the desired performance.

10.3.1 DESIGN OF DISTRIBUTED EVENT-TRIGGERED CONTROLLER

Before moving on, we require the assumptions as below:

Assumption 10.7. The sampler is clock-driven in each subsystem. Let h_i denote the upper bound of sampling intervals for the i-th subsystem,

$$t_{k+1}^i - t_k^i \leq h_i, k \in \mathbb{N},$$

where $h_i > 0$.

Assumption 10.8. For each subsystem, it is assumed that the sensor-controller channel is closed via network-based communications.

Assumption 10.9. The zero-order-hold (ZOH) is event-triggered, and it makes use of the latest sampling data and holds them until the next transmitting data are received.

In the NCSs, the traditionally time-triggering implementation is undesirable because of limited bandwidths. Here, for the purpose of reducing the data transmissions, inspired by the work in [2], the distributed event-triggering fuzzy controller is proposed as follows:

Controller Rule $\mathscr{R}_i^s$: **IF** $\hat{\zeta}_{i1}(t_k^i)$ is $\mathscr{F}_{i1}^s$, $\hat{\zeta}_{i2}(t_k^i)$ is $\mathscr{F}_{i2}^s$, $\cdots$, and $\hat{\zeta}_{ig}(t_k^i)$ is $\mathscr{F}_{ig}^s$, **THEN**

$$u_i(t) = K_{ii}^s \hat{x}_i(t_k^i) + \sum\nolimits_{j=1,j\neq i}^{N} K_{ij}^s \hat{x}_j(t_k^i), t \in [z_k^i, z_{k+1}^i), s \in \mathscr{I}_i, \tag{10.125}$$

where $i \in \mathscr{N}$ and $K_{ii}^s \in \Re^{n_{ui} \times n_{xi}}, K_{ij}^s \in \Re^{n_{ui} \times n_{xj}}$ are controller gains to be designed; $\hat{\zeta}_i(z_k^i)$ and $\hat{x}_i(z_k^i)$ represent the renewal signals in the i-th ZOH.

Likewise, the total event-based distributed fuzzy controller is given by

$$u_i(t) = K_{ii}(\hat{\mu}_i)\hat{x}_i(t_k^i) + \sum\nolimits_{j=1,j\neq i}^{N} K_{ij}(\hat{\mu}_i)\hat{x}_j(t_k^j), t \in [z_k^i, z_{k+1}^i), \tag{10.126}$$

where $K_{ii}(\hat{\mu}_i) := \sum_{s=1}^{r_i} \hat{\mu}_i^s \left[\hat{\zeta}_i(t_k^i)\right] K_{ii}^s, K_{ij}(\hat{\mu}_i) := \sum_{s=1}^{r_i} \hat{\mu}_i^s \left[\hat{\zeta}_i(t_k^i)\right] K_{ij}^s, \sum_{s=1}^{r_i} \hat{\mu}_i^s \left[\hat{\zeta}_i(t_k^i)\right] = 1$. In the following, we will denote $\hat{\mu}_{is} := \hat{\mu}_{is}\left[\hat{z}_i(t_k^i - \tau_i)\right]$ for brevity.

Note: If $K_{ij}^s \equiv 0$ for any $s \in \mathscr{I}_i, i \in \mathscr{N}$, the distributed controller (10.126) degrades to the decentralized one used in (10.64). The simulation part will show that the supplemental feedbacks including the interconnected information provided in (10.126) for the local controllers guarantee better stabilization results compared to those for decentralized ones.

Note: The premise variables of the distributed fuzzy controller (10.126) are subject to event-based control and network-induced time delays. Hence, the asynchronous premise variables between $\zeta_i(t)$ and $\hat{\zeta}_i(t)$ are more practical. As reported in [9], if the information between $\zeta_i(t)$ and $\hat{\zeta}_i(t)$ is unavailable, this condition leads to generating a linear controller, which reduces the stabilization ability of the designed tool.

Here, an event-triggered scheme operates as follows:

$$\text{EBM: both } x_i(t_k^i) \text{ and } \zeta_i(t_k^i) \text{ are sent} \Leftrightarrow \| x_i(t_k^i) - \hat{x}_i(t_{k-1}^i) \| > \sigma_i \| x_i(t_k^i) \|, \tag{10.127}$$

where $\sigma_i \geq 0$ is a properly chosen design parameter.

In accordance with the operating condition (10.127), the event-based strategy in the i-th subsystem is established:

$$\hat{x}_i\left(t_k^i\right) = \begin{cases} x_i\left(t_k^i\right), \text{if } \left\|x_i\left(t_k^i\right) - \hat{x}_i\left(t_{k-1}^i\right)\right\| > \sigma_i \left\|x_i\left(t_k^i\right)\right\|, \\ \hat{x}_i\left(t_{k-1}^i\right), \text{if } \left\|x_i\left(t_k^i\right) - \hat{x}_i\left(t_{k-1}^i\right)\right\| \leq \sigma_i \left\|x_i\left(t_k^i\right)\right\|, \end{cases} \tag{10.128}$$

$$\hat{\zeta}_i\left(t_k^i\right) = \begin{cases} \zeta_i\left(t_k^i\right), \text{if } x_i\left(t_k^i\right) \text{ is sent,} \\ \hat{\zeta}_i\left(t_{k-1}^i\right), \text{if } x_i\left(t_k^i\right) \text{ is not sent.} \end{cases} \tag{10.129}$$

For simplicity, we define

$$x_i(v) = x_i(t_k^i) - x_i(t), t \in [z_k^i, z_{k+1}^i),$$
$$e_i(t) = \hat{x}_i(t_k^i) - x_i(t_k^i), t \in [z_k^i, z_{k+1}^i). \tag{10.130}$$

Submitting (10.130) into the fuzzy controller (10.126) and combined with the fuzzy system ((10.62)), the closed-loop fuzzy control system is

$$\begin{aligned}\dot{x}_i(t) = {} & \bar{A}_{ii}(\mu_i, \hat{\mu}_i)x_i(t) + B_i(\mu_i)K_{ii}(\hat{\mu}_i)x_i(v) + B_i(\mu_i)K_{ii}(\hat{\mu}_i)e_i(t) \\ & + \sum\nolimits_{j=1, j\neq i}^{N} B_i(\mu_i)K_{ij}(\hat{\mu}_i)x_j(v) + \sum\nolimits_{j=1, j\neq i}^{N} B_i(\mu_i)K_{ij}(\hat{\mu}_i)e_j(t) \\ & + \sum_{j=1, i\neq j}^{N} \bar{A}_{ij}(\mu_i, \hat{\mu}_i)x_j(t),\end{aligned} \tag{10.131}$$

where $\bar{A}_{ii}(\mu_i, \hat{\mu}_i) = A_{ii}(\mu_i) + B_i(\mu_i)K_{ii}(\hat{\mu}_i), \bar{A}_{ij}(\mu_i, \hat{\mu}_i) = A_{ij}(\mu_i) + B_i(\mu_i)K_{ij}(\hat{\mu}_i)$.

Note: When the premise variables on controller (10.126) are not considered in sampled-data measurement, the premise variables between the fuzzy system and the fuzzy controller are synchronous. Here, we just consider the asynchronous sampled-data measurement. However, the obtained result can be easily extend to the synchronous case.

Note: The premise variable $z_i(t)$ undergoes time-driven sensors and event-driven ZOHs, and is implemented by the proposed fuzzy controller in (10.126). Hence the premise variable spaces in asynchronous form between $z_i(t)$ and $z_i(t_k^i)$ are more practical.

Note: As pointed out in [9], when the information between μ_{il} and $\hat{\mu}_{il}$ is unavailable, the condition $\mu_{il} \neq \hat{\mu}_{il}$ generally leads to a linear controller instead of a fuzzy one, which degrades the stabilization ability of the controller. When the data on μ_{il} and $\hat{\mu}_{il}$ is available, the design conservatism can be improved, and obtaining the corresponding fuzzy controller.

Note: t_k^i is relative to the clock on the i-th subsystem. In other words, the sampled-data clocks can be different among all subsystems.

Now, we introduce the following Lyapunov function:

$$\begin{aligned}V(t) &= \sum_{i=1}^{N} V_i(t) \\ &= \sum_{i=1}^{N} x_i^T(t) P_i x_i(t) + \sum_{i=1}^{N} h_i^2 \int_{t_k^i}^{t} \dot{x}_i^T(\alpha) Q_i \dot{x}_i(\alpha) d\alpha \\ &\quad - \sum_{i=1}^{N} \frac{\pi^2}{4} \int_{t_k^i}^{t} \left[x_i(\alpha) - x_i(t_k^i)\right]^T Q_i \left[x_i(\alpha) - x_i(t_k^i)\right] d\alpha,\end{aligned} \tag{10.132}$$

where $\{P_i, Q_i\} \in \Re^{n_{xi} \times n_{xi}}$ are positive definite symmetric matrices.

Using Wirtinger's inequality in [11] we can see that $V(t) > 0$. Based on the new model in (10.131) and the Lyapunov function in (10.132), a sufficient condition for devising a decentralized sampled-data controller can be given as below:

Lemma 10.5: Stability Analysis of Distributed Event-Triggered Controller

The closed-loop fuzzy system in (10.62) using a distributed event-triggered fuzzy controller (10.126), is asymptotically stable, if there exist the symmetric positive definite matrices $\{P_i, W_{i1}, W_{i2}, W_{i3}, Z_i\} \in \Re^{n_{xi} \times n_{xi}}$, $K_{ii}(\hat{\mu}_i) \in \Re^{n_{ui} \times n_{xi}}$, $K_{ij}(\hat{\mu}_i) \in \Re^{n_{ui} \times n_{xj}}$, and the positive scalars $\{h_i, \sigma_i\}$, such that the following matrix inequalities hold:

$$\begin{aligned} &\Theta_i + \mathrm{Sym}\left(\mathbb{G}_i \mathbb{A}_i(\mu_i, \hat{\mu}_i)\right) + \sum_{\substack{j=1 \\ j \neq i}}^{N} \mathbb{G}_i \bar{A}_{ij}(\mu_i, \hat{\mu}_i) W_{i1} \bar{A}_{ij}^T(\mu_i, \hat{\mu}_i) \mathbb{G}_i^T \\ &+ \sum_{\substack{j=1 \\ j \neq i}}^{N} \mathbb{G}_i B_i(\mu_i) K_{ij}(\hat{\mu}_i) W_{i2} K_{ij}^T(\hat{\mu}_i) B_i^T(\mu_i) \mathbb{G}_i^T \\ &+ \sum_{\substack{j=1 \\ j \neq i}}^{N} \mathbb{G}_i B_i(\mu_i) K_{ij}(\hat{\mu}_i) W_{i3} K_{ij}^T(\hat{\mu}_i) B_i^T(\mu_i) \mathbb{G}_i^T < 0, \end{aligned} \tag{10.133}$$

where

$$\Theta_i = \begin{bmatrix} h_i^2 Q_i & P_i & 0 & 0 \\ \star & \sum_{\substack{j=1 \\ j \neq i}}^{N} W_{j1}^{-1} + \sigma_i^2 Z_i & \sigma_i^2 Z_i & 0 \\ \star & \star & -\frac{\pi^2}{4} Q_i + \sigma_i^2 Z_i + \sum_{\substack{j=1 \\ j \neq i}}^{N} W_{j2}^{-1} & 0 \\ \star & \star & \star & -Z_i + \sum_{\substack{j=1 \\ j \neq i}}^{N} W_{j3}^{-1} \end{bmatrix},$$

$$\mathbb{A}_i(\mu_i, \hat{\mu}_i) = \begin{bmatrix} -I & \bar{A}_{ii}(\mu_i, \hat{\mu}_i) & B_i(\mu_i) K_{ii}(\hat{\mu}_i) \end{bmatrix}. \tag{10.134}$$

■

Proof. By taking the time derivative of $V(t)$ in (10.132),

$$\dot{V}(t) = \sum_{i=1}^{N} \{2x_i^T(t) P_i \dot{x}_i(t) + h_i^2 \dot{x}_i^T(t) Q_i \dot{x}_i(t) - \frac{\pi^2}{4} x_i^T(v) Q_i x_i(v)\}. \tag{10.135}$$

Define the matrix $\mathbb{G}_i \in \Re^{3n_{xi} \times n_{xi}}$ and $\chi_i(t) = \begin{bmatrix} \dot{x}_i^T(t) & x_i^T(t) & x_i^T(v) & e_i^T(t) \end{bmatrix}^T$, and it follows from (10.131) that

$$0=\sum_{i=1}^{N}2\chi_i^T(t)\mathbb{G}_i\mathbb{A}_i(\mu_i,\hat{\mu}_i)\chi_i(t)+\sum_{i=1}^{N}2\chi_i^T(t)\mathbb{G}_i\sum_{j=1,i\neq j}^{N}\bar{A}_{ij}(\mu_i,\hat{\mu}_i)x_j(t)$$
$$+\sum_{i=1}^{N}2\chi_i^T(t)\mathbb{G}_i\sum_{\substack{j=1\\ j\neq i}}^{N}B_i(\mu_i)K_{ij}(\hat{\mu}_i)x_j(v)+\sum_{i=1}^{N}2\chi_i^T(t)\mathbb{G}_i\sum_{\substack{j=1\\ j\neq i}}^{N}B_i(\mu_i)K_{ij}(\hat{\mu}_i)e_j(t), \tag{10.136}$$

where $\mathbb{A}_i(\mu_i,\hat{\mu}_i)=\begin{bmatrix} -I & \bar{A}_{ii}(\mu_i,\hat{\mu}_i) & B_i(\mu_i)K_{ii}(\hat{\mu}_i) & B_i(\mu_i)K_{ii}(\hat{\mu}_i)\end{bmatrix}$.

Note that

$$2\bar{x}^T\bar{y}\leq\bar{x}^TM^{-1}\bar{x}+\bar{y}^TM\bar{y}, \tag{10.137}$$

where $\{\bar{x},\bar{y}\}\in\Re^n$ and symmetric matrix $M>0$.

By introducing matrix $0<W_{i1}=W_{i1}^T\in\Re^{n_{xi}\times n_{xi}}$, $0<W_{i2}=W_{i2}^T\in\Re^{n_{xi}\times n_{xi}}$, and $0<W_{i3}=W_{i3}^T\in\Re^{n_{xi}\times n_{xi}}$, and using the relation of (10.137),

$$\sum_{i=1}^{N}2\chi_i^T(t)\mathbb{G}_i\sum_{\substack{j=1\\ j\neq i}}^{N}\bar{A}_{ij}(\mu_i,\hat{\mu}_i)x_j(t)$$
$$\leq\sum_{i=1}^{N}\sum_{\substack{j=1\\ j\neq i}}^{N}\chi_i^T(t)\mathbb{G}_i\bar{A}_{ij}(\mu_i,\hat{\mu}_i)W_{i1}\bar{A}_{ij}^T(\mu_i,\hat{\mu}_i)\mathbb{G}_i^T\chi_i(t)+\sum_{i=1}^{N}\sum_{\substack{j=1\\ j\neq i}}^{N}x_j^T(t)W_{i1}^{-1}x_j(t)$$
$$=\sum_{i=1}^{N}\sum_{\substack{j=1\\ j\neq i}}^{N}\chi_i^T(t)\mathbb{G}_i\bar{A}_{ij}(\mu_i,\hat{\mu}_i)W_{i1}\bar{A}_{ij}^T(\mu_i,\hat{\mu}_i)\mathbb{G}_i^T\chi_i(t)+\sum_{i=1}^{N}\sum_{\substack{j=1\\ j\neq i}}^{N}x_i^T(t)W_{j1}^{-1}x_i(t), \tag{10.138}$$

and

$$\sum_{i=1}^{N}2\chi_i^T(t)\mathbb{G}_i\sum_{\substack{j=1\\ j\neq i}}^{N}B_i(\mu_i)K_{ij}(\hat{\mu}_i)x_j(v)$$
$$\leq\sum_{i=1}^{N}\sum_{\substack{j=1\\ j\neq i}}^{N}\chi_i^T(t)\mathbb{G}_iB_i(\mu_i)K_{ij}(\hat{\mu}_i)W_{i2}K_{ij}^T(\hat{\mu}_i)B_i^T(\mu_i)\mathbb{G}_i^T\chi_i(t)+\sum_{i=1}^{N}\sum_{\substack{j=1\\ j\neq i}}^{N}x_j^T(v)W_{i2}^{-1}x_j(v)$$
$$=\sum_{i=1}^{N}\sum_{\substack{j=1\\ j\neq i}}^{N}\chi_i^T(t)\mathbb{G}_iB_i(\mu_i)K_{ij}(\hat{\mu}_i)W_{i2}K_{ij}^T(\hat{\mu}_i)B_i^T(\mu_i)\mathbb{G}_i^T\chi_i(t)+\sum_{i=1}^{N}\sum_{\substack{j=1\\ j\neq i}}^{N}x_i^T(v)W_{j2}^{-1}x_i(v), \tag{10.139}$$

and

$$\begin{aligned}
&\sum_{i=1}^{N} 2\chi_i^T(t)\mathbb{G}_i \sum_{\substack{j=1\\j\neq i}}^{N} B_i(\mu_i)K_{ij}(\hat{\mu}_i)e_j(t)\\
&\le \sum_{i=1}^{N}\sum_{\substack{j=1\\j\neq i}}^{N} \chi_i^T(t)\mathbb{G}_iB_i(\mu_i)K_{ij}(\hat{\mu}_i)W_{i3}K_{ij}^T(\hat{\mu}_i)B_i^T(\mu_i)\mathbb{G}_i^T\chi_i(t)+\sum_{i=1}^{N}\sum_{\substack{j=1\\j\neq i}}^{N} e_j^T(t)W_{i3}^{-1}e_j(t)\\
&= \sum_{i=1}^{N}\sum_{\substack{j=1\\j\neq i}}^{N} \chi_i^T(t)\mathbb{G}_iB_i(\mu_i)K_{ij}(\hat{\mu}_i)W_{i3}K_{ij}^T(\hat{\mu}_i)B_i^T(\mu_i)\mathbb{G}_i^T\chi_i(t)+\sum_{i=1}^{N}\sum_{\substack{j=1\\j\neq i}}^{N} e_i^T(t)W_{j3}^{-1}e_i(t).
\end{aligned} \tag{10.140}$$

On the other hand, we can define $0 < Z_i = Z_i^T \in \Re^{n_{xi}\times n_{xi}}$. It follows from the event-triggered strategy in (10.128) and (10.129) that

$$\begin{aligned}
e_i^T(t)Z_ie_i(t) &\le \sigma_i^2 x_i^T(t_k^i)Z_ix_i(t_k^i)\\
&= \sigma_i^2[x_i(t)+x_i(v)]^T Z_i(\star).
\end{aligned} \tag{10.141}$$

It follows from (10.135)-(10.141) that the result on (10.133) can be obtained directly and is not LMI-based. When the asynchronized information of μ_{il} and $\hat{\mu}_{il}$ is unknown, the designed result generally leads to the linear controller instead of the fuzzy one [9]. From a practical perspective, obtaining a priori knowledge of μ_{il} and $\hat{\mu}_{il}$ is possible. Thus, we assume that the asynchronized condition is subject to

$$\underline{\rho}_{il} \le \frac{\hat{\mu}_{il}}{\mu_{il}} \le \bar{\rho}_{il}, \tag{10.142}$$

where $\underline{\rho}_{il}$ and $\bar{\rho}_{il}$ are positive scalars.

It follows from (10.133) and (10.142) that the design result on the distributed event-triggered fuzzy controller can be summarized as below:

Theorem 10.9: Design of Decentralized Event-Triggered Fuzzy Control Using Asynchronized Method

The closed-loop fuzzy system in (10.62) using a distributed event-triggered fuzzy controller (10.126), is asymptotically stable, if there exist the symmetric positive definite matrices $\{\bar{P}_i, W_{i1}, W_{i2}, W_{01}, W_{02}, Z_i\} \in \Re^{n_{xi}\times n_{xi}}$, $W_{01} \le W_{i1}$, $W_{02} \le W_{i2}$, and matrices $G_i \in \Re^{n_{xi}\times n_{xi}}$, $M_{ils} = M_{isl}^T \in \Re^{4n_{xi}\times 4n_{xi}}$, $\bar{K}_{iis} \in \Re^{n_{ui}\times n_{xi}}$, $\bar{K}_{ijs} \in \Re^{n_{ui}\times n_{xj}}$, and the positive scalars $\{h_i, \underline{\rho}_{il}, \bar{\rho}_{il}\}$, such that for all $(l,s) \in \mathcal{L}_i$, the following LMIs hold:

$$\bar{\rho}_{il}\Sigma_{ill} + M_{ill} < 0, \tag{10.143}$$

$$\underline{\rho}_{il}\Sigma_{ill} + M_{ill} < 0, \tag{10.144}$$

$$\bar{\rho}_{is}\Sigma_{ils} + \bar{\rho}_{il}\Sigma_{isl} + M_{ils} + M_{isl} < 0, \tag{10.145}$$

$$\underline{\rho}_{is}\Sigma_{ils} + \underline{\rho}_{il}\Sigma_{isl} + M_{ils} + M_{isl} < 0, \tag{10.146}$$

$$\underline{\rho}_{is}\Sigma_{ils} + \bar{\rho}_{il}\Sigma_{isl} + M_{ils} + M_{isl} < 0, \tag{10.147}$$

$$\bar{\rho}_{is}\Sigma_{ils} + \underline{\rho}_{il}\Sigma_{isl} + M_{ils} + M_{isl} < 0, \tag{10.148}$$

$$\begin{bmatrix} M_{i11} & \cdots & M_{i1r} \\ \vdots & \ddots & \vdots \\ M_{ir1} & \cdots & M_{irr} \end{bmatrix} > 0, \tag{10.149}$$

where

$$\Sigma_{ils} = \begin{bmatrix} \tilde{\Theta}_{i1} + \mathrm{Sym}\left(\bar{\mathbb{I}}_i \bar{\mathbb{A}}_{ils}\right) & \bar{G}_{i1} & \bar{G}_{i2} & \tilde{\Theta}_i \\ \star & -\frac{1}{(N-1)}W_{01} + \sigma_i^2 \bar{Z}_i & \sigma_i^2 \bar{Z}_i & 0 \\ \star & \star & -\frac{1}{(N-1)}W_{02} + \sigma_i^2 \bar{Z}_i & 0 \\ \star & \star & \star & \tilde{\mathbb{W}} \end{bmatrix},$$

$$\tilde{\Theta}_i = \begin{bmatrix} \bar{G}_{i3} & \bar{\Theta}_{i2ls} & \bar{\Theta}_{i3ls} \end{bmatrix}, \tilde{\mathbb{W}} = \mathrm{diag}\left\{-\frac{W_{02}}{(N-1)}, \mathbb{W}_1 - \mathbb{G}_i - \mathbb{G}_i, \mathbb{W}_2 - \mathbb{G}_i - \mathbb{G}_i\right\},$$

$$\tilde{\Theta}_{i1} = \begin{bmatrix} h_i^2 \bar{Q}_i & \bar{P}_i & 0 & 0 \\ \star & \sigma_i^2 \bar{Z}_i & \sigma_i^2 \bar{Z}_i & 0 \\ \star & \star & -\frac{\pi^2}{4}\bar{Q}_i + \sigma_i^2 \bar{Z}_i & 0 \\ \star & \star & \star & -\bar{Z}_i \end{bmatrix}, \bar{G}_{i1} = \begin{bmatrix} 0 \\ G_i^T \\ 0 \\ 0 \end{bmatrix},$$

$$\bar{G}_{i2} = \begin{bmatrix} 0 \\ 0 \\ G_i^T \\ 0 \end{bmatrix}, \bar{G}_{i3} = \begin{bmatrix} 0 \\ 0 \\ 0 \\ G_i^T \end{bmatrix}, \bar{\mathbb{I}}_i = \begin{bmatrix} I \\ I \\ 0 \\ 0 \end{bmatrix},$$

$$\bar{\Theta}_{i2ls} := \underbrace{\begin{bmatrix} \bar{\Theta}_{i12ls} & \cdots & \bar{\Theta}_{ij2ls} & \cdots & \bar{\Theta}_{iN2ls} \end{bmatrix}}_{N-1}, \bar{\Theta}_{ij2ls} = \bar{\mathbb{I}}_i \left(A_{ijl} G_i + B_{il} \bar{K}_{ijs}\right),$$

$$\bar{\Theta}_{i3ls} := \underbrace{\begin{bmatrix} \bar{\Theta}_{i13ls} & \cdots & \bar{\Theta}_{ij3ls} & \cdots & \bar{\Theta}_{iN3ls} \end{bmatrix}}_{N-1}, \bar{\Theta}_{ij3ls} = \sqrt{2}\bar{\mathbb{I}}_i B_{il} \bar{K}_{ijs},$$

$$\bar{\mathbb{A}}_{ils} = \begin{bmatrix} -G_i & A_{iil} G_i + B_{il} K_{iis} G_i & B_{il} K_{iis} G_i & B_{il} K_{iis} G_i \end{bmatrix}. \tag{10.150}$$

In that case, the proposed sampled-data fuzzy controller gains can be calculated by

$$K_{iis} = \bar{K}_{iis} G_i^{-1}, K_{ijs} = \bar{K}_{ijs} G_i^{-1}, s \in \mathcal{L}_i. \tag{10.151}$$

■

Proof. Define $W_{01} \leq W_{i1}$ and $W_{02} \leq W_{i2}$ and $W_{i3} \equiv W_{i2}, i \in \mathscr{N}$,

$$W_{01}^{-1} \geq W_{i1}^{-1}, W_{02}^{-1} \geq W_{i2}^{-1}, \tag{10.152}$$

which implies $(N-1)\,W_{01}^{-1} \geq \sum_{\substack{j=1 \\ j\neq i}}^{N} W_{j1}^{-1}$ and $(N-1)\,W_{02}^{-1} \geq \sum_{\substack{j=1 \\ j\neq i}}^{N} W_{j2}^{-1}$, respectively.

It is easy to see that the following inequality holds, which implies (10.133),

$$\begin{bmatrix} \bar{\mathbb{A}}_i(\mu_i,\hat{\mu}_i) & \bar{I}_1 & \bar{I}_2 & \bar{I}_3 \\ \star & -\frac{1}{(N-1)}W_{01} & 0 & 0 \\ \star & \star & -\frac{1}{(N-1)}W_{02} & 0 \\ \star & \star & \star & -\frac{1}{(N-1)}W_{02} \end{bmatrix} < 0, \tag{10.153}$$

where

$$\bar{\mathbb{A}}_i(\mu_i,\hat{\mu}_i) = \bar{\Theta}_{i1} + \mathrm{Sym}\left(\mathbb{G}_i\mathbb{A}_i(\mu_i,\hat{\mu}_i)\right) + \sum_{\substack{j=1 \\ j\neq i}}^{N} \bar{\Theta}_{ij2}(\mu_i,\hat{\mu}_i)W_{i1}\bar{\Theta}_{ij2}^{T}(\mu_i,\hat{\mu}_i)$$
$$+2\sum_{\substack{j=1 \\ j\neq i}}^{N} \bar{\Theta}_{ij3}(\mu_i,\hat{\mu}_i)W_{i2}\bar{\Theta}_{ij3}^{T}(\mu_i,\hat{\mu}_i),$$
$$\bar{\Theta}_{i1} = \begin{bmatrix} h_i^2Q_i & P_i & 0 & 0 \\ \star & \sigma_i^2Z_i & \sigma_i^2Z_i & 0 \\ \star & \star & -\frac{\pi^2}{4}Q_i+\sigma_i^2Z & 0 \\ \star & \star & \star & -Z_i \end{bmatrix},$$
$$\bar{I}_1 = \begin{bmatrix} 0 \\ I \\ 0 \\ 0 \end{bmatrix}, \bar{I}_2 = \begin{bmatrix} 0 \\ 0 \\ I \\ 0 \end{bmatrix}, \bar{I}_3 = \begin{bmatrix} 0 \\ 0 \\ 0 \\ I \end{bmatrix},$$
$$\bar{\Theta}_{ij2}(\mu_i,\hat{\mu}_i) = \mathbb{G}_i\bar{A}_{ij}(\mu_i,\hat{\mu}_i), \bar{\Theta}_{ij3}(\mu_i,\hat{\mu}_i) = \mathbb{G}_iB_i(\mu_i)K_{ij}(\hat{\mu}_i). \tag{10.154}$$

It follows from (10.143) that

$$h_i^2Q_i - \mathrm{Sym}\{G_i\} < 0, i \in \mathscr{N} \tag{10.155}$$

which implies that $G_i, i \in \mathscr{N}$ are nonsingular matrices.

We further define

$$\left\{\begin{array}{l} \mathbb{G}_i=\left[\begin{array}{cccc} G_i^{-1} & G_i^{-1} & 0 & 0 \end{array}\right]^T, \Gamma_1:=\operatorname{diag}\left\{\begin{array}{cccccccc} G_i & G_i & G_i & G_i & I & I & \mathbb{I} & \mathbb{I} \end{array}\right\}, \\ \mathbb{I}:=\operatorname{diag}\underbrace{\left\{\begin{array}{ccccc} I & \cdots & I & \cdots & I \end{array}\right\}}_{N-1}, \bar{P}_i=G_i^T P_i G_i, \bar{Q}_i=G_i^T Q_i G_i, \\ \bar{Z}_i=G_i^T Z_i G_i, \bar{K}_{ijs}=K_{ijs} G_i, \mathbb{W}_1:=\operatorname{diag}\underbrace{\left\{\begin{array}{ccccc} W_{i1} & \cdots & W_{i1} & \cdots & W_{i1} \end{array}\right\}}_{N-1}, \\ \mathbb{W}_2:=\operatorname{diag}\underbrace{\left\{\begin{array}{ccccc} W_{i2} & \cdots & W_{i2} & \cdots & W_{i2} \end{array}\right\}}_{N-1}, \\ \bar{\Theta}_{i2}(\mu_i,\hat{\mu}_i):=\underbrace{\left[\begin{array}{ccccc} \bar{\Theta}_{i12}(\mu_i,\hat{\mu}_i) & \cdots & \bar{\Theta}_{ij2}(\mu_i,\hat{\mu}_i) & \cdots & \bar{\Theta}_{iN2}(\mu_i,\hat{\mu}_i) \end{array}\right]}_{N-1}, \\ \bar{\Theta}_{i3}(\mu_i,\hat{\mu}_i):=\underbrace{\left[\begin{array}{ccccc} \bar{\Theta}_{i13}(\mu_i,\hat{\mu}_i) & \cdots & \bar{\Theta}_{ij3}(\mu_i,\hat{\mu}_i) & \cdots & \bar{\Theta}_{iN3}(\mu_i,\hat{\mu}_i) \end{array}\right]}_{N-1}, \\ \bar{\Theta}_{ij2}(\mu_i,\hat{\mu}_i)=\mathbb{G}_i\bar{A}_{ij}(\mu_i,\hat{\mu}_i), \bar{\Theta}_{ij3}(\mu_i,\hat{\mu}_i)=\sqrt{2}\mathbb{G}_i B_i(\mu_i)K_{ij}(\hat{\mu}_i). \end{array}\right. \tag{10.156}$$

By substituting (10.156) into (10.153), and using the Schur complement lemma, one has

$$\left[\begin{array}{cccc} \bar{\Theta}_{i1}+\operatorname{Sym}(\mathbb{G}_i\mathbb{A}_i(\mu_i,\hat{\mu}_i)) & \mathbb{I}_1 & \mathbb{I}_2 & \tilde{\Theta}_i \\ \star & -\frac{1}{(N-1)}W_{01} & 0 & 0 \\ \star & \star & -\frac{1}{(N-1)}W_{02} & 0 \\ \star & \star & \star & \Lambda \end{array}\right]<0, \tag{10.157}$$

where $\tilde{\Theta}_i=\left[\begin{array}{ccc} \mathbb{I}_3 & \bar{\Theta}_{i2}(\mu_i,\hat{\mu}_i) & \bar{\Theta}_{i3}(\mu_i,\hat{\mu}_i) \end{array}\right], \Lambda=\operatorname{diag}\left\{-\frac{W_{02}}{N-1}, -\mathbb{W}_1^{-1}, -\mathbb{W}_2^{-1}\right\}$.

By performing a congruence transformation by Γ_1, and extracting the fuzzy membership functions,

$$\sum_{l=1}^{r_i}\sum_{s=1}^{r_i}\mu_{il}\hat{\mu}_{is}\Sigma_{ils}<0, \tag{10.158}$$

where Σ_{ils} is defined in (10.150).

By taking the relation in (10.142) and using the asynchronous method proposed in [9], the inequality in (10.158) holds if the inequalities (10.143)-(10.149) hold, thus completing this proof.

It is worth pointing output that the number of LMIs on Theorem 10.9 is large. The existing relaxation technique $\sum_{l=1}^{r_i}[\mu_{il}]^2\Sigma_{ill}+\sum_{l=1}^{r_i}\sum_{l<s\leq r_i}\mu_{il}\mu_{is}\Sigma_{ls}<0$ is no longer applicable to fuzzy controller synthesis, since $\mu_{is}\neq\hat{\mu}_{is}$. Similarly to the relaxation technique in [10], it is assumed that

$$|\mu_{il}-\hat{\mu}_{il}|\leq\delta_{il}, l\in\mathscr{L}_i, \tag{10.159}$$

where δ_{il} is a positive scalar. If $\Sigma_{ils} + M_{il} \geq 0$, where M_{il} is a symmetric matrix, one obtains

$$\sum_{l=1}^{r_i}\sum_{s=1}^{r_i} \mu_{il}\hat{\mu}_{is}\Sigma_{ils} = \sum_{l=1}^{r_i}\sum_{s=1}^{r_i} \mu_{il}\mu_{is}\Sigma_{ils} + \sum_{l=1}^{r_i}\sum_{s=1}^{r_i} \mu_{il}\left(\hat{\mu}_{is} - \mu_{is}\right)\left(\Sigma_{ils} + M_{il}\right)$$
$$\leq \sum_{l=1}^{r_i}\sum_{s=1}^{r_i} \mu_{il}\mu_{is}\left[\Sigma_{ils} + \sum_{f=1}^{r_i} \delta_{if}\left(\Sigma_{ilf} + M_{il}\right)\right]. \quad (10.160)$$

Therefore, upon defining $\Sigma_{ils} = \Phi_{ils} + \sum_{s=1}^{r_i} \delta_{is}\left(\Phi_{ils} + M_{il}\right)$, the existing relaxation technique from [16] can be applied to (10.160).

Based on the assumption in (10.159), and the result of (10.158) and (10.160), the design result on fuzzy control is proposed as below:

Theorem 10.10: Design of Distributed Event-Triggered Fuzzy Control Using Synchronized Method

The closed-loop fuzzy system in (10.62) using a distributed event-triggered fuzzy controller (10.126), is asymptotically stable, if there exist the symmetric positive definite matrices $\{\bar{P}_i, W_i, W_0\} \in \Re^{n_{xi}\times n_{xi}}$, $W_0 \leq W_i$, and matrices $G_i \in \Re^{n_{xi}\times n_{xi}}$, $M_{il} = M_{il}^T \in \Re^{4n_{xi}\times 4n_{xi}}$, $\bar{K}_{is} \in \Re^{n_{ui}\times n_{xi}}$, and the positive scalars $\{h_i, \delta_{il}\}$, such that for all $(l,s) \in \mathcal{L}_i$, the following LMIs hold:

$$\bar{\Sigma}_{ill} < 0, l \in \mathcal{L}_i \quad (10.161)$$

$$\bar{\Sigma}_{ils} + \bar{\Sigma}_{isl} < 0, 1 \leq l < s \leq r_i \quad (10.162)$$

where

$$\bar{\Sigma}_{ils} = \Sigma_{ils} + \sum_{f=1}^{r_i} \delta_{if} \left(\Sigma_{ilf} + M_{il}\right)$$

$$\Sigma_{ils} = \begin{bmatrix} \tilde{\Theta}_{i1} + \mathrm{Sym}\left(\mathbb{I}_i \bar{\mathbb{A}}_{ils}\right) & \bar{G}_{i1} & \bar{G}_{i2} & \tilde{\Theta}_i \\ \star & -\frac{1}{(N-1)} W_{01} & 0 & 0 \\ \star & \star & -\frac{1}{(N-1)} W_{02} & 0 \\ \star & \star & \star & \Lambda \end{bmatrix},$$

$$\tilde{\Theta}_i = \begin{bmatrix} \bar{G}_{i3} & \bar{\Theta}_{i2ls} & \bar{\Theta}_{i3ls} \end{bmatrix}, \Lambda = \mathrm{diag}\left\{ -\frac{W_{02}}{N-1}, \mathbb{W}_1 - \mathbb{G}_i - \mathbb{G}_i, \mathbb{W}_2 - \mathbb{G}_i - \mathbb{G}_i \right\},$$

$$\tilde{\Theta}_{i1} = \begin{bmatrix} h_i^2 \bar{Q}_i & \bar{P}_i & 0 & 0 \\ \star & \sigma_i^2 \bar{Z}_i & \sigma_i^2 \bar{Z}_i & 0 \\ \star & \star & -\frac{\pi^2}{4} \bar{Q}_i + \sigma_i^2 \bar{Z} & 0 \\ \star & \star & \star & -Z_i \end{bmatrix}, \bar{G}_{i1} = \begin{bmatrix} 0 \\ G_i^T \\ 0 \\ 0 \end{bmatrix},$$

$$\bar{G}_{i2} = \begin{bmatrix} 0 \\ 0 \\ G_i^T \\ 0 \end{bmatrix}, \bar{G}_{i3} = \begin{bmatrix} 0 \\ 0 \\ 0 \\ G_i^T \end{bmatrix}, \mathbb{I}_i = \begin{bmatrix} I \\ I \\ 0 \\ 0 \end{bmatrix},$$

$$\bar{\Theta}_{i2ls} := \underbrace{\begin{bmatrix} \bar{\Theta}_{i12ls} & \cdots & \bar{\Theta}_{ij2ls} & \cdots & \bar{\Theta}_{iN2ls} \end{bmatrix}}_{N-1}, \bar{\Theta}_{ij2ls} = \mathbb{I}_i \left(A_{ijl} G_i + B_{il} \bar{K}_{ijs}\right),$$

$$\bar{\Theta}_{i3ls} := \underbrace{\begin{bmatrix} \bar{\Theta}_{i13ls} & \cdots & \bar{\Theta}_{ij3ls} & \cdots & \bar{\Theta}_{iN3ls} \end{bmatrix}}_{N-1}, \bar{\Theta}_{ij3ls} = \sqrt{2} \mathbb{I}_i B_{il} \bar{K}_{ijs},$$

$$\bar{\mathbb{A}}_{ils} = \begin{bmatrix} -G_i & A_{iil} G_i + B_{il} K_{iis} G_i & B_{il} K_{iis} G_i & B_{il} K_{iis} G_i \end{bmatrix}. \tag{10.163}$$

In that case, the controller gains can be calculated by

$$K_{iis} = \bar{K}_{iis} G_i^{-1}, K_{ijs} = \bar{K}_{ijs} G_i^{-1}, s \in \mathscr{L}_i. \tag{10.164}$$

■

Note that when the asynchronized information of μ_{il} and $\hat{\mu}_{il}$ is unknown, the design result on a distributed event-triggered linear controller can be directly derived as below:

Theorem 10.11: Distributed Event-Triggered Linear Controller Design

The closed-loop fuzzy system in (10.62) using a distributed event-triggered linear controller $u_i(t) = K_{ii} x_i(t_k^i) + \sum_{j=1, j \neq i}^{N} K_{ij} x_j(t_k^i), t \in [t_k^i, t_{k+1}^i)$, is asymptotically stable, if there exist the symmetric positive definite matrices $\{\bar{P}_i, W_{i1}, W_{i2}, W_{01}, W_{02}\} \in \Re^{n_{xi} \times n_{xi}}$,

$W_{01} \leq W_{i1}$, $W_{02} \leq W_{i2}$, and matrices $G_i \in \Re^{n_{xi} \times n_{xi}}$, $\bar{K}_{ii} \in \Re^{n_{ui} \times n_{xi}}$, $\bar{K}_{ij} \in \Re^{n_{ui} \times n_{xj}}$, and the positive scalars $\{h_i, \sigma_i\}$, such that for all $l \in \mathscr{L}_i$, the following LMIs hold:

$$\begin{bmatrix} \tilde{\Theta}_{i1} + \mathrm{Sym}\left(\bar{\mathbb{I}}_i \bar{\mathbb{A}}_{il}\right) & \bar{G}_{i1} & \bar{G}_{i2} & \tilde{\Theta}_i \\ \star & -\frac{1}{(N-1)} W_{01} & 0 & 0 \\ \star & \star & -\frac{1}{(N-1)} W_{02} & 0 \\ \star & \star & \star & \Lambda \end{bmatrix} < 0, \tag{10.165}$$

where

$$\tilde{\Theta}_i = \begin{bmatrix} \bar{G}_{i3} & \bar{\Theta}_{i2l} & \bar{\Theta}_{i3l} \end{bmatrix}, \Lambda = \mathrm{diag}\left\{ -\frac{W_{02}}{N-1}, \mathbb{W}_1 - \mathbb{G}_i - \mathbb{G}_i, \mathbb{W}_2 - \mathbb{G}_i - \mathbb{G}_i \right\},$$

$$\tilde{\Theta}_{i1} = \begin{bmatrix} h_i^2 \bar{Q}_i & \bar{P}_i & 0 & 0 \\ \star & \sigma_i^2 \bar{Z}_i & \sigma_i^2 \bar{Z}_i & 0 \\ \star & \star & -\frac{\pi^2}{4}\bar{Q}_i + \sigma_i^2 \bar{Z} & 0 \\ \star & \star & \star & -Z_i \end{bmatrix}, \bar{G}_{i1} = \begin{bmatrix} 0 \\ G_i^T \\ 0 \\ 0 \end{bmatrix},$$

$$\bar{G}_{i2} = \begin{bmatrix} 0 \\ 0 \\ G_i^T \\ 0 \end{bmatrix}, \bar{G}_{i2} = \begin{bmatrix} 0 \\ 0 \\ 0 \\ G_i^T \end{bmatrix}, \bar{\mathbb{I}}_i = \begin{bmatrix} I \\ I \\ 0 \\ 0 \end{bmatrix},$$

$$\bar{\Theta}_{i2l} := \underbrace{\begin{bmatrix} \bar{\Theta}_{i12l} & \cdots & \bar{\Theta}_{ij2l} & \cdots & \bar{\Theta}_{iN2l} \end{bmatrix}}_{N-1}, \bar{\Theta}_{ij2l} = \bar{\mathbb{I}}_i \left(A_{ijl} G_i + B_{il} \bar{K}_{ij}\right),$$

$$\bar{\Theta}_{i3l} := \underbrace{\begin{bmatrix} \bar{\Theta}_{i13l} & \cdots & \bar{\Theta}_{ij3l} & \cdots & \bar{\Theta}_{iN3l} \end{bmatrix}}_{N-1}, \bar{\Theta}_{ij3l} = \sqrt{2} \bar{\mathbb{I}}_i B_{il} \bar{K}_{ij},$$

$$\bar{\mathbb{A}}_{il} = \begin{bmatrix} -G_i & A_{iil} G_i + B_{il} K_{ii} G_i & B_{il} \bar{K}_{ii} & B_{il} \bar{K}_{ii} \end{bmatrix}. \tag{10.166}$$

In that case, the proposed linear controller gains can be calculated by

$$K_{ii} = \bar{K}_{ii} G_i^{-1}, K_{ij} = \bar{K}_{ii} G_j^{-1}. \tag{10.167}$$

■

10.4 SIMULATION STUDIES

Consider a system using a DC-DC converter and a permanent-magnet synchronous generator (PMSG) using an AC-DC converter. Their dynamic models are respectively described by the following differential equations [18]

$$\begin{cases} \dot{v}_{PV} = \frac{1}{C_{PV}}\left(\phi_{PV} - \phi_L u\right), \\ \dot{\phi}_L = \frac{1}{L}\left(R_0\left(\phi_0 - \phi_L\right) - R_L \phi_L - v_0\right) \\ \qquad + \frac{1}{L}\left(V_D + v_{PV} - R_M \phi_L\right) u - \frac{V_D}{L}, \\ \dot{v}_0 = \frac{1}{C_0}\left(\phi_L - \phi_0\right), \end{cases}$$

and [19]

$$\begin{cases} L_d \dot{\phi}_{ds} = -R_s \phi_{ds} + \omega L_q(i_{qs}) \phi_{qs} + v_d, \\ L_q(i_{qs}) \dot{\phi}_{qs} = -R_s \phi_{qs} - 0.5\omega_g P_n L_d \phi_{ds} - \omega \psi_m + v_q, \\ C_s \dot{v}_{dc} = \frac{3}{2}(d_s \phi_{ds} + d_q \phi_{qs}) - \phi_0, \end{cases}$$

where in the solar PV system, v_{PV}, ϕ_L, and v_0 are the PV array voltage, the current on the inductance L, and the voltage of the capacitance C_0, respectively; R_0, R_L, and R_M are the internal resistance on the capacitance C_0, the inductance L, and the power MOSFET, respectively; V_D is the forward voltage of the power diode; ϕ_0 is the measurable load current. In the PMSG system, ψ_m and R_s denote the magnet flux linkage and stator resistance, respectively; ϕ_{ds}, v_d, L_d and $\phi_{qs}, v_q, L_q(i_{qs})$ are the current, voltage, inductance in d-axis and in q-axis, respectively; P_n and ω_g are the number of poles and the rotor speed, respectively; d_s and d_q are the duty-ratio signals, ω is the electrical angular velocity, v_{dc} is the voltage of the capacitance C_s.

Now, let us consider a direct current (DC) microgrid with two solar PV subsystems and two PMSGs. The dynamic model can be given by

$$\begin{cases} \dot{v}_{PV(i)} = \frac{1}{C_{PV(i)}} \frac{\phi_{PV(i)}}{v_{PV(i)}} v_{PV(i)} - \frac{1}{C_{PV(i)}} \phi_{L(i)} u_{(i)}, \\ \dot{\phi}_{L(i)} = \left(-\frac{1}{L_{(i)}} R_{0(i)} - \frac{1}{L_{(i)}} R_{L(i)} - \frac{V_{D(i)}}{L_{(i)} \phi_{L(i)}} \right) \phi_{L(i)} \\ \quad + \left(\frac{R_{0(i)}}{L_{(i)}(R_{line(i)} + R_{load})} - \frac{1}{L_{(i)}} \right) v_{0(i)} \\ \quad - \sum_{\substack{j=1 \\ j \neq i}}^{N} \frac{R_{0(i)} R_{load}}{L_{(i)}(R_{line(i)} + R_{load})} \phi_{0(j)} \\ \quad + \frac{1}{L_{(i)}} \left(V_{D(i)} + v_{PV(i)} - R_{M(i)} \phi_{L(i)} \right) u_{(i)}, \\ \dot{v}_{0(i)} = \frac{1}{C_{0(i)}} \phi_{L(i)} - \frac{1}{C_{0(i)}(R_{line(i)} + R_{load})} v_{0(i)} \\ \quad + \sum_{\substack{j=1 \\ j \neq i}}^{N} \frac{R_{load}}{C_{0(i)}(R_{line(i)} + R_{load})} \phi_{0(j)}, i = \{1, 2\} \end{cases}$$

and

$$\begin{cases} \dot{\phi}_{ds(i)} = \frac{-R_{s(i)}}{L_{d(i)}} \phi_{ds(i)} + \frac{0.5\omega_{g(i)} P_{n(i)} L_{q(i)}(i_{qs(i)})}{L_{d(i)}} \phi_{qs(i)}, \\ \dot{\phi}_{qs(i)} = \frac{-R_{s(i)}}{L_{q(i)}(i_{qs(i)})} \phi_{qs(i)} - \frac{0.5\omega_{g(i)} P_{n(i)} L_{d(i)}}{L_{q(i)}(i_{qs})} \phi_{ds(i)}, \\ \dot{v}_{dc(i)} = -\frac{1}{C_{s(i)}(R_{load} + R_{line(i)})} v_{dc(i)} \\ \quad + \sum_{\substack{j=1 \\ j \neq i}}^{N} \frac{R_{load}}{C_{s(i)}(R_{load} + R_{line(i)})} \phi_{0(j)} \\ \quad + \frac{3}{2C_{s(i)}} u_{(i)}, i = \{3, 4\} \end{cases}$$

where $R_{line(i)}$ denotes the line resistance on the i-th subsystem, R_{load} is the consumer load.

In this simulation, the values of the parameters are $V_{D(1)} = 9.1$ V, $C_{PV(1)} = 0.0101$ F, $C_{0(1)} = 0.0472$ F, $L_{(1)} = 0.0516$ H, $R_{L(1)} = 1.7\ \Omega$, $R_{0(1)} = 1.2\ \Omega$, $R_{M(1)} = 0.8$

Ω; $V_{D(2)} = 9.2$ V, $C_{PV(2)} = 0.0108$ F, $C_{0(2)} = 0.0411$ F, $L_{(2)} = 0.0514$ H, $R_{L(2)} = 1.8$ Ω, $R_{0(2)} = 1.1$ Ω, $R_{M(2)} = 0.85$ Ω; $R_{s(3)} = 8.9$ Ω, $L_{d(3)} = 0.0629$ H, $P_{n(3)} = 12$, $C_{s(3)} = 0.0418$ F; $R_{s(4)} = 9$ Ω, $L_{d(4)} = 0.0684$ H, $P_{n(4)} = 12$, $C_{s(4)} = 0.0431$ F, $R_{load} = 18$ Ω, $R_{line(i)} = 1.3$ Ω. Assume that $\phi_{L(i)} = 0.25\phi_{0(i)}, i = \{1,2\}$, and $\phi_{qs(i)} = 0.25\phi_{0(i)}, i = \{3,4\}$, we can approximate the DC microgrid by a T-S model based on the following fuzzy rules:

Plant Rule $\mathscr{R}_i^1$: **IF** $(v_{PV(i)}, \phi_{PV(i)})$ is $(9.5, 0.36)$ and $\phi_{L(i)}$ is 0.25, **THEN**

$$\dot{x}_i(t) = A_{i1}x_i(t) + B_{i1}u_i(t) + \sum_{\substack{j=1 \\ j\neq i}}^{4} \bar{A}_{ij}x_j(t), i = \{1,2\},$$

Plant Rule $\mathscr{R}_i^2$: **IF** $(v_{PV(i)}, \phi_{PV(i)})$ is $(9.5, 0.36)$ and $\phi_{L(i)}$ is 0.29, **THEN**

$$\dot{x}_i(t) = A_{i2}x_i(t) + B_{i2}u_i(t) + \sum_{\substack{j=1 \\ j\neq i}}^{4} \bar{A}_{ij}x_j(t), i = \{1,2\},$$

Plant Rule $\mathscr{R}_i^3$: **IF** $(v_{PV(i)}, \phi_{PV(i)})$ is $(12.3, 0.42)$ and $\phi_{L(i)}$ is 0.25, **THEN**

$$\dot{x}_i(t) = A_{i3}x_i(t) + B_{i3}u_i(t) + \sum_{\substack{j=1 \\ j\neq i}}^{4} \bar{A}_{ij}x_j(t), i = \{1,2\},$$

Plant Rule $\mathscr{R}_i^4$: **IF** $(v_{PV(i)}, \phi_{PV(i)})$ is $(12.3, 0.42)$ and $\phi_{L(i)}$ is 0.29, **THEN**

$$\dot{x}_i(t) = A_{i4}x_i(t) + B_{i4}u_i(t) + \sum_{\substack{j=1 \\ j\neq i}}^{4} \bar{A}_{ij}x_j(t), i = \{1,2\},$$

Plant Rule $\mathscr{R}_i^1$: **IF** $\omega_{g(i)}$ is 74 and $L_{q(i)}$ is 0.0515, **THEN**

$$\dot{x}_i(t) = A_{i1}x_i(t) + B_{i1}u_i(t) + \sum_{\substack{j=1 \\ j\neq i}}^{4} \bar{A}_{ij}x_j(t), i = \{3,4\},$$

Plant Rule $\mathscr{R}_i^2$: **IF** $\omega_{g(i)}$ is 74 and $L_{q(i)}$ is 0.0541, **THEN**

$$\dot{x}_i(t) = A_{i2}x_i(t) + B_{i2}u_i(t) + \sum_{\substack{j=1 \\ j\neq i}}^{4} \bar{A}_{ij}x_j(t), i = \{3,4\},$$

Plant Rule $\mathscr{R}_i^3$: **IF** $\omega_{g(i)}$ is 80 and $L_{q(i)}$ is 0.0515, **THEN**

$$\dot{x}_i(t) = A_{i3}x_i(t) + B_{i3}u_i(t) + \sum_{\substack{j=1 \\ j\neq i}}^{4} \bar{A}_{ij}x_j(t), i = \{3,4\},$$

Plant Rule $\mathscr{R}_i^4$: **IF** $\omega_{g(i)}$ is 80 and $L_{q(i)}$ is 0.0541, **THEN**

$$\dot{x}_i(t) = A_{i4}x_i(t) + B_{i4}u_i(t) + \sum_{\substack{j=1\\ j\neq i}}^{4} \bar{A}_{ij}x_j(t), i = \{3,4\},$$

where $\{A_{il}, B_{il}, \bar{A}_{ij}\}, l = \{1,2,3,4\}$ is listed in [20].

Figure 10.1 shows that the open-loop DC microgrid is unstable under the initial conditions $x_1(0) = [2,2,4]^T, x_2(0) = [2.5,1.9,4.5]^T, x_3(0) = [3,1.8,5]^T$, and $x_4(0) = [3.5,1.7,4.5]^T$. Our aim is to design a decentralized event-triggered controller in the form of (10.64) such that the resulting closed-loop fuzzy control system is asymptotically stable with fewer data transmissions. Assume that $\tau_i = 0.17$, $\bar{\eta}_i = 0.18$, by applying Lemma 10.4 with $\bar{Q}_i > 0$, we indeed obtain a feasible solution with the maximum triggered parameter $\sigma_{imax} = 0.032$. Consider the condition that $\bar{Q}_i$ may be negative definite, and by using Lemma 10.4, the maximum triggered parameter can be further improved to $\sigma_{imax} = 0.038$, and the corresponding linear controller gains are

$$\begin{aligned} K_1 &= \begin{bmatrix} 0.1706 & 0.0001 & -0.0001 \end{bmatrix}, \\ K_2 &= \begin{bmatrix} 0.1773 & 0.0001 & -0.0001 \end{bmatrix}, \\ K_3 &= \begin{bmatrix} 0.0091 & 0.0058 & -0.1211 \end{bmatrix}, \\ K_4 &= \begin{bmatrix} 0.0152 & -0.0045 & -0.1250 \end{bmatrix}. \end{aligned}$$

Now, let us assume that the information in the asynchronous variables μ_{il} and $\hat{\mu}_{il}$ is known and satisfies that $\underline{\rho}_{il} = 0.5$ and $\bar{\rho}_{il} = 2$. By applying Theorem 10.7, the triggered parameter σ_i is further improved to 0.041, and the corresponding fuzzy controller gains are

$$\begin{aligned} K_{11} &= \begin{bmatrix} 0.1843 & 0.0001 & \text{-}0.0001 \end{bmatrix}, \\ K_{12} &= \begin{bmatrix} 0.1573 & 0.0001 & 0.0001 \end{bmatrix}, \\ K_{13} &= \begin{bmatrix} 0.1659 & 0.0001 & \text{-}0.0001 \end{bmatrix}, \\ K_{14} &= \begin{bmatrix} 0.1417 & 0.0001 & 0.0001 \end{bmatrix} \end{aligned}$$

for the first subsystem, and

$$\begin{aligned} K_{21} &= \begin{bmatrix} 0.1869 & 0.0001 & \text{-}0.0001 \end{bmatrix}, \\ K_{22} &= \begin{bmatrix} 0.1592 & 0.0001 & 0.0001 \end{bmatrix}, \\ K_{23} &= \begin{bmatrix} 0.1678 & 0.0001 & \text{-}0.0001 \end{bmatrix}, \\ K_{24} &= \begin{bmatrix} 0.1429 & 0.0001 & 0.0001 \end{bmatrix}, \end{aligned}$$

for the second subsystem, and

$$\begin{aligned} K_{31} &= \begin{bmatrix} 0.0018 & \text{-}0.0149 & \text{-}0.0271 \end{bmatrix}, \\ K_{32} &= \begin{bmatrix} 0.0018 & \text{-}0.0148 & \text{-}0.0270 \end{bmatrix}, \\ K_{33} &= \begin{bmatrix} 0.0018 & \text{-}0.0148 & \text{-}0.0270 \end{bmatrix}, \\ K_{34} &= \begin{bmatrix} 0.0018 & \text{-}0.0146 & \text{-}0.0268 \end{bmatrix} \end{aligned}$$

for the third subsystem, and

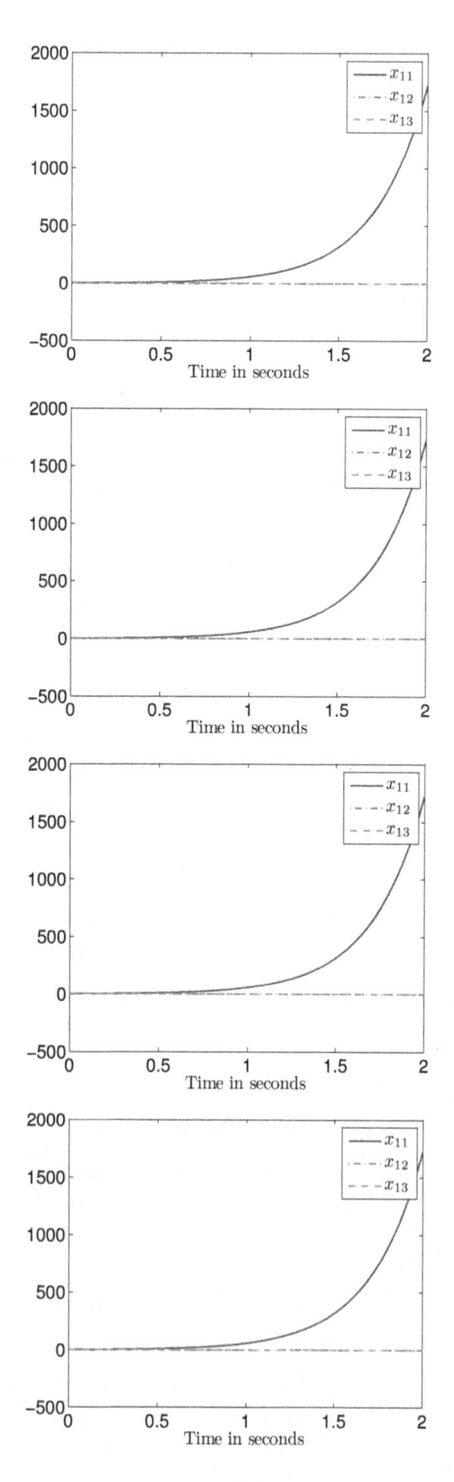

Figure 10.1 State responses for open-loop DC microgrid.

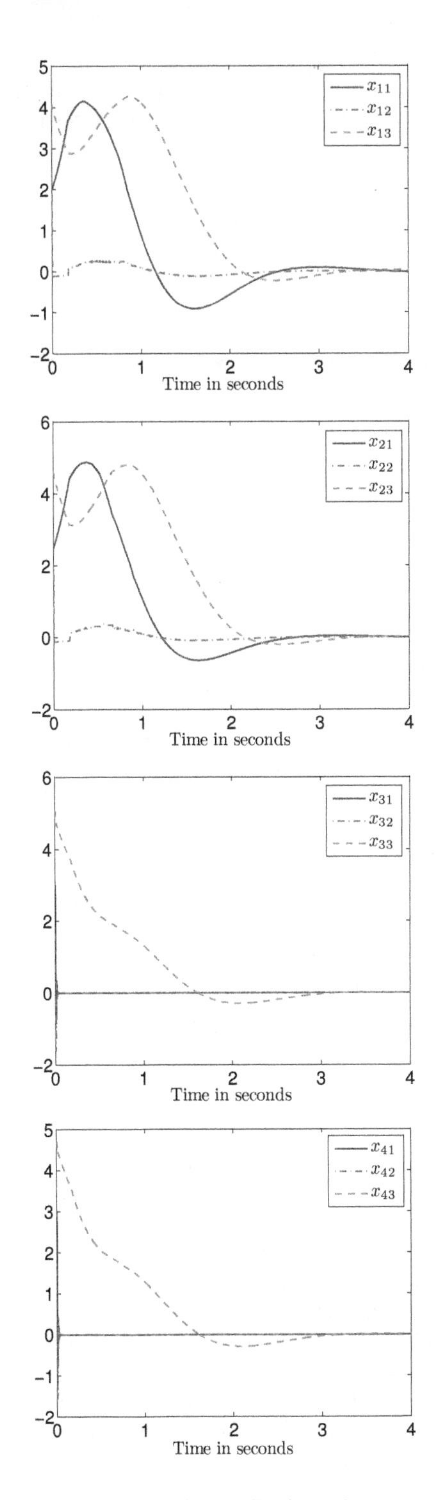

Figure 10.2 State responses for closed-loop DC microgrid.

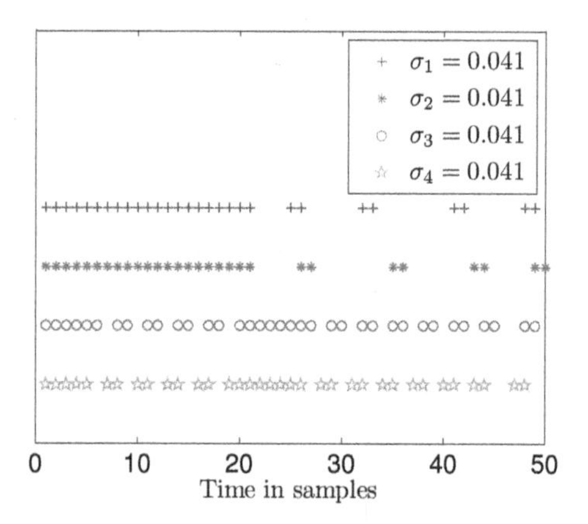

Figure 10.3 Event-triggered times for $\sigma_i = 0.041$.

$$
\begin{aligned}
K_{41} &= \begin{bmatrix} \text{-0.0125} & \text{-0.0319} & \text{-0.0283} \end{bmatrix}, \\
K_{42} &= \begin{bmatrix} \text{-0.0124} & \text{-0.0317} & \text{-0.0282} \end{bmatrix}, \\
K_{43} &= \begin{bmatrix} \text{-0.0124} & \text{-0.0317} & \text{-0.0282} \end{bmatrix}, \\
K_{44} &= \begin{bmatrix} \text{-0.0124} & \text{-0.0318} & \text{-0.0282} \end{bmatrix}
\end{aligned}
$$

for the fourth subsystem.

Figure 10.1 indicates state responses for open-loop DC microgrid. However, based on the above solutions, Figure 10.2 shows convergence to zero for closed-loop DC microgrid. Figure 10.3 shows that the number of transmissions reduces from 50 to 29 in PV subsystem 1, and from 50 to 29 in PV subsystem 2, and from 50 to 36 in PMSG subsystem 3, and from 50 to 35 in PMSG subsystem 4, respectively.

10.5 REFERENCES

1. Cheng, J., Ju H. Park, Zhang, L., and Zhu, Y. (2018). An asynchronous operation approach to event-triggered control for fuzzy Markovian jump systems with general switching policies. IEEE Transactions on Fuzzy Systems, 26(1), 6-18.
2. Heemels, W. P. M. H. and Donkers, M. C. F. (2013). Model-based periodic event-triggered control for linear systems. Automatica, 49(3), 698-711.
3. Donkers, M. C. F. and Heemels, W. P. M. H. (2012). Output-based event-triggered control with guaranteed-gain and improved and decentralized event-triggering. IEEE Transactions on Automatic Control, 57(6), 1362-1376.
4. Yue, D. , Tian, E., and Han, Q. L. (2013). A delay system method for designing event-triggered controllers of networked control systems. IEEE Transactions on Automatic Control, 58(2), 475-481.

5. Hristu-Varsakelis, D. and Kumar, P. (2002). Interrupt-based feedback control over a shared communication medium. In Proceedings of the 2002 IEEE Conference on Decision and Control, 3223-3228.
6. Wang, X. and Lemmon, M. D. (2011). Event-triggering in distributed networked control systems. IEEE Transactions on Automatic Control, 56(3), 586-601.
7. Tabuada, P. and Wang, X. W. X. (2006). Preliminary results on state-triggered scheduling of stabilizing control tasks. IEEE Conference on Decision and Control.
8. Moarref, M. and Rodrigues, L. (2014). Stability and stabilization of linear sampled-data systems with multi-rate samplers and time driven zero order holds. Automatica, 50(10), 2685-2691.
9. Arino, C. and Sala, A. (2008). Extensions to "Stability analysis of fuzzy control systems subject to uncertain grades of membership". IEEE Transactions on Systems, Man, and Cybernetics, Part B, 38(2), 558-563.
10. Fridman, E. (2010). A refined input delay approach to sampled-data control. Automatica, 46(2), 421-427.
11. Liu, K. and Fridman, E. (2012). Wirtinger's inequality and Lyapunov-based sampled-data stabilization. Automatica, 48(1), 102-108.
12. Gu, K. (2002). An integral inequality in the stability problem of time-delay systems. In Proceedings of the 2000 IEEE Conference on Decision and Control, 2805-2810.
13. Xu, S., Lam, J., Zhang, B., and Zou, Y. (2015). New insight into delay-dependent stability of time-delay systems. International Journal of Robust and Nonlinear Control, 25(7), 961-970.
14. Zhang, D., Han, Q. L., and Jia, X. (2017). Network-based output tracking control for a class of T-S fuzzy systems that can not be stabilized by nondelayed output feedback controllers. IEEE Transactions on Cybernetics, 45(8), 1511-1524.
15. Zhang, H., Yu, G., Zhou, C., and Dang, C. (2013). Delay-dependent decentralised $\mathscr{H}_\infty$ filtering for fuzzy interconnected systems with time-varying delay based on Takagi-Sugeno fuzzy model. IET Control Theory and Applications, 7(5), 720-729.
16. Feng, G. (2006). A survey on analysis and design of model-based fuzzy control systems. IEEE Transactions on Fuzzy Systems, 14(5), 676-697.
17. Zhong, Z.X. and Lin C.M. (2017). Large-Scale Fuzzy Interconnected Control Systems Design and Analysis, Pennsylvania: IGI Global.
18. Chiu C. and Ouyang, Y. (2011). Robust maximum power tracking control of uncertain photovoltaic systems: A unified T-S fuzzy model-based approach. IEEE Transactions on Control Systems and Technology, 19(6): 1516-1526.
19. Boukhezzar, B. and Siguerdidjane, H. (2010). Comparison between linear and nonlinear control strategies for variable speed wind turbines. Control Engineering Practice, 18(12), 1357-1368.
20. Zhong, Z., Lin, C. M., Shao, Z., and Xu, M. (2018). Decentralized event-triggered control for large-scale networked fuzzy systems. IEEE Transactions on Fuzzy Systems, 26(1), 29-45.

11 Estimation and Compensation for TDS Attacks

Cyber-physical systems (CPSs) represent integrations of cyber and physical subsystems. They perform realtime monitoring by utilizing a combination of sensors, actuators and computation and communication modules. Due to the increasing developments in distributed sensing, action and networking technologies, CPS technology has found use in large-scale networks and is called networked CPS (NCPS) [1].

Power networks are large-scale spatially distributed systems located over widespread areas. Reliable monitoring and control systems are essential for guaranteeing the safety and stability of power networks. [2]. Traditionally, this is done in a centralized control framework using a single controller. Its core methodology for control of power networks requires the whole data measurements and powerful processing capabilities. In recent years, control techniques with higher sampling rates have been developed. Such centralized control schemes reduce system reliability and increase a network's sensitivity to a single point of failure [3]. Currently, distributed control has been increasing developed for many large-scale systems, and some efforts on distributed control of power systems have advanced [4].

Power systems are large-scale nonlinear operations [5]. Traditionally, simplified linear models have been generally utilized in synchronous generators for a long time. Due to their inherently simple designs, linear control methodologies were widely used in utilities. However, such linear control results only provide asymptotic stability in a small region of the equilibrium and attenuate the impacts on small disturbances [6, 7]. Recently, researchers have proposed cyber-physical systems (CPSs) approach to the design of power systems [8]. Due to the openness of system operation and communication protocols, CPSs are extremely vulnerable to cyber attacks, which can produce catastrophic consequences. Thus, security is a major challenge for implementing CPSs [9]. Recently, many researchers have extensively investigated security issues from different perspectives. One of the main goals is to evaluate the vulnerabilities of CPSs to cyber attacks. Much of the literature focuses on false data injection attacks [10], denial-of-service (DoS) attacks [11], time delay switch (TDS) attacks [11], deception attacks [12], against particular systems. More recently, NCPS has been used in modern power systems to monitor and control distributed systems over a wide area and has raised many important research perspectives. This makes modern power systems depending on networks vulnerable to attacks, particularly high-profile cyber attacks [13].

Extensive uncertainties and disturbances impact many utility engineering applications and adversely affect performances of closed-loop systems [14]. Recently, it has been recognized that the finite time stable systems had two advantages: i) faster convergence speed around the equilibrium point; ii) better disturbance rejection performance [15]. Imposing bounds on initial conditions prevents system states from exceeding certain thresholds within certain time intervals. In that case, the system is said to be finite-time stable. In practice, compared with Lyapunov asymptotic stability, FTS is more suitable for analyzing the transient dynamics of the controlled system within a specified finite interval. It ensures that state variables fall within a specified threshold and thus avoid excitations and saturations of nonlinear dynamics during transient events [16].

This chapter focuses on nonlinear power networks subject to the TDS attacks. The main purpose is to establish an effective method to countervail the negative influence from TDS attacks. To more effectively monitor system states and vulnerability to TDS attacks, we propose an augmented observer system to ensure that the estimation error system is the finite-time boundedness (FTB). The elimination of negative effects, from the TDS attacks to the system performance, is achieved by means of a compensation control. Finally, the transient performance of power networks is exhaustively tested by a numerical simulation.

11.1 TDS ATTACK OF LOCAL COMPONENTS

Consider a cyber-physical power system, its nonlinear dynamics are given by

$$\dot{x}(t) = A(t)x(t) + B(t)u(t) + D(t)d(t), \tag{11.1}$$

where $x(t) \in \Re^{n_x}$ and $u(t) \in \Re^{n_u}$ are the system state signal and the control input signal, respectively, $\{A(t), B(t), D(t)\}$ is a set of the time-varying parameter matrices. In this section, the T-S fuzzy modeling approach is introduced to describe the original nonlinear system (11.1). In the T-S fuzzy modeling, we choose several operating points for a considered nonlinear system. The corresponding local subsystems are then obtained for each operating point of the identification or linearization technique, where subsystems are connected with a fuzzy inference rule together with a membership function. Thus, the nonlinear system (11.1) can be characterized via the following T-S fuzzy model:

Plant Rule $\mathscr{R}^l$: **IF** z_1 is $\mathscr{F}_1^l$ and z_2 is $\mathscr{F}_2^l$ and $\cdots$ and z_g is $\mathscr{F}_g^l$, **THEN**

$$\dot{x}(t) = A_l x_i(t) + B_l u(t) + D_l d(t), l \in \mathscr{L} := \{1, 2, \ldots, r\} \tag{11.2}$$

where $\mathscr{R}^l$ is the l-th fuzzy inference rule; $\mathscr{F}_\phi^l$ $(\phi = 1, 2, \ldots, g)$ are the fuzzy sets; r is the number of inference rules; $z(t) \triangleq [z_1, z_2, \ldots, z_g]$ represents the measurable variables; $\{A_l, B_l, D_l\}$ is the l-th local linear subsystem.

Denote $\mathscr{F}^l \triangleq \prod_{\phi=1}^{g} \mathscr{F}_\phi^l$ as the inferred fuzzy set, and $\mu_l[z(t)]$ as the normalized membership function,

$$\mu_l[z(t)] \triangleq \frac{\prod_{\phi=1}^{g} \mu_{l\phi}[z_\phi(t)]}{\sum_{\varsigma=1}^{r} \prod_{\phi=1}^{g} \mu_{\varsigma\phi}[z_\phi(t)]} \geq 0, \sum_{l=1}^{r} \mu_l[z(t)] = 1. \tag{11.3}$$

We denote $\mu_l \triangleq \mu_l[z(t)]$ for brevity.

The T-S fuzzy dynamic model could be obtained via the fuzzy blending, that is

$$\dot{x}(t) = A(\mu)x(t) + B(\mu)u(t) + D(\mu)d(t), \tag{11.4}$$

where

$$A(\mu) \triangleq \sum_{l=1}^{r} \mu_l A_l,\ B(\mu) \triangleq \sum_{l=1}^{r} \mu_l B_l,\ D(\mu) \triangleq \sum_{l=1}^{r} \mu_l D_l. \tag{11.5}$$

This assumes that the attacker can only target output measurements between the plant and the controller. The attacker can also launch the relay-based or timestamp-based or noise-based attacks. Moreover, it may be impossible to obtain the exact knowledge about TDS attacks [11]. Here, under the relay-based or timestamp-based attacks, the system output is given by

$$y(t) = Cx(t-\varepsilon), \tag{11.6}$$

where ε is the unknown delay that can be either time-varying or constant one.

11.1.1 REACHABLE SET ESTIMATION FOR TRACKING CONTROL

This subsection will propose an efficient method to avoid the negative influence from the TDS attack. An augmented fuzzy observer is first developed for the fuzzy system (11.4) utilizing simultaneous estimation of system state and attack perturbation. Then, based on the estimation results, a compensation technique is adopted to avoid the impact of the TDS attack.

11.1.2 OBSERVER DESIGN FOR SYSTEM STATE AND DELAY PERTURBATION

First, we define the perturbation of the TDS attack as

$$\omega(t) = C[x(t-\varepsilon) - x(t)]. \tag{11.7}$$

The system output (11.6) thus becomes

$$y(t) = Cx(t) + \omega(t). \tag{11.8}$$

Motivated by [17], combined with (11.4) and (11.8), let $\mathscr{X}(t) \triangleq \begin{bmatrix} x^T(t) & \omega^T(t) \end{bmatrix}^T$, and introduce an auxiliary variable $\sigma(\mu)$,

$$\begin{cases} E\dot{\mathscr{X}}(t) = \bar{A}(\mu)\mathscr{X}(t) + \bar{B}(\mu)u(t) + \bar{D}(\mu)d(t) + \bar{B}_\omega(\mu)\omega(t), \\ y(t) = \bar{C}\mathscr{X}(t), \end{cases} \tag{11.9}$$

where

$$\left\{\begin{array}{l} E \triangleq \begin{bmatrix} \mathbf{I} & 0 \\ 0 & 0 \end{bmatrix}, \bar{A}(\mu) \triangleq \begin{bmatrix} A(\mu) & 0 \\ 0 & -\sigma(\mu)\mathbf{I} \end{bmatrix}, \bar{B}(\mu) \triangleq \begin{bmatrix} B(\mu) \\ 0 \end{bmatrix}, \\ \bar{D}(\mu) \triangleq \begin{bmatrix} D(\mu) \\ 0 \end{bmatrix}, \bar{B}_{\omega}(\mu) \triangleq \begin{bmatrix} 0 \\ \sigma(\mu)\mathbf{I} \end{bmatrix}, \bar{C} \triangleq \begin{bmatrix} C & \mathbf{I} \end{bmatrix}. \end{array}\right. \tag{11.10}$$

The augmented state vector $\mathscr{X}(t)$ in (11.9) is composed of the state $x(t)$ and the disturbance term $\omega(t)$ induced by the TDS attack. Thus, the estimation of system state together with the perturbation of the TDS attack will be implemented once the state estimation of $\mathscr{X}(t)$ is acquired.

Based on the system (11.9), an augmented observer is constructed with the form of

$$\left\{\begin{array}{l} S\dot{\mathscr{G}}(t) = \left(\bar{A}(\mu) - K(\mu)\bar{C}\right)\mathscr{G}(t) + \bar{B}(\mu)u(t) + \bar{B}_{\omega}(\mu)y(t), \\ \hat{\mathscr{X}}(t) = \mathscr{G}(t) + S^{-1}\bar{L}y(t), \end{array}\right. \tag{11.11}$$

where $\bar{L} \triangleq \begin{bmatrix} 0 & L^T \end{bmatrix}^T$, and $\mathscr{G}(t)$ is an auxiliary state vector. From the expressions in (11.9) and (11.11), as well as the estimation error, which is defined by $\mathscr{E}(t) \triangleq \mathscr{X}(t) - \hat{\mathscr{X}}(t)$, we define

$$S \triangleq E + \bar{L}\bar{C}, \bar{B}_{\omega}(\mu) \triangleq -\bar{A}(\mu)S^{-1}\bar{L}, \tag{11.12}$$

which implies that S is a non-singular matrix. It is clear that

$$\bar{C}S^{-1}\bar{L} = \begin{bmatrix} C & \mathbf{I} \end{bmatrix} \begin{bmatrix} \mathbf{I} & 0 \\ -C & L^{-1} \end{bmatrix} \begin{bmatrix} 0 \\ L \end{bmatrix} = \mathbf{I}. \tag{11.13}$$

In terms of (11.9)-(11.13), one can obtain

$$S\dot{\mathscr{E}}(t) = \left(\bar{A}(\mu) - K(\mu)\bar{C}\right)\mathscr{E}(t) + \bar{D}(\mu)d(t) + \bar{B}_{\omega}(\mu)\omega(t). \tag{11.14}$$

Noting that S is non-singular, the above error system (11.14) can be expressed by

$$\dot{\mathscr{E}}(t) = S^{-1}\left(\bar{A}(\mu) - K(\mu)\bar{C}\right)\mathscr{E}(t) + \bar{B}_{dw}(\mu)\bar{\omega}(t), \tag{11.15}$$

where $\bar{B}_{dw}(\mu) \triangleq \begin{bmatrix} S^{-1}\bar{D}(\mu) & S^{-1}\bar{B}_{\omega}(\mu) \end{bmatrix}, \bar{\omega}(t) \triangleq \begin{bmatrix} d^T(t) & \omega^T(t) \end{bmatrix}^T$.

Here, we further assume that the disturbance is bounded, which satisfies

$$\bar{\omega}^T(t)\bar{\omega}(t) \leq \tilde{\omega}^2, \tag{11.16}$$

where $\tilde{\omega} > 0$ is a scalar.

In this subsection, the main purpose is to search for a set as small as possible to limit the following reachable set:

$$\mathbb{S} \triangleq \{\mathscr{E}(t) \in \Re^{n_x+n_y} \,|\, \mathscr{E}(t) \text{ and } \bar{\omega}(t) \text{ such that (11.15) and (11.16) are satisfied}, t \geq 0\}. \tag{11.17}$$

An ellipsoid that bounds the reachable set of closed-loop error system (11.15) is expressed by

$$\mathbb{E} \triangleq \left\{\mathscr{E} \mid \mathscr{E}^T P \mathscr{E} > 1,\ \mathscr{E} \in \Re^{(n_x+n_y)}\right\}, \tag{11.18}$$

where $P = P^T > 0$.

Note: It is easy to see from (11.15) that $\bar{\omega}^T(t)\bar{\omega}(t) = d^T(t)d(t) + \omega^T(t)\omega(t)$. The assumption that $\bar{\omega}^T(t)\bar{\omega}(t) \leq \tilde{\omega}^2$ in (11.16) may be conservative. Alternatively, we can define $\bar{\omega}(t) \equiv d(t)$, which implies that the smaller scalar $\tilde{\omega}$ can be found. In addition, the perturbation of the TDS attack $\omega(t)$ depends on the system state $x(t)$, thus there always exists a scalar ρ such that the inequality $\omega^T(t)\omega(t) \leq \rho x^T(t)x(t)$ holds. In that case, it is expected to obtain the less design conservatism. However, the scalar ρ is required to be known a priori.

Next, a sufficient condition is provided to ascertain that the reachable set of closed-loop error system (11.15) can be limited with the help of the intersection of ellipsoids in (11.18).

Theorem 11.1: Reachable Set Estimation of Observer System

A fuzzy observer in (11.11) can ensure that the reachable set of closed-loop error system (11.15) can be limited with the help of the intersection of ellipsoids in (11.18), if there exist the matrix $P \in \Re^{(n_x+n_y)\times(n_x+n_y)}$, $P = P^T > 0$, the gain matrices $\bar{K}_l \in \Re^{(n_x+n_y)\times n_y}$, and the positive scalars $\{\tilde{\omega}, \gamma\}$, such that for $\forall\ l \in \mathscr{L}$, the following LMIs hold:

$$\begin{bmatrix} Sym\left\{PS^{-1}\bar{A}_l - \bar{K}_l\bar{C}\right\} + \gamma P & P\bar{B}_{dw} \\ \star & -\frac{\gamma}{\tilde{\omega}^2}\mathbf{I} \end{bmatrix} > 0, \tag{11.19}$$

where $S \triangleq E + \bar{L}\bar{C}$, $\bar{B}_{dw} \triangleq \begin{bmatrix} S^{-1}\bar{D}_l & S^{-1}\bar{B}_\omega \end{bmatrix}$, and the corresponding fuzzy observer gains are given by

$$K_l = SP^{-1}\bar{K}_l,\ l \in \mathscr{L}. \tag{11.20}$$

■

Proof. We consider the following two steps to complete the proof. In particular, Step I will derive the stability results of the resulting closed-loop error system (11.15), and the reachable set estimation will proceed in Step II.

Step I. First, a Lyapunov candidate function is considered:

$$V(t) = \mathscr{E}^T(t)P\mathscr{E}(t), \tag{11.21}$$

where $P \in \Re^{(n_x+n_y)\times(n_x+n_y)}$ is the positive definite symmetric matrix.

Taking the time-derivative of $V(t)$ gives

$$\begin{aligned} \dot{V}(t) &= 2\mathscr{E}^T(t)P\dot{\mathscr{E}}(t) \\ &= 2\mathscr{E}^T(t)P\left[S^{-1}\left(\bar{A}(\mu) - K(\mu)\bar{C}\right)\mathscr{E}(t) + \bar{B}_{dw}(\mu)\bar{\omega}(t)\right]. \end{aligned} \tag{11.22}$$

It follows from (11.16), (11.21) and (11.22) that

$$
\begin{aligned}
J(t) &= \dot{V}(t) + \gamma V(t) - \frac{\gamma}{\tilde{\omega}^2}\bar{\omega}^T(t)\bar{\omega}(t) \\
&= 2\mathscr{E}^T(t)P\left[S^{-1}\left(\bar{A}(\mu) - K(\mu)\bar{C}\right)\mathscr{E}(t) + \bar{B}_{dw}(\mu)\bar{\omega}(t)\right] + \gamma V(t) - \frac{\gamma}{\tilde{\omega}^2}\bar{\omega}^T(t)\bar{\omega}(t) \\
&= \begin{bmatrix} \mathscr{E}(t) \\ \bar{\omega}(t) \end{bmatrix}^T \begin{bmatrix} \mathrm{Sym}\left\{PS^{-1}\left(\bar{A}(\mu) - K(\mu)\bar{C}\right)\right\} + \gamma P & P\bar{B}_{dw}(\mu) \\ \star & -\frac{\gamma}{\tilde{\omega}^2}\mathbf{I} \end{bmatrix} \begin{bmatrix} \mathscr{E}(t) \\ \bar{\omega}(t) \end{bmatrix}.
\end{aligned}
\tag{11.23}
$$

When $\bar{\omega}(t) \equiv 0$, it is straightforward to see that the inequality in (11.19) implies $\dot{V}(t) < 0$. Therefore, the asymptotic stability of closed-loop error system (11.15) is verified.

Step II. Since the inequality in (11.19) holds, one can obtain $J(t) < 0$. Then, by multiplying both its sides with $e^{\gamma t}$,

$$
\begin{aligned}
\frac{d\left(e^{\gamma t}V(t)\right)}{dt} &= e^{\gamma t}\dot{V}(t) + e^{\gamma t}\gamma V(t) \\
&< e^{\gamma t}\frac{\gamma}{\tilde{\omega}^2}\bar{\omega}^T(t)\bar{\omega}(t).
\end{aligned}
\tag{11.24}
$$

Now, by performing the integral of (11.24) from 0 to $T > 0$,

$$
\begin{aligned}
e^{\gamma T}V(T) &< \int_0^T e^{\gamma t}\frac{\gamma}{\tilde{\omega}^2}\bar{\omega}^T(t)\bar{\omega}(t)dt \\
&< e^{\gamma T} - 1.
\end{aligned}
\tag{11.25}
$$

Thus, for any time $T > 0$, one can obtain $V(T) < 1$.

Since $V(T) < 1$, it implies that

$$
\begin{aligned}
V(t) &= \mathscr{E}^T(t)P\mathscr{E}(t) \\
&< 1.
\end{aligned}
\tag{11.26}
$$

Hence, the reachable set (11.18) can be obtained directly, which completes the proof.

Here, the main purpose is to consider the fuzzy observer with the form of (11.11) such that the "smallest" bound for the reachable set (11.18) can be determined. To achieve this, a simple optimization algorithm has been developed in [18], i.e. maximizing δ subject to $\delta\mathbf{I} < P$. By using Schur complement to (11.26), one can readily solve the following optimization problem:

Algorithm 11.1.

$$
\min \bar{\delta}, \text{ s.t. } \begin{bmatrix} \bar{\delta}\mathbf{I} & \mathbf{I} \\ \star & P \end{bmatrix} \succ 0 \text{ and } (11.19),
$$

where $\bar{\delta} = \delta^{-1}$.

Note: A quadratic Lyapunov function $V(t) = \mathscr{E}^T(t)P\mathscr{E}(t)$ is considered in (11.21). It is clear that if $P \equiv \sum_{l=1}^{r}\mu_l P_l$, the function (11.21) turns to the fuzzy-basis-dependent Lyapunov function $V(t) = \mathscr{E}^T(t)P(\mu)\mathscr{E}(t)$. However, it requires that the time-derivative of μ_l is known a priori, which may be impractical.

11.1.3 COMPENSATION MECHANISM FOR THE PERTURBATION OF TDS ATTACK

Based on the augmented observer (11.11), we construct the following fuzzy controller

$$\begin{aligned} u(t) &= -F(\mu)\hat{x}(t) \\ &= -\left[\begin{array}{cc} F(\mu) & 0 \end{array}\right]\hat{\mathscr{X}}(t) \\ &= -F(\mu)x(t)+\mathscr{F}(\mu)\mathscr{E}(t), \end{aligned} \tag{11.27}$$

where $\mathscr{F}(\mu) \triangleq \left[\begin{array}{cc} F(\mu) & 0 \end{array}\right]$.

Motivated by [22], combined with (11.4) and (11.27),

$$E\dot{\bar{x}}(t) = \tilde{A}(\mu)\bar{x}(t) + \tilde{B}(\mu)\bar{d}(t), \tag{11.28}$$

where

$$\left\{ \begin{array}{l} E \triangleq \left[\begin{array}{cc} \mathbf{I} & 0 \\ 0 & 0 \end{array}\right], \tilde{A}(\mu) \triangleq \left[\begin{array}{cc} A(\mu) & B(\mu) \\ -F(\mu) & -\mathbf{I} \end{array}\right], \tilde{B}(\mu) \triangleq \left[\begin{array}{cc} D(\mu) & 0 \\ 0 & \mathscr{F}(\mu) \end{array}\right], \\ \bar{x}(t) \triangleq \left[\begin{array}{cc} x^T(t) & u^T(t) \end{array}\right]^T, \bar{d}(t) \triangleq \left[\begin{array}{cc} d^T(t) & \mathscr{E}^T(t) \end{array}\right]^T. \end{array} \right. \tag{11.29}$$

Based on Algorithm 11.1, it can be known that $\mathscr{E}^T\mathscr{E} \leq \bar{\delta}$. Thus, we can further assume that the disturbance is limited, which satisfies

$$\bar{d}^T(t)\bar{d}(t) \leq \tilde{d}^2, \tag{11.30}$$

where $\tilde{d}$ is a positive scalar.

In this subsection, the purpose is to search for a region as small as possible to limit the following reachable set:

$$\tilde{\mathbb{S}} \triangleq \left\{ x(t) \in \Re^{n_x} \left| x(t) \text{ and } \bar{d}(t) \text{ such that (11.28) and (11.30) are satisfied, } t \geq 0 \right. \right\}. \tag{11.31}$$

An ellipsoid to limit the reachable set of closed-loop control system (11.28) is determined by

$$\mathbb{E} \triangleq \left\{ x \mid x^T\tilde{P}_{(1)}x < 1,\ x \in \Re^{n_x} \right\}, \tag{11.32}$$

where $\tilde{P}_{(1)} = \tilde{P}_{(1)}^T > 0$.

Here, a sufficient condition is presented to ascertain the boundness of reachable set belonging to the closed-loop control system (11.28). It can be confined with the help of the intersection of ellipsoids in (11.32).

Theorem 11.2: Reachable Set Estimation of Compensation Control System

A fuzzy controller (11.27) can ensure that the reachable set of system (11.28) can be limited with the help of intersecting multiple ellipsoids in (11.18), if there exist matrices $\tilde{P}_{(1)} \in \Re^{n_x \times n_x}$ satisfying $\tilde{P}_{(1)} = \tilde{P}_{(1)}^T > 0$, $\tilde{P}_{(2)} \in \Re^{n_u \times n_x}$, $\tilde{P}_{(3)} \in \Re^{n_u \times n_u}$,

$J \in \Re^{n_u \times n_x}$, $\tilde{F}_l \in \Re^{n_u \times n_x}$, and the positive scalars $\{\tilde{d}, \gamma\}$, such that for $\forall\ l \in \mathscr{L}$, the following LMIs hold:

$$\begin{bmatrix} Sym\left\{\begin{bmatrix} A_l^T \tilde{P}_{(1)} - \tilde{F}_l J & -\tilde{F}_l \\ B^T \tilde{P}_{(1)} - \tilde{P}_{(3)} J & -\tilde{P}_{(3)} \end{bmatrix}\right\} + \gamma \begin{bmatrix} \tilde{P}_{(1)} & 0 \\ 0 & 0 \end{bmatrix} & \star \\ \begin{bmatrix} D_l^T \tilde{P}_{(1)} & 0 \\ \tilde{F}_l J & \tilde{F}_l \\ 0 & 0 \end{bmatrix} & -\frac{\gamma}{\tilde{d}^2}\mathbf{I} \end{bmatrix} > 0, \quad (11.33)$$

where the corresponding controller gains are given by

$$F_l = \left[\tilde{F}_l \tilde{P}_{(3)}^{-1}\right]^T, l \in \mathscr{L}. \quad (11.34)$$

■

Proof. A Lyapunov candidate function is considered as

$$V(t) = \bar{x}^T(t) E^T \tilde{P} \bar{x}(t), \quad (11.35)$$

where

$$\tilde{P} \triangleq \begin{bmatrix} \tilde{P}_{(1)} & 0 \\ \tilde{P}_{(2)} & \tilde{P}_{(3)} \end{bmatrix}, \quad (11.36)$$

with $0 < \tilde{P}_{(1)} = \tilde{P}_{(1)}^T \in \Re^{n_x \times n}$, $\tilde{P}_{(2)} \in \Re^{n_u \times n_x}$, $\tilde{P}_{(3)} \in \Re^{n_u \times n_u}$.

Then, based on the descriptor system in (11.28), the time-derivative of $V(t)$ is calculated as

$$\begin{aligned} \dot{V}(t) &= \dot{\bar{x}}^T(t) E^T \tilde{P} \bar{x}(t) + \bar{x}^T(t) E^T \tilde{P} \dot{\bar{x}}(t) \\ &= 2\left[\tilde{A}(\mu)\bar{x}(t) + \tilde{B}(\mu)\bar{d}(t)\right]^T \tilde{P}\bar{x}(t). \end{aligned} \quad (11.37)$$

It derives from (11.28), (11.30) and (11.37) that

$$\begin{aligned} J(t) &= \dot{V}(t) + \gamma V(t) - \frac{\gamma}{\tilde{d}^2} \bar{d}^T(t)\bar{d}(t) \\ &= 2\left[\tilde{A}(\mu)\bar{x}(t) + \tilde{B}(\mu)\bar{d}(t)\right]^T \tilde{P}\bar{x}(t) + \gamma V(t) - \frac{\gamma}{\tilde{d}^2}\bar{d}^T(t)\bar{d}(t) \\ &= \begin{bmatrix} \bar{x}(t) \\ \bar{d}(t) \end{bmatrix}^T \begin{bmatrix} \mathrm{Sym}\{\tilde{A}^T(\mu)\tilde{P}\} + \gamma E^T \tilde{P} E & \star \\ \tilde{B}^T(\mu)\tilde{P} & -\frac{\gamma}{\tilde{d}^2}\mathbf{I} \end{bmatrix} \begin{bmatrix} \bar{x}(t) \\ \bar{d}(t) \end{bmatrix}. \end{aligned} \quad (11.38)$$

Similar to the proof in Theorem 11.1, $J(t) < 0$ implies

$$\begin{aligned} \bar{x}^T(t) E^T \tilde{P} \bar{x}(t) &= x^T(t) \tilde{P}_{(1)} x(t) \\ &< 1. \end{aligned} \quad (11.39)$$

In order to cast (11.38) into a solvable LMI, we further define $\tilde{P}_{(2)} = \tilde{P}_{(3)}J, \tilde{F}(\mu) = F^T(\mu)\tilde{P}_{(3)}$, where $J \in \Re^{n_u \times n_x}$. Thus, the following inequality means that $J(t) < 0$,

$$\begin{bmatrix} \text{Sym}\left\{\begin{bmatrix} A^T(\mu)\tilde{P}_{(1)} - \tilde{F}(\mu)J & -\tilde{F}(\mu) \\ B^T\tilde{P}_{(1)} - \tilde{P}_{(3)}J & -\tilde{P}_{(3)} \end{bmatrix}\right\} + \gamma\begin{bmatrix} \tilde{P}_{(1)} & 0 \\ 0 & 0 \end{bmatrix} & \star \\ \begin{bmatrix} B_d^T\tilde{P}_{(1)} & 0 \\ \begin{bmatrix} \tilde{F}(\mu)J \\ 0 \end{bmatrix} & \begin{bmatrix} \tilde{F}(\mu) \\ 0 \end{bmatrix} \end{bmatrix} & -\frac{\gamma}{d^2}\mathbf{I} \end{bmatrix} > 0. \tag{11.40}$$

By extracting the fuzzy membership functions, the inequality (11.33) can be directly derived. Thus, the closed-loop system (11.28) is stable, which completes the proof.

In comparison with Theorem 11.1, the main aim here is to establish a fuzzy controller (11.27) such that the "smallest" bound of the reachable set given in (11.32) can be obtained. As discussed in [18], a typical optimization algorithm is to maximize $\hat{\delta}$ satisfying $\hat{\delta}\mathbf{I} < \tilde{P}_{(1)}$, that is

Algorithm 11.2.

$$\min \tilde{\delta}, \text{ s.t. } \begin{bmatrix} \tilde{\delta}\mathbf{I} & \mathbf{I} \\ \star & \tilde{P}_{(1)} \end{bmatrix} \succ 0 \text{ and } (11.33),$$

where $\tilde{\delta} = \hat{\delta}^{-1}$.

11.1.4 DESIGN PROCEDURE FOR REACHABLE SET ESTIMATION

The detailed calculating steps of the reachable set estimation for the presented T-S fuzzy systems are summarized in the following:

i) Use the T-S fuzzy model method to describe the nonlinear system (11.4);

ii) Construct the augmented T-S fuzzy system with the form of (11.9);

iii) Choose an auxiliary gain matrix L for the observer (11.11), and the auxiliary variable σ_l of the augmented fuzzy system in the form of (11.9);

iv) Solve LMIs in (11.19) to obtain matrices P and K_l, and use Algorithm 11.1 to calculate the reachable set estimation $\bar{\delta}$. Repeatedly apply Algorithm 11.1 to obtain the minimum reachable set estimation $\bar{\delta}$ via adjusting σ_l and L;

v) Implement the fuzzy observer in the form of (11.9), and produce the estimation of state $x(t)$ and delay perturbation $\omega(t)$ at the same time;

vi) Solve LMIs in (11.33) to obtain matrices $\tilde{P}$ and F_l, and use Algorithm 11.2 to calculate the minimum reachable set estimation $\hat{\delta}$. Apply the designed controller (11.27) to system (11.5), which satisfies that the reachable set of closed-loop control system (11.28) can be limited with the help of the intersection of ellipsoids in (11.32).

11.2 TDS ATTACK OF POWER NETWORKS

11.2.1 FUZZY MODELING OF POWER NETWORKS

In power systems, several generators and loads are dynamically interconnected. Thus, they can be seen as an example of complex networks, where each bus can be regarded as a node in the network. Assume that in the network all the buses are connected to synchronous machines. Its nonlinear model for the active power flow in a transmission line connected between bus i and bus j, is described by the following swing equation [19],

$$m_i\ddot{\delta}_i + d_i\dot{\delta}_i - P_{mi}(t) = -\sum_{j=1,i\neq j}^{N_i} P_{ij}(\delta_i,\delta_j), \tag{11.41}$$

where $i \in \mathscr{N}_i := \{1,2,\ldots,N_i\}$, N_i is the neighbourhood set of bus i, where bus j and i share a transmission line or communication link. N is the number of all bus nodes; δ_i is the phase angle in the i-th bus; m_i is the inertia coefficient of motors, and d_i is the damping coefficient of generators; P_{mi} is the mechanical input power; P_{ij} is the active power flow from bus i to j.

We define $\dot{\delta}_i = \omega_i$, and follow from (11.41) that

$$\begin{aligned} \dot{\delta}_i &= \omega_i, \\ \dot{\omega}_i &= m_i^{-1}d_i\omega_i - m_i^{-1}P_{mi}(t) - \sum_{j=1,i\neq j}^{N_i} m_i^{-1}P_{ij}(\delta_i,\delta_j). \end{aligned} \tag{11.42}$$

Assume that there are no power losses nor ground admittances. Then, the active power flow P_{ij} is given by

$$P_{ij}(\delta_i,\delta_j) = H_{ij}^{(1)}\cos(\delta_i-\delta_j) + H_{ij}^{(2)}\sin(\delta_i-\delta_j), \tag{11.43}$$

where $H_{ij}^{(1)} = |V_i|\,|V_j|\,G_{ij}, H_{ij}^{(2)} = |V_i|\,|V_j|\,F_{ij}, G_{ij}$ and F_{ij} are the branch conductance and susceptance between bus i and bus j, respectively; $V_i = |V_i|\,e^{j\delta_i}$ denotes the complex voltage of bus i where j represents the imaginary unit.

It is noted that the term $P_{ij}(\delta_i,\delta_j)$ is nonlinear. In this section, the T-S fuzzy modeling approach is introduced to describe the original nonlinear system (11.41). In the T-S fuzzy modeling, we choose several operating points for a considered nonlinear system. The corresponding local subsystems are then obtained for each operating point in virtue of the identification or linearization technique, where subsystems are connected with a fuzzy inference rule together with a membership function. Thus, define $z_{ij}(\delta_i,\delta_j) = P_{ij}(\delta_i,\delta_j)/\delta_j$ as the fuzzy premise variables, the nonlinear system (11.41) can be characterized via the following T-S fuzzy model:

Plant Rule $\mathscr{R}_i^l$: **IF** $z_{ij}(\delta_i,\delta_j)$ is $\mathscr{T}_i^l$, **THEN**

$$\dot{x}_i(t) = A_{ii}x_i(t) + B_iu_i(t) + \sum_{j=1,i\neq j}^{N_i} A_{ij}^l(\delta_j)x_j(t),\, l \in \mathscr{L}_i \tag{11.44}$$

where $\mathscr{L}_i := \{1,2,\ldots,r_i\}$, $\mathscr{R}_i^l$ is the l-th fuzzy inference rule; $\mathscr{T}_\phi^l$ $(\phi = 1,2,\ldots,g)$ are the fuzzy sets; r_i is the number of inference rules; $z_{ij}(t)$ represents the measurable variables; $\left\{A_{ii}, A_{ij}^l(\delta_j)\right\}$ is the l-th local linear subsystem for the i-th node. Similar to (11.44), $A_{ij}^{l}(\delta_j)$ is described by $A_{ij}^{lp}, p \in \mathscr{L}_j$ with if-then rules, and the system parameters are as below:

$$A_{ii} = \begin{bmatrix} 0 & 1 \\ 0 & m_i^{-1}d_i \end{bmatrix}, B_i = \begin{bmatrix} 0 \\ -m_i^{-1} \end{bmatrix}, A_{ij}^{lp} = \begin{bmatrix} 0 & 0 \\ z_{ij}^{lp} & 0 \end{bmatrix}, x_i(t) = \begin{bmatrix} \delta_i & \omega_i \end{bmatrix}^T. \tag{11.45}$$

Denote $\mu_i^l[z_{ij}(\delta_i)]$ and $\mu_i^p[z_{ij}(\delta_j)]$ as the normalized membership function. By fuzzy blending, the i-th global T-S fuzzy dynamic model is obtained by

$$\begin{cases} \dot{x}_i(t) = A_{ii}x_i(t) + B_iu_i(t) + \sum\limits_{j=1, i\neq j}^{N_i} A_{ij}(\mu_i, \mu_j)x_j(t), \\ y_i(t) = C_ix_i(t), \end{cases} \tag{11.46}$$

where $A_{ij}(\mu_i, \mu_j) := \sum\limits_{l=1}^{r_i}\sum\limits_{p=1}^{r_j} \mu_i^l \mu_j^p A_{ij}^{lp}, C_i = \begin{bmatrix} 1 & 0 \end{bmatrix}$.

11.2.2 TDS ATTACKS

Assume that the attacker can only target output measurements between the plant and the controller. The attacker can launch the relay-based or timestamp-based or noise-based attacks. Moreover, it may be not possible to obtain the exact knowledge about TDS attacks [11]. Here, under the relay-based or timestamp-based attacks, the system output is given by

$$y_i(t) = C_ix_i(t-\varepsilon), \tag{11.47}$$

where ε denotes the unknown time delays.

Now, we define the unknown TDS attack as

$$\omega_i(t) = C_i[x_i(t-\varepsilon) - x_i(t)]. \tag{11.48}$$

The system output (11.47) thus becomes

$$y_i(t) = C_ix_i(t) + \omega_i(t). \tag{11.49}$$

Note: The work of [11] proposes three typical types of TDS attacks: Relay-based attacks, timestamp-based ones, and noise-based ones. In (11.49), we propose a unified framework to describe three types of TDS attacks in output measurement.

Moreover, we define where $\omega(t) = \begin{bmatrix} \omega_1^T(t) & \omega_2^T(t) & \cdots & \omega_N^T(t) \end{bmatrix}^T$ and assume that the TDS attack $\omega_i(t)$ is a norm-bounded square intergrable signal over $[0, T]$, defined as below:

$$\mathrm{W}_{[0,T],\delta} \triangleq \{\omega(t) \in \mathscr{L}_2[0,T], \omega_i^T(t)\omega_i(t) \leq \delta_i\}, \tag{11.50}$$

where δ_i is a known positive scalar, and $\delta = \sum\limits_{i=1}^{N} \delta_i$.

The aim of this work is to estimate the TDS attack $\omega_i(t)$ and design compensation controller for the power network (11.46) such that the resulting closed-loop control system is finite time boundedness (FTB) with respect to the specified parameters $(c_1, c_2, [0,T], R, \mathbb{W}_{[0,T],\delta})$. First, the definition of FTB for nonlinear systems generalized from [20] is recalled as below:

Definition 11.1. Consider the following nonlinear dynamic system:

$$\dot{x}(t) = f(x, u, \omega).$$

Given a time interval $[0,T]$, two positive scalars a_1, a_2, with $a_1 < a_2$, and a weighted matrix $R > 0$. System (11.46) with $u(t) = 0$ is said to be FTB with respect to $(a_1, a_2, [0,T], R, \mathbb{W}_{[0,T],\delta})$, if

$$x^T(0) R x(0) \leq a_1 \Longrightarrow x^T(t) R x(t) < a_2, \forall t \in [0,T],$$

for all $\omega(t) \in \mathbb{W}_{[0,T],\delta}$.

In the next section, we introduce an augmented observer for realizing the estimation for both system state and TDS attack induced by hackers. Then, we propose a compensation control and give conditions for when the solutions exist.

11.2.3 OBSERVER DESIGN FOR TDS ATTACKS

This subsection will construct an augmented observer for realizing the estimation for both system state and TDS attack induced by hackers, such that the estimation error is the FTB with respect to the specified parameters $(b_1, b_2, [0,T], R, \mathbb{W}_{[0,T],\delta_i})$. Motivated by [17], combined with (11.46) and (11.49), let $\mathscr{X}_i(t) \triangleq \begin{bmatrix} x_i^T(t) & \omega_i^T(t) \end{bmatrix}^T$, and introduce an auxiliary matrix variable σ_i,

$$\begin{cases} E_i \dot{\mathscr{X}}_i(t) = \bar{A}_{ii} \mathscr{X}_i(t) + \bar{B}_i u_i(t) + \sum_{j=1, i \neq j}^{N_i} \bar{A}_{ij}(\mu_i, \mu_j) x_j(t) + \bar{D}_i \omega_i(t), \\ y_i(t) = \bar{C}_i \mathscr{X}_i(t), \end{cases} \tag{11.51}$$

where

$$\begin{cases} E_i \triangleq \begin{bmatrix} \mathbf{I} & 0 \\ 0 & 0 \end{bmatrix}, \bar{A}_{ii} \triangleq \begin{bmatrix} A_{ii} & 0 \\ 0 & -\sigma_i \mathbf{I} \end{bmatrix}, \bar{B}_i \triangleq \begin{bmatrix} B_i \\ 0 \end{bmatrix}, \\ \bar{D}_i \triangleq \begin{bmatrix} 0 \\ \sigma_i \mathbf{I} \end{bmatrix}, \bar{A}_{ij}(\mu_i, \mu_j) = \begin{bmatrix} A_{ij}(\mu_i, \mu_j) \\ 0 \end{bmatrix}, \bar{C}_i \triangleq \begin{bmatrix} C_i & \mathbf{I} \end{bmatrix}. \end{cases} \tag{11.52}$$

The augmented state vector $\mathscr{X}_i(t)$ in (11.51) is composed of the state $x_i(t)$ and the TDS attack $\omega_i(t)$ induced by hackers. Thus, the estimation of system state together with the attack perturbation will be implemented once the state estimation of $\mathscr{X}_i(t)$ is acquired.

Based on the system (11.46), an augmented observer is constructed with the form of

$$\begin{cases} S_i\dot{\mathscr{G}}_i(t) = \left(\bar{A}_{ii} - K_{ii}\bar{C}_i\right)\mathscr{G}_i(t) + \bar{B}_i u_i(t) + \sum_{j=1,i\neq j}^{N_i} \bar{A}_{ij}\left(\mu_i,\mu_j\right)\hat{x}_j(t) - \bar{D}_i y_i(t), \\ \hat{\mathscr{X}}_i(t) = \mathscr{G}_i(t) + S_i^{-1}\bar{L}_i y_i(t), \end{cases} \tag{11.53}$$

where $\bar{L}_i \triangleq \begin{bmatrix} 0 & L_i^T \end{bmatrix}^T$, and $\mathscr{G}_i(t)$ is an auxiliary state vector.

Now, we further define

$$\begin{aligned} &\mathscr{E}_i(t) \triangleq \mathscr{X}_i(t) - \hat{\mathscr{X}}_i(t), \mathscr{E}_i(t) = \begin{bmatrix} \mathscr{E}_{(i1)}^T & \mathscr{E}_{(i2)}^T \end{bmatrix}^T, \\ &\hat{\mathscr{X}}_i(t) = \begin{bmatrix} \hat{x}_i^T(t) & \hat{\omega}_i^T(t) \end{bmatrix}^T, S_i \triangleq E_i + \bar{L}_i\bar{C}_i, \bar{D}_i \triangleq -\bar{A}_{ii}S_i^{-1}\bar{L}_i, \end{aligned} \tag{11.54}$$

which implies that S_i is a non-singular matrix. It is clear that

$$\bar{C}_i S_i^{-1}\bar{L}_i = \begin{bmatrix} C_i & \mathbf{I} \end{bmatrix} \begin{bmatrix} \mathbf{I} & 0 \\ -C_i & L_i^{-1} \end{bmatrix} \begin{bmatrix} 0 \\ L_i \end{bmatrix} = \mathbf{I}. \tag{11.55}$$

It follows from (11.51)-(11.55) that

$$S_i\dot{\mathscr{E}}_i(t) = \left(\bar{A}_{ii} - K_{ii}\bar{C}_i\right)\mathscr{E}_i(t) + \sum_{j=1,i\neq j}^{N_i} \bar{A}_{ij}\left(\mu_i,\mu_j\right)\mathscr{E}_{(j1)} + \bar{D}_i\omega_i(t). \tag{11.56}$$

Noting that S_i is non-singular, the above error system (11.56) can be expressed by

$$\dot{\mathscr{E}}_i(t) = \mathscr{A}_{ii}\mathscr{E}_i(t) + \sum_{j=1,i\neq j}^{N_i} S_i^{-1}\bar{A}_{ij}\left(\mu_i,\mu_j\right)\mathscr{E}_{(j1)} + \bar{\mathscr{I}}\omega_i(t), \tag{11.57}$$

where $\mathscr{A}_{ii} = S_i^{-1}\left(\bar{A}_{ii} - K_{ii}\bar{C}_i\right), \bar{\mathscr{I}} = \begin{bmatrix} 0 \\ \sigma_i L_i^{-1} \end{bmatrix}$.

Note: It is easy to see from (11.57) that the effect from $\omega_i(t)$ to the error dynamics can be effectively attenuated by selecting a small scalar σ_i or a large matrix L_i.

Note: It is also seen from (11.57) that a large matrix S_i will lead to a weakening interconnection, which easily leads to a feasible solution.

The following lemma presents a sufficient condition for guaranteeing FTB of the error system (11.57), which is fundamental to obtain the main results of this work [21].

Theorem 11.3: FTB of Observer Design for TDS Attacks

Consider the positive scalars $(b_1, b_2, 0, T, \delta)$ and the positive-definite matrices $\{R, M_i\}$. The error system in (11.57) is the FTB with respect to

$(b_1, b_2, [0,T], R, \omega_{[0,T],\delta})$, if the following relationships hold:

$$\begin{bmatrix} Sym\{P_iS_i^{-1}\bar{A}_{ii} - \bar{K}_{ii}\bar{C}_i\} + \Phi_i & P_i\bar{\mathscr{I}} & \Psi_i^{lp} \\ \star & -\beta_i\mathbf{I} & 0 \\ \star & \star & -M_i \end{bmatrix} < 0, \tag{11.58}$$

$$b_1 < b^* < b_2, \tag{11.59}$$

$$\frac{\bar{\rho}_P b_1 + T\beta\delta}{\underline{\rho}_P e^{-\beta T}} < b^*, \tag{11.60}$$

where

$$\Phi_i = \sum_{j=1, i\neq j}^{N} \begin{bmatrix} M_j & 0 \\ 0 & 0 \end{bmatrix} - \beta P_i, S_i \triangleq E_i + \bar{L}_i\bar{C}_i, \bar{A}_{ii} = \begin{bmatrix} A_{ii} & 0 \\ 0 & -\sigma_i\mathbf{I} \end{bmatrix},$$

$$M_i = diag\underbrace{\{\begin{matrix} M_1 & M_2 & \cdots & M_N \end{matrix}\}}_{N-1}, \Psi_i^{lp} = \underbrace{\{\begin{matrix} \Psi_{i1}^{lp} & \Psi_{i2}^{lp} & \cdots & \Psi_{iN}^{lp} \end{matrix}\}}_{N-1},$$

$$\Psi_{ij}^{lp} = \nu_{ij}P_iS_i^{-1}\bar{A}_{ij}^{lp}, \bar{A}_{ij}^{lp} = \begin{bmatrix} A_{ij}^{lp} \\ 0 \end{bmatrix}, \bar{C}_i = \begin{bmatrix} C_i & \mathbf{I} \end{bmatrix}, \bar{\mathscr{I}} = \begin{bmatrix} 0 \\ \sigma_i L_i^{-1} \end{bmatrix}. \tag{11.61}$$

Furthermore, the corresponding fuzzy observer gains are given by

$$K_{ii} = S_iP_i^{-1}\bar{K}_{ii}. \tag{11.62}$$

■

Proof. Consider the following Lyapunov functional:

$$\begin{aligned} V(t) &= \sum_{i=1}^{N} V_i(t) \\ &= \sum_{i=1}^{N} \mathscr{E}_i^T(t) P_i \mathscr{E}_i(t), \forall t \in [0,T] \end{aligned} \tag{11.63}$$

where $0 < P_i = P_i^T \in \Re^{n_{xi}\times n_{xi}}$. Taking the time derivative of the system (11.63),

$$\dot{V}(t) = \sum_{i=1}^{N} 2\mathscr{E}_i^T(t)P_i\left[\mathscr{A}_{ii}\mathscr{E}_i(t) + \bar{\mathscr{I}}\omega_i(t)\right] + \sum_{i=1}^{N}\sum_{j=1,i\neq j}^{N_i} 2\mathscr{E}_i^T(t)P_iS_i^{-1}\bar{A}_{ij}(\mu_i,\mu_j)\mathscr{E}_{(j1)}. \tag{11.64}$$

Note that the relationship $N_i \subset N$ holds. In order to implement the observer proposed in (11.53), we introduce the scalar $\nu_{ij} \in [0,1]$ to describe the set $N_i \subset N$. Thus,

$$\sum_{i=1}^{N}\sum_{j=1,i\neq j}^{N_i} \mathscr{E}_i^T(t)P_iS_i^{-1}\bar{A}_{ij}(\mu_i,\mu_j)\mathscr{E}_{(j1)} = \sum_{i=1}^{N}\sum_{j=1,i\neq j}^{N} \nu_{ij}\mathscr{E}_i^T(t)P_iS_i^{-1}\bar{A}_{ij}(\mu_i,\mu_j)\mathscr{E}_{(j1)}, \tag{11.65}$$

where the scalar $\nu_{ij} \equiv 1$ holds for $\mathscr{N}_i \subset \mathscr{N}$, otherwise $\nu_{ij} \equiv 0$.

Note that

$$2\bar{x}^T\bar{y} \leq \kappa^{-1}\bar{x}^T\bar{x} + \kappa\bar{y}^T\bar{y}, \tag{11.66}$$

where $\{\bar{x},\bar{y}\} \in \Re^n$ and scalar $\kappa > 0$. Define the matrix $0 < M_i = M_i^T$,

$$\begin{aligned}
&\sum_{i=1}^{N}\sum_{j=1,i\neq j}^{N} \nu_{ij}\mathscr{E}_i^T(t)P_iS_i^{-1}\bar{A}_{ij}(\mu_i,\mu_j)\mathscr{E}_{(j1)} \\
&\leq \sum_{i=1}^{N}\sum_{j=1,i\neq j}^{N} \nu_{ij}\mathscr{E}_i^T(t)P_iS_i^{-1}\bar{A}_{ij}(\mu_i,\mu_j)M_i^{-1}(\star) + \sum_{i=1}^{N}\sum_{j=1,i\neq j}^{N} \mathscr{E}_{(j1)}^T M_i\mathscr{E}_{(j1)} \\
&= \sum_{i=1}^{N}\sum_{j=1,i\neq j}^{N} \nu_{ij}\mathscr{E}_i^T(t)P_iS_i^{-1}\bar{A}_{ij}(\mu_i,\mu_j)M_i^{-1}(\star) + \sum_{i=1}^{N}\sum_{j=1,i\neq j}^{N} \mathscr{E}_{(i1)}^T M_j\mathscr{E}_{(i1)}.
\end{aligned} \tag{11.67}$$

Now, define the following index

$$\begin{aligned}
J(t) &= \sum_{i=1}^{N} J_i(t) \\
&= \sum_{i=1}^{N} \left[\dot{V}_i(t) - \beta V_i(t) - \beta\omega_i^T(t)\omega_i(t)\right].
\end{aligned} \tag{11.68}$$

Combined with (11.63)-(11.68), $J(t) < 0$ can be satisfied if the following inequality holds,

$$\begin{bmatrix} P_i\mathscr{A}_{ii} + \mathscr{A}_{ii}^T P_i + \Phi_i(\mu_i,\mu_j) & P_i\bar{\mathscr{I}} \\ \star & -\beta_i\mathbf{I} \end{bmatrix} < 0, \tag{11.69}$$

where $\Phi_i(\mu_i,\mu_j) = \sum_{j=1,i\neq j}^{N} \nu_{ij}P_iS_i^{-1}\bar{A}_{ij}(\mu_i,\mu_j)M_i^{-1}(\star) + \sum_{j=1,i\neq j}^{N} \begin{bmatrix} M_j & 0 \\ 0 & 0 \end{bmatrix} - \beta P_i$.

Based on the Schur complement,

$$\begin{bmatrix} \Gamma_i + \Phi_i & P_i\bar{\mathscr{I}} & \Psi_i(\mu_i,\mu_j) \\ \star & -\beta_i\mathbf{I} & 0 \\ \star & \star & -\mathbb{M}_i \end{bmatrix} < 0, \tag{11.70}$$

where

$$\begin{aligned}
&\Gamma_i = \mathrm{Sym}\left\{P_iS_i^{-1}\left(\bar{A}_{ii} - K_{ii}\bar{C}_i\right)\right\}, \Phi_i = \sum_{j=1,i\neq j}^{N} \begin{bmatrix} M_j & 0 \\ 0 & 0 \end{bmatrix} - \beta_iP_i, \Psi_{ij}^{lp} = \nu_{ij}P_iS_i^{-1}\bar{A}_{ij}^{lp}, \\
&\mathbb{M}_i = \mathrm{diag}\underbrace{\{\, M_1 \;\; M_2 \;\; \cdots \;\; M_N \,\}}_{N-1}, \Psi_i(\mu_i,\mu_j) = \underbrace{\left\{\Psi_{i1}^{lp} \; \Psi_{ij}^{lp} \; \cdots \; \Psi_{iN}^{lp}\right\}}_{N-1}.
\end{aligned} \tag{11.71}$$

By extracting the fuzzy premise variables, the inequality in (11.58) can be directly obtained. It is easy to see that the result on (11.58) implies $J(t) < 0$, that is

$$\dot{V}(t) < \beta V(t) + \beta \omega(t)\omega(t). \tag{11.72}$$

By multiplying both sides of expression (11.72) by $e^{-\beta t}$ and integrating the successive inequality from 0 to t with $t \in [0, T]$, we have

$$\begin{aligned} e^{-\beta t} V(t) &< V(0) + \beta \int_0^t e^{-\beta \theta} \omega^T(\theta)\omega(\theta) d\theta \\ &\leq \bar{\rho}_P c_1 + T\beta\delta, \end{aligned} \tag{11.73}$$

where $\bar{\rho}_P = \lambda_{\max}\left(R^{-\frac{1}{2}} P R^{-\frac{1}{2}}\right), P = \text{diag}\{P_1\ P_2 \cdots P_N\}$.

In addition, it follows from (11.63) that

$$e^{-\beta t} V(t) \geq \underline{\rho}_P e^{-\beta T} \mathscr{E}^T(t) R \mathscr{E}(t), \tag{11.74}$$

where $\underline{\rho}_P = \lambda_{\min}\left(R^{-\frac{1}{2}} P R^{-\frac{1}{2}}\right)$. It is easy to see from (11.64) that

$$\mathscr{E}^T(t) R \mathscr{E}(t) \leq \frac{\bar{\rho}_P c_1 + T\beta\delta}{\underline{\rho}_P e^{-\beta T}}. \tag{11.75}$$

Define $T^* \leq T$. Note that $\mathscr{E}^T(t) R \mathscr{E}(t) < b^*$ for all $t \in [0, T^*]$. According to Definition 11.1, the error system in (11.57) is the FTB with respect to $(b_1, b_2, [0, T], R, \mathbb{W}_{[0,T],\delta})$. This completes the proof.

Note: A quadratic Lyapunov function $V(t) = \mathscr{E}^T(t) P \mathscr{E}(t)$ is considered in (11.63). It is clear that if $P \equiv \sum_{l=1}^r \mu_l P_l$, the function (11.63) turns to the fuzzy-basis-dependent Lyapunov function $V(t) = \mathscr{E}^T(t) P(\mu) \mathscr{E}(t)$. However, it requires that the time-derivative of μ_l is known a priori, which may be unpractical.

11.2.4 COMPENSATION CONTROL FOR TDS ATTACKS

This subsection describes the design of a compensation controller such that TDS attacks can be attenuated in the FTB. Now, we construct the distributed compensation controller as below:

$$\begin{aligned} u_i(t) &= -F_{ii}\hat{x}_i(t) - \sum_{j=1, i\neq j}^{N_i} F_{ij}(\mu_i, \mu_j)\hat{x}_j(t) \\ &= -\mathscr{F}_{ii}\hat{\mathscr{X}}_i(t) - \sum_{j=1, i\neq j}^{N_i} \mathscr{F}_{ij}(\mu_i, \mu_j)\hat{\mathscr{X}}_j(t) \\ &= -F_{ii}x_i(t) + \mathscr{F}_{ii}\mathscr{E}_i(t) - \sum_{j=1, i\neq j}^{N_i} F_{ij}(\mu_i, \mu_j)x_j(t) + \sum_{j=1, i\neq j}^{N_i} \mathscr{F}_{ij}(\mu_i, \mu_j)\mathscr{E}_j(t), \end{aligned} \tag{11.76}$$

where $\mathscr{F}_{ii} \triangleq \begin{bmatrix} F_{ii} & 0 \end{bmatrix}, \mathscr{F}_{ij}(\mu_i,\mu_j) \triangleq \begin{bmatrix} F_{ij}(\mu_i,\mu_j) & 0 \end{bmatrix}$.

Submitting (11.76) into (11.46) yields the following closed-loop control system:

$$\dot{x}_i(t) = (A_{ii} - B_i F_{ii}) x_i(t) + \sum_{j=1, i\neq j}^{N_i} (A_{ij}(\mu_i,\mu_j) - B_i F_{ij}(\mu_i,\mu_j)) x_j(t)$$
$$+ B_i \mathscr{F}_{ii} \mathscr{E}_i(t) + \sum_{j=1, i\neq j}^{N_i} B_i \mathscr{F}_{ij}(\mu_i,\mu_j) \mathscr{E}_j(t). \tag{11.77}$$

Note: The interconnections are decoupled when $A_{ij}(\mu_i,\mu_j) - B_i F_{ij}(\mu_i,\mu_j) \equiv 0$. Here, we can choose $F_{ij}(\mu_i,\mu_j) = \sum_{l=1}^{r_i} \sum_{p=1}^{r_j} \mu_i^l \mu_j^p \begin{bmatrix} m_i z_{ij}^{lp} & 0 \end{bmatrix}$. In that case, the closed-loop control system declines to

$$\dot{x}_i(t) = (A_{ii} - B_i F_{ii}) x_i(t) + B_i \mathscr{F}_{ii} \mathscr{E}_i(t) + \sum_{j=1, i\neq j}^{N_i} B_i \mathscr{F}_{ij}(\mu_i,\mu_j) \mathscr{E}_j(t).$$

Based on Theorem 11.1, it can be known that $\mathscr{E}^T(t) R \mathscr{E}(t) < b^*$ for all $t \in [0, T^*]$. Thus, we have $\mathscr{E}^T(t) \mathscr{E}(t) < \frac{b^*}{\lambda_{\min}(R)}$. Here, a sufficient condition for the proposed compensation controller, which attenuates the TDS attack with respect to $(c_1, c_2, [0,T], R, \mathscr{E}_{[0,T], \frac{b^*}{\lambda_{\min}(R)}})$, is given by:

Theorem 11.4: FTB of Compensation Control System

Given the positive scalars, positive-definite matrices and $0 < \tilde{P}_{i(1)} = \tilde{P}_{i(1)}^T \in \Re^{n_x \times n}$, $\tilde{P}_{i(2)} \in \Re^{n_u \times n_x}$, $\tilde{P}_{i(3)} \in \Re^{n_u \times n_u}$, $J_i \in \Re^{n_u \times n_x}$, a fuzzy controller (11.76) can ensure that the resulting closed-loop system is the FTB with respect to $(c_1, c_2, [0,T], R, \mathscr{E}_{[0,T], \frac{b^*}{\lambda_{\min}(R)}})$, if the following inequalities hold for all $i \in \mathscr{N}, l \in \mathscr{L}_i$, $p \in \mathscr{L}_j$:

$$\begin{bmatrix} Sym\{P_i(A_{ii} - B_i F_{ii})\} - \beta P_i & P_i B_i \mathscr{F}_{ii} & \Upsilon_i^{lp} \\ \star & -\beta \mathbf{I} + \sum_{j=1, i\neq j}^{N} U_j & 0 \\ \star & \star & -U_i \end{bmatrix} < 0, \tag{11.78}$$

$$c_1 < c^* < c_2, \tag{11.79}$$

$$\frac{\bar{\rho}_P c_1 + T\beta \frac{b^*}{\lambda_{\min}(R)}}{\underline{\rho}_P e^{-\beta T}} < c^*, \tag{11.80}$$

where $U_i = diag\underbrace{\{ U_1 \quad U_2 \quad \cdots \quad U_N \}}_{N-1}, \Upsilon_i^{lp} = \underbrace{\left\{ \Upsilon_{i1}^{lp} \; \Upsilon_{i2}^{lp} \; \cdots \; \Upsilon_{iN}^{lp} \right\}}_{N-1}, \Upsilon_{ij}^{lp} = \nu_{ij} P_i B_i \mathscr{F}_{ij}^{lp}$.

■

Proof. The proof is similar to Theorem 11.3, thus is deleted.

11.2.5 DESIGN PROCEDURE FOR ATTENUATING TDS ATTACKS

The detailed calculating steps of attenuating TDS attacks for the considered power systems are summarized in the following:

a) Use the T-S fuzzy model method to describe the nonlinear power system (11.46);

b) Choose a smaller value for the scalar σ_i, and construct the augmented T-S fuzzy system with the form of (11.51);

c) Choose a larger value for the auxiliary gain matrix L_i, and construct the observer (11.53);

d) We give β and solve (11.58) of Theorem 11.3 to obtain matrices P_i and K_{ii}, given $R, T, \delta, \mathscr{E}(t)$ and use (11.60) of Theorem 11.3 to calculate the bounding b^*;

e) Calculate the controller gain matrix F_{ij}^{lp}, and construct the compensation controller (11.76);

f) Choose the controller gain matrix F_{ii}, and use Theorem 11.4 to obtain matrices P_i, given $R, T, \frac{b^*}{\lambda_{\min}(R)}, x(0)$, and calculate the bounding c^*;

g) Implement the obtained fuzzy observer and compensation controller.

11.3 SIMULATION STUDIES

This section considers a power network with three buses. We assume that the phase angle δ_i is measurable but ω_i is immeasurable. Choose the model parameters $m_1 = 0.1, m_2 = 0.2, m_3 = 0.1, d_1 = -8, d_2 = -16, d_3 = -10$.

Attenuating TDS attacks in the nonlinear power network can be achieved as below.

a) In this simulation, by linearizing the phase angle around $\delta_i = 1$ and $\delta_i = 8$, then the nonlinear system is expressed by the T-S fuzzy model with the following parameters:

$$A_1 = \begin{bmatrix} 0 & 1 \\ 0 & -80 \end{bmatrix}, B_1 = \begin{bmatrix} 0 \\ -10 \end{bmatrix}, A_2 = \begin{bmatrix} 0 & 1 \\ 0 & -80 \end{bmatrix}, B_2 = \begin{bmatrix} 0 \\ -5 \end{bmatrix},$$
$$A_3 = \begin{bmatrix} 0 & 1 \\ 0 & -100 \end{bmatrix}, B_3 = \begin{bmatrix} 0 \\ -10 \end{bmatrix}, A_{ij}^{l1} = \begin{bmatrix} 0 & 0 \\ 1 & 0 \end{bmatrix}, A_{ij}^{l2} = \begin{bmatrix} 0 & 0 \\ 8 & 0 \end{bmatrix},$$
$$C_i = \begin{bmatrix} 1 & 0 \end{bmatrix}, (i,j) \in \{1,2,3\}, l \in \{1,2\}.$$

b) Choose the scalar $\sigma_i = 0.2$, and the matrix $\bar{L}_i = \begin{bmatrix} 0 \\ 0 \\ 14 \end{bmatrix}$, and construct the fuzzy observer in the form of (11.53).

c) Given $\beta = 0.04$, solve (11.58) of Theorem 11.3 with constraints on controller gains to obtain matrices P_i and K_{ii} as below:

$$P_1 = \begin{bmatrix} 0.0522 & 0 & 0.0501 \\ 0 & 0.0005 & 0 \\ 0.0501 & 0 & 0.0550 \end{bmatrix}, P_2 = \begin{bmatrix} 0.0235 & 0 & 0.0213 \\ 0 & 0.0004 & 0 \\ 0.0213 & 0 & 0.0252 \end{bmatrix},$$

$$P_3 = \begin{bmatrix} 0.0713 & 0 & 0.0687 \\ 0 & 0.0006 & 0 \\ 0.0687 & 0 & 0.0776 \end{bmatrix},$$

$$K_{11} = \begin{bmatrix} 0.2829 \\ 0.3102 \\ 0.1060 \end{bmatrix}, K_{22} = \begin{bmatrix} 0.4485 \\ 0.4071 \\ 0.2424 \end{bmatrix}, K_{33} = \begin{bmatrix} 0.2140 \\ 0.2300 \\ 0.1082 \end{bmatrix}.$$

Given the scalars $b_1 = 3, T = 0.1, \sigma_1 = 1.5, \sigma_2 = 2, \sigma_3 = 4.5$, and the matrix $R = \text{diag}\{1,1,1\}$, use (11.60) of Theorem 11.3 to calculate the bounding $b^* = 1155$;

d) Assume that $A_{ij}(\mu_i,\mu_j) - B_i F_{ij}(\mu_i,\mu_j) \equiv 0$, and calculate the controller gain matrix $F_{1j}^{l1} = [-0.1 \ \ 0]$, $F_{1j}^{l2} = [-0.8 \ \ 0]$, $F_{2j}^{l1} = [-0.2 \ \ 0]$, $F_{2j}^{l2} = [-1.6 \ \ 0]$,$F_{3j}^{l1} = [-0.1 \ \ 0]$, $F_{3j}^{l2} = [-0.8 \ \ 0]$. Then construct the compensation controller (11.76);

e) Given the matrix P_i as below:

$$P_1 = \begin{bmatrix} 0.0277 & 0.0006 \\ 0.0006 & 0.0117 \end{bmatrix}, P_2 = \begin{bmatrix} 0.0368 & 0.0010 \\ 0.0010 & 0.0232 \end{bmatrix},$$

$$P_3 = \begin{bmatrix} 0.0342 & 0.0010 \\ 0.0010 & 0.0215 \end{bmatrix},$$

and the scalar $\beta = 0.04$, use Theorem 11.4 to obtain the controller gain matrix F_{ii} as below:

$$F_{11} = \begin{bmatrix} 0.4917 & 0.1038 \end{bmatrix}, F_{22} = \begin{bmatrix} 0.7350 & 0.0603 \end{bmatrix},$$
$$F_{33} = \begin{bmatrix} 0.4661 & 0.0680 \end{bmatrix}.$$

Given $R = \text{diag}\{1,1\}, T = 0.1, \frac{b^*}{\lambda_{\min}(R)} = 1155, c_1 = 17$, and calculate the bounding $c^* = 451$;

f) Assume that attackers target output measurements from 0.05 s to 0.1 s as shown in Figure 11.1. Implement the obtained observer and compensation controller. Figure 11.2 shows that the closed-loop control system is the FTB with respect to $(c_1, c_2, [0,T], R, \mathscr{E}_{[0,T]}, \frac{b^*}{\lambda_{\min}(R)})$.

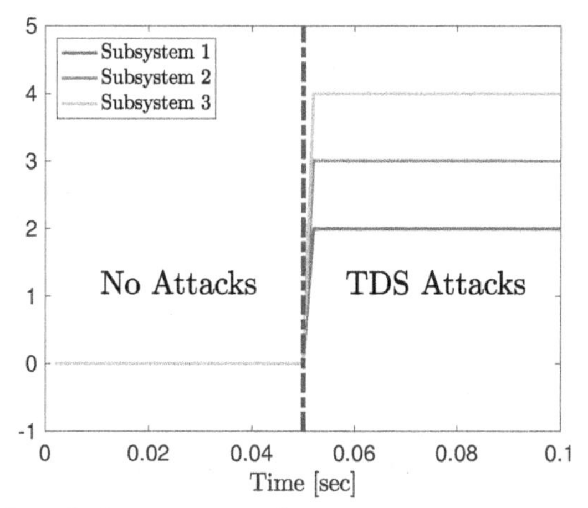

Figure 11.1 TDS attacks for power networks.

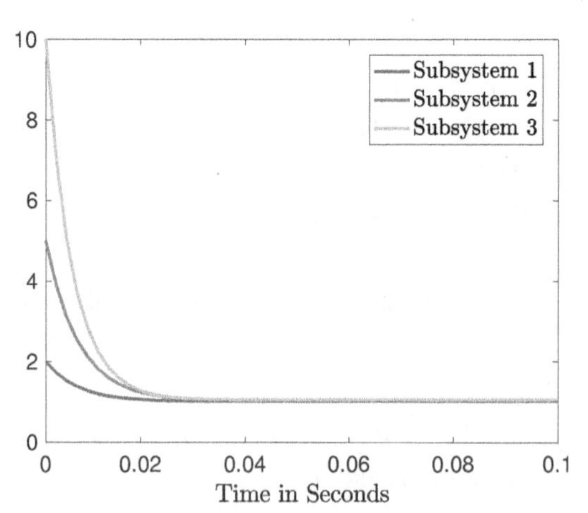

Figure 11.2 Response of $x_i^T(t) R_i x_i(t)$.

11.4 REFERENCES

1. Cao, X., Cheng, P., Chen, J., and Sun, Y. (2013). An online optimization approach for control and communication codesign in networked cyber-physical systems. IEEE Transactions on Industrial Informatics, 9(1), 439-450.
2. Cao, X., Cheng, P., Chen, J., Ge, S. S., Cheng, Y., and Sun, Y. (2014). Cognitive radio based state estimation in cyber-physical systems. IEEE Journal on Selected Areas in Communications, 32(3), 489-502.
3. Chen, J., Cao, X., Cheng, P., Xiao, Y., and Sun, Y. (2010). Distributed collaborative control for industrial automation with wireless sensor and actuator networks. IEEE Transactions on Industrial Electronics, 57(12), 4219-4230.
4. Zhong, Z., Lin, C. M., Shao, Z., and Xu, M. (2018). Decentralized event-triggered control for large-scale networked fuzzy systems. IEEE Transactions on Fuzzy Systems, 26(1), 29-45.
5. Zhong, Z., Zhu, Y., and Lam, H. K. (2017). Asynchronous piecewise output-feedback control for large-scale fuzzy systems via distributed event-triggering schemes. IEEE Transactions on Fuzzy Systems, 26(3), 1688-1703.
6. Cao, X., Cheng, P., Chen, J., and Sun, Y. (2013). An online optimization approach for control and communication codesign in networked cyber-physical systems. IEEE Transactions on Industrial Informatics, 9(1), 439-450.
7. Yan, R., Dong, Z. Y., Saha, T. K., and Majumder, R. (2010). A power system nonlinear adaptive decentralized controller design. Automatica, 46(2), 330-336.
8. Khaitan, S. K. and Mccalley, J. D. (2013). Cyber physical system approach for design of power grids: A survey. In Proceedings of the 2013 IEEE Power and Energy Society General Meeting.
9. Zhang, H., Qi, Y., Wu, J., Fu, L., and He, L. (2018). DoS attack energy management against remote state estimation. IEEE Transactions on Control of Network Systems, 5(1), 383-394.
10. Zhang, H., Meng, W., Qi, J., Wang, X., and Zheng, W. X. (2018). Distributed load sharing under false data injection attack in inverter-based microgrid. IEEE Transactions on Industrial Electronics, 66(2), 1543-1551.
11. Sargolzaei, A., Yen, K. K., Abdelghani, M. N., Sargolzaei, S., and Carbunar, B. (2017). Resilient design of networked control systems under time delay switch attacks, application in smart grid. IEEE Access, 5, 15901-15912.
12. Zhang, H., Cheng, P., Wu, J., Shi, L., and Chen, J. (2014). Online deception attack against remote state estimation. IFAC Proceedings Volumes, 47(3), 128-133.
13. Zhang, H. and Zheng, W. X. (2018). Denial-of-service power dispatch against linear quadratic control via a fading channel. IEEE Transactions on Automatic Control, 63(9), 3032-3039.
14. Basin, M. V., Yu, P., and Shtessel, Y. B. (2017). Hypersonic missile adaptive sliding mode control using finite- and fixed-time observers. IEEE Transactions on Industrial Electronics, 65(1), 930-941.
15. Li S., Sun, H., Yang J., and Yu X. (2015). Continuous finite-time output regulation for disturbed systems under mismatching condition. IEEE Transactions on Automatic Control, 60(1), 277-282.
16. Basin, M., Panathula, C. B., Shtessel, Y. B., and Rodriguez-Ramirez, P. (2016). Continuous finite-time higher-order output regulators for systems with unmatched unbounded disturbances. IEEE Transactions on Industrial Electronics, 63(8), 5036-5043.

17. Gao, Z. (2015). Estimation and compensation for Lipschitz nonlinear discrete-time systems subjected to unknown measurement delays. IEEE Transactions on Industrial Electronics, 62(9), 5950-5961.
18. Fridman, E. and Shaked, U. (2003). On reachable sets for linear systems with delay and bounded peak inputs. Automatica, 39(11), 2005-2010.
19. Pan, W. , Yuan, Y., Sandberg, H., GoncAlves, J., and Stan, G. B. (2015). Online fault diagnosis for nonlinear power systems. Automatica, 55, 27-36.
20. Elbsat, M. N. and Yaz, E. E. (2013). Robust and resilient finite-time bounded control of discrete-time uncertain nonlinear systems. Automatica, 49(7), 2292-2296.
21. Zhong Z., Zhu Y., Lin C., and Huang T. (2019). A fuzzy control framework for interconnected nonlinear power networks under TDS attack: Estimation and compensation. Journal of the Franklin Institute, doi.org/10.1016/j.jfranklin.2018.12.012.
22. Chadli M., Karimi H., and Shi P. (2014). On stability and stabilization of singular uncertain Takagi-Sugeno fuzzy systems. Journal of the Franklin Institute, 351(3), 1453-1463.

Index

A

B

C

E